Battery Technologies

Battery Technologies

Materials and Components

Edited by Jianmin Ma

WILEY-VCH

Editor

Prof. Jianmin Ma
University of Electronic Science and Technology of China
School of Materials and Energy
No. 2006, Xiyuan Ave
West Hi-Tech Zone
611731 Chengdu
China

Cover Image: © Immersion Imagery/Shutterstock

Library of Congress Card No.: applied for

British Library Cataloguing-in-Publication Data
A catalogue record for this book is available from the British Library.

Bibliographic information published by the Deutsche Nationalbibliothek
The Deutsche Nationalbibliothek lists this publication in the Deutsche Nationalbibliografie; detailed bibliographic data are available on the Internet at <http://dnb.d-nb.de>.

Print ISBN: 978-3-527-34858-9
ePDF ISBN: 978-3-527-83004-6
ePub ISBN: 978-3-527-83003-9
oBook ISBN: 978-3-527-83005-3

Typesetting Straive, Chennai, India

Printed on acid-free paper

Printed and bound in Great Britain by
CPI Group (UK) Ltd, Croydon, CR0 4YY

C9783527348589_270824

Contents

Preface

We are living in a society driven by electrics. We can use mobile phones, internet, personal computers, and other portable electronics everywhere. This is attributed to the batteries used, which can supply electrics for these applications. In China, we can take electric vehicle taxies in bigger cities very conveniently and more families own the electric vehicles for their daily use. This also depends on the batteries. In our cities, we can also employ batteries to store electric energy at night and then use in heavy load, which is also called "peak cut". Therefore, batteries are becoming a necessary part in our life.

Batteries can be thought as controlled reactions which should be reversibility, high ratio of transferred charges/reactant weight, green process, and low cost. At present, it is very hot research topic on batteries and batteries are very huge fields, i.e. systems from organic electrolytes to aqueous electrolytes, charged ions from cations (e.g., Li^+, Na^+, K^+, Zn^{2+}, Mg^{2+}, Ca^{2+}, and Al^{3+}) to anions (e.g. OH^-, F^-, and Cl^-), electrode materials from gases (e.g. O_2, CO_2, and N_2) to liquids, and (organic/inorganic or metals/non-metals) solids. Therefore, it is very interesting and attractive to carry out the research of batteries. In 2019, Prof. John B. Goodenough, Professor M. Stanley Whittingham, and Professor Akira Yoshino were awarded to Noble Chemistry Medals due to their contributions to the development of lithium-ion batteries. Under this background, more and more people have been joining the research and development of rechargeable batteries.

To make more people to get the frontier knowledges for battery fields, I edited this book together with some friends. They are. The book includes the wide range of contents, i.e. lithium-ion batteries, lithium–air batteries, lithium–sulfur batteries, sodium-ion batteries, sodium–air batteries, zinc-ion batteries, zinc–air batteries, aluminum-ion batteries, aluminum–air batteries, and dual ion batteries. The battery systems and their components are talked. Although there are many topics to be talked as introduced before, we could not write all parts in one book. So, we hope that readers can understand this. In future, we will add more information into this book if there is any opportunity.

I firmly think that this book will be useful handbook for batteries to every reader. Finally, we hope that this book will enrich the knowledge of students from all levels, also help researchers to know more about batteries, and make the workers in battery industries to know more about this field.

School of Materials and Energy *Jianmin Ma*
University of Electronic Science and Technology of China
Chengdu
611731
P.R. China

1

Li-Ion Battery

Ruiping Liu

China University of Mining & Technology (Beijing), Department of Materials Science and Engineering, 100083, Beijing, China

1.1 Introduction

1.1.1 History of the Lithium-Ion Battery

Lithium is the lightest metal in nature, with an atomic weight of 6.94, a density of 0.534 $g\,cm^{-3}$, and a standard electrode potential of −3.045 V, which is the lowest potential among all the metal electrodes. In the 1970s, the first lithium metal battery was prepared by using titanium sulfide as the cathode and lithium metal as the anode. However, the lithium dendrites caused by the uneven deposition and distribution of lithium during charging will cause a large irreversible loss of active lithium, and even short-circuit of the battery, thus the battery is prohibited from charging [1]. In the 1980s, it is found that the lithium ions can be reversibly and freely embedded into the graphite materials, and soon, the battery was successfully prepared by using graphite as the anode material. Lithium ions can be freely and reversibly inserted and extracted between the cathode and anode, which is visually called "rocking chair battery" and later named "lithium ion battery" [2]. Both of the lithium-ion batteries and lithium metal batteries can work by the insertion and extraction of lithium ions at the electrode. However, compared with the lithium metal battery, the lithium metal can be replaced by other active materials to solve the safety problem of lithium metal as the anode material in lithium-ion battery [3]. As the first company to commercialize lithium-ion batteries, Sony Corporation has done much research work [4, 5]. Currently, commercial lithium-ion batteries mainly use transition metal lithium salts as the positive electrode Li_xM_2 (M represents a transition metal such as Co, Mn, Ni, Fe, etc.), and inexpensive and excellent conductive porous graphite as the negative electrode. They are widely used in digital products, grid energy storage, electric vehicles (EVs), hybrid electric vehicles (HEVs), and etc. [6–9].

1.1.2 Basic Structure of Lithium-Ion Battery

The composition of lithium-ion battery is shown in Figure 1.1. Lithium-ion battery is mainly composed of the following four parts: cathode, anode, electrolyte, and

Battery Technologies: Materials and Components, First Edition. Edited by Jianmin Ma.

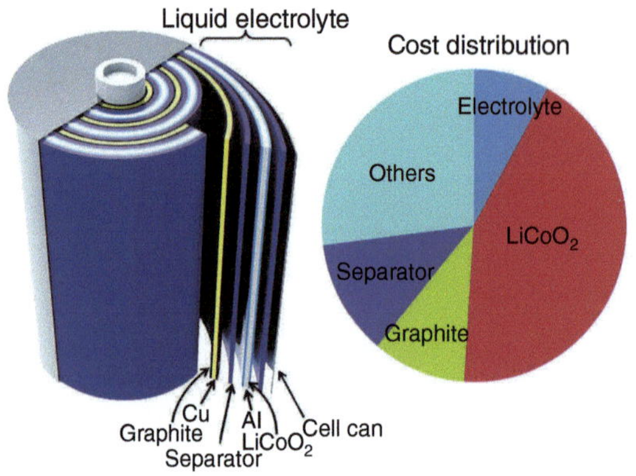

Figure 1.1 Structural illustration and relative cost of each component. Source: Zhou et al. [10]. Reproduced with permission of John Wiley Sons.

separator. The main purpose of cathode materials is to provide lithium ions for the whole battery system. At present, the main positive materials are Li_2M (M = Co, Ni, Mn, and other transition metals) with layered structure, ternary materials (Li [Co, Ni, Mn]$_2$), $LiMn_2O_4$, and $LiMPO_4$ (M = Fe, Co, Ni, Mn, and so on) with spinel structure. The main commercial cathode material of lithium-ion battery is $LiCoO_2$. The cost of the material can account for about half of the total cost of lithium-ion battery. Its theoretical capacity is 274 mAh g^{-1}, and the discharge voltage is 3.6 V [5].

The anode is generally prepared by uniformly loading the active material together with the conductive agent (generally carbon black) and the binder on the current collector, and it is the key part of lithium-ion battery. Currently, the commonly used collector is copper foil with the thickness of 7–15 μm.

The electrolyte in lithium-ion batteries is an important medium for the free transportation of lithium ions between the cathode and anode. The electrolyte is generally composed of lithium salts ($LiPF_6$, $LiClO_4$, and $LiBF_4$) and organic solvents. The common organic solvents are ethylene carbonate (EC), propylene carbonate (PC), dimethyl carbonate (DMC), and diethyl carbonate (DEC).

The function of the separator in the lithium-ion battery is to prevent the anode and cathode from contacting and thus avoiding the short circuit of the battery. The most commonly used membranes are polymer films, including polypropylene (PP) and polyethylene (PE). Generally, the strength of the separator is improved by three-layer structure, and the lithium ions can pass through the separator smoothly.

1.1.3 Working Mechanisms of Lithium-Ion Battery

The working principle of lithium-ion battery is a simply process of lithium-ion continuously embedding and detaching between the cathode and anode. Its essence is a kind of concentration battery. Figure 1.2 shows the working principle of lithium-ion battery [11]. During charging process, the oxidation reaction of the cathode materials is taken place and the lithium ion transfers from the cathode to the anode. The lithium ion is embedded in the anode material after passing through the electrolyte and separator. At the same time, the electron reaches the anode through the external circuit. During the discharge, the lithium ion is removed from the anode and

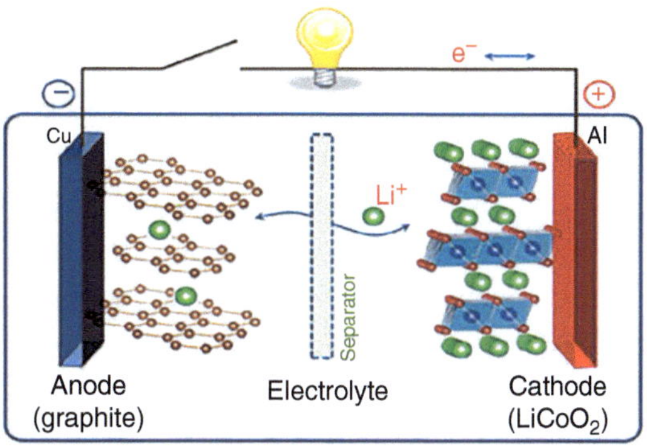

Figure 1.2 The working principle of lithium-ion battery. Source: Wang et al. [11] Reproduced with permission of Elsevier.

transferred to the cathode and embedded in the cathode materials, accompanied by the electron transfer in the external circuit.

Taking porous graphite as the anode materials, lithium cobaltate ($LiCoO_2$) as the cathode materials, the electrochemical reactions during the charge and discharge process are briefly described as follows:

$$LiCo_2 + 6C \leftrightharpoons Li_{1-x}Co_2 + Li_xC_6 \tag{1.1}$$

$$\text{Cathode}: LiCo_2 \leftrightharpoons xLi^+ + e^- + Li_{1-x}Co_2 + Li_xC_6 \tag{1.2}$$

$$\text{Anode}: xLi^+ + e^- + 6C \leftrightharpoons Li_xC_6 \tag{1.3}$$

1.1.4 Characteristics of Lithium-Ion Batteries

The advantages of lithium ion batteries are mainly as follows:

- Compared with normal chemical batteries, the lithium-ion battery has a large specific capacity and energy density. The volume of the lithium-ion battery is 20–50% of that of a chemical battery with the same capacity. At this stage, the actual specific energy of the lithium ion battery is 150–200 Wh kg^{-1}, and the specific energy of the lithium ion battery can eventually reach 250–300 Wh kg^{-1}.
- Lithium-ion battery allows a wide working range. Under room temperature, the discharge capacity of the battery accounts for more than 85% of the overall theoretical capacity after one month in an open circuit. Lithium-ion batteries can be discharged steadily in a wide temperature range (−20 to 55 °C).
- Lithium-ion battery can be recycled and used many times. The lithium-ion batteries currently used have avoided the problems of internal lithium dendrite short circuits that cause damage and potential safety hazards. The remaining capacity of the lithium-ion battery with carbon material as the anode is still more than 60% of the theoretical capacity after 1200 cycles, which is much higher than that of other types of batteries.
- Compared with lithium metal battery, lithium-ion battery possesses the resistance characteristics of short circuit, overcharge, overdischarge, and impact. It can be quickly charged and discharged at a current density of 1 C.

- No memory effect exists in lithium-ion battery, and it can be repeatedly charged and discharged.
- Lithium-ion battery can be packed with small size and lightweight.

The shortcomings of lithium-ion batteries are mainly manifested as follows:

- Since the electrolyte of lithium-ion battery mainly consists of organic component, the conductivity is lower than that of a chemical battery on the market, so the corresponding internal circuit impedance is greater than that of a chemical battery.
- The voltage platform of the lithium-ion battery changes greatly (about 40%) during the discharge process, and there is no relatively more stable discharge platform. For a device that requires stable power supply, it is impossible to maximize its efficiency, and also due to the large change in the voltage platform of the lithium-ion battery, it is also difficult to estimate the remaining capacity.
- The cost of battery composition materials is high. The main cost of lithium-ion batteries comes from the relatively expensive cathode material $LiCo_2O_4$.
- A comprehensive battery management system is needed to prevent the lithium-ion battery from overcharging.

1.2 Cathode Materials for Lithium-Ion Batteries

Cathode material is one of the key components of lithium-ion battery, which determines the working voltage, capacity, and cycle life of the battery. At present, the potential cathode materials mainly include layered-structural cathode materials ($LiCoO_2$, $LiMnO_2$, $LiMn_2O_4$, $LiNi_xCo_yMn_zO_2$, and $LiNi_{0.8}Co_{0.15}Al_{0.05}O_2$), spinel structural materials ($LiMn_2O_4$ and $LiNi_{0.5}Mn_{1.5}O_4$), polyanionic materials ($LiFePO_4$, $Li_3V_2(PO_4)_3$, and Li_2FeSiO_4), etc. The ideal cathode material should have the following characteristics: (i) high capacity; (ii) high oxidation reduction potential; (iii) good chemical and thermal stability; (iv) high ionic and electronic conductivity; (v) high safety; (vi) low price.

1.2.1 Layer-Structured Cathode Materials

Common layer-structured cathode materials mainly include lithium cobalt oxide ($LiCoO_2$), lithium nickel oxide ($LiNiO_2$), and lithium manganese oxide ($LiMnO_2$), all of which exhibit the crystal structure of α-$NaFeO_2$. Among them, $LiCoO_2$ with rhombohedral structure is the first commercialized layered cathode material, and the O atom is formed hexagonal dense accumulation according to the order of ABCABC (Figure 1.3). It is first reported by Goodenough in 1981, the theoretical specific capacity of $LiCoO_2$ is 274 $mAh\,g^{-1}$, and the average working voltage is 3.9 V (vs. Li/Li^+). However, during the actual charging process, the layer structure of Co—O octahedron will become unstable and gradually transition to spinel phase when the amount of lithium removal is more than 50% and thus results in the decrease of battery capacity. Thus, the actual reversible specific capacity of $LiCoO_2$ can only be up to 140 $mAh\,g^{-1}$. In addition, the high cost and toxicity of Co limit its application in large-scale energy storage system.

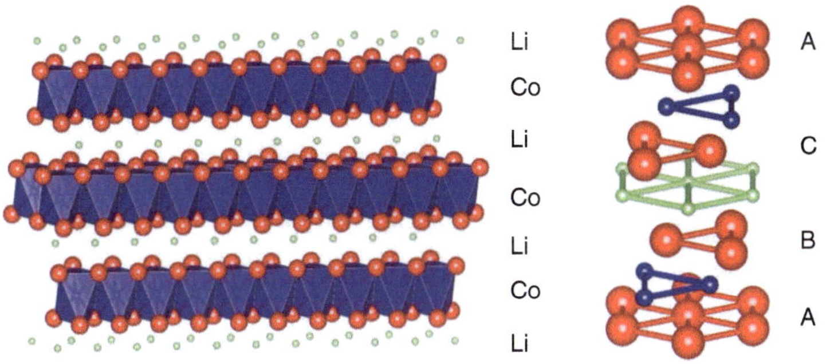

Figure 1.3 Crystal structure diagram of $LiCoO_2$ (red: oxygen, purple: cobalt, green: lithium). Source: Erickson et al [12] Reproduced with permission of IOP Publishing.

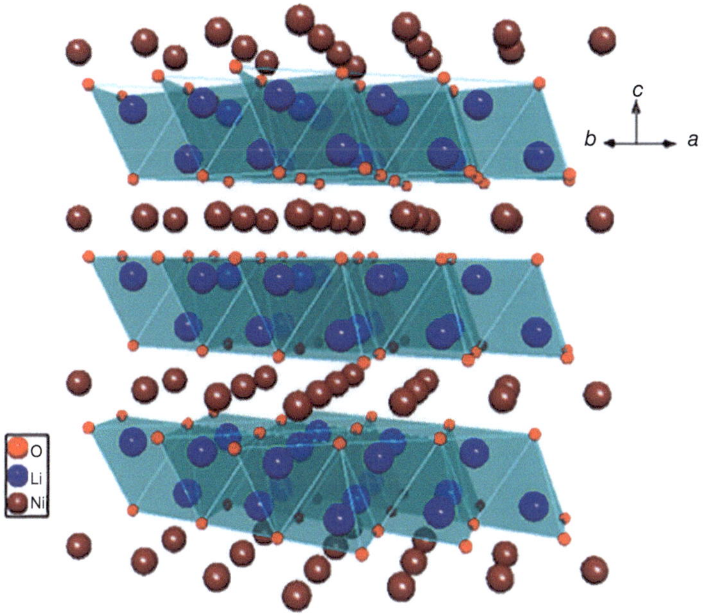

Figure 1.4 Crystal structure diagram of $LiNiO_2$ (red: oxygen, brown: nickel, blue: lithium). Source: Manthiram et al [13]. Reproduced with permission of Elsevier.

As shown in Figure 1.4, the layered $LiNiO_2$ cathode materials have a theoretical specific capacity of 274 mAh g^{-1}, but only 140 mAh g^{-1} can be used in the actual charging and discharging process. During the charging process, lithium ions are removed from the interlayer of $LiNiO_2$, due to the approach of the Ni^{2+} ion radius (0.69) and Li^+ (0.76), it is very easy to occupy the lithium position and cause the mixed arrangement of lithium nickel, which will lead to the destruction of the original layered structure, and some lithium ions cannot be re-embedded into $LiNiO_2$ during discharging and thus cause the rapid attenuation of reversible capacity [14]. In addition, it is very difficult to synthesize and unstable at high temperature, the structure will change from hexagonal phase to cubic phase.

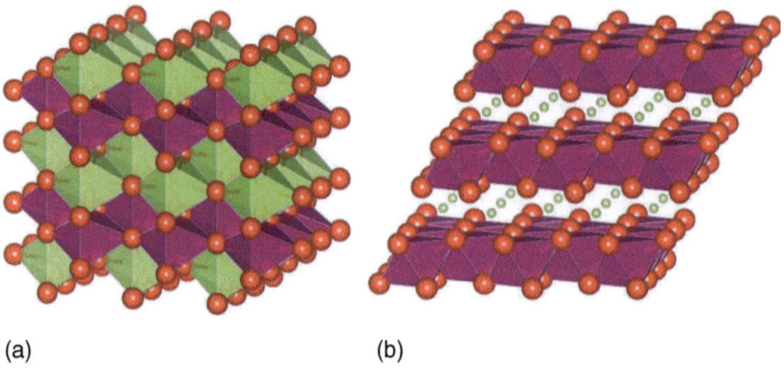

Figure 1.5 (a) Orthorhombic $LiMnO_2$ (*Pmmn*), (b) monoclinic $LiMnO_2$ (*C2/m*); (red: oxygen, pink: manganese, green: lithium). Source: Erickson et al [12]. Reproduced with permission of IOP Publishing.

The theoretical specific capacity of $LiMnO_2$ is 285 mAh g^{-1}, and the actual specific capacity is about 165–195 mAh g^{-1}. As shown in Figure 1.5, the layer-structured $LiMnO_2$ belongs to metastable phase, in which Mn^{3+} is prone to disproportionation to form Mn^{2+} and Mn^{4+}. Meanwhile, Mn^{3+} is easy to dissolve and occupy the position of Li^+ during the charging process, finally forming Jahn–Teller distortion and causing the damage of layered structure [15]. In addition, it is difficult to synthesize layer-structured $LiMnO_2$ because it is easy to transform into rhombic or spinel structure; thus, the application of pure $LiMnO_2$ as cathode materials is difficult.

As cathode materials of lithium-ion batteries, $LiCoO_2$, $LiNiO_2$, and $LiMnO_2$ exhibit their own shortcomings as cathode, and they cannot be widely used in the field of power batteries. However, it is possible to obtain new cathode materials with high capacity, high voltage, and high cycle stability by preparing Li–Ni–Co–Mn–O ternary materials, in which $LiCoO_2$ has a higher electron and ion transfer rate to ensure the rate performance, the higher capacity of $LiNiO_2$ can provide the cathode a high capacity, and $LiMnO_2$ mainly plays a role in stabilizing the structure. Theoretically, the three components can form any uniform solid solution while maintaining the original layered structure. In the ternary system, Ni, Co, and Mn elements can be randomly distributed in the transition metal layer, and Li still occupies the original position. Liu et al. [16] prepared ternary cathode materials with excellent electrochemical performance for the first time in 1999. Among them, the high capacity of nickel-rich ternary cathode material (more than 60% Ni in the composition) makes it a potential cathode material for the current development of high-energy-density power battery. Therefore, the current research field is mainly focused on the area near the $LiNiO_2$ component. As shown in Figure 1.6, it mainly includes the common $LiNi_{1/3}Co_{1/3}Mn_{1/3}O_2$ (NCM333), $LiNi_{0.5}Co_{0.2}Mn_{0.3}O_2$ (NCM523), $LiNi_{0.6}Co_{0.2}Mn_{0.2}O_2$ (NCM622), and $LiNi_{0.8}Co_{0.1}Mn_{0.1}O_2$ (NCM811). However, these kinds of nickel-rich cathode materials exhibit some drawbacks for the commercialization, including capacity fading during long-term cycling, poor rate capacity, thermal instability, short storage lifetime at elevated temperature, high residual lithium, gas evolution, serious safety concern, Li/Ni cation mixing, and so on. Thereby, some crucial mechanisms are proposed to explain these problems in

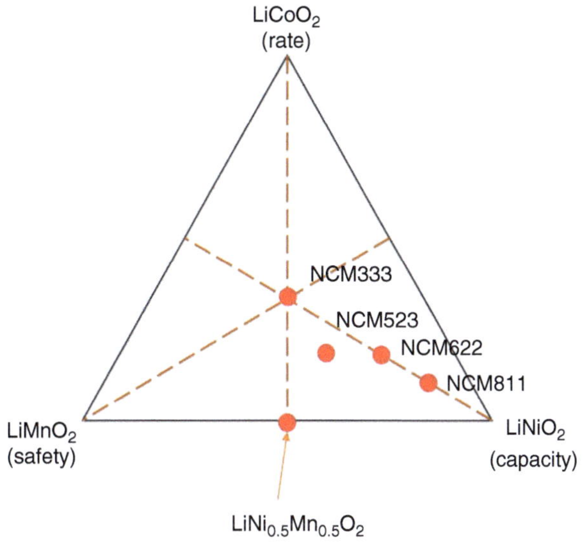

Figure 1.6 Phase diagrams of $LiNiO_2$, $LiCoO_2$, and $LiMnO_2$ ternary systems and some representative components. Source: Erickson et al [12]. Reproduced with permission of IOP Publishing.

literature: (i) side reaction of electrolyte catalyzed by the delithiated NCM at voltages above 4.3 V with a concomitant oxygen release; (ii) dissolution of the transition metal ions corroded by HF acid from the electrolyte; (iii) layered-to-spinel phase transformation and occurrence of NiO-type phase; (iv) formation of micro-crack and particle fracture originated from internal strain, expansion, and contraction of lattice volume during cycling. The micro-cracks enlarged along the parasitic reaction area between active materials and the electrolyte and accelerated the fracture of particles and decomposition of the electrolyte. Finally, the pulverization of bulk cathode particles occurred. The pulverized and separated particles cannot participate in electrochemical reaction, resulting in the capacity decay (Figure 1.7).

In addition to the common $LiNi_xCo_yMn_{1-x-y}O_2$ (NCM) ternary cathode materials, there is also a special high-capacity ternary cathode material, which is $LiNi_{0.8}Co_{0.15}Al_{0.05}O_2$ (NCA) [18]. The actual specific capacity is as high as 220 mAh g^{-1}. A small amount of Al can stabilize the layered structure. Compared with NCM, NCA has higher structural stability. However, the amphoteric characteristics of Al element make it difficult to synthesize the stable NCA precursor. At present, the NCA cathode materials have been successfully developed by Panasonic in Japan and Tesla in USA, and the energy density of 300 wh kg^{-1} of single cell can be obtained by matching the NCA with silicon/carbon anode.

1.2.2 Spinel-Structured Cathode Materials

$LiMn_2O_4$ with spinel structure belongs to cubic system, and the space group is *Fd*-3*m*. Among them, O atoms are arranged in cubic dense packing, Li ions are in the tetrahedral sites, and Mn occupies half of the octahedral sites (Figure 1.8). Thus, $LiMn_2O_4$ has three-dimensional lithium-ion diffusion channel and excellent rate performance, which is suitable for high-power lithium-ion batteries [19, 20]. The

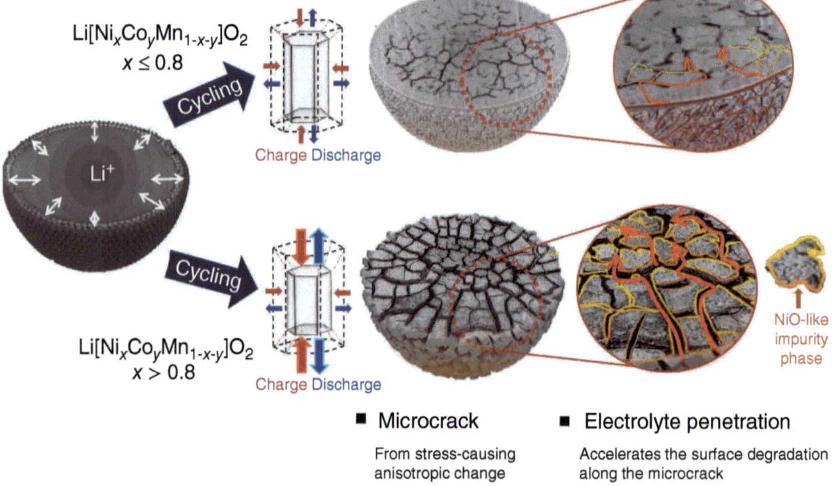

Figure 1.7 Capacity fading scheme of Ni-rich cathode materials. Source: Ryu et al. [17]. Reproduced with permission of American Chemical Society.

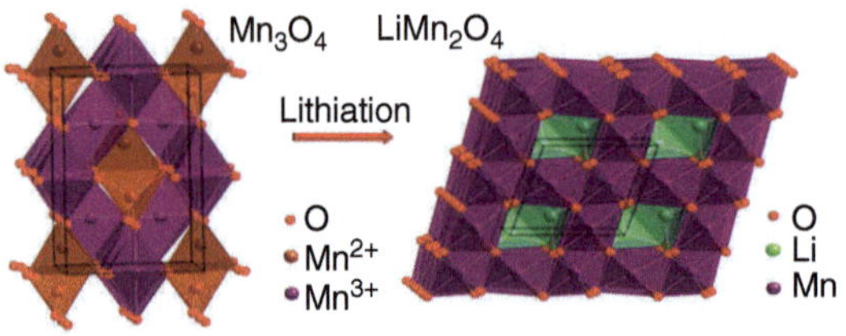

Figure 1.8 Crystal structures of spinel Mn_3O_4 and spinel $LiMn_2O_4$. Source: Adapted from Xia et al [19].

theoretical specific capacity of $LiMn_2O_4$ is 148 mAh g^{-1}, and the working voltage is 4.0 V (vs. Li/Li$^+$). However, Jahn–Teller distortion caused by entering the octahedron sites of lithium ions will occur during the charging and discharging process, which will affect the reversible capacity and cycle performance of the material. Therefore, the actual specific capacity of $LiMn_2O_4$ can only be 110–120 mAh g^{-1}. In addition, Mn is easy to dissolve in the electrolyte, and the self-discharge will lead to the decline of reversible capacity of the battery.

$LiNi_{0.5}Mn_{1.5}O_4$ can be formed by replacing part of Mn in $LiMn_2O_4$ with Ni. During charging and discharging process, the valence state of Mn in $LiNi_{0.5}Mn_{1.5}O_4$ remains unchanged and only Ni^{2+} and Ni^{4+} change, which will ensure the high structural stability of the materials. At the same time, the working voltage is also increased from the original 4.0 to 4.7 V, which helps to improve the energy density of the battery [21]. According to the different positions occupied by Ni and Mn, $LiNi_{0.5}Mn_{1.5}O_4$ can be divided into *Fd*-3*m* and *P*4$_3$32 space group. In comparison, $LiNi_{0.5}Mn_{1.5}O_4$ with *Fd*-3*m* space group has better structural stability and cycle performance. However, the ordinary organic electrolyte will decompose when the voltage is charged to 4.7 V, resulting in the rapid degradation of the battery capacity deterioration of cycle performance.

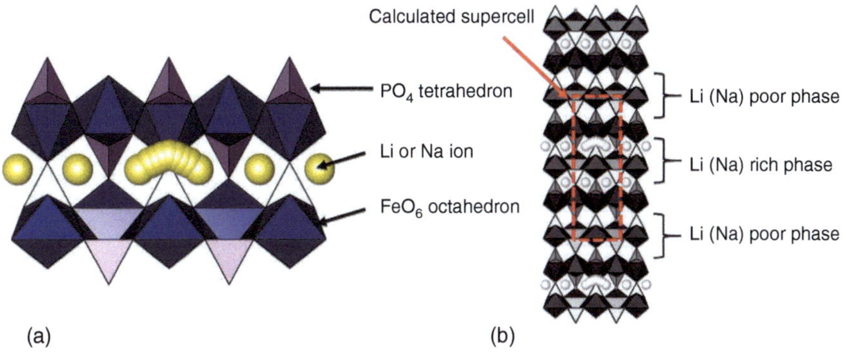

Figure 1.9 Crystal structure diagram of $LiFePO_4$. Source: Nakayama et al [22]. Reproduced with permission of Elsevier.

1.2.3 Olivine-Structured Cathode Materials

Lithium iron phosphate ($LiFePO_4$) with olivine structure has been widely used in power batteries for its excellent cycling performance and safety [22]. It was first reported by Goodenough et al. $LiFePO_4$ of orthorhombic system belongs to the *Pnma* space group, and its theoretical specific capacity is 170 mAh g^{-1}, the working voltage is about 3.4 V (vs. Li/Li^+). During the charging and discharging process, Li^+ occupied in octahedral sites can be transmitted in one-dimensional channel, accompanied by the phase transition of $LiFePO_4$ and $FePO_4$. Due to the strong bonding energy of polyanion PO_4^{3-}, it can still maintain the structural stability when Li ions are completely removed from $LiFePO_4$. However, due to the low ion and electric conductivity of pure $LiFePO_4$, it is necessary to optimize the structure of $LiFePO_4$ for practical application. To sum up, $LiFePO_4$ with low cost and mature synthesis technology is suitable for energy storage system with low energy density requirements (Figure 1.9).

1.3 Anode Materials for LIBs

For cathode materials, the commonly used materials such as lithium cobaltic acid, lithium iron phosphate, and lithium manganate have been well developed, and the capacity is close to the theoretical value. Sulfur and other high-capacity cathode materials are also under development. Therefore, in addition to properly increasing the capacity of cathode materials, people are looking for high-capacity anode materials to replace the traditional graphite anode, which is also one of the most effective strategies to improve the energy density of the lithium-ion battery.

The anode materials of lithium-ion battery can react with lithium ion during the charging process to form lithium-containing compounds, and also lithium ions can be effectively removed during the discharging process. The requirements for an ideal anode material are as follows:

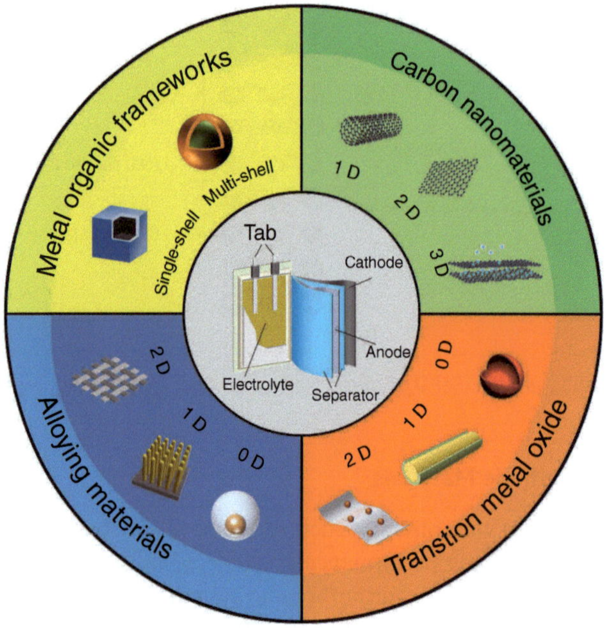

Figure 1.10 Schematic diagram of several representative nanostructured anode materials for LIBs. Source: Cheng et al. [23]. Reproduced with permission of Elsevier.

- low operating voltage;
- high electron and ion transfer rate;
- high capacity;
- high stability

Due to its high stability, graphite with the theoretical specific capacity of 372 mAh g^{-1} is commercially used as anode material, and the actual specific capacity can only be 320–350 mAh g^{-1}. With the increasing requirement of the energy density for lithium-ion battery, the low capacity of graphite cannot meet the requirements of the battery. Therefore, newly high-capacity anode materials are highly desired to replace graphite to improve the energy density of the battery (Figures 1.10 and 1.11). According to the mechanism of lithium storage, the anode materials mainly studied at present can be divided into the following three categories: that is, (i) intercalation anode material (lithium ion is reversibly intercalated/exfoliated in the gap between material layers), mainly including common graphite like carbon materials and lithium titanate, (ii) alloy-type anode materials (lithium ion reacts with anode material to form alloy for lithium storage), including silicon, germanium, tin, aluminum, magnesium, etc., and (iii) conversion anode materials (lithium ions react reversibly with metal oxides, nitrides, or phosphates to form metal atoms and lithium-containing compounds including transition metal oxides, sulfides, nitrides, etc.).

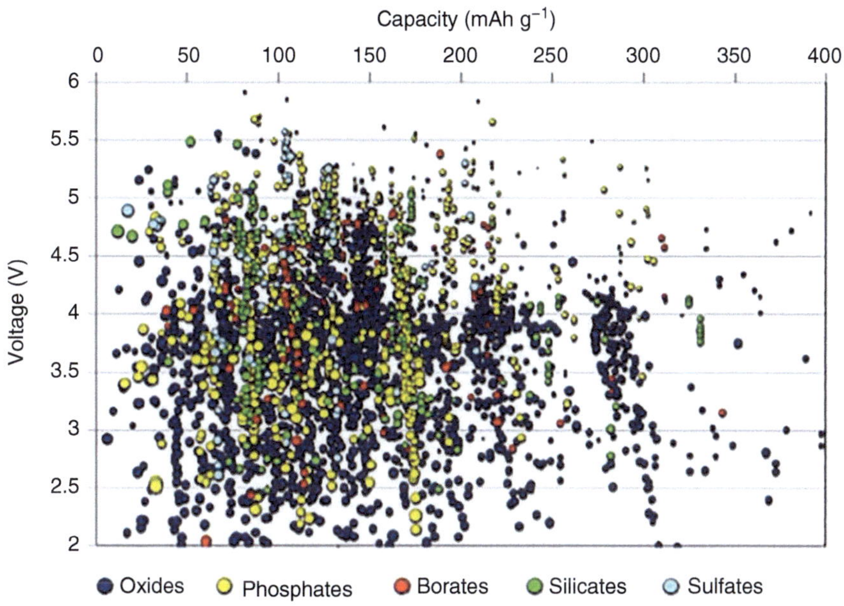

Figure 1.11 The calculated relationship between the LIBs voltage and the theoretical capacity of different compounds for screening materials. Source: Cheng et al. [23]. Reproduced with permission of Elsevier.

1.3.1 Intercalation Anode Materials

Intercalation anode materials mainly rely on the lithium ion diffusion to the gap of layered structure for lithium storage, including carbon-based materials and spinel structured lithium titanate ($Li_4Ti_5O_{12}$). According to the graphite degree of the carbon-based materials, it can be divided into layered graphite, soft carbon, and hard carbon. As a typical carbon, graphite, which mainly includes artificial graphite and natural modified graphite, was commercially used as anode material in 1990s. As shown in Figure 1.12a, during the charging process, lithium ion can be inserted into the interlayer of graphite to form LiC_6, which will impart the graphite with the theoretical specific capacity of 372 mAh g^{-1}. Meanwhile, the voltage of lithiation platform is low. Due to its excellent cycle stability and low cost, graphite anodes are widely used in commercial lithium-ion batteries. However, the poor compatibility between graphite and organic electrolyte will cause the solvent co-insertion during charging, which affects the performance of the battery. In addition, the lower redox potential of carbon materials (close to that of lithium precipitation) is prone to cause the precipitation of lithium dendrite at a high rate or under the condition of overcharge, which leads to short circuit of battery and serious safety problems.

As shown in Figure 1.12b, three lithium ions can be reversibly intercalated into the spinel structured lithium titanate ($Li_4Ti_5O_{12}$) to form $Li_7Ti_9O_{12}$ phase, and the theoretical specific capacity is 175 mAh g^{-1}, the working voltage is about

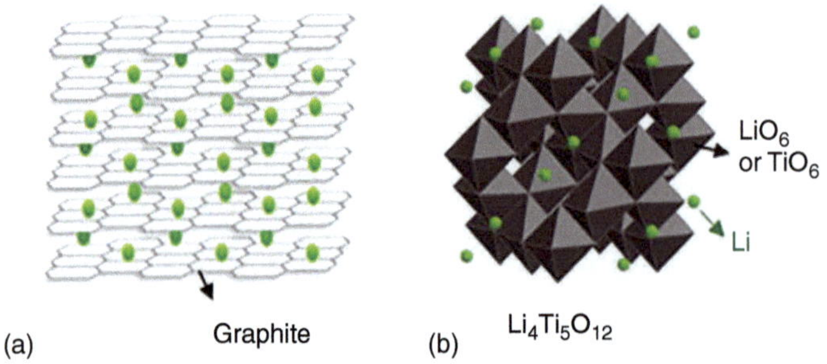

Figure 1.12 Crystal structures of (a) lithiated graphite and (b) $Li_4Ti_5O_{12}$. Source: Nitta et al. [24]. Reproduced with permission of Elsevier.

1.55 V (vs. Li/Li^+). The formation of lithium dendrites can be effectively avoided by a higher lithium intercalation/de-intercalation platform, thus improving the safety of battery [25]. At the same time, no structural change of $Li_4Ti_5O_{12}$ occurred during the whole charging and discharging process, and the volume change is less than 0.2%, which is called a "zero strain material" with ultrahigh structural and cyclic stability. However, the poor ions and electric conductivity of $Li_4Ti_5O_{12}$ will result in the poor rate performance of the battery [26]. In general, the intercalation anode materials are of low cost, high safety, and long cycle life. However, the low theoretical specific capacity makes it suitable for energy storage systems with low energy density requirements.

Compared with the ordered materials such as graphite or lithium titanate, amorphous carbon with disordered structure can be thought to be a large number of graphite particles with high crystallinity that are dispersed in the amorphous carbon matrix. Amorphous carbon mainly can be prepared by pyrolysis of the organics at high temperature, and the graphitization degree can be increased with the increase of carbonization temperature. It is worth noting that amorphous carbon can be divided into soft carbon and hard carbon according to the degree of graphitization, as shown in Figure 1.13. Among them, soft carbon refers to the amorphous carbon materials, which can be graphitized above 2000 °C. Soft carbon with low crystallinity, large crystal surface spacing, high irreversible capacity, and good compatibility with electrolyte mainly includes mesophase carbon microsphere, coke, carbon fiber grown in gas phase, and pitch-based carbon fiber, etc. Therefore, soft carbon is generally not directly used as the anode material for lithium-ion battery. Hard carbon refers to the amorphous carbon materials, which are difficult to graphitize even at 3000 °C, and it can be prepared by pyrolysis of polymer (such as polyvinyl alcohol and polyethylene, etc.) and biomass materials (such as shell, animal shell, and cotton, etc.). As anode materials for lithium-ion battery, hard carbon with the high specific capacity of 1000 mAh g^{-1} is favorable for the intercalation of lithium ions without significant expansion of the structure [28]. The layer spacing of amorphous carbon is generally larger than that of traditional graphite, and more lithium ions can be stored in the process of lithium intercalation. In addition, defects in amorphous carbon can absorb lithium ions stably and

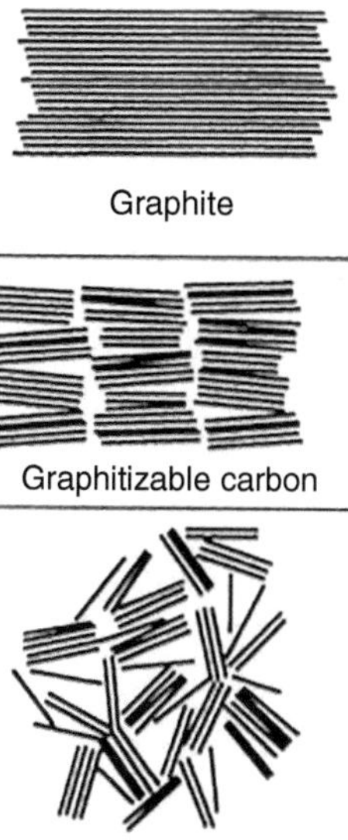

Figure 1.13 Schematics of morphologies of graphite, graphitizable carbon (soft carbon), and nongraphitizable carbon (hard carbon). Source: Nishi et al. [27]. Reproduced with permission of John Wiley & Sons.

improve the lithium storage performance. In conclusion, amorphous carbon with high capacity possesses a wide range of applications as the anode of lithium-ion batteries.

1.3.2 Alloy Anode Materials

Alloy anode materials mainly include silicon (Si), germanium (Ge), tin (Sn), antimony (Sb), aluminum (Al), and magnesium (Mg). Table 1.1 compares the relevant parameters of alloy anode materials in IVA family and graphite. For C, Si, Ge, and Sn, their fully lithium embedded states are LiC_6, $Li_{4.4}Si$, $Li_{4.4}Ge$, and $Li_{4.4}Sn$, respectively. Compared with carbon-based materials, one atom in the alloy anode can react with many lithium ions and show high capacity. In addition, the alloy anode also has a suitable lithiation/delithiation platform, which is considered to be the most promising alternative anode material to carbon materials for the new generation of battery. For example, Si and Ge can achieve high theoretical specific capacities of 4200 and 1625 mAh g^{-1} by forming $Li_{4.4}Si$ and $Li_{4.4}Ge$ alloys, respectively. However, there are also some problems behind the high capacity of alloy anode materials. Due to the large volume change of active substances during cycling, the active material will be powdered and even peeled off from the current collector and lose electric contact with the conductive network, and the reversible capacity and the cycle stability of the battery will be deteriorated. In addition, the poor conductivity of alloy anode also limits its capacity. In order to realize the commercialization of alloy anode, it is necessary to optimize their structure and composition to improve its electrochemical performance.

Table 1.1 Comparison of Group IVA elements as anode materials for lithium-ion batteries.

Materials	C	Si	Ge	Sn
Bulk density ($g\,cm^{-3}$)	2.25	2.33	5.32	7.36
Lithiated phase	LiC_6	$Li_{4.4}Si$	$Li_{4.4}Ge$	$Li_{4.4}Sn$
Theoretical gravimetric capacity ($mAh\,g^{-1}$)	372	4200	1625	994
Theoretical volumetric capacity ($mAh\,cm^{-3}$)	837	9781	8645	7316
Voltage (vs. Li/Li^+)	0.05	0.4	0.5	0.6
Volume change (%)	12	400	272	259

Source: Liu et al. [29]. Reproduced with permission of Wiley-VCH.

1.3.3 Conversion Anode Materials

The conversion anode materials mainly include oxides, nitrides, phosphates, and sulfides of various transition metals. As an anode material for lithium-ion battery, it has a high theoretical specific capacity and is considered to be a new type of anode material with application prospect. Among the conversion anode materials A_xB_y (A = Co, Fe, Ni, Cu, Mn, Cr, Mo, etc., B = O, S, N, P, etc.), transition metal atom A can react with multiple lithium ions and thus impart the materials with high specific capacity [30]. Take metal oxide A_xO_y as an example, Figure 1.14 shows a schematic illustration of the conversion reaction of transition metal oxides for LIB anodes. From Eq. (1.4), $2y$ lithium ions can be stored per formula unit of metal oxide through a conversion reaction, causing a structural change and amorphization of transition metal oxides that involves large volume expansion. At the end of lithiation, nanoscale transition metal clusters are embedded in the lithium oxide (Li_2O) matrix. During delithiation, these transition metal clusters are oxidized to form amorphous transition metal oxide. The Li_2O, which is inert to lithium, will cause the loss of the first reversible capacity, resulting in the lower initial coulomb efficiency of the material. The high working voltage of these materials (0.5–1.5 V, vs. Li/Li^+) can avoid the formation of lithium dendrites and ensure the high safety of batteries. However, in order to obtain high energy density of battery, it is necessary to match high-voltage cathode materials, which limits the practical application of conversion anode. In addition, A_xB_y anode material is also accompanied by a certain volume change during cycling, which will inevitably cause the collapse and destruction of electrode

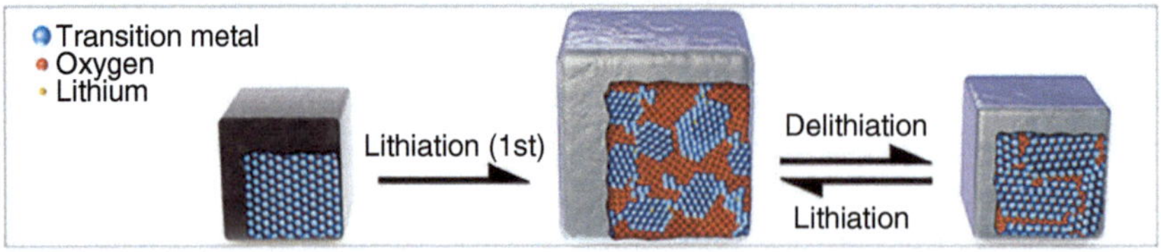

Figure 1.14 Working mechanisms of the conversion anode materials. Source: Yu et al. [31]. Reproduced with permission of John Wiley Sons.

structure, eventually leading to the decline of electrode capacity and the shortening of battery life. It is worth noting that the density of most transition metal elements is relatively high, which will increase the proportion of the anode material in the whole battery and reduce the overall energy density of the battery.

$$M_xO_y + 2yLi^+ + 2ye^- \leftrightharpoons xM^0 + yLi_2O \text{ (M transition metals)} \quad (1.4)$$

Taking tin oxide (SnO_2) as an example, it is considered as one of the most potential anode materials for the next-generation lithium-ion batteries because of its low operating voltage, high specific and volumetric capacity, high abundance, and low cost [32, 33]. However, the SnO_2-based anode materials suffer from three main issues during the charge/discharge process. The first one is the large volume change (~400%) during the lithiation and delithiation process, resulting in severe electrode pulverization and even peel off from the current collector and thus fast capacity fading during cycling [34, 35]. The second one is the poor electric conductivity of the SnO_2-based anode materials, which will reduce the charge transfer and lead to the low rate capability [36]. The last one is the poor initial Columbic efficiency due to the irreversible conversion reaction during the initial lithiation process, which finally results in the additional cathode material consumption [37, 38]. It is well known that both the conversion reaction and alloying reaction between SnO_2 and Li^+ exist during lithium storage process, as shown below (Eq. (1.6)).

$$SnO_2 + 4Li^+ + 4e^- \rightarrow Sn + 2Li_2O \quad (1.5)$$

$$Sn + xLi^+ + xe^- \rightarrow Li_xSn \quad (1.6)$$

The conversion reaction (Eq. (1.5)) is partially reversible, while the alloying reaction (Eq. (1.6)) is fully reversible. The common theoretical capacity of SnO_2 anode materials, which can be calculated from the maximum uptake of the Li ions of 4.4 mol per unit of Sn in Eq. (1.6), is 782 mAh g^{-1}. Assuming that Eq. (1.5) is reversible, the capacity contribution from Eq. (1.5) will be 712 mAh g^{-1}, and in this case, the total theoretical capacity of SnO_2 anode materials will reach as high as 1494 mAh g^{-1}. It can be concluded that the electrochemical performance of the SnO_2 anode materials largely depends on the reversibility of the two reactions, and thus it is crucial to improve the reaction reversibility of the reactions during charge/discharge process.

Three strategies can be adopted to solve the above issues (Figure 1.15). One is to encapsulate SnO_2 NPs in robust physical barriers of carbonaceous materials (e.g. amorphous carbon [C], carbon nanotube [CNT], and graphene [Gr], etc.), conductive polymers (e.g. polypyrrole [PPy], polydopamine [PDA], and polyaniline [PANI], etc.), or inorganic materials (e.g. TiO_2, Au, NiTi, vanadium carbide [V_2C] MXene). These physical barriers can effectively hinder Sn coarsening by disjoining SnO_2 particles, while simultaneously improving the structural stability and electrical conductivity of the electrodes. Moreover, during the synthesis process, the physical barriers can function as loading matrixes that promote the generation of well-dispersed SnO_2 NPs. The second one is to construct hierarchical, porous, or hollow-structured SnO_2 architectures containing plenty of voids inside the particles

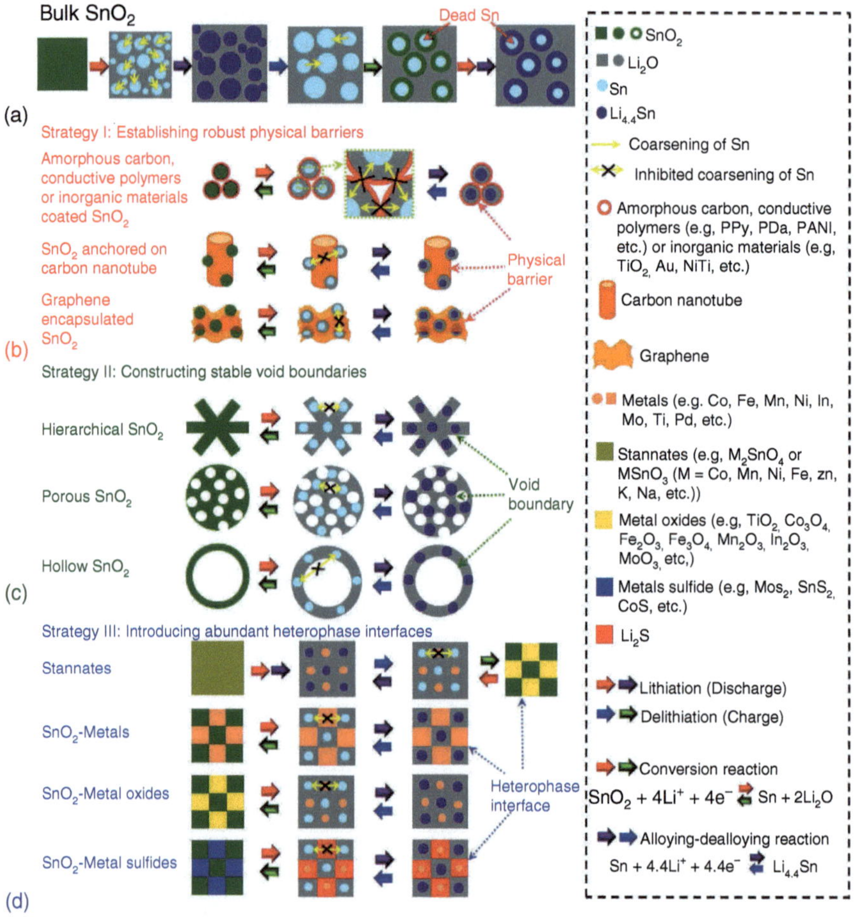

Figure 1.15 Schematic diagram showing the structure evolutions of (a) bulk SnO_2, (b) SnO_2-based nanocomposites with robust physical barriers, (c) hierarchical/porous/hollow-structured SnO_2 with stable void boundaries, and (d) stannates and heterogeneous materials/SnO_2 hybrids with abundant heterophase interfaces. Source: Zhao et al. [39]. Reproduced with permission of John Wiley Sons.

that divide particles into nanosized subunits separated by high-specific-area void boundaries. The voids not only promote electrolyte infiltration, shortening the lithium transfer distance, but also buffer the volume changes of SnO_2 particles, preventing particle pulverization and stabilizing the void boundaries that hinder Sn coarsening among the divided subunits. The last one is to fabricate stannates (e.g. M_2SnO_4 or $MSnO_3$, M = Mn, Co, Zn, Ba, etc.), doping SnO_2 with metal ions (Mx^+, M = Zn, Fe, In, W, etc.), or mixing SnO_2 with heterogeneous materials (e.g. metals (Ms, M = Au, Co, Pb, Fe, Mn, Ag, etc.), metal oxides, or sulfides (MOs or MSs, M = Co, Fe, Mn, Ni, Cu, Zn, V, Mo, Ti, etc.). The introduced Mx^+, Ms., MOs, or MSs generate abundant heterophase interfaces in cycled SnO_2-based electrodes that divide SnO_2 into individual nanocrystalline domains and impede Sn coarsening between the isolated SnO_2 domains. Furthermore, some transition metal elements (e.g. Co, Fe, Ni, Mn, etc.) that exhibit high conversion reaction reversibility with Li_2O can promote the decomposition of Li_2O and facilitate the reverse conversion reaction from Sn and Li_2O to SnO_2, improving the capacity.

1.3.4 Lithium Metal Anode

Lithium metal anode has been thought to be the holy grail to realize the highest energy density owing to the high theoretical capacity (3860 mAh g^{-1}, about 10 times that of the graphite anode) and the low electrochemical potential (−3.04 V vs. the standard hydrogen electrode) of Li metal anode [40–42]. Nevertheless, Li dendrites formed during cycling hinder the development of lithium metal anode, which leads to low Columbic efficiency, poor cycle stability, and even short-circuiting-related safety hazards [43, 44]. As illustrated in Figure 1.16, the lithium foil contacts and reacts with electrolyte to form a solid electrolyte interface (SEI) during cycling due to its high activity, and the SEI between the lithium and electrolyte is inhomogeneous and relatively fragile. In the stage of lithium deposition, Li^+ ions are more likely to deposit on the bumps located on the surface of the lithium foil due to the accumulation of electrons, forming dendritic lithium metal. The generation of dendrites and the volume change of the electrode during cycling lead to the rupture of the SEI, and the fresh lithium is again exposed to the electrolyte and reacts with it to form new SEI. This process continues throughout the battery cycling, constantly consuming electrolyte and lithium, resulting in low Columbic efficiency. The lithium dendrites may break from the matrix and be wrapped by the SEI [45], accompanied by losing their activity and forming "dead lithium," at the same time, the SEI also becomes thicker and thicker, resulting in higher and higher interface resistance. More seriously, the gathering of dendrite is highly possible to pierce the membrane, causing a short circuit inside the battery, leading to ignition or explosion

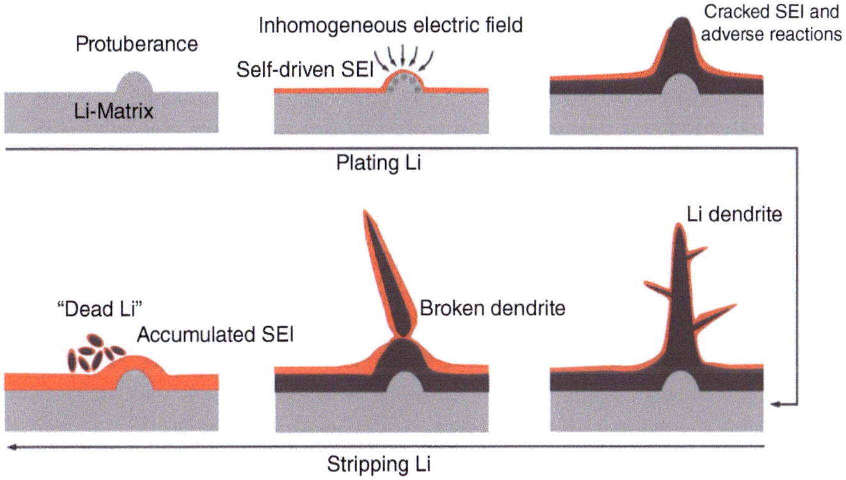

Figure 1.16 Scheme of dilemma for Li metal anode in rechargeable batteries.

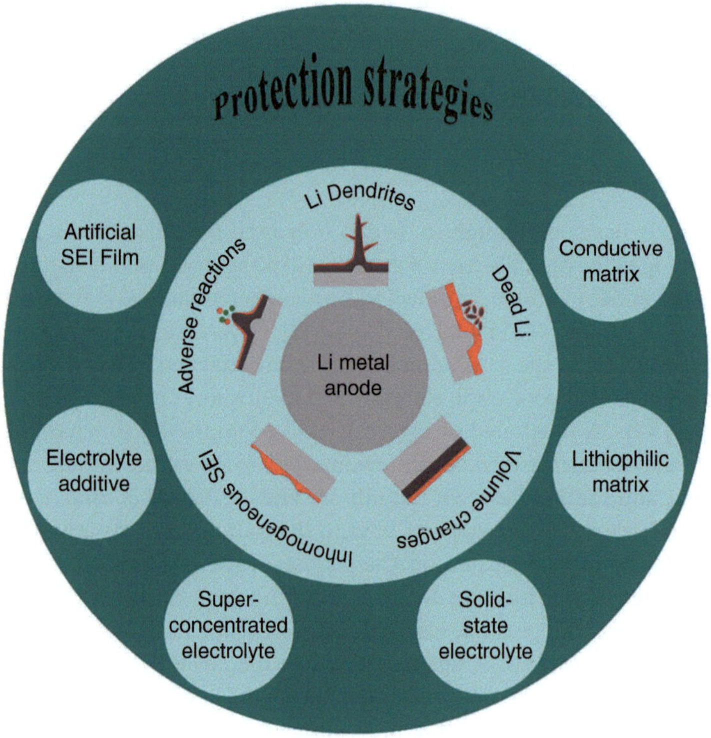

Figure 1.17 Solutions for Li metal anode in rechargeable batteries.

of the battery. Therefore, suppressing dendrite growth and mitigating electrode's volume change are fundamental methods for improving the performance of lithium metal anode [46–51].

According to previous reports, the current strategies for addressing the issues of lithium metal anodes are as follows (Figure 1.17): (i) designing a stable and uniform artificial SEI to replace the fragile native SEI, and thus achieving homogeneous Li deposition [52, 53]; (ii) developing electrolyte additives to help uniform Li deposition or stabilize SEI [54–56]; (iii) employing high-modulus solid-state electrolyte (SSE) to inhibit the growth of lithium dendrites [57–59]; (iv) replacing lithium metal with lithium alloy to suppress dendritic lithium formation [60, 61]; (v) constructing novel structured Li metal anode by nanotechnology to regulate Li ions plating/stripping behavior and mitigate the volume change during repeated cycling [62–65]. Each of the above methods has its advantages and disadvantages. For example, although the artificial SEI cannot reduce the overall energy density of the batteries, most of them are not strong enough to withstand the constant volume change and may be ruptured after a long cycling. Adding additives into electrolyte is easy to operate on a large scale; however, those additives will be continuously consumed during cycling. SSEs are considered particularly promising because of their inherent safety characteristics and potential to prevent dendritic deposition of the lithium, while the problems of low ionic conductivity and large interfacial impedance hinder the commercialization of SSE [66]. Alloyed anodes can efficiently suppress the Li dendritic formation, but they will increase the mass of the electrode, resulting in lower energy density of battery. Structured anodes can improve the performance of lithium metal anode in several aspects, but they should present good mechanical properties, and almost all of them meet the trouble of severe interfacial side reactions.

1.4 Electrolyte

The electrolyte acts as a bridge between cathode and anode to transport Li^+. The basic requirements of electrolyte with excellent performance include: (i) high ionic conductivity and low viscosity; (ii) low melting point, high boiling point, and wide temperature range; (iii) wide electrochemical window and good chemical stability; (iv) good compatibility with positive and negative materials. The above requirements are necessary for the electrolyte to work continuously and stably in the lithium-ion battery and guarantee the high performance of the lithium-ion battery [13]. The compatibility between electrolyte and cathode and anode will also affect the performance of lithium-ion battery. The reduction of electrolyte on the surface of carbon anode results in the formation of solid electrolyte interface, referred as SEI film. The oxidation reaction on the surface of metal oxide leads to the formation of cathode electrolyte interface film, which is referred as CEI film. The formation of stable interfacial film is conducive to improve the cycle stability of batteries.

1.4.1 Liquid Electrolyte

The electrolyte consists of lithium salt, solvent, and functional additive, and the three components and their working principles are described in detail below.

1.4.1.1 Lithium Salts

Lithium salt mainly provides a large amount of Li^+ for electrolyte, but its anion is also an important factor affecting the physical and chemical properties of electrolyte. Proper lithium salt can effectively improve the energy density, broaden the electrochemical window, improve the cycle life of the battery, and broaden the working temperature range of the electrolyte. Lithium salt should exhibit the following characteristics: (i) it is easy to dissolve in organic solvent with high solubility; (ii) it is easy to ionize, so that the electrolyte can be highly ionic conductive; (iii) good thermal stability and not easy to decompose; (iv) good oxidation–reduction stability; (v) noncorrosive to Al and Cu collector; (vi) easy to prepare, low cost, environmentally friendly. At present, the widely used lithium salts mainly include $LiPF_6$, $LiBF_4$, lithium hexafluoride arsenate ($LiAsF_6$), lithium perchlorate ($LiClO_4$), and other inorganic salts, as well as the organic lithium salts such as LiBOB, LiODFB, lithium trifluoromethylsulfonate ($Li(CF_3SO_3)$), and lithium trifluoromethylsulfonimide [LiTFSI]) [67–69].

(1) $LiPF_6$. At present, the commercial lithium salt is mainly $LiPF_6$, which possesses the following advantages: high solubility and conductivity; effective film formation on the electrode surface; passivation of the positive collector to prevent its dissolution; wide electrochemical stability window. However, the thermal stability of $LiPF_6$ is poor, and it is easy to decompose into PF_5 and LiF at high temperature, and the decomposition products are easy to hydrolyze to produce HF, which will damage the structure and performance of the battery and pollute the environment [70].
(2) $LiBF_4$. $LiBF_4$, which exhibits higher thermal stability than $LiPF_6$, is not sensitive to water and is not easy to generate HF. It has small charge transfer impedance at low temperature. It is the most widely used lithium salt in industry except $LiPF_6$ [71]. Compared with $LiPF_6$, $LiBF_4$ has a small anion radius and is easy to associate, so its solubility in organic carbonate solvent is low and its conductivity is not high. Moreover, $LiBF_4$ has a poor film-forming property and poor compatibility with the electrode, so it is usually used in combination with other lithium salts [72].
(3) LiBOB. LiBOB is a new type of boron-based lithium salt. Due to its excellent performance, its research is more in-depth. LiBOB has high decomposition temperature, good thermal stability, easy to form film on the negative electrode surface, and has a good passivation effect on Al collector [70]. The lithium-ion battery with LiBOB can be cycled well at 60 and 70 °C, while lithium-ion battery with $LIPF_6$ cannot work under the same conditions [67]. LiBOB is mainly sensitive to water, and its solubility in common carbonate solvent is small. Although it is easy to form a film, its impedance is large, which seriously affects the low-temperature performance and rate performance of the battery [69]. It is rarely used alone and is generally used as an additive.
(4) LiODFB. LiODFB is another new type of boron-based lithium salt. It combines the structure and advantages of LiBOB and $LiBF_4$. LiODFB is not only similar to LiBOB, but also can form stable SEI film on the anode surface. Compared with LiBOB, the SEI film formed by LiODFB has smaller impedance and higher stability [73]. LiODFB can passivate aluminum foil and inhibit the oxidation of electrolyte, so it has good compatibility with cathode materials. However, the high impedance of LiODFB at low temperature limits its application at low temperature.
(5) LiTFSI. Because of its high conductivity (comparable to $LiPF_6$) and high thermal decomposition temperature (>360 °C), the stable interfacial film on the cathode surface and low charge transfer resistance at low temperature can be obtained. It has attracted wide attention in the field of high and low-temperature batteries [74]. However, LiTFSI is corrosive to the current collector.

1.4.1.2 Organic Solvent

The properties of organic solvents also affect the properties of electrolyte. The following properties should be considered in the selection of solvents for wide-temperature electrolyte: (i) high flash point and low vapor pressure, which are related to the safety

Table 1.2 Some physical properties of organic solvents.

	Statue	Solvent	Melting point (°C)	Boiling point (°C)	Flash point (°C)	Dielectric constant (25 °C)	Viscosity (25 °C)	D.N.	A.N.
Carbonate	Ring	EC	36.4	248	150	89.78	1.90 (40 °C)	16.4	
		PC	−48.8	242	135	64.92	2.53	15.1	18.3
		BC	−53	240		53	3.2		
		VC	22	162	73				
	Chain	DMC	4.6	91	15	3.107	0.59 (20 °C)	16	
		DEC	−74.3	126	33	2.805	0.75	14.6	
		EMC	−53	110	23	2.958	0.64		
Carboxylate	Ring	γBL	−43.5	204	101	39	1.73	18	18.2
	Chain	EA	−84	77	−4	6.02	0.45	17	
		MF	−99	32	−32	8.5	0.33		
		MA	−98	57		6.7	3.64	16.5	
		PA	−92.5	101.6	14				
		EP	−73.9	99.1	12.2		0.9 (15 °C)		
		MB	−84	102	11		0.6		

EC: Ethylene carbonate, PC: Propylene carbonate, BC: Butene carbonate, VC: Vinyl carbonate, DMC: Dimethyl carbonate, DEC: Diethyl carbonate, EMC: Ethyl carbonate, γ BL: γ-butyrolactone, EA: Ethyl acetate, MF: Methyl formate, MA: Methyl acetate, PA: Propyl acetate, EP: Ethyl propionate, MB: Methyl butyrate.

of the battery at high temperature; (ii) low melting point and high boiling point, the solvent is in a liquid state in a wide temperature range; (iii) high dielectric constant, which is favorable for the dissolution of lithium salt and ionize into free ions; (iv) wide electrochemical window. At present, the main solvent used in the electrolyte of lithium-ion battery is carbonate solvent. The common organic solvents are listed in Table 1.2.

Carbonate mainly includes two types, that is, ring carbonate with high dielectric constant and viscosity, and chain carbonate with low dielectric constant and viscosity. Carbonate solvents are widely used in lithium-ion batteries because of their good electrochemical stability, high flash point, and low melting point. The ring carbonate solvents (EC and PC) with a higher dielectric constant and flash point can effectively dissolve and ionize lithium salts. EC is an indispensable component in electrolyte solvent system because of its good film-forming property. The ionic conductivity of EC is higher than that of the corresponding PC-based electrolyte, but EC is a solid at room temperature, which is often mixed with chain carbonate (mainly DMC, DEC, and EMC [Ethyl carbonate]) with lower melting point; however, the low-temperature performance of the electrolyte system containing EC is poor. The lower melting point of PC can widen the lower temperature limit of electrolyte. However, PC and Li^+ are easy to be co-embedded into graphite. In general, PC needs to be used together with solvents with good film-forming properties, such as EC, or film-forming additives. It can be seen from the data in Table 1.2 that the polarity of the solvent with high dielectric constant is larger, which is conducive to the dissolution and ionization of lithium salt, meanwhile, the viscosity and melting point of the solvent are often high, which will affect the low-temperature performance of the electrolyte, while the relative dielectric constant of the solvent with low viscosity is also lower. Therefore, it is not easy to meet all the requirements of electrolyte for one solvent. Generally, the ring carbonate and chain carbonate are used together to widen the application temperature range of electrolyte.

Due to the low oxidation potential of ether-based electrolyte, it is rarely used in lithium-ion batteries. Sulfone solvents are often used in high-voltage batteries because of their wide electrochemical window. However, the melting point of sulfone is generally high, so it cannot be used in wide-temperature electrolyte system. Carboxylate solvent has low melting point and can be used as a cosolvent in low-temperature electrolyte system [75]. However, carboxylate with low flash point and high vapor pressure may reduce the safety of electrolyte at high temperature, which is not suitable for wide-temperature electrolyte [76].

1.4.1.3 Functional Additives

Additives are also one of the essential components in the electrolyte system and are considered to be the most economical and effective component to improve the cycle performance and life of the battery. The amount of additives, whether calculated by mass or volume fraction, shall not exceed 5%. There are many kinds of electrolyte additives, such as lithium salt, solvent, and even polymer. According to its function, it can be divided into anode film-forming additive, cathode film-forming additive, anti-overcharge and overdischarge protection additive, and flame retardant additive. The film-forming additives of electrodes affect the interface properties between the electrolyte and the electrode.

(1) Anode film-forming additive. During the first charge of the battery, the electrolyte will reduce and decompose on the surface of anode, thus forming SEI film on the surface of anode. PC is an ideal solvent for widening the lower temperature limit of electrolyte because of its low melting point. However, PC and Li^+ are easy to be co-embedded between graphite layers, so it is necessary to add film-forming agent to form stable SEI membrane.
    At present, vinyl carbonate (VC) is widely used as anode film-forming additive, due to the unsaturated C=C double bond in VC molecular structure and the lower energy of the lowest space orbit (LUMO), which can preferentially reduce the electrolyte, form a stable SEI film, and improve the capacity of cell. It is reported that the initial capacity of cell can be increased from 208 to 334 mAh g^{-1}. It is shown that the mechanical properties of SEI film can be improved by adding VC in the electrolyte [76]. Fluoroethylene carbonate (FEC) is also a commonly used anode film-forming additive [77]. F atom can

be introduced into EC to form C—F bond by adding FEC. Due to the strong electron absorption ability of C—F bond, the LUMO energy of FEC is lower than that of EC. Similar to VC, FEC can preferentially reduce to form stable and low-impedance SEI film, so as to improve the compatibility between electrode and electrolyte.

Boron compounds have been extensively studied because of their ability to improve the interfacial membrane of the battery. For example, the LiBOB mentioned above also has the advantage of low LUMO energy than that of organic carbonates, which can be used as an additive to give priority to film formation, inhibit the decomposition of electrolyte, and stabilize the interface impedance. LiODFB formed by introducing F atom with stronger electronegativity to LiBOB has similar film-forming performance with LiBOB, and SEI film with higher stability and lower impedance can be formed on the surface of anode [78].

(2) Cathode film-forming additive. When Li^+ is embedded/removed from the cathode materials, the valence state of the central transition metal atom will change, which makes the lattice constant change. During the charging process, when a large amount of Li^+ comes out of the cathode, the change of lattice constant may lead to the phase transformation of crystal structure and even destroy the crystal structure, resulting in the dissolution of metal ions. The electrolyte will be oxidized and decomposed under the catalysis of transition metal ions at high potential. Therefore, in order to protect the stability of the cathode materials, the cathode film-forming additive should also be added. The formation mechanism of CEI film is that the electrolyte is oxidized on the surface of cathode. According to this mechanism, as long as the molecules with higher energy of the highest electron occupied orbit (HOMO) than that of the electrolyte solvent are selected and added to the electrolyte, they will be preferentially oxidized, so as to improve the compatibility between the electrolyte and cathode.

It has been reported that biphenyl and thiophene possess high orbital energy (HOMO) and are prone to oxidative decomposition. The capacity retention of $LiCoO_2$ cells can be significantly improved by adding 0.1% biphenyl and thiophene [79]. It is shown that biphenyl and thiophene can form interfacial film on the surface of cathode, which will inhibit the oxidation decomposition of the electrolyte and improve the cycle performance of the battery. In addition, LiBOB and LiODFB can also help to form film on the surface of cathode, finally improving the cycling performance of the battery [80].

1.4.2 Solid Electrolyte

Solid electrolyte, also known as fast ionic conductor or super ionic conductor [67], can be divided into polymer electrolyte and inorganic solid electrolyte (Figures 1.18 and 1.19). According to crystalline state, it can be divided into crystalline electrolyte and amorphous electrolyte. Crystalline electrolyte refers to inorganic solid electrolyte, while amorphous electrolyte includes polymer electrolyte and glassy inorganic solid electrolyte. The solid electrolyte for lithium-ion battery should meet

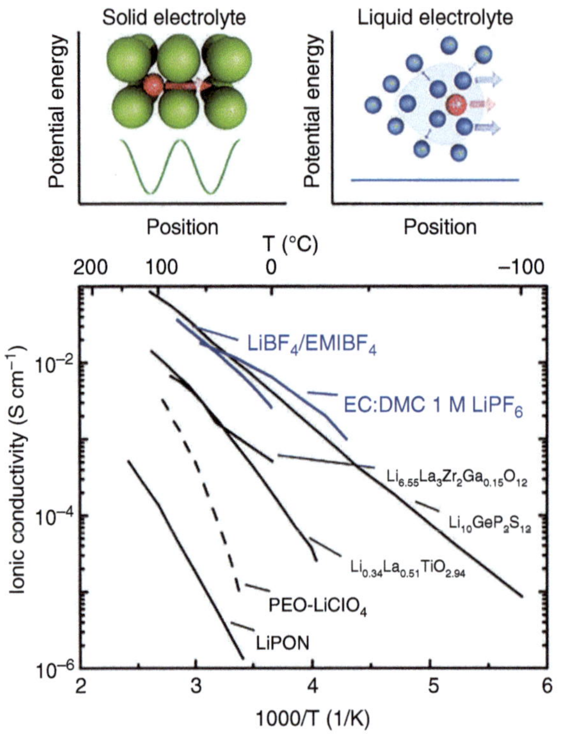

Figure 1.18 Reported total lithium-ion conductivity (unless otherwise mentioned) as a function of temperature. Source: Bachman et al. [81]. Reproduced with permission of American Chemical of Society.

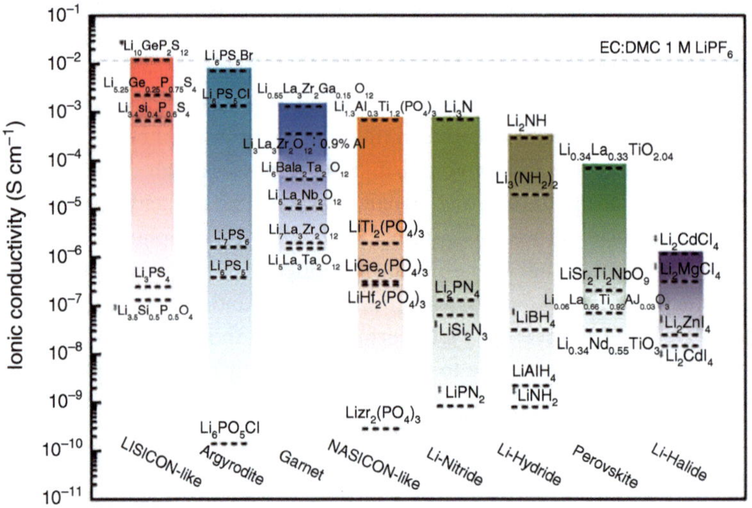

Figure 1.19 Reported total ionic conductivity of solid-state lithium-ion conductors at room temperature. Source: Bachman et al. [81]. Reproduced with permission of American Chemical of Society.

the following requirements [82]: good lithium-ion conductivity; very low electronic conductivity; no or very small grain boundary resistance; good chemical stability, no reaction with electrode materials and Li; high electrochemical decomposition voltage; green environmental protection, low price and easy to prepare.

1.4.2.1 Polymer Electrolyte

Polymer electrolyte is a kind of polymer/lithium salts composite, which is based on the amorphous structure of PEO, polyvinyl butyral (PVB), polyacrylonitrile (PAN), polyvinylidene fluoride (PVDF), polymethylmethacrylate (PMMA), and so on. The conductivity comes from the migration or transition of lithium ions in the spiral channel formed by polymer chain (Figure 1.20). At the early stage, the polymer electrolytes were mainly formed by PEO and alkali metal salts, and the ionic conductivity was only 10^{-8} S cm^{-1} [84]. Later, the researchers found that the room temperature conductivity of the material was obviously improved to 10^{-4}–10^{-3} S cm^{-1} by introducing the organic solution into polymer electrolyte to form gel polymer electrolyte [85, 86]. Decreasing the glass transition temperature and increasing the number of carrier (Li^+) are the main ways to improve the ionic conductivity of materials, include doping, adjusting polymer structure, organic solvent plasticizing, and changing the structure and concentration of lithium salt [86]. However, due to its poor thermal stability and mechanical properties, its development and application are seriously limited.

1.4.2.2 Li_3N and its Derivatives

Li_3N is the first reported inorganic solid electrolyte with high ionic conductivity at room temperature. Its crystal structure is hexagonal system, which contains Li_2N layer and pure Li layer perpendicular to the *c* axis (Figure 1.21). The conductivity of Li_3N crystal is anisotropic. The conductivity of ions perpendicular to the *c*-axis

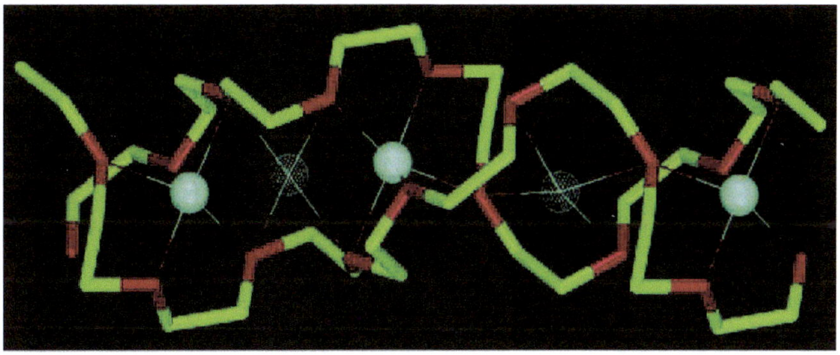

Figure 1.20 Schematic diffusion pathway of the Li^+ cations in PEO_6:$LiPF_6$. Thin lines indicate coordination around the Li^+ cation; solid blue spheres, lithium in the crystallographic five-coordinate site (note that the fifth thin line is very short in this view); meshed blue spheres, lithium in the intermediate four coordinate sites; green, carbon; red, oxygen. Source: Stoeva et al. [83]. Reproduced with permission of American Chemical Society.

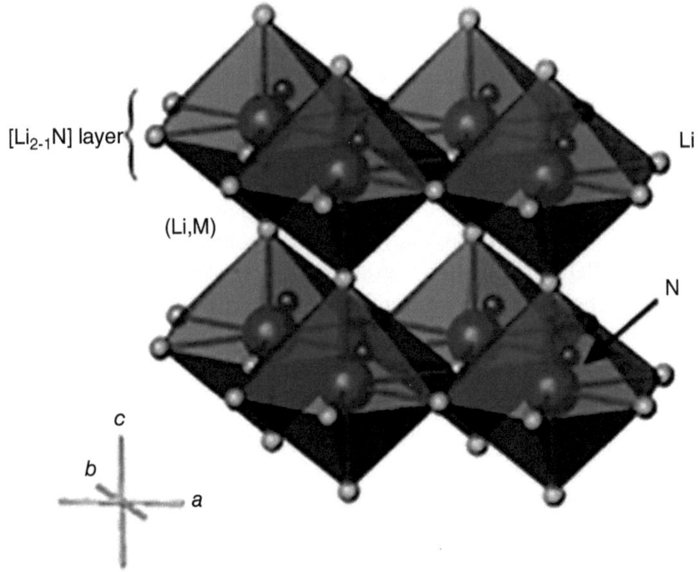

Figure 1.21 Structure of $Li_{3-x-y}M_xN$. Polyhedral representation showing layers of edge-sharing $NLi_6(Li, M)_2$ hexagonal bipyramids linked by vertexes along the *c* axis. Source: Gregory et al. [87]. Reproduced with permission of American Chemical Society.

can reach 10^{-3} S cm^{-1}, but the conductivity parallel to the *c*-axis is very low. The decomposition voltage of Li_3N is only 0.45 V, and its chemical stability is poor (sensitive to air and flammable in water, etc.), which seriously limits its practical application. In order to improve the performance of Li_3N, LiX (X = Cl^{-1}, Br^{-1}, I^{-1}) was added to form Li_3N–LiX eutectic. The stability and decomposition voltage of Li_3N–LiX eutectic were improved obviously (>2.5 V), but the ionic conductivity decreased to 10^{-5}–10^{-6} S cm^{-1}. The conductivity of $Li_9N_2Cl_3$ was improved by replacing part of Li^+ with metal cations such as Na^+, K^+, Rb^+, CS^+, Mg^{2+}, Ba^{2+}, and Al^{3+} [87].

1.4.2.3 Perovskite Solid Electrolyte

The general structural formula of perovskite solid electrolyte is $Li_{3x}La_{2/3-x}TiO_3$ ($0.04 < X < 0.17$, LLTO) (Figure 1.22) [88]. $Li_{0.34}La_{0.51}TiO_{2.91}$ with the room-temperature ionic conductivity of 10^{-3} S cm^{-1} was successfully synthesized by Ingaguma and Chen Liquan et al. Although this kind of material has high ionic conductivity, when contacting with Li, Ti^{4+} in the structure is reduced to Ti^{3+}, it makes the material to show high electronic conductivity and become a mixed conductor of electrons and lithium ions, leading to the failure of battery, which seriously limits its practical application. The properties of the electrolyte materials can be improved by substitution of some elements. For example, Sr^{2+} is used to replace part of Li^+ and La^{3+} in the structure of ABO_3, the vacancy concentration of Li is increased, the cell volume is increased, and the bottleneck of Li^+ transport is enlarged. The bulk lithium ionic conductivity at 25 °C can be as high as 1.5×10^{-3} S cm^{-1}, while the

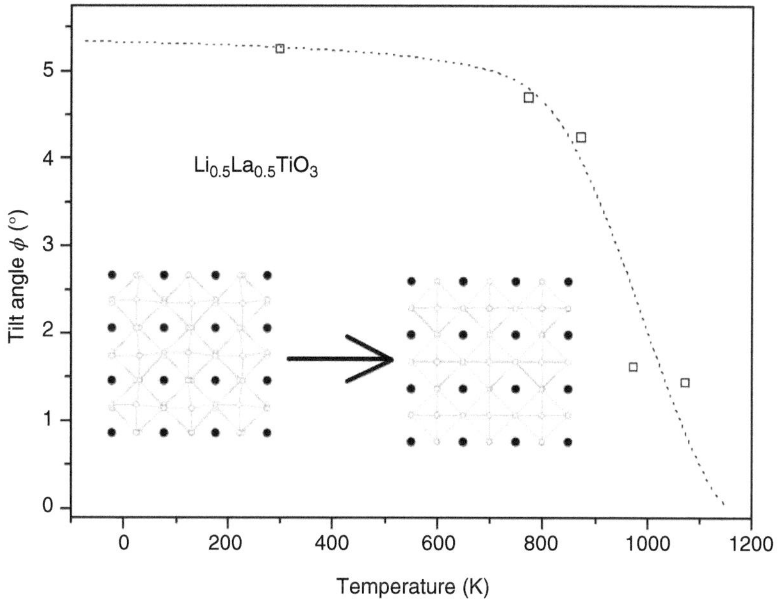

Figure 1.22 Structure of perovskite solid electrolyte. Source: Varez et al. [88]. Reproduced with permission of American Chemical Society.

total conductivity is $5.5 \times 10^{-4}\ S\ cm^{-1}$ [89]. The properties can also be improved by doping at B position with metal elements (SN, Zr, Mn, Ge, Al, etc.).

1.4.2.4 LISICON

$Li_{14}Zn(GeO_4)_4$ reported by Hong at Massachusetts Institute of technology is the first LISICON (lithium super ionic conductor) compound. The ionic conductivity can reach $0.13\ S\ cm^{-1}$ at 300 °C and $\sim 10^{-7}\ S\ cm^{-1}$ at room temperature. Then, Bruce and West studied the properties of $Li_{2+2x}Zn_{1-x}GeO_4$ [90]. LISICON can be considered as the solid solution of Li_4GeO_4 and Zn_2GeO_4, and the structure is similar to that of γ-Li_3PO_4. At the same time, the solid solutions of Li_4XO_4 (X = Si, Sc, Ge, Ti) and Li_3YO_4 (Y = P, As, V, Cr), with the general formula of $Li_{3+x}Y_{1-x}X_xO_4$, have also been studied (Figure 1.23) [91, 92]. The volume of crystal cell increased significantly by substituting O^{2-} with S^{2-} of larger radius, and the size of ion transport channel was increased, in addition, the binding of structural framework to Li^+ was weakened due to the strong polarity of S^{2-}, making the conductivity of Li_2S–GeS_2–P_2S_5 reach $10^{-3}\ S\ cm^{-1}$, which is equivalent to that of liquid organic electrolyte [93, 94].

1.4.2.5 NASICON

NASICON is the sodium fast ionic conductor, which firstly refers to solid solution of $Na_{1+x}Zr_2Si_xP_{3-x}O_{12}$ ($x = 2$), $Li_3Zr_2Si_2PO_4$ was obtained by replacing Na^+ in the structure with Li^+; however, due to the larger radius of Na^+ than that of Li^+, its ionic conductivity is very low (Figure 1.24). The channel that is suitable for Na^+ transportation is too large for Li^+ [96, 97]. Based on this, the $LiTi_2(PO_4)_3$ was obtained

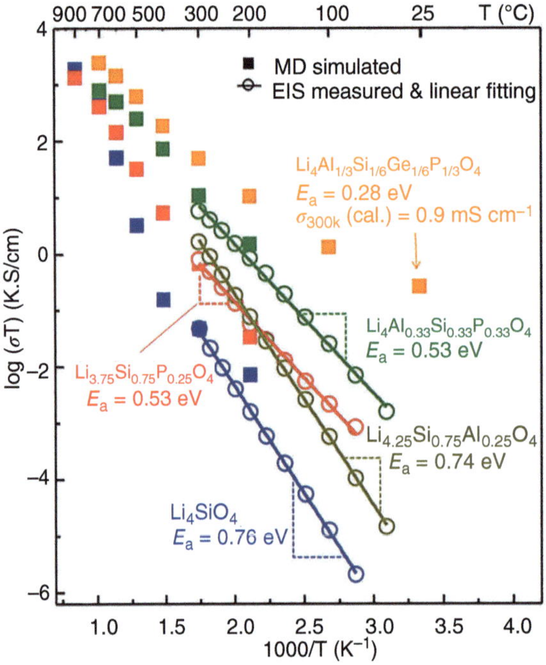

Figure 1.23 Li^+ ionic conductivity (σ), Arrhenius plots for Li_4SiO_4 (blue), $Li_{3.75}Si_{0.75}P_{0.25}O_4$ (red), $Li_{4.25}Si_{0.75}Al_{0.25}O_4$ (dark yellow), $Li_4Al_{0.33}Si_{0.33}P_{0.33}O_4$ (green), and $Li_4Al_{1/3}Si_{1/6}Ge_{1/6}P_{1/3}O_4$ (orange). MD simulated values are shown in solid squares. Conductivity values deduced from electrochemical impedance spectroscopy (EIS) measurements are shown in circles. Linear fits for experimental values are plotted in solid lines. The activation energies are derived in the temperature range of 50–300 °C. Source: Deng et al. [90]. Reproduced with permission of American Chemical Society.

by replacing Zr^{4+} in the structure with Ti^{4+} with a smaller ion radius, and the ionic conductivity was significantly improved. In addition, Ti^{4+} can be replaced by Al^{3+}, In^{3+}, Ga^{3+}, La^{3+}, and Y^{3+}, which can also effectively improve the ionic conductivity of the material. It is difficult to obtain single dense phase of the solid electrolyte with NASICON structure. The large resistance of the grain boundary is an important factor for the low conductivity of the material. Moreover, the poor chemical compatibility between NASICON and lithium must be overcome before commercialization.

1.4.2.6 Garnet

$Li_5La_3M_2O_{12}$ (M = Ta, Nb) with garnet structure found by Thangadurai and weppner et al. [98] is another type of lithium-ion solid electrolyte. Figure 1.25 shows the crystal structure of $Li_5La_3M_2O_{12}$ [99], which belongs to the cubic system, and the space group is *Ia-3d*. At room temperature, it has high conductivity of ~10^{-6} S cm^{-1} [100] and high decomposition voltage (6 V vs. Li/Li$^+$). In the garnet crystal structure, lithium ions occupy tetrahedral and octahedral positions, respectively, while tetrahedron and adjacent octahedron are coplanar. When the transition lithium ions enter the tetrahedron sites, due to coplanar connection, the lithium ions will generate

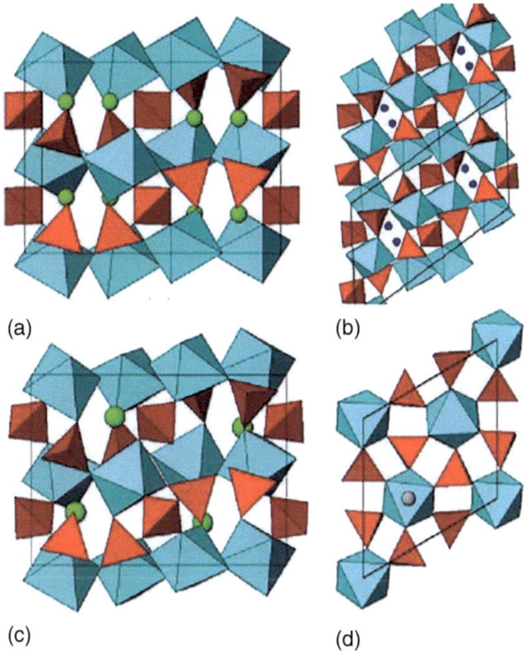

Figure 1.24 Structures of NASICON polymorphs: (a) orthorhombic (Pbna), (b) monoclinic ($P2_1/c$), (c) triclinic, and (d) Corundum-like. Source: Anantharamulu et al. [95]. Reproduced with permission of Springer Nature.

strong electrostatic repulsion with the lithium ions at the octahedron site, thus, part of the lithium ions at the octahedron site will transport to the tetrahedral position or octahedral position connected by common edges, and the above lithium ion will interact with the adjacent lithium ions, thus forming a so-called synergistic effect, and finally lithium ions can migrate rapidly in the crystal by this synergistic effect. The ionic conductivity of the materials can be improved by substitution of La^{2+} with Ca^{2+}, Sr^{2+}, Ba^{2+}, Eu^{3+}, K^{+}, or M with In^{3+}, Zr^{4+}, Gd^{3+}, and Y^{3+} [101, 102].

1.4.2.7 Glassy Inorganic Solid Electrolyte

Amorphous fast ionic conductor, also known as ionic conducting glass, is characterized by a long-range disordered structure of rigid skeleton. Compared with the crystal fast ionic conductor, the amorphous state itself belongs to the high defect structure, which is conducive to the migration of ions. The chemical composition is continuously adjustable, which makes it easy to explore and find new materials in a wide range of components. The macroscopic properties are isotropic, the preparation and processing are relatively simple, and the production cost is low. On the other hand, like other amorphous materials, there are inherent disadvantages, such as thermodynamic instability, spontaneous crystallization, and so on.

Glass itself belongs to an irregular network structure with different sizes of channels, which is easy to block cations with larger radius, while for lithium ions with smaller radius, conduction in the glass network will not be blocked. Therefore, the conductivity of glassy solid electrolyte is high, which can reach 10^{-3} S cm^{-1} at room temperature. However, due to its metastable state, it is unstable at high temperature and easy to crystallize, which will decrease the strength and conductivity of the

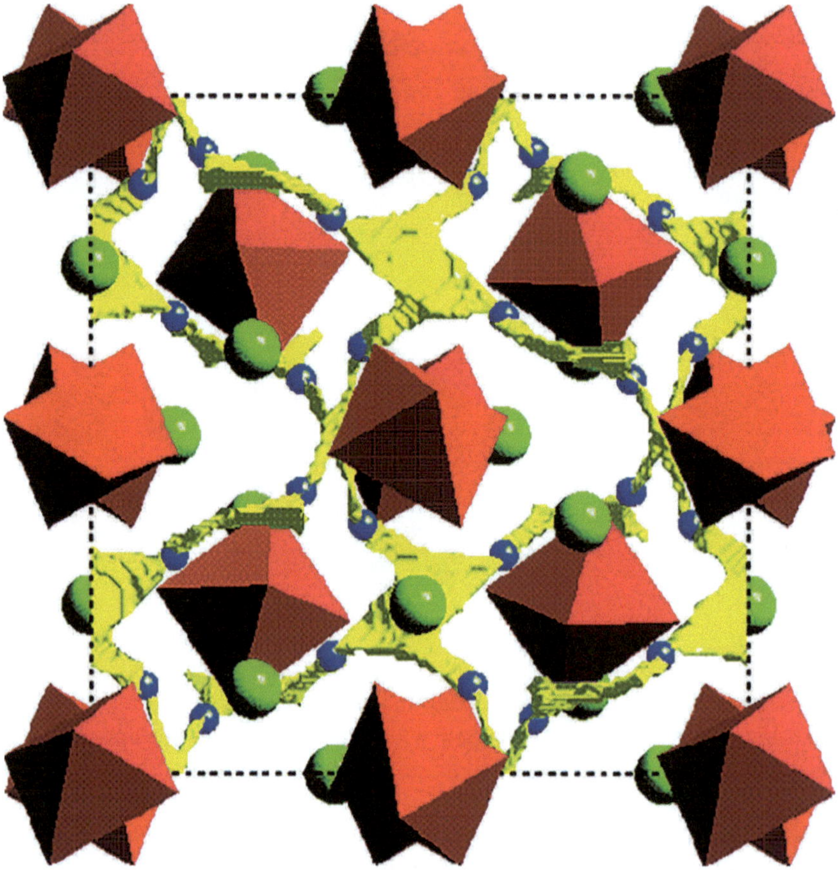

Figure 1.25 Crystal structure of garnet-like $Li_5La_3M_2O_{12}$ (MO_6: octahedral, lanthanum: large solid circles; Li(I): empty circles; Li(II): small solid circles). The lithium's oxygen coordination environment is shown on the right-hand side. Source: Thangadurai et al. [98]. Reproduced with permission of American Chemical Society.

material. At present, glass-based inorganic lithium-ion solid electrolyte is mainly divided into oxide glassy electrolyte and sulfide glassy electrolyte.

The oxide glassy electrolyte is mainly composed of network-modified oxides (Li_2O) and network oxides (such as SiO_2, P_2O_5, B_2O_3, etc.), and in general, the conductivity is about 10^{-6} S cm^{-1}. The concentration of Li^+ can be increased by increasing the content of Li_2O or adding some lithium salt, so as to improve the conductivity of the material. LiPON with the improved ionic conductivity can be formed by introducing nitrogen into Li_2O–P_2O_5 oxide glass solid electrolyte [92]; meanwhile, the thermal stability, the hardness of the glass, and the ability to resist the corrosion of water and salt solution have also been significantly improved. In addition, the introduction of ceramic crystal phase to form glass ceramic composite electrolyte can also improve the ionic conductivity of the material.

The ion conductivity of sulfide glassy electrolyte prepared by using S^{2-} with larger ion radius and higher polarity to replace O^{2-} in the oxide glass electrolyte system can reach 10^{-3} S cm^{-1} [103]. The developing trend of solid-state sulfide electrolytes is shown in Figure 1.26. However, the environmental and equipment conditions during the preparation process are crucial due to the intrinsic sensitivity of the raw materials to air and water.

1.5 Separators

Battery separators are electric insulating membranes placed between cathodes and anodes. They have complex three-dimensional porous structures. When liquid electrolyte fills in, the separator allows rapid transfer of lithium ions between cathode and anode, which are necessary to complete the circuit of the battery. Moreover, separators are electrically insulating. Thus, the cathodes and anodes are physically isolated apart, which otherwise would lead to internal short-circuiting.

Although separators are not active components in batteries, it can be seen from the above discussion that the separator is very important for the lithium-ion battery system, and its structure and performance directly affect the performance of the whole battery [105]. Some key parameters influencing the battery safety are as follows.

- Thickness. The thickness controls the mechanical strength of the separators and the impedance of the lithium-ion battery. The thicker the separators, the better the mechanical strength, the stronger the piercing resistance. However, the increase of thickness will also increase the internal resistance of lithium-ion battery, reduce the utilization of active substances, and reduce the battery capacity. The thickness of separators is generally required to be less than 25 μm.
- Permeability. The permeability refers to the time required for gas to pass through the separators per unit area under a certain pressure, generally using Gurley value to evaluate. The smaller the Gurley value is, the shorter the time for gas to pass through the separators and the faster the gas speed is, the larger the pore and porosity are. The pressure drop method is generally used to determine the Gurley value of the separators. For commercial polyene separator, the Gurley value is generally less than 750 seconds. For a specific separator, the permeability of the separator is positively related to the internal resistance. It is suggested that the pore structure of the separator affects the transport of lithium ions, so a low Gurley value represents the high air permeability and low resistance.
- Pore size and distribution. The pore size of the separator is strictly required, which must be submicron. This is because the self-discharge of lithium-ion battery is serious when the pore size of the separator is too high, while the small pore size will affect the lithium ion transport.
- Porosity. The porosity of separator is crucial for lithium-ion battery, which directly affects the core of the battery (transport of lithium ion and storage of electrolyte). The separator with high porosity has better lithium ion permeability, and at the same time, it can also provide the storage site for the electrolyte of the battery.
- Mechanical strength. During the LIB battery fabrication process, separators are wound with the electrodes under tension. Thus, there are some basic

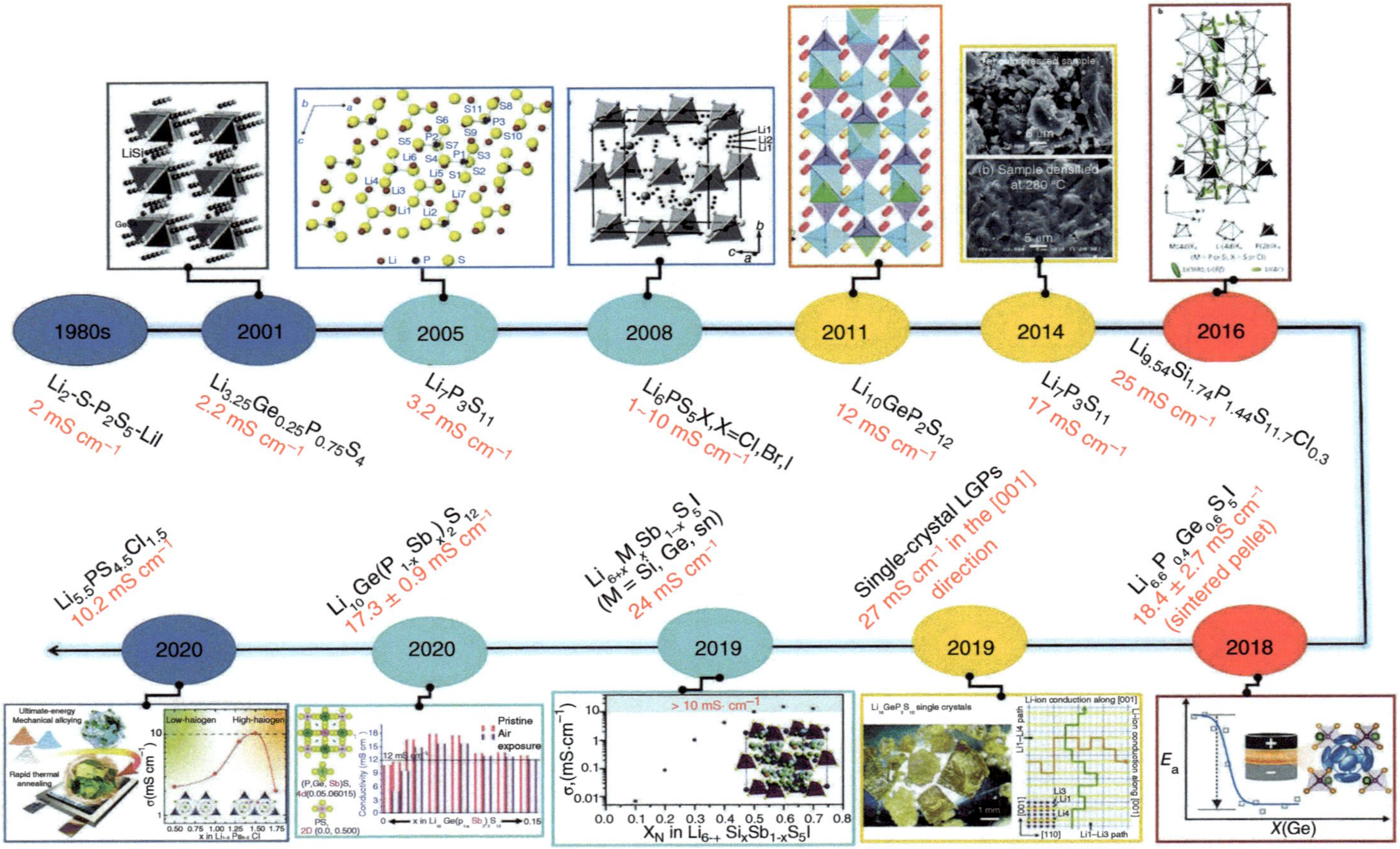

Figure 1.26 Developing trend of solid-state sulfide electrolytes. Source: Changhong Wang et al. [104] Reproduced with permission of Royal Society of Chemistry.

requirements for the mechanical strength of a separator: a reasonable tensile strength to sustain the stress during battery assembly (take the Celgard 2325 as an example, the machine direction Tensile Strength is as high as 1700 $kgf\,cm^{-2}$); a high puncture strength to avoid penetration of electrode material through the separator (empirically, the puncture strength should be at least >300 $g\,mil^{-1}$), a good mix penetration strength to avoid loose electrode particles penetration and short cell, which requires >100 $kgf\,mil^{-1}$ for separators in LIBs [106]. It should be noted that the selection of battery materials should be considered based on the circumstances where they will be practically used. The possibilities of harsh conditions such as battery crush and mechanical drop off should be given considerable attention.

- Thermal stability. Most of the battery separator membranes are polymeric materials, which will shrink and wrinkle apparently above a certain temperature. Thus, the thermal shrinkage at the early stage of battery thermal runaway process should be minimized, otherwise the cathode and anode of the battery will physically contact and eventually lead to thermal runaway. The requirement is generally <5% thermal shrinkage after 60 minutes at 90 °C (in a vacuum) [107].
- Wettability. The separators should wet out quickly and completely in typical battery electrolytes. Incomplete and nonuniform wetting of the separator could result in heterogeneous lithium ion flux and thus the possible plating of lithium dendrites, which may short-circuit the battery.
- Ionic conductivity. The ionic conductivity of the separators refers to the ionic flow energy of the separators after being fully wetted by the electrolyte. The performance of lithium-ion battery mainly depends on the ionic conductivity of the electrolyte in the separator. The ionic conductivity of the common organic liquid electrolyte at room temperature is 10^{-4}–10^{-3} $S\,cm^{-1}$. The separator prevents the electrodes from contacting; however, the volume occupied by the separator will undoubtedly reduce the capacity of electrolyte between electrodes, resulting in the reduction of the effective conductivity of electrolyte and increase of the resistance.
- Chemical and electrochemical stability. The separator is soaked by electrolyte for a long time, and it needs to exist stably in the battery for a long time without shrinkage or swelling by solvent, etc. At the same time, it cannot be degraded by the strong oxidizing electrolyte, and it is strictly electrochemical inert. The material determines the stability of the separator (solvent resistance, strong oxidation resistance, electrical resistance, and chemical stability). The electrochemical stability of the separator can be measured by linear voltammetry.

It is difficult to simultaneously meet all the requirements. Most of the commercial LIB separators are polyolefin membranes, e.g. PE and/or PP. And some other separator materials with different chemical components are also emerging in the literatures. The scheme in Figure 1.27 describes the primary considerations for the separator modification in improving the battery safety.

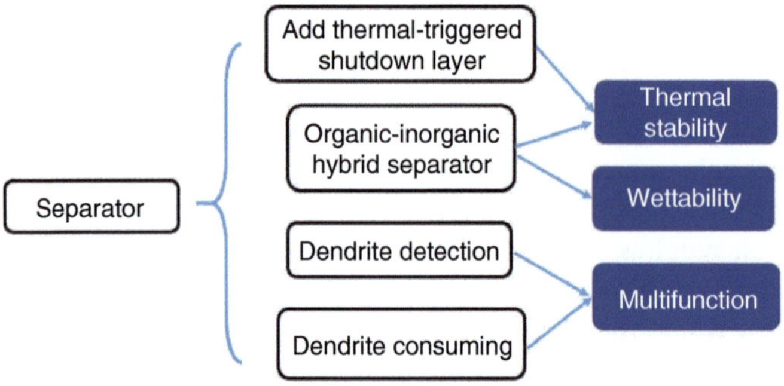

Figure 1.27 Scheme describing the primary considerations for the separator modification in improving the battery safety. Source: Yuan and Liu [108]. Reproduced with permission of Elsevier.

1.5.1 Polyolefin Separator

Due to its good insulation, low density, high mechanical strength, chemical resistance, and electrochemical corrosion resistance, almost all the separators for commercial lithium-ion batteries are made of semicrystalline polyolefin-based materials, such as PP, PE, and composite membranes of PP/PE/PP. The thickness of polyolefin separator used in 3 C electronic products is generally not greater than 25 μm. In order to meet the high power demand of HEV and EV, the thickness of separator is 32 and 40 μm, respectively. However, the polyolefin separators still have shortcomings. Firstly, due to the high crystallinity and low polarity of the polyolefin, the surface energy of the polyolefin separator is low, while the polarity of the electrolyte is high, and the affinity between the separator and the electrolyte is poor, so it is not easy to be infiltrated by the electrolyte, thus it makes the electrolyte easy to leak. Secondly, the porosity of polyolefin separator prepared by melt-drawing process is low, and the liquid absorption rate of the separator also decreases, which is not conducive to the migration of lithium ion and finally affects the electrochemical performance of the battery (rate performance and cycle stability). Thirdly, the thermal stability of polyolefin separator is limited, which may lead to temperature rise and even out of control of heat during overcharge, the separator began to shrink itself before shutting down the electrochemical reaction of the battery, resulting in the contact of positive and negative electrodes, causing short circuit and even explosion, thus posing a huge threat to the safety of lithium-ion battery. Among them, the major issue of polyolefin separators is their thermal instability. PP and PE are the most commonly used polyolefin separators, whereas their thermal stability is poor. The melting point (T_m) is just ~165 °C for PP and ~135 °C for PE [106], respectively. As temperatures approach their melting points, the separator shrinks dramatically in dimension. The internal short circuit of the battery happens and is further exaggerated afterward by the "positive feedback loop" (Figure 1.28), leading to thermal runaway. Several strategies have been employed to alleviate the safety issue of polyolefin-based separator.

A popular strategy is to employ multilayered structure to shut down the conduction pathway of lithium ions through the separator in case of overheating.

Separators with a PP/PE/PP trilayer structure have already been well commercialized (Figure 1.29a). When the internal temperature of the battery increases above ~130 °C, the porous middle PE layer partially melts, closing the pores inside the separator and preventing conduction of lithium ions in the liquid electrolytes (Figure 1.29b), while the PP layer provides mechanical support to maintain the overall dimensional stability, thus avoiding internal short-circuiting. The trilayer structure of the separator indeed has enhanced the safety of the LIBs. However, it does not always function well. In practical application, the heat in the cell could accumulate very quickly under harsh conditions and the internal temperature climbs up so fast that the thermal shutdown effect in PP/PE/PP can only last for a short duration due to the small melting temperature gap between PP and PE. Thus, the melting of the separator is still inevitable in some practical circumstances.

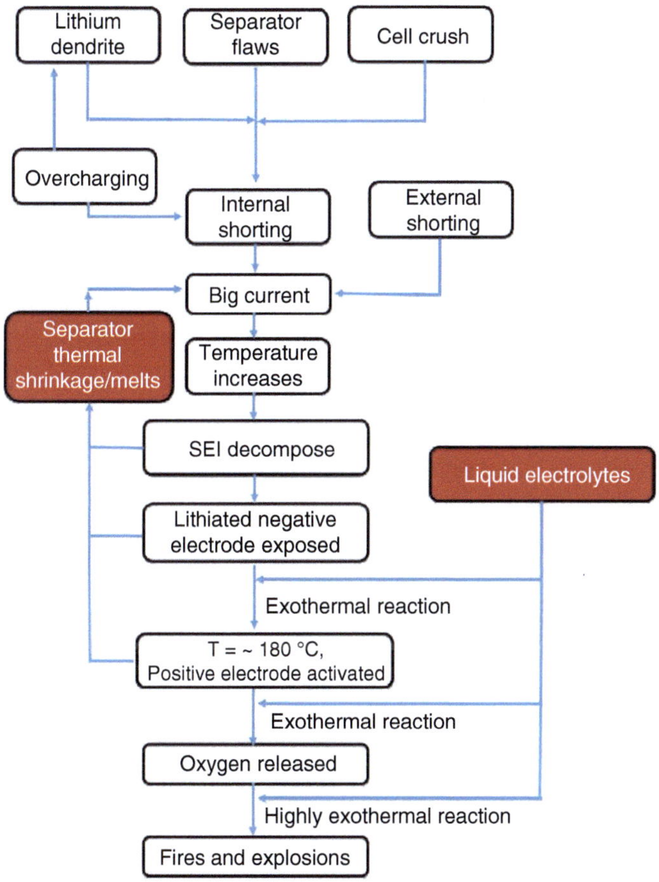

Figure 1.28 Flowchart illustrating the thermal runaway process. The key role of separator and liquid electrolyte is highlighted. Source: Yuan and Liu [108]. Reproduced with permission of Elsevier.

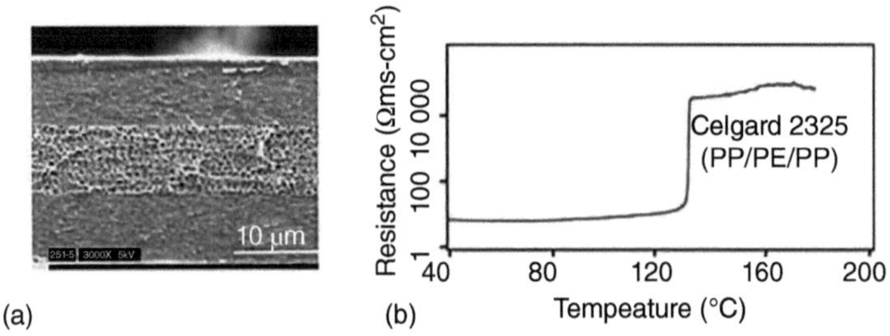

Figure 1.29 (a) Cross-section scanning electron micrographs of Celgard 2325 (PP/PE/PP) separator used in LIBs; (b) internal impedance (at 1 kHz) of Celgard 2325 (PP/PE/PP) separator as a function of temperature; heating rate: 60 °C min^{-1}. Source: Zhang et al. [106]. Reproduced with permission of American Chemical Society.

1.5.2 Polymers with High Melting Points for Separators

Porous membranes of poly(esters) [109], cellulose [110], polyimide [111], and other analogous have been demonstrated to be effective strategies to improve the thermal stability of separators (Figure 1.30a). For example, it has been reported that polyimide is a kind of thermosetting polymer widely considered as a promising alternative to polyolefin due to its excellent thermal stability (stable over 400 °C), high tensile strength, good electrolyte wettability, and flame retardancy, etc. (Figure 1.30b). The key step for preparing this kind of separator is to make interpenetrated micro/nanopores as lithium ion transportation channel across these polymer membranes. In a recently reported work, recyclable LiBr was utilized as the template for nanopores creation. The fabrication was carried out at the intermediate polyamic acid stage of the polyimide, which shows much better processability. LiBr salt, used as the template, was mixed with the intermediates, and then the final mixture was casted into thin films after the chemical condensation reaction was finished. Uniform and interconnected nanopores across the thin membrane can be created by simply removing LiBr in the water bath. Once dissolved, LiBr can be further recycled. This facile synthesis method offers an exciting possibility for a PI separator both in lab-scale and potential manufacturing in industry.

1.5.3 Inorganic Composite Separators

The inorganic composite separator is based on a highly ordered porous structure matrix, which is coated with a layer of inorganic ceramic particles in the presence of high molecular organic adhesive. Due to its large specific surface area and good hydrophilicity, the inorganic ceramic particles exhibit excellent affinity with organic electrolyte solvents (ethylene carbonate [EC], propylene carbonate [PC], and butyrolactone [GBL]); meanwhile, rigid inorganic particles can improve the thermal stability of the separator. The most common composition of ceramic coating is mainly composed of nonmetallic oxides such as SiO_2 and Al_2O_3 with hydroxyl groups on the surface, which are bonded with PVDF–HFP or PVDF. The characteristics of these nonmetallic oxides are that the particles are tightly stacked and the existing gap

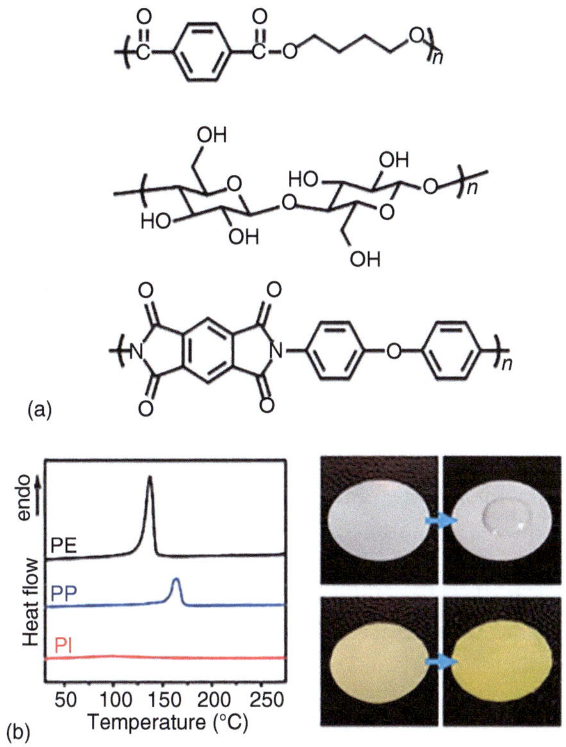

Figure 1.30 (a) Molecular structure of some high-melting temperature polymers as separator materials with low shrinkage at high temperatures. Top: polyimide (PI). Middle: cellulose. Bottom: poly(butylene) terephthalate. (b) Left: comparison of the differential scanning calorimetry (DSC) spectra of the PI with the PE and PP separator; the PI separator shows excellent thermal stability at the temperatures ranging from 30 to 275 °C. Right: digital camera photos comparing the wettability of a commercial separator and the as-synthesized PI separator with a propylene carbonate electrolyte. Source: Lin et al. [111]. Reproduced with permission of American Chemical Society.

provides a developed porous structure without affecting the air permeability and porosity of the separator, the thermal stability and liquid absorption of the separator can be improved. The most classical one is prepared by Kim et al. [112], who used PVDF–HFP as binder and coated SiO_2 particles on both sides of a commercial PE separator. It was found that the composite separators with SiO_2: PVDF–HFP of 9 : 1 possessed better porous structure, good liquid absorption, and high ionic conductivity and excellent thermal stability.

It should be noted that although numerous new materials have been reported in the literature, which are considered to be promising to replace the commercial polyolefin separators, few of them have been really commercialized. For practical applications, a good balance of the different parameters of the separators should be considered, such as mechanical strength, melting point, process ability, and cost.

1.6 Conclusions and Perspective

Energy density is the main issue of the batteries for commercial application. Lithium-ion batteries with high energy density are highly desired for the electric vehicles, portable devices, and large-scale energy storage techniques. Materials are crucial for the development of lithium-ion batteries, although various technologies have been developed to improve the electrochemical performance of the electrode materials, electrolytes, and separator, there are still many difficulties that limit their commercialization; thus, new materials design and full understanding of the charge/discharge mechanisms of the batteries are the two main challenges in the future. Incremental developments can be expected in the following four aspects:

Firstly, a thorough understanding of the deposition behavior of lithium as well as the formation mechanism and components of SEI film is the basis for the development of next-generation lithium-ion batteries. The advanced characterization methods such as freeze electron microscopy, environmental scanning electron microscopy, and in situ techniques need to be further developed to reveal the essence of the lithium-ion batteries.

Secondly, the commercialized graphite-based anode materials limit their wider application due to their relatively low theoretical capacity, easy to form lithium dendrite, and low lithium-ion intercalation potential. Therefore, it is highly desirable to explore the substitute anode materials with high electrochemical performance. Ideally, the optimized architecture, geometry, and compositions of materials will impart the electrode with high stability and high energy density simultaneously.

Thirdly, the low energy density is the main reason for the limited application of lithium-ion batteries at present, and the development lag of cathode materials is the bottleneck of the improvement of energy density of lithium-ion batteries. Lithium-rich manganese-based materials have high discharge capacity and wide voltage window, which are considered as the main candidates for the next generation of high-performance lithium-ion battery cathode materials. Therefore, how to solve the issues such as low first coulomb efficiency, poor cycling performance, continuous voltage attenuation, and so on, which hinder the practical application of materials, needs to be further researched.

Fourthly, the traditional lithium ion separator is difficult to adapt for the lithium-ion batteries with lithium metal as the anode and flexible lithium-ion batteries with bending. Therefore, further efforts should be made in the polymer electrolyte system with spectral suitability, such as the gel electrolyte based on the covalent self-cross-linking system and ionic liquid gel electrolytes based on the curved cross-linking system.

Last but not least, all solid-state lithium-ion batteries are thought to be the most promising candidate for the next-generation lithium-ion batteries with high energy density. Compared with organic electrolyte, solid electrolyte has lower ionic conductivity and better machining performance. Therefore, how to realize the thin-film electrolyte to promote the development of all solid-state thin-film battery will be an important research direction in the future.

Acknowledgments

The authors gratefully acknowledge financial supports from the talents project of Beijing Municipal Committee Organization Department (No. 2018000021223ZK21), the Fundamental Research Funds for the Central Universities (No. 2021JCCXJD01 and 2021YJSJD01), Key R & D, and transformation projects in Qinghai Province (2021-HZ-808) and Hebei Province (21314401D).

References

1 Wang, J., Song, W.L., Wang, Z.Y. et al. (2015). Facile fabrication of binder-free metallic tin nanoparticle/carbon nanofiber hybrid electrodes for lithium-ion batteries. *Electrochim. Acta* 153: 468–475.
2 Ji, X.X., Huang, X.T., Liu, J.P. et al. (2010). Carbon-coated SnO(2) nanorod array for lithium-ion battery anode material. *Nanoscale Res. Lett.* 5: 649–653.
3 Xue, X.Y., Chen, Z.H., Xing, L.L. et al. (2011). SnO_2/alpha-MoO_3 core-shell nanobelts and their extraordinarily high reversible capacity as lithium-ion battery anodes. *Chem. Commun.* 47: 5205–5207.
4 Thomas, R. and Rao, G.M. (2015). SnO_2 nanowire anchored graphene nanosheet matrix for the superior performance of Li-ion thin film battery anode. *J. Mater. Chem. A* 3: 274–280.
5 Zhou, D., Song, W.L., and Fan, L.Z. (2015). Hollow core-shell SnO_2/C fibers as highly stable anodes for lithium-ion batteries. *Acs Appl. Mater. Inter.* 7: 21472–21478.
6 Mauger, A., Armand, M., Julien, C.M., and Zaghib, K. (2017). Challenges and issues facing lithium metal for solid-state rechargeable batteries. *J. Power Sources* 353: 333–342.
7 Patil, A., Choi, J.W., and Yoon, S.J. (2006). Review of issue and challenges facing rechargeable nanostructured lithium batteries. *IEEE Nmdc 2006: IEEE Nanotechnology Materials and Devices Conference 2006, Proceedings*, 196–197.
8 Patil, A., Patil, V., Shin, D.W. et al. (2008). Issue and challenges facing rechargeable thin film lithium batteries. *Mater. Res. Bull.* 43: 1913–1942.
9 Tarascon, J.M. and Armand, M. (2001). Issues and challenges facing rechargeable lithium batteries. *Nature* 414: 359–367.
10 Zhou, L.M., Zhang, K., Hu, Z. et al. (2018). Recent developments on and prospects for electrode materials with hierarchical structures for lithium-ion batteries. *Adv. Energy Mater.* 8: 1701415.
11 Wang, Y., Niu, S.S., and Lu, S. (2015). Controlled-synthesis and lithium storage properties of SnO_2 porous core-shell spheres and core-in-double-shell spheres. *Mater. Lett.* 157: 209–211.
12 Erickson, E.M., Schipper, F., Penki, T.R. et al. (2017). Review-recent advances and remaining challenges for lithium ion battery cathodes: II. Lithium-rich, xLi subset of 2 subset of MnO subset of 3 subset of (1-x)LiNi subset of a subset

of co subset of b subset of Mn subset of c subset of O subset of 2 subset of. *J. Electrochem. Soc.* 164: A6341–A6348.

13 Manthiram, A., Song, B., and Li, W. (2017). A perspective on nickel-rich layered oxide cathodes for lithium-ion batteries. *Energy Storage Materials* 6: 125–139.

14 Wang, G.X., Zhong, S., Bradhurst, D.H. et al. (1998). Synthesis and characterization of LiNiO compounds as cathodes for rechargeable lithium batteries. *J. Power Sources* 76: 141–146.

15 Cho, J.P., Kim, T.J., and Park, B. (2002). The effect of a metal-oxide coating on the cycling behavior at 55 degrees C in orthorhombic $LiMnO_2$ cathode materials. *J. Electrochem. Soc.* 149: A288–A292.

16 Liu, Z.L., Yu, A.S., and Lee, J.Y. (1999). Synthesis and characterization of $LiNi_{1-x-y}Co_xMn_yO_2$ as the cathode materials of secondary lithium batteries. *J. Power Sources* 81: 416–419.

17 Ryu, H.-H., Park, K.-J., Yoon, C.S., and Sun, Y.-K. (2018). Capacity fading of Ni-rich $Li[Ni_xCo_yMn_{1-x-y}]O_2$ ($0.6 \leq x \leq 0.95$) cathodes for high-energy-density lithium-ion batteries: bulk or surface degradation? *Chem. Mater.* 30: 1155–1163.

18 Mukherjee, P., Faenza, N.V., Pereira, N. et al. (2018). Surface structural and chemical evolution of layered $LiNi_{0.8}Co_{0.15}Al_{0.050}O_2$ (NCA) under high voltage and elevated temperature conditions. *Chem. Mater.* 30: 8431–8445.

19 Xia, H., Xia, Q.Y., Lin, B.H. et al. (2016). Self-standing porous $LiMn_2O_4$ nanowall arrays as promising cathodes for advanced 3D microbatteries and flexible lithium-ion batteries. *Nano Energy* 22: 475–482.

20 Lee, M.J., Lee, S., Oh, P. et al. (2014). High performance $LiMn_2O_4$ cathode materials grown with epitaxial layered nanostructure for Li-ion batteries. *Nano Lett.* 14: 993–999.

21 Chen, R.J., Zhao, T.L., Zhang, X.X. et al. (2016). Advanced cathode materials for lithium-ion batteries using nanoarchitectonics. *Nanoscale Horiz.* 1: 423–444.

22 Nakayama, M., Yamada, S., Jalem, R., and Kasuga, T. (2016). Density functional studies of olivine-type $LiFePO_4$ and $NaFePO_4$ as positive electrode materials for rechargeable lithium and sodium ion batteries. *Solid State Ionics* 286: 40–44.

23 Cheng, H., Shapter, J.G., Li, Y., and Gao, G. (2021). Recent progress of advanced anode materials of lithium-ion batteries. *J. Energy Chem.* 57: 451–468.

24 Nitta, N., Wu, F.X., Lee, J.T., and Yushin, G. (2015). Li-ion battery materials: present and future. *Mater. Today* 18: 252–264.

25 Tian, Q.H., Chen, P., Zhang, Z.X., and Yang, L. (2017). Achievement of significantly improved lithium storage for novel clew-like $Li_4Ti_5O_{12}$ anode assembled by ultrafine nanowires. *J. Power Sources* 350: 49–55.

26 Zhang, Y.X., Luo, Y., Chen, Y. et al. (2017). Enhanced rate capability and low-temperature performance of $Li_4Ti_5O_{12}$ anode material by facile surface fluorination. *ACS Appl. Mater. Inter.* 9: 17146–17155.

27 Nishi, Y. (2001). The development of lithium ion secondary batteries. *Chem. Rec.* 1: 406–413.

28 Endo, M., Kim, C., Nishimura, K. et al. (2000). Recent development of carbon materials for Li ion batteries. *Carbon* 38: 183–197.

29 Liu, D.Q., Liu, Z.J., Li, X.W. et al. (2017). Group IVA element (Si, Ge, Sn)-based alloying/dealloying anodes as negative electrodes for full-cell lithium-ion batteries. *Small* 13: 1702000.

30 Guo, J.X., Zhu, H.F., Sun, Y.F. et al. (2017). Pie-like free-standing paper of graphene paper@Fe_3O_4 nanorod array@carbon as integrated anode for robust lithium storage. *Chem. Eng. J.* 309: 272–277.

31 Yu, S.H., Lee, S.H., Lee, D.J. et al. (2016). Conversion reaction-based oxide nanomaterials for lithium ion battery anodes. *Small* 12: 2146–2172.

32 Lou, X.W., Li, C.M., and Archer, L.A. (2009). Designed synthesis of coaxial SnO_2@carbon hollow nanospheres for highly reversible lithium storage. *Adv. Mater.* 21: 2536.

33 Wang, C., Zhou, Y., Ge, M.Y. et al. (2010). Large-scale synthesis of SnO_2 nanosheets with high lithium storage capacity. *J. Am. Chem. Soc.* 132: 46–47.

34 Jin, Y.H., Min, K.M., Seo, S.D. et al. (2011). Enhanced Li storage capacity in 3 nm diameter SnO_2 nanocrystals firmly anchored on multiwalled carbon nanotubes. *J. Phys. Chem. C* 115: 22062–22067.

35 Liang, J., Yu, X.Y., Zhou, H. et al. (2014). Bowl-like SnO_2@carbon hollow particles as an advanced anode material for lithium-ion batteries. *Angew. Chem. Int. Ed.* 53: 12803–12807.

36 Ren, J.G., Yang, J.B., Abouimrane, A. et al. (2011). SnO_2 nanocrystals deposited on multiwalled carbon nanotubes with superior stability as anode material for Li-ion batteries. *J. Power Sources* 196: 8701–8705.

37 Hu, R.Z., Chen, D.C., Waller, G. et al. (2016). Dramatically enhanced reversibility of Li_2O in SnO_2-based electrodes: the effect of nanostructure on high initial reversible capacity. *Energy Environ. Sci.* 9: 595–603.

38 Zhou, X.S., Wan, L.J., and Guo, Y.G. (2013). Binding SnO_2 nanocrystals in nitrogen-doped graphene sheets as anode materials for lithium-ion batteries. *Adv. Mater.* 25: 2152–2157.

39 Zhao, S., Sewell, C.D., Liu, R. et al. (2019). SnO_2 as advanced anode of alkali-ion batteries: inhibiting Sn coarsening by crafting robust physical barriers, void boundaries, and heterophase interfaces for superior electrochemical reaction reversibility. *Adv. Energy Mater.*: 1902657.

40 Lang, J., Long, Y., Qu, J. et al. (2019). One-pot solution coating of high quality LiF layer to stabilize Li metal anode. *Energy Storage Mater.* 16: 85–90.

41 Adair, K.R., Iqbal, M., Wang, C. et al. (2018). Towards high performance Li metal batteries: nanoscale surface modification of 3D metal hosts for pre-stored Li metal anodes. *Nano Energy* 54: 375–382.

42 Wang, Q., Yang, C., Yang, J. et al. (2018). Stable Li metal anode with protected interface for high-performance Li metal batteries. *Energy Storage Mater.* 15: 249–256.

43 Yun, Q.B., He, Y.B., Lv, W. et al. (2016). Chemical dealloying derived 3D porous current collector for Li metal anodes. *Adv. Mater.* 28: 6932.

44 Xu, W., Wang, J.L., Ding, F. et al. (2014). Lithium metal anodes for rechargeable batteries. *Energy Environ. Sci.* 7: 513–537.

45 Cheng, X.B., Zhang, R., Zhao, C.Z., and Zhang, Q. (2017). Toward safe lithium metal anode in rechargeable batteries: a review. *Chem. Rev.* 117: 10403–10473.

46 Huang, K., Li, Z., Xu, Q. et al. (2019). Lithiophilic CuO nanoflowers on Ti-mesh inducing lithium lateral plating enabling stable lithium-metal anodes with ultrahigh rates and ultralong cycle life. *Adv. Energy Mater.* 9.

47 Li, T., Shi, P., Zhang, R. et al. (2019). Dendrite-free sandwiched ultrathin lithium metal anode with even lithium plating and stripping behavior. *Nano Res.* 12: 2224–2229.

48 Jiao, S., Zheng, J., Li, Q. et al. (2018). Behavior of lithium metal anodes under various capacity utilization and high current density in lithium metal batteries. *Joule* 2: 110–124.

49 Zou, P., Wang, Y., Chiang, S.W. et al. (2018). Directing lateral growth of lithium dendrites in micro-compartmented anode arrays for safe lithium metal batteries. *Nat. Commun.* 9: 464.

50 Goodenough, J.B. and Kim, Y. (2010). Challenges for rechargeable Li batteries. *Chem. Mater.* 22: 587–603.

51 Goodenough, J.B. and Park, K.S. (2013). The Li-ion rechargeable battery: a perspective. *J. Am. Chem. Soc.* 135: 1167–1176.

52 Dong, H.Y., Xiao, X.L., Jin, C. et al. (2019). High lithium-ion conductivity polymer film to suppress dendrites in Li metal batteries. *J. Power Sources* 423: 72–79.

53 Cheng, X.B., Yan, C., Chen, X. et al. (2017). Implantable solid electrolyte interphase in lithium-metal batteries. *Chem-Us* 2: 258–270.

54 Tao, R., Bi, X.X., Li, S. et al. (2017). Kinetics tuning the electrochemistry of lithium dendrites formation in lithium batteries through electrolytes. *ACS Appl. Mater. Inter.* 9: 7003–7008.

55 Li, X., Zheng, J., Ren, X. et al. (2018). Dendrite-free and performance-enhanced lithium metal batteries through optimizing solvent compositions and adding combinational additives. *Adv. Energy Mater.* 8.

56 Zhang, X.-Q., Cheng, X.-B., Chen, X. et al. (2017). Fluoroethylene carbonate additives to render uniform Li deposits in lithium metal batteries. *Adv. Funct. Mater.* 27.

57 Wan, H., Peng, G., Yao, X. et al. (2016). Cu_2ZnSnS_4/graphene nanocomposites for ultrafast, long life all-solid-state lithium batteries using lithium metal anode. *Energy Storage Mater.* 4: 59–65.

58 Bouchet, R., Maria, S., Meziane, R. et al. (2013). Single-ion BAB triblock copolymers as highly efficient electrolytes for lithium-metal batteries. *Nat. Mater.* 12: 452–457.

59 Liu, R., Wu, Z., He, P. et al. (2019). A self-standing, UV-cured semi-interpenetrating polymer network reinforced composite gel electrolytes for dendrite-suppressing lithium ion batteries. *J. Materiomics* 5: 185–194.

60 A. Varzi, L. Mattarozzi, S. Cattarin, P. Guerriero, S. Passerini, (2018). *3D porous Cu-Zn alloys as alternative anode materials for Li-ion batteries with superior low T performance. Adv. Energy Mater.*, 8. https://doi.org/10.1002/aenm.201701706

61 Peng, Z., Ren, F.H., Yang, S.S. et al. (2019). A highly stable host for lithium metal anode enabled by Li_9Al_4-Li_3N-AlN structure. *Nano Energy* 59: 110–119.
62 Guo, F., Wang, Y., Kang, T. et al. (2018). A Li-dual carbon composite as stable anode material for Li batteries. *Energy Storage Mater.* 15: 116–123.
63 Liu, B., Zhang, L., Xu, S. et al. (2018). 3D lithium metal anodes hosted in asymmetric garnet frameworks toward high energy density batteries. *Energy Storage Mater.* 14: 376–382.
64 Yue, X.-Y., Wang, W.-W., Wang, Q.-C. et al. (2018). CoO nanofiber decorated nickel foams as lithium dendrite suppressing host skeletons for high energy lithium metal batteries. *Energy Storage Mater.* 14: 335–344.
65 Wang, L., Zhu, X., Guan, Y. et al. (2018). ZnO/carbon framework derived from metal-organic frameworks as a stable host for lithium metal anodes. *Energy Storage Mater.* 11: 191–196.
66 Zhao, Q., Liu, X., Stalin, S. et al. (2019). Solid-state polymer electrolytes with in-built fast interfacial transport for secondary lithium batteries. *Nat. Energy* 4: 365–373.
67 Aravindan, V., Gnanaraj, J., Madhavi, S., and Liu, H.K. (2011). Lithium-ion conducting electrolyte salts for lithium batteries. *Chem.-Eur. J.* 17: 14326–14346.
68 Kita, F., Sakata, H., Sinomoto, S. et al. (2000). Characteristics of the electrolyte with fluoro organic lithium salts. *J. Power Sources* 90: 27–32.
69 Edman, L., Doeff, M.M., Ferry, A. et al. (2000). Transport properties of the solid polymer electrolyte system P(EO)(n)LiTFSI. *J. Phys. Chem. B* 104: 3476–3480.
70 Campion, C.L., Li, W.T., and Lucht, B.L. (2005). Thermal decomposition of $LiPF_6$-based electrolytes for lithium-ion batteries. *J. Electrochem. Soc.* 152: A2327–A2334.
71 Zhang, S.S., Xu, K., and Jow, T.R. (2002). Study of $LiBF_4$ as an electrolyte salt for a Li-ion battery. *J. Electrochem. Soc.* 149: A586–A590.
72 Ding, M.S. and Jow, T.R. (2004). How conductivities and viscosities of PC-DEC and PC-EC solutions of $LiBF_4$, $LiPF_6$, LiBOB, Et_4NBF_4, and Et_4NPF_6 differ and why. *J. Electrochem. Soc.* 151: A2007–A2015.
73 Zhou, H.M., Liu, F.R., and Li, J. (2012). Preparation, thermal stability and electrochemical properties of LiODFB. *J. Mater. Sci. Technol.* 28: 723–727.
74 Zhang, S.S., Xu, K., and Jow, T.R. (2006). Study of the charging process of a LiCoO(2)-based Li-ion battery. *J. Power Sources* 160: 1349–1354.
75 Zhang, S.S., Xu, K., and Jow, T.R. (2002). A new approach toward improved low temperature performance of Li-ion battery. *Electrochem. Commun.* 4: 928–932.
76 Herreyre, S., Huchet, O., Barusseau, S. et al. (2001). New Li-ion electrolytes for low temperature applications. *J. Power Sources* 97-8: 576–580.
77 Liao, L.X., Cheng, X.Q., Ma, Y.L. et al. (2013). Fluoroethylene carbonate as electrolyte additive to improve low temperature performance of $LiFePO_4$ electrode. *Electrochim. Acta* 87: 466–472.
78 Zhang, S.S. (2006). An unique lithium salt for the improved electrolyte of Li-ion battery. *Electrochem. Commun.* 8: 1423–1428.

79 Chung, G.C., Kim, H.J., Yu, S.I. et al. (2000). Origin of graphite exfoliation – an investigation of the important role of solvent cointercalation. *J. Electrochem. Soc.* 147: 4391–4398.

80 Lee, D.J., Im, D., Ryu, Y.G. et al. (2013). Phosphorus derivatives as electrolyte additives for lithium-ion battery: the removal of O-2 generated from lithium-rich layered oxide cathode. *J. Power Sources* 243: 831–835.

81 Bachman, J.C., Muy, S., and Grimaud, A. (2016). Inorganic solid-state electrolytes for lithium batteries: mechanisms and properties governing ion conduction. *Chem. Rev.* 116: 140–162.

82 Takada, K. (2013). Progress and prospective of solid-state lithium batteries. *Acta Mater.* 61: 759–770.

83 Stoeva, Z., Martin-Litas, I., Staunton, E. et al. (2003). Ionic conductivity in the crystalline polymer electrolytes PEO_6: $LiXF_6$, X = P, As, Sb. *J. Am. Chem. Soc.* 125: 4619–4626.

84 Shin, J.H., Henderson, W.A., Scaccia, S. et al. (2006). Solid-state Li/$LiFePO_4$ polymer electrolyte batteries incorporating an ionic liquid cycled at 40 degrees C. *J. Power Sources* 156: 560–566.

85 Nakano, H., Dokko, K., Sugaya, J.I. et al. (2007). All-solid-state micro lithium-ion batteries fabricated by using dry polymer electrolyte with micro-phase separation structure. *Electrochem. Commun.* 9: 2013–2017.

86 Stephan, A.M. (2006). Review on gel polymer electrolytes for lithium batteries. *Eur. Polym. J.* 42: 21–42.

87 Gregory, D.H., O'Meara, P.M., Gordon, A.G. et al. (2002). Structure of lithium nitride and transition-metal-doped derivatives, $Li_{3-x-y}MxN$ (M = Ni, Cu): a powder neutron diffraction study. *Chem. Mater.* 14: 2063–2070.

88 Varez, A., Fernández-Díaz, M.T., Alonso, J.A., and Sanz, J. (2005). Structure of fast ion conductors $Li_3xLa_{2/3-x}TiO_3$ deduced from powder neutron diffraction experiments. *Chem. Mater.* 17: 2404–2412.

89 Morata-Orrantia, A., Garcia-Martin, S., and Alario-Franco, M.A. (2003). Optimization of lithium conductivity in La/Li titanates. *Chem. Mater.* 15: 3991–3995.

90 Deng, Y., Eames, C., Fleutot, B. et al. (2017). Enhancing the lithium ion conductivity in lithium superionic conductor (LISICON) solid electrolytes through a mixed polyanion effect. *ACS Appl. Mater. Interfaces* 9: 7050–7058.

91 Knauth, P. (2009). Inorganic solid Li ion conductors: an overview. *Solid State Ionics* 180: 911–916.

92 Song, S.W., Choi, H., Park, H.Y. et al. (2010). High rate-induced structural changes in thin-film lithium batteries on flexible substrate. *J. Power Sources* 195: 8275–8279.

93 Hassoun, J., Verrelli, R., Reale, P. et al. (2013). A structural, spectroscopic and electrochemical study of a lithium ion conducting $Li_{10}GeP_2S_{12}$ solid electrolyte. *J. Power Sources* 229: 117–122.

94 Ong, S.P., Mo, Y.F., Richards, W.D. et al. (2013). Phase stability, electrochemical stability and ionic conductivity of the Li10 +/− 1MP2X12 (M = Ge, Si, Sn, Al or P, and X = O, S or Se) family of superionic conductors. *Energy Environ. Sci.* 6: 148–156.

95 Anantharamulu, N., Rao, K.K., Rambabu, G. et al. (2011). A wide-ranging review on Nasicon type materials. *J. Mater. Sci.* 46: 2821–2837.

96 Jackman, S.D. and Cutler, R.A. (2013). Stability of NaSICON-type $Li_{1.3}Al_{0.3}Ti_{1.7}P_3O_{12}$ in aqueous solutions. *J. Power Sources* 230: 251–260.

97 Swamy, D.T., Babu, K.E., and Veeraiah, V. (2013). Evidence for high ionic conductivity in lithium-lanthanum titanate, Li0 center dot 29La0 center dot $57TiO_3$. *B Mater. Sci.* 36: 1115–1119.

98 Thangadurai, V., Adams, S., and Weppner, W. (2004). Crystal structure revision and identification of Li^+-ion migration pathways in the garnet-like $Li_5La_3M_2O_{12}$ (M = Nb, Ta) oxides. *Chem. Mater.* 16: 2998–3006.

99 Thangadurai, V. and Weppner, W. (2005). Li(6)ALa(2)Ta(2)O(12) (A=Sr, Ba): Novel garnet-like oxides for fast lithium ion conduction. *Adv. Funct. Mater.* 15: 107–112.

100 Thangadurai, V., Kaack, H., and Weppner, W.J.F. (2003). Novel fast lithium ion conduction in garnet-type $Li_5La_3M_2O_{12}$ (M = Nb, Ta). *J. Am. Ceram. Soc.* 86: 437–440.

101 Narayanan, S., Baral, A.K., and Thangadurai, V. (2016). Dielectric characteristics of fast Li ion conducting garnet-type $Li_{5+2x}La_3Nb_{2-x}YxO_{12}$ (x = 0.25, 0.5 and 0.75). *Phys. Chem. Chem. Phys.* 18: 15418–15426.

102 Ahmad, M.M. and Al-Jaafari, A. (2015). Concentration and mobility of mobile Li^+ ions in $Li_6BaLa_2Ta_2O_{12}$ and $Li_5La_3Ta_2O_{12}$ garnet lithium ion conductors. *J. Mater. Sci.-Mater. El* 26: 8136–8142.

103 Ohtomo, T., Hayashi, A., Tatsumisago, M. et al. (2013). All-solid-state lithium secondary batteries using the 75Li(2)S center dot 25P(2)S(5) glass and the 70Li(2)S center dot 30P(2)S(5) glass-ceramic as solid electrolytes. *J. Power Sources* 233: 231–235.

104 Wang, C., Liang, J., Zhao, Y. et al. (2009). All-solid-state lithium batteries enabled by sulfide electrolytes: from fundamental research to practical engineering design. *Energy Environ. Sci.* 14: 2577–2619.

105 Lee, H., Yanilmaz, M., Toprakci, O. et al. (2014). A review of recent developments in membrane separators for rechargeable lithium-ion batteries. *Energy Environ. Sci.* 7: 3857–3886.

106 Zhang, P.A.Z.J. (2004). Battery separators. *Chem. Rev.* 104: 4419–4462.

107 Zhang, S.S. (2007). A review on the separators of liquid electrolyte Li-ion batteries. *J. Power Sources* 164: 351–364.

108 Yuan, M.Q. and Liu, K. (2020). Rational design on separators and liquid electrolytes for safer lithium-ion batteries. *J. Energy Chem.* 43: 58–70.

109 Orendorff, C.J., Lambert, T.N., Chavez, C.A. et al. (2013). Polyester separators for lithium-ion cells: improving thermal stability and abuse tolerance. *Adv. Energy Mater.* 3: 314–320.

110 Li, L., Yu, M., Jia, C. et al. (2017). Cellulosic biomass-reinforced polyvinylidene fluoride separators with enhanced dielectric properties and thermal tolerance. *ACS Appl. Mater. Inter.* 9: 20885–20894.

111 Lin, D.C., Zhuo, D., Liu, Y.Y., and Cui, Y. (2016). All-integrated bifunctional separator for Li dendrite detection via novel solution synthesis of a thermostable polyimide separator. *J. Am. Chem. Soc.* 138: 11044–11050.

112 Kim, K.J., Kim, J.H., Park, M.S. et al. (2012). Enhancement of electrochemical and thermal properties of polyethylene separators coated with polyvinylidene fluoride-hexafluoropropylene co-polymer for Li-ion batteries. *J. Power Sources* 198: 298–302.

2

Li-O_2 Battery

Zhijia Zhang[1], Jun Wang[2], Shaofei Zhang[3], Shihao Sun[1], and Xia Ma[1]

[1] Tiangong University, School of Material Science and Engineering, State Key Laboratory of Separation Membrane and Membrane Processes, Tianjin Municipal Key Laboratory of Advanced Fibers and Energy Storage, No. 399 BinShuiXi Road, Tianjin, 300387, China
[2] Shandong University, Key Laboratory for Liquid-Solid Structural Evolution and Processing of Materials (Ministry of Education), 27 Shanda Nanlu, Jinan, 250061, China
[3] Hebei University of Science & Technology, School of Materials Science and Engineering, Hebei Key Laboratory of Material Near-Net Forming Technology, No. 26 Yuxiang Street, Shijiazhuang, 050018, China

2.1 Li-O_2 Battery

2.1.1 Introduction

Lithium–air batteries have attracted worldwide attention in recent years, due to its ultrahigh theoretical energy density (3600 Wh kg^{-1}) [1–3]. As early as 1987, Semkow and Sammells had been engaged in Li–O_2 batteries and developed a kind of ZrO_2 solid electrolyte operated at 650–800 °C [4]. The first room temperature Li–O_2 battery, with a polymer electrolyte, was constructed by Abraham and Jiang in 1996 [5]. Since the early attempts, extensive research on lithium–air batteries has been conducted and the fundamental understanding of the Li–O_2 reaction has acquired a series of significant progress [6–8]. Currently, lithium–air batteries can be divided into four types according to the type of the electrolyte employed in batteries: aprotic, aqueous, hybrid aqueous/aprotic, and all-solid-state batteries [9–11]. The research efforts have been mainly devoted to the aprotic Li–O_2 battery due to its extremely high energy density and simple battery structures. Hence, this chapter introduces lithium–air battery, mainly based on aprotic Li–O_2 battery.

The structure of the aprotic Li–O_2 battery consists of cathode, electrolyte, separators, and anode, which is similar to a lithium-ion battery. The cathode of Li–O_2 battery is different from that of lithium-ion battery, which is in direct contact with air or oxygen. And the working mechanism of Li–O_2 battery is very different from that of lithium-ion battery (rocking chair-type lithium intercalation/deintercalation) [12], which uses the e^- gain/loss of oxygen in the ambient to achieve energy storage

Battery Technologies: Materials and Components, First Edition. Edited by Jianmin Ma.

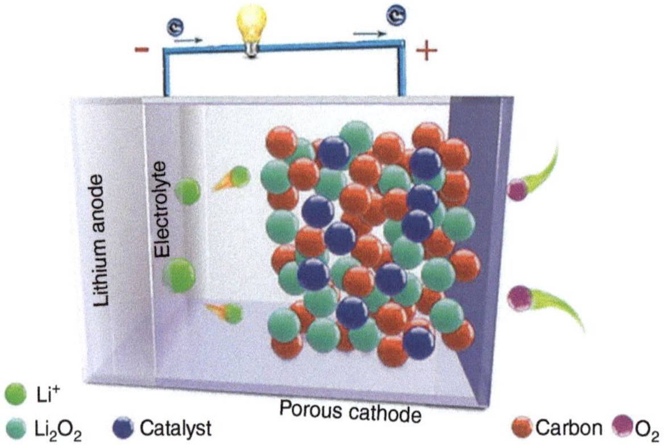

Figure 2.1 Schematic of the charge/discharge process of a typical aprotic Li–O_2 battery. Source: From Shu et al. [13]. Reproduced with permission of John Wiley and Sons.

and conversion. It can be said that Li-O_2 battery is a semi-open system (Figure 2.1). During the process of discharging, the lost electrons of lithium are oxidized into lithium ions and then diffused to the catalytic cathode under the control of electric field. At the same time, oxygen obtains electrons and composes with lithium ions to form lithium oxides on catalytic cathode. The electrochemical reaction is opposite in the charging process. Ideally, Li_2O_2 is the product of Li-O_2 battery discharge process, and its electrochemical reactions could be generally written as follows [13–15]:

$$\text{Cathode: } 2\text{Li} + \text{O}_2 + 2\text{e}^- \leftrightarrow \text{Li}_2\text{O}_2 \tag{2.1}$$

$$\text{Anode: } 2\text{Li} \leftrightarrow 2\text{Li}^+ + 2\text{e}^- \tag{2.2}$$

$$\text{Total: } 2\text{Li} + \text{O}_2 \leftrightarrow \text{Li}_2\text{O}_2\ (E_{\text{ocv}} = 2.90\ \text{V vs. Li/Li}^+) \tag{2.3}$$

In fact, the Li-O_2 batteries are still in the primary stage of basic research. Li-O_2 batteries had some unavoidable problems in practical application, such as harsh working environment, low energy efficiency, and strong dependence on lithium metal [16–18]. These disadvantages are mainly due to the active nature of lithium metal itself and the slow kinetics of oxygen reduction reaction (ORR) and oxygen evolution reaction (OER) [19, 20]. Hence, a large part of the research work is devoted to the development of novel cathode materials with high ORR and OER activities to improve kinetics and tolerate lithium activity. Meanwhile, the electrodes need to have a special nano-sized architecture with enough pore structure for accelerating the ions transmission. This chapter focuses on the high-efficiency cathode materials with various nanostructures. Although the cathode plays an important role in Li-O_2 batteries, its other components are not satisfactory, so that, the anode, electrolyte, and other related components of Li-O_2 batteries are also introduced in detail. In addition, the influence of nitrogen, carbon dioxide, and water in the air on the Li-O_2 batteries is also concerned, which helps to think about the development of Li-O_2 batteries to lithium–air battery. The author thinks that only when all units work together, can the lithium–air battery with practical prospect be obtained.

2.1.2 Cathode Materials

As we all know, cathode is one of the most important components in the electrochemical reaction of Li–O_2 batteries. Li–O_2 batteries possess special semi-open structure and complex mechanism of lithium–oxygen reaction, so that this section summarizes three key characteristics of cathode: efficient catalytic performance, good conductivity, and appropriate porosity [21–23]. First of all, the active material of cathode must have efficient catalytic properties to control the mild ORR/OER reactions, so as to keep the overpotential of Li—O_2 battery at a low level. Second, good conductivity provides stable electron supply for fast electrochemical reactions. Third, high porosity can provide active sites and confined space for discharge products and then provide reliable guarantee for stable electrochemical reaction. Meanwhile, high porosity is the basis of rapid oxygen transport and lithium ion migration. Here we divide cathode materials into three types: carbon-based catalytic materials, non-noble metal-based catalytic materials, and noble metal-based catalytic materials. We give a detailed description of various typical cathode materials in this section.

2.1.2.1 Carbon-Based Materials

Carbon-based materials exhibit numerous advantages, including low density, low prices, good conductivity, and tunable structure. It has always been a research hotspot. Therefore, carbon-based materials have received a lot of research and application in energy storage systems, including lithium/sodium-ion batteries, Zn–MnO_2 batteries, and supercapacitors [24–26]. Meanwhile, the heteroatom-doped carbon could show effectively bifunctional catalytic activity (ORR/OER), which makes it an ideal cathode material for Li–O_2 batteries [27]. We will focus only on carbon-based materials in this section.

Activated Carbon Carbon black, a conductive additive, is widely used in the preparation of electrodes from active materials in batteries due to the characteristics of low cost, high conductivity, and high specific surface area (up to 3000 $m^2\ g^{-1}$) [28]. The high specific surface area of carbon black is the main reason for its selection as cathode of Li–O_2 battery. At present, there are many kinds of carbon black used as catalytic materials for Li–O_2 batteries, such as ketjenblack (KB), XC-72, super P, and so on [21, 29, 30].

Lin et al. evaluated the accommodating capacity of Li–O_2 batteries attributed to the high porosity utilization of KB, which is dominating the solvent pathway during the formation of Li_2O_2 (Figure 2.2) [29]. During the discharge process, Li_2O_2 can format along the KB surface which is generated through the surface path and the solvent path perpendicular to the surface. In the initial stage of oxidization process, both the solvent ion channels and surface pathways are available. With the discharge going on, the part of the electrode surface path is covered by Li_2O_2 quickly, while the surface favorable to the solvent path can continue to deposit Li_2O_2. Since the Li_2O_2 nanosheet becomes dominant through the solvent path, the passage for Li^+/O_2 transport channel can be kept smooth for further growth of Li_2O_2, resulting

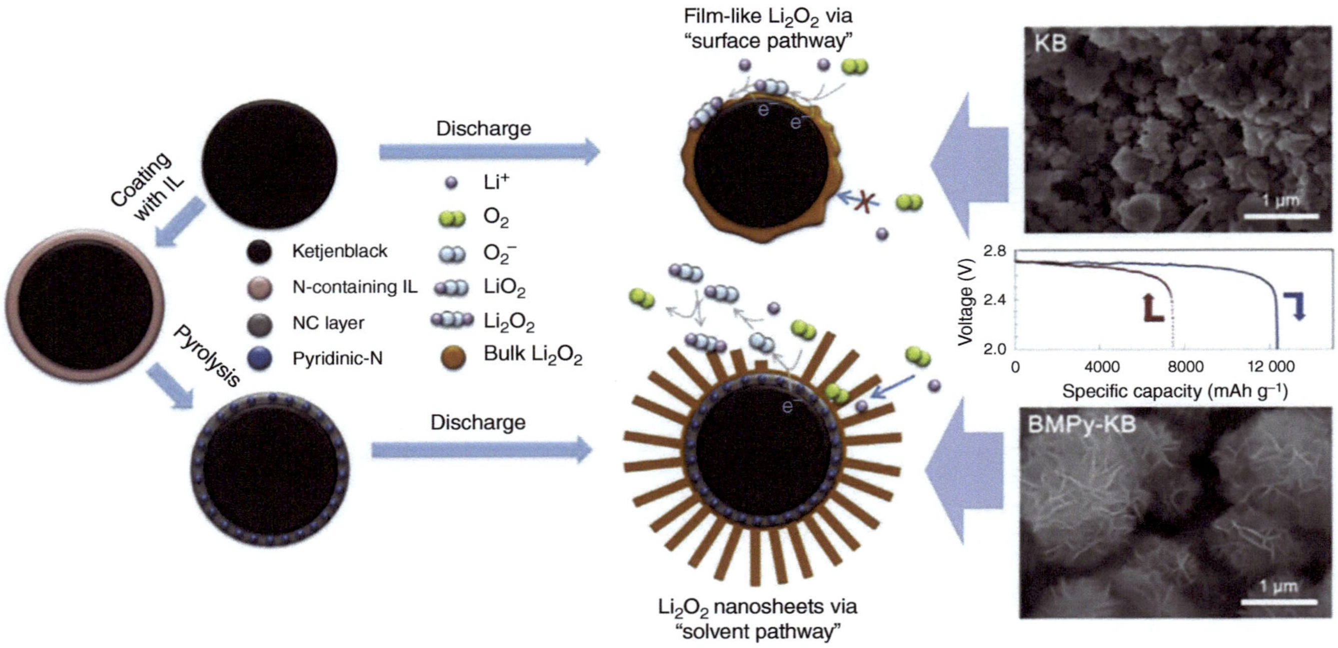

Figure 2.2 Schematic illustration of the Li_2O_2 growth mode and discharge capacity. Source: From Lin et al. [29]. Reproduced with permission of Elsevier.

in an increase in discharge capacity and possibly a decrease in the charging platform after discharge process (Figure 2.2). Although the carbon with ultrahigh specific surface area could afford numerous active sites for Li_2O_2, their geometric shape makes them easy to agglomerate and O_2 is difficult to penetrate into the carbon pores. This restricts the flow of active substances to the site of catalytic activity in subsequent cycles, resulting in premature battery failure.

Novel Carbon Materials As mentioned previously, the catalytic cathode of Li-O_2 battery needs appropriate porosity and high specific surface area. Nanodesign of materials is an effective method to increase the specific surface area. Meanwhile, nanodesign of materials provides convenient conditions for the construction of various three-dimensional (3D) porous structures. The preparation of carbon nanomaterials has been extensively studied, such as one-dimensional carbon nanotubes (1D CNTs) and carbon nanofibers (CNFs), two-dimensional (2D) graphene, and 3D carbon foam [31–33].

The synergistic combination of catalytic and conductive properties of CNTs cathode materials can effectively improve the cycling performance and discharge specific capacity of the Li-O_2 battery (Figure 2.3) [34]. In particular, surface defects can promote the electrocatalytic activity of O_2. At the same time, O_2 is preferentially adsorbed on the defects of the oxidized CNTs, which leads to the formation of defect Li_2O_2. The modified CNTs have a unique low-voltage plateau at about 3.5 V and a reversible capacity of 500 mAh g^{-1}.

The reaction environment around the cathode of lithium-O_2 battery is complex, including three phases, namely, gas phase (O_2), liquid phase (electrolyte), and solid phase (current collector), and the reaction is accompanied by the appearance and disappearance of solid Li_2O_2. Therefore, the air cathode of lithium-O_2 battery needs enough space to ensure that the change of Li_2O_2 does not affect the oxygen/electrolyte/electron transport path. Moreover, the air cathode needs a large number of evenly distributed catalytic active sites to achieve efficient formation and decomposition of Li_2O_2. 2D carbon materials with high conductivity have been proved to be excellent air cathode materials due to their load-carrying capacity and good structural stability [35–38]. The Li-O_2 battery with modified graphene exhibits high reversible capacity (85%) and stable cycle life (>100 cycles) under deep discharge/charging conditions, with a capacity of 17 000 mAh g^{-1} (Figure 2.4). The modified graphene exhibited excellent 3D Li^+/O_2 transfer characteristics and metal-free catalytic activity [39].

Three-dimensional carbon materials can better meet the requirements of air cathode because of its good conductivity and good layered porous structure [40–42]. It is noted that compared with the reported graphene-based macromaterials, such as nickel foam-based 3D graphene prepared by electrochemical methods, the 3D graphene monolithic materials fabricated directly by hydrothermal method with self-assembly have the advantages of adjustable pore structure, low density, and convenient preparation [43]. The modified 3D graphene can not only be used as an excellent self-supporting cathode but also avoid the negative effects of binder and

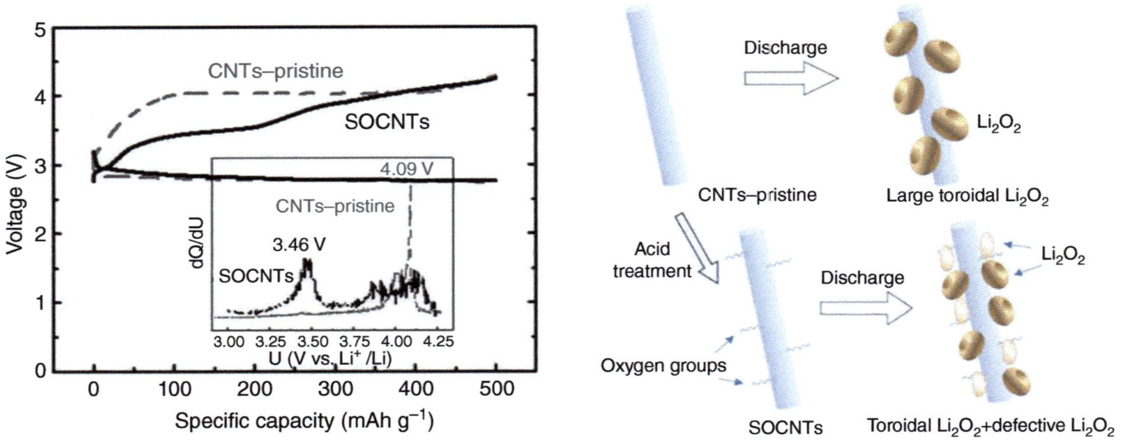

Figure 2.3 Schematic illustrations of the charge/discharge curves and Li_2O_2 growth mode. Source: Qin et al [34]. Reproduced with permission of Elsevier.

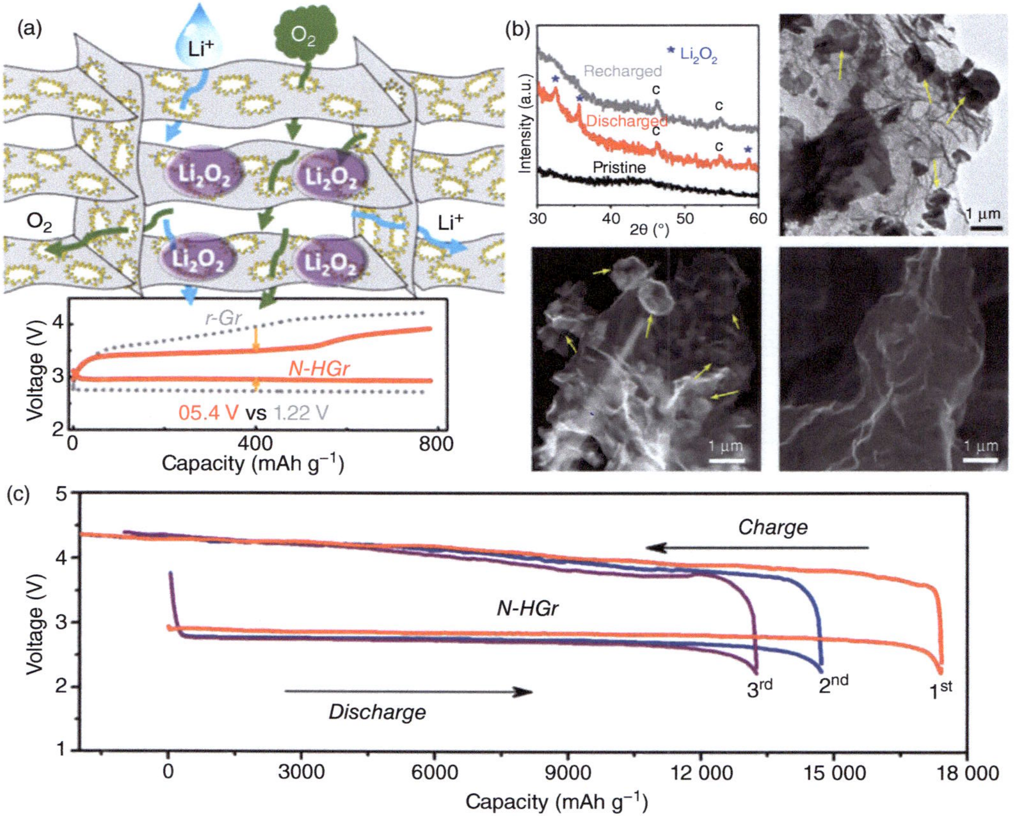

Figure 2.4 (a) Schematic illustration of the graphene for high-performance $Li{-}O_2$ batteries and Li_2O_2 growth mode. (b) Phase composition and morphology of grapheme. (c) Full discharge/charge curves of different graphene within the voltage range of 2.2–4.4 V under 100 mA g^{-1}. Source: From Shui et al. [39]. Reproduced with permission of American Chemical Society.

conductive additive. Moreover, it can be used as a new type of conductive framework catalyst to further fill the specific surface area and pore volume.

Figure 2.5 shows nitrogen-doped 3D graphene (3DG) with improved catalytic activity. As a new type catalyst for Li–O_2 battery, its unique 3D structure increases its catalytic active sites and pore volume. The obtained macro- and mesopores are conducive to the storage of discharge products, electrolyte transport, and oxygen diffusion. Compared with the 3D graphene, N-3DG has higher specific capacity and better cycle stability [44, 45]. The 3D porous structure effectively improves the catalytic activity sites of ORR and further optimizes the air cathode structure, which makes oxygen diffusion, electrolyte penetration, and discharge product (Li_2O_2) storage more effective. Therefore, when N-3DG is directly used as an independent cathode of Li–O_2 battery, its specific capacity reaches 7300 mAh g^{-1}, which has a longer 21 cycle life (21 cycles) than that of 3DG (2250 mAh g^{-1}, 8 cycles).

Atypical Carbon Materials with Special Structure In the previous report, the formation of Li_2O_2, the key product of air cathode, is different, including its size, structure, and crystallinity. Therefore, it is difficult to further understand the key mechanism in the charging/discharging process of Li–O_2 battery. To some extent, this problem is due to the unclear definition of the morphology and pore structure of carbon carriers. In view of this, it is valuable to study the effect of cathode materials with clear structure on lithium–air battery. A series of carbon materials with special structure have been systematically and deeply studied and have shown unique performance on lithium–air battery (Figure 2.6), including inverse replica of face-centered cubic structure, mesoporous carbon nanotube architecture, onion-like carbon structure, etc. [46–49].

Those special carbon structures make the operation mechanism of Li–O_2 battery further improved. For example, the possible cathodic reaction mechanism in discharge/charge processes is proposed based on onion-like carbon structure [48]. On onion-like carbon structure with high specific surface area, the nucleation rate exceeds the growth rate, and the density of active center is high, resulting in amorphous Li_2O_2 with thin film morphology as the main product (D1 stage in Figure 2.7), which has better ionic and electronic conductivity compared with thick lithium oxide.

Su et al. believe that the possible reaction process related to the first stage might be the LiO_2 formation on the catalyst surface through the solid solution path from the outside of Li_2O_2, where the species of LiO_2 is disproportionate to evolve O_2. In the second stage (>3.6 V), there is no residual film-like product (Figure 2.7, stage C1). Therefore, it can be easily concluded that ring Li_2O_2 decomposition resulted in a high charge potential. This reaction may be related to the formation of bulk Li_2O_2, which generates O_2 through two-phase transformation.

2.1.2.2 Noble Metal-Based Materials

Noble metals have good catalytic properties because of their physical and chemical properties. In Li–O_2 battery, noble metals have low lithium metal nucleation potential, which can guide the uniform formation of Li metal on the catalyst surface and

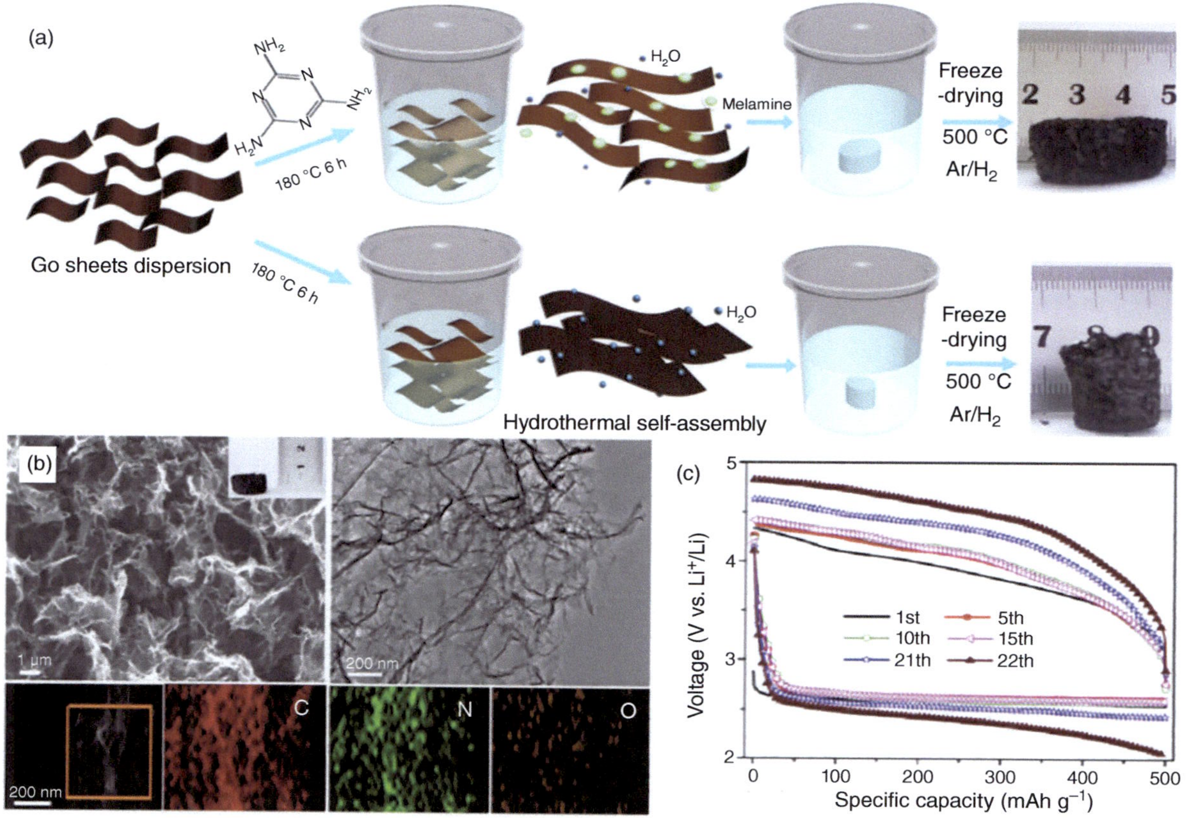

Figure 2.5 (a) Schematic illustration of the preparation processes of nitrogen-doped 3D graphene and 3D graphene. (b) Morphology and (c) electrochemical properties of nitrogen-doped 3D graphene. Source: From He et al. [44]. Reproduced with permission of Elsevier.

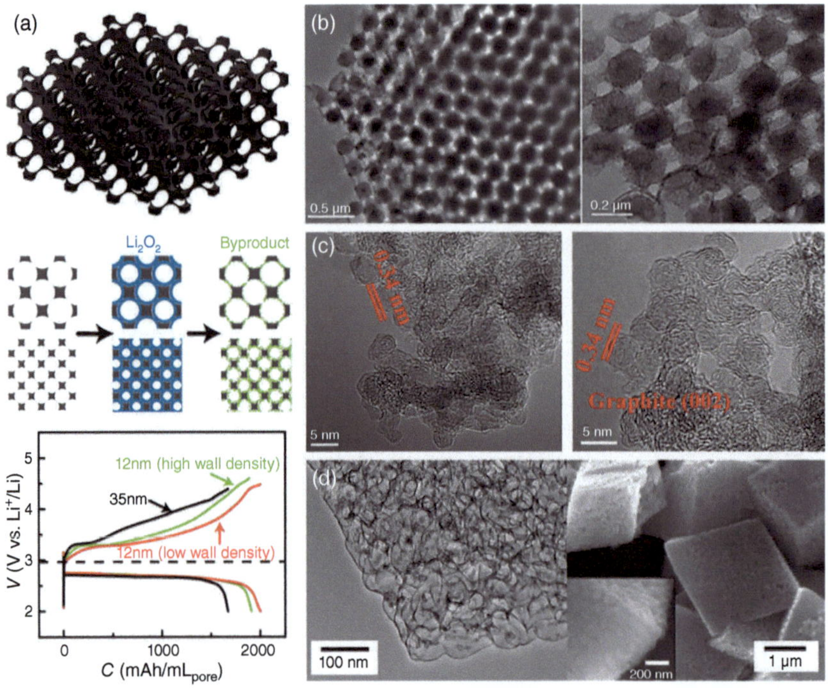

Figure 2.6 (a) Structure and electrochemical performance of three dimensionally ordered mesoporous carbon. Source: From Xie et al. [46]. Reproduced with permission of Wiley-VCH. (b) TEM images of as-prepared 3D ordered meso−/macroporous carbon sphere arrays. Source: From Guo et al. [47]. Reproduced with permission of Wiley-VCH. (c) High resolution transmission electron microscope (HRTEM) images of onion-like carbon (a) and N-doped onion-like carbon (b). Source: From Shu et al. [48]. Reproduced with permission of Royal Society of Chemistry. (d) Scanning electron microscope (SEM) and transmission electron microscope (TEM) images of mesoporous carbon nanotubes. Source: From Sun et al. [49]. Reproduced with permission of Wiley-VCH.

efficiently catalyze the deposition/decomposition of Li_2O_2 on the catalytic cathode. Because of the high cost of noble metals, the research focuses on the synthesis of composite catalysts with high-efficiency catalytic properties and reducing the proportion of noble metals as far as possible [50–53].

Xu's group reported a direct growth of copper nanoscale arrays with nanostructured Au surfaces on foam copper substrates (Au/Cu@FCu), which can be used as the anode substrate and the cathode of Li–O_2 battery [50]. The Au/Cu@FCu electrode with significant catalytic activity can tune the formation behavior of Li metal on anode and realize high efficient reversible decomposition of Li_2O_2 in cathode (Figure 2.8). The results show that lithium forms uniformly on the surface of Au/Cu@FCu anode without lithium dendrite. The Coulombic efficiency is more than 96%, and the cycle life is long and stable, more than 970 hours. At the same time, the overpotential of the gold cathode is very low (0.64 V) and the specific capacity is as high as 27 270 mAh g^{-1} compared with the carbon-free cathode of Li–O_2 battery reported so far.

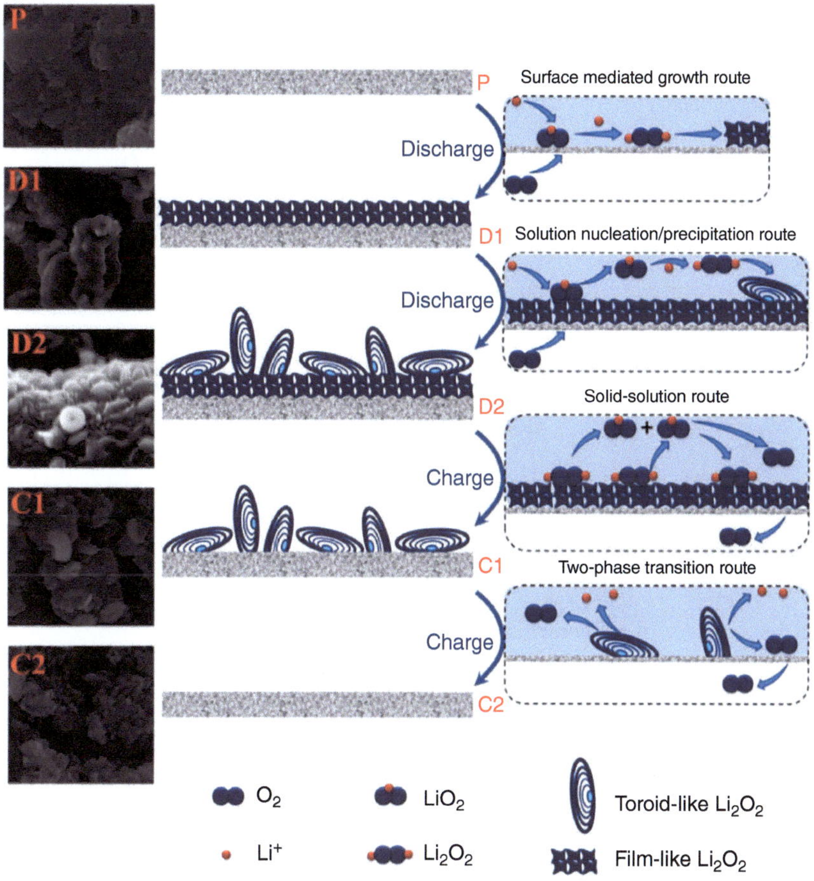

Figure 2.7 Schematic diagram of ORR and OER mechanism under different redox reactions. Source: From Shu et al. [48]. Reproduced with permission of Royal Society of Chemistry.

Chou's group improved the utilization rate of noble metals by preparing porous AgPd–Pd composite nanotubes (AgPd–Pd NTs) [52]. AgPd–Pd NTs are a kind of excellent bifunctional catalyst for ORR and OER in Li–O_2 batteries. The porous nanotube (NT) architecture exhibits good cycle stability and excellent energy efficiency, which contribute to the rapid diffusion of O_2 and electrolyte in NTs (Figure 2.9). And the porous NT structure forms a conductive network during the charge and discharge processes. The electrode of AgPd–Pd NTs shows stable discharge and charging capacity in Li–O_2 batteries. The discharge voltage is higher than 2.5 V, and the charging voltage is lower than 4.1 V, which can last for 100 cycles (Figure 2.9c).

2.1.2.3 Non-noble Metal-Based Materials

It is still a great challenge to develop catalysts based on earth's rich elements without affecting their performance. In recent years, the development of non-noble metal-based cathodes for Li–O_2 batteries has become a research hotspot. Although

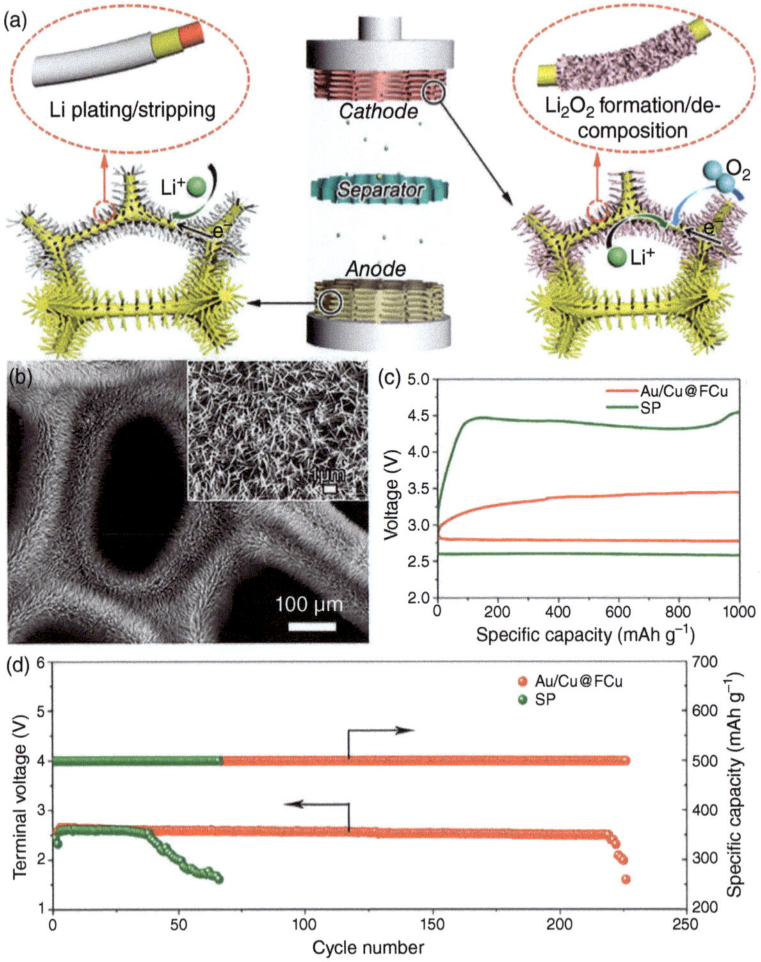

Figure 2.8 (a) Schematic diagram of Au/Cu@FCu preparation process. (b) Field emission scanning electron microscope (FESEM) of $Cu(OH)_2$@FCu. (c) and (d) Electrochemical performance of as-prepared Au/Cu@FCu electrode. Source: From Luo et al. [50]. Reproduced with permission of American Chemical Society.

a series of progress has been made on non-noble metal-based cathodes in recent years, it has been a great challenge to design and develop catalysts with high rate and stable cycle performance of Li–O_2 batteries [54–58].

Transition Metals Except for noble metals, some transition metals have good catalytic performance in Li–O_2 batteries. The bimetallic catalyst composed of two different transition metals has higher catalytic activity because the bimetallic catalyst not only inherits the essential catalytic properties of each component but also has higher catalytic efficiency than the single metal catalyst due to the strong synergy between metal atoms [59–61].

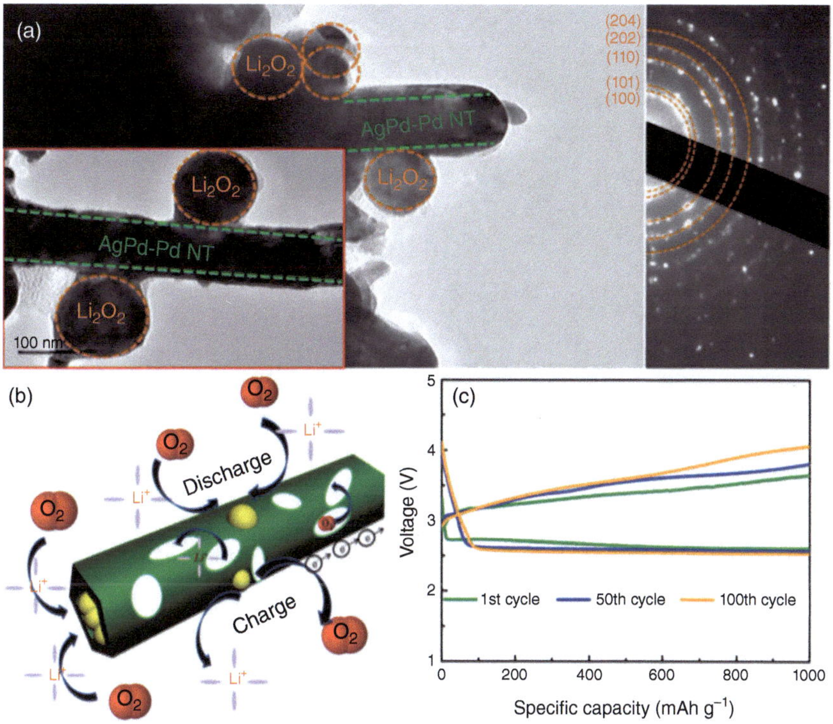

Figure 2.9 (a) TEM and selected area electron diffraction (SAED) patterns of AgPd Pd NTs catalyst after full discharge. (b) Schematic diagram of discharge/charge process. (c) Discharge/charge patterns of AgPd Pd NTs at selected cycles. Source: From Luo et al. [52]. Reproduced with permission of Wiley-VCH.

A carbon-free nickel/ruthenium (Ni/Ru) air cathode was developed using a three-dimensional ultralight porous Ni substrate [62]. First, three-dimensional scaffolds were fabricated by template plating and selective corrosion. Ru nanoparticles were coated on the scaffold to form a Ni/Ru air cathode which exhibited a high specific capacity with 2410 mAh g^{-1} at 150 mA g_{Ru}^{-1} (Figure 2.10). The Ni/Ru air cathode-specific capacity is 306 times higher than that of commercial nickel foam because the 3D cathode has super lightweight, large specific surface area, and good electronic conductivity.

Zhou's group reported the preparation of cobalt/copper bimetallic nanoparticles (CoCu/graphene) based on graphene and its performance in Li–O_2 batteries, as shown in Figure 2.11 [63]. The CoCu/graphene electrode exhibits high initial discharge capacity (14 821 mAh g^{-1}) at 200 mA g^{-1} and a low average charging voltage (about 4.0 V). In addition, the CoCu/graphene electrode exhibits superior high rate performance (7955 mAh g^{-1} at 800 mA g^{-1}), high cycle stability (122 cycles at 200 mA g^{-1} with the cutoff specific capacity of 1000 mAh g^{-1}), and excellent Coulombic efficiency (92% at 200 mA g^{-1}). These excellent electrochemical properties are mainly due to the synergistic effect of transition metals Co and Cu, which can improve the kinetics of ORR and OER evolution at the same time. A good

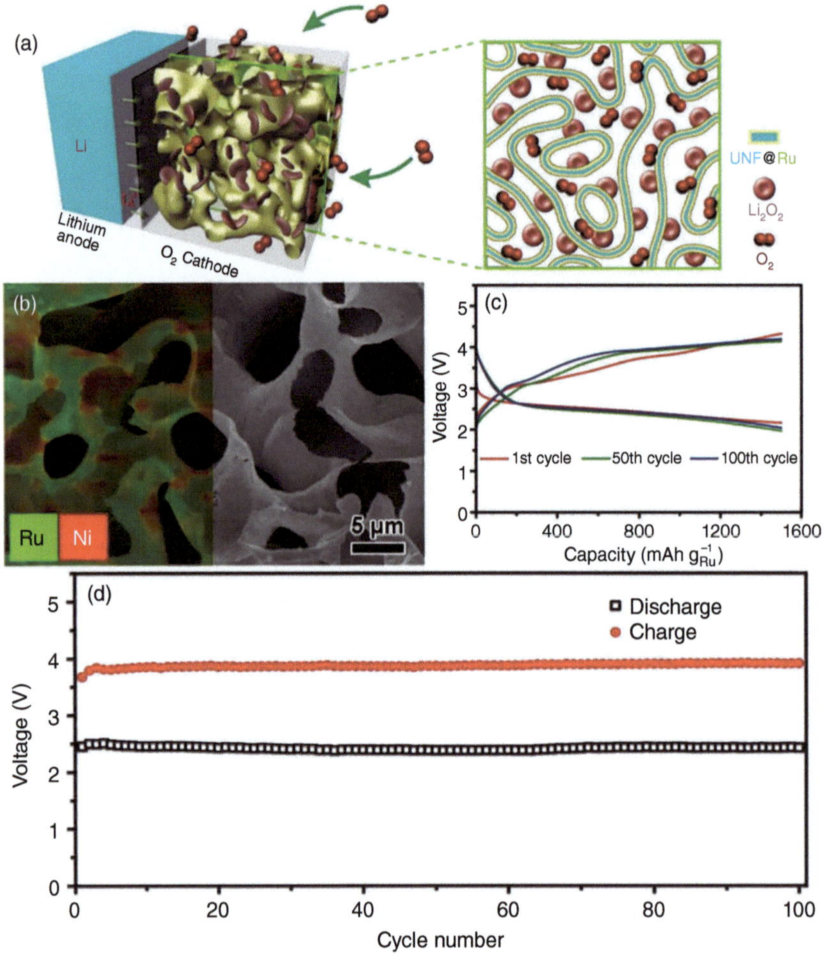

Figure 2.10 (a) Working mechanism of Ni/Ru cathode. (b) energy dispersive X-ray spectroscopy (EDX) elemental mapping and SEM image. (c) and (d) Electrochemical performance of as-prepared Ni/Ru electrode. Source: From Liu et al. [62]. Reproduced with permission of Wiley-VCH.

composite material ensures that the nanowall-like Li_2O_2 is uniformly formed on CoCu/graphene, replace of the typical ring-shaped Li_2O_2, which promotes the efficient reversible formation and decomposition of Li_2O_2.

Transition Metal Oxides The development of catalyst cathode from transition metal to transition oxide can further reduce the cost and contribute to the industrialization of Li–O_2 batteries. A variety of transition metal oxides, such as Fe_2O_3 [64], Fe_3O_4 [65], NiO [66], CuO [67], and Co_3O_4 [68], have been studied as cathode catalysts for Li–O_2 batteries. The transition group oxide electrodes not only show high capacity but also have high cycle stability and good bifunctional catalytic performance.

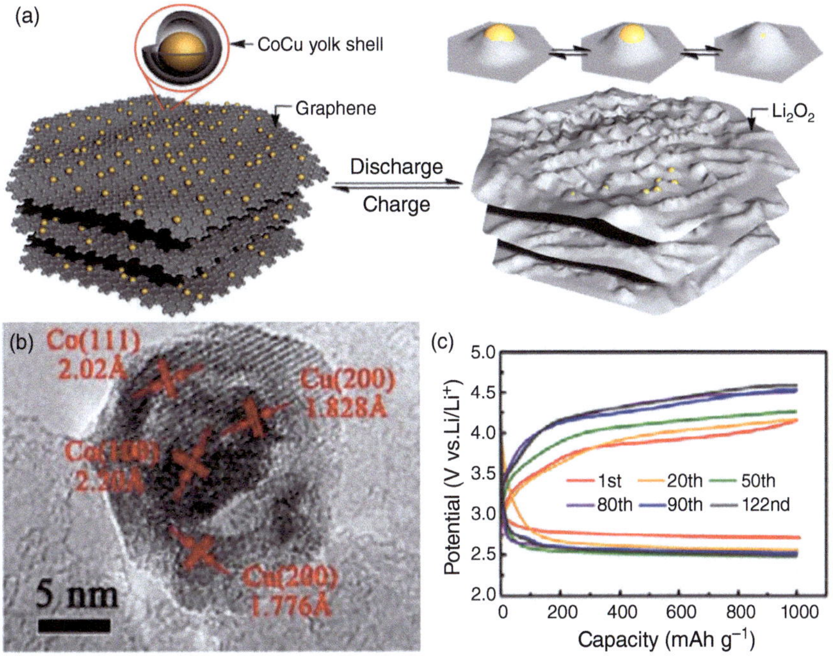

Figure 2.11 (a) Schematic diagram of CoCu/graphene cathode in charge/discharge process of Li–O_2 battery. (b) HRTEM image of CoCu/graphene yolk shell structure. (c) Cycle performance of CoCu/graphene cathode at 200 mA g^{-1} with capacity cut-off of 1000 mAh g^{-1} and discharge end voltage of 2.5 V. Source: From Chen et al. [63]. Reproduced with permission of Royal Society of Chemistry.

Liu's group proposed a strategy to increase the catalytic property of CoO by introducing dotted carbon species and O vacancies [69]. Carbon-dotted defective Co with O vacancy (CoO/C) was successfully prepared by sintering the $Co(AC)_2{\cdot}4H_2O$ (Figure 2.12), which was mediated by ethanol. Compared with commercial CoO and pure O vacancy CoO, the cycle performance, initial capacity, thermal stability, and rate performance of CoO/C catalytic electrode are obviously improved, which is due to the synergistic effect of point carbon species and O vacancies on ORR and OER. O vacancies can improve the flowability of Li^+/e^-, and provide nucleation sites for Li_2O_2. In addition, carbon-dotted species increase the CoO conductivity and make the ORR and OER process more stable.

Gu et al. developed a metal oxide cathode catalyst by using the surface anchoring of high selenium oxide ($SeO_4{}^{2-}$) clusters to control the chemical bond properties [70]. The bond competition between (Se^{6+}–O) and (Mn—O) bonds is especially beneficial to inhibit the formation of Mn^{3+} and improve the oxygen electron density of α-MnO_2 nanowires (NW) as shown in Figure 2.13. The selenate anchored α-MnO_2-NW electrode has good cathode catalytic activity for Li–O_2 batteries attributed to the increasing charge transfer kinetics and reversible Li_2O_2 decomposition. This study emphasizes that the surface modification with high valence

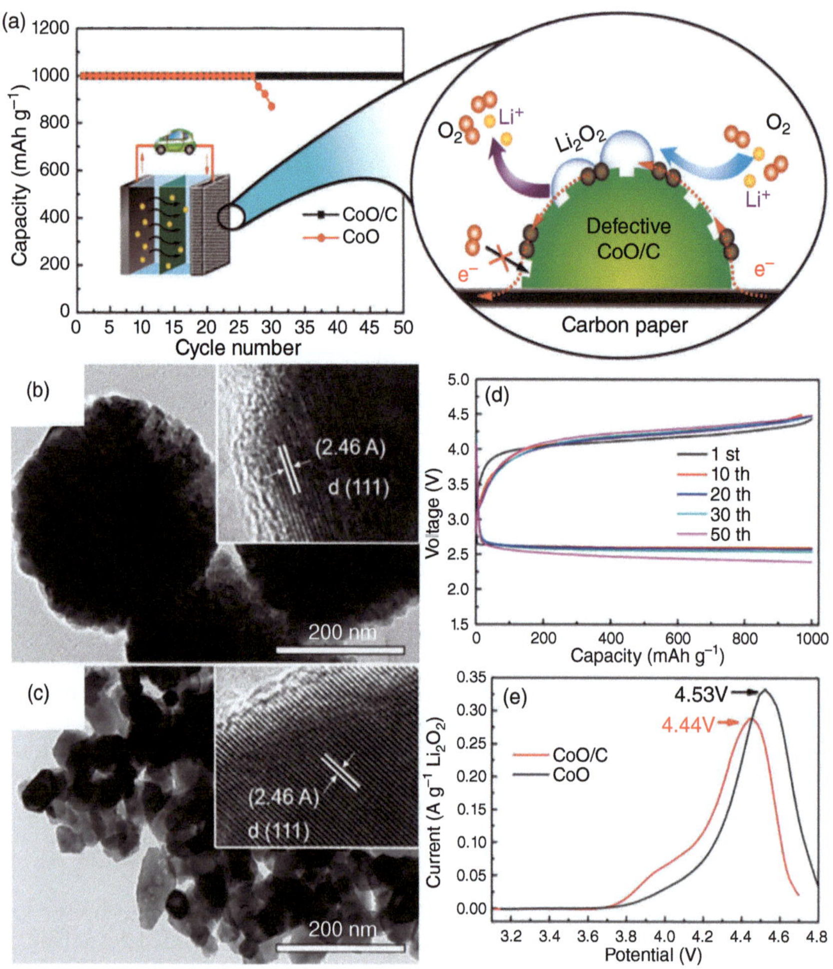

Figure 2.12 (a) The synergetic effect diagram shows the dotted carbon and oxygen vacancies in CoO/C on ORR and OER. The HRTEM images of (c) carbon-dotted defective CoO and (d) commercial CoO. (d) Discharge/charge curves of CoO/C with selected cycles. (e) Linear sweep voltammetric curves of Li_2O_2-containing batteries catalyzed by CoO and CoO/C with scanning rate of 0.1 $mV\,s^{-1}$. Source: From Gao et al. [69]. Reproduced with permission of American Chemical Society.

oxidation clusters can effectively increase the electrochemical property of nanosized metal oxides in Li–O_2 batteries.

Transition Metal Carbides, Sulfides, and Nitrides Transition metal carbides [71], sulfides [72], and nitrides [73] have also been found to have catalytic properties in Li–O_2 batteries. Among these transition metal compounds, Mo_2C is a very attractive material because of its multivalent state, high electrochemical activity, and low cost. Nazar' group report a passivated nanostructured metal carbide (Mo_2C), which has a very low oxidation potential of Li_2O_2 (3.2 V) [74]. Due to the similar redox potential of Mo_2C, the passivated conductive interface between Li_2O_2 and Li_xMoO_3 formed

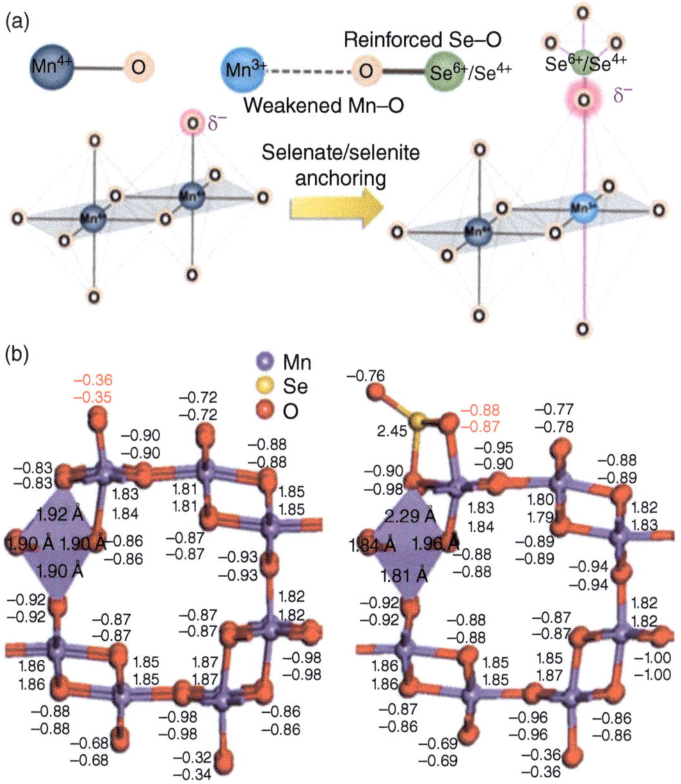

Figure 2.13 (a) The schematic of beneficial effects of SeO_4^{2-} anchoring o α-MnO_2-NW. (b) The surface O charge states of α-MnO_2-NW (left) and SeO_4^{2-} anchored α-MnO_2-NW (right) were calculated by Bader charge analysis. Source: From Gu et al. [70]. Reproduced with permission of Wiley-VCH.

on carbide is oxidized simultaneously (Figure 2.14a,b). This process explains the low charging potential by releasing Li_xMoO_3 into the electrolyte, keeping the surface of each cycle alive. On the other hand, a layer of amorphous MoS_2 film was deposited on the 3D conductive carbon scaffold as cathode of Li–O_2 battery. As can be seen from Figure 2.14a,b, the energy efficiency was increased to about 83% and the cycle life was 190 times [75]. In this report, a pressure tuned stop flow atomic layer deposition method was used to deposit amorphous MoS_2 (ALD–MoS_2) layer with thickness of about 5 nm on the CNT forest-covered graphite foam. The 3D integrated cathode is very effective for both ORR and OER. It can be concluded from the first principle calculation that ALD–MoS_2 effectively reduces the catalytic barrier.

In addition, Kim and coworkers reported a kind of cathode catalytic electrode using cobalt nitride (Co_4N) CNF film as cathode catalyst electrode of Li–O_2 battery [76]. In this report, Co(OH)F nanorods were grown by hydrothermal method and then nitride to obtain brush-like Co_4N nanorods (Figure 2.15). Co_4N modified CNF (Co_4N/CNF) catalytic cathode shows excellent battery performance with stable cycle

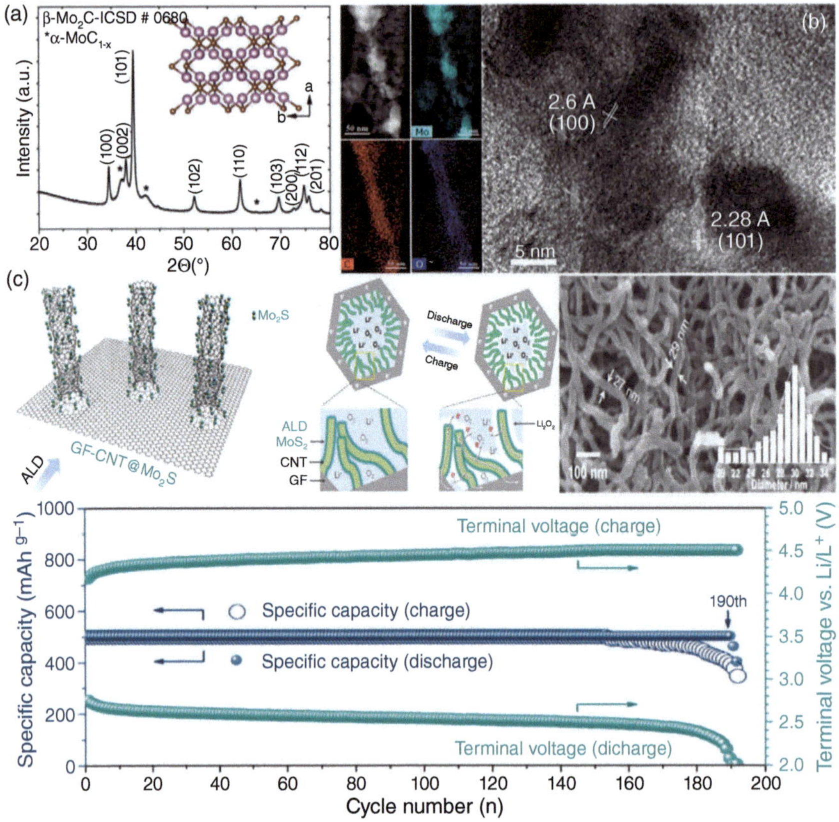

Figure 2.14 (a) The X-ray diffraction (XRD) pattern of the as-prepared Mo_2C nanofibers, TEM images of the same region of a nanofiber with the electron energy loss spectroscopy (EELS) and TEM images. Source: From Kundu et al. [74]. Reproduced with permission of American Chemical Society. (b) Schematic diagrams of the fabrication and charge/discharge process for GF–CNT@MoS_2 cathode of Li–O_2 battery and the corresponding SEM image and electrochemical performance. Source: From Song et al. [75]. Reproduced with permission of Wiley-VCH, Weinheim.

over 177 cycles. Co_4N nanorods provide a great deal of reaction sites and fast electron transport paths for continuous network CNF during charge–discharge process. The thin oxide layer formed on the surface of Co_4N nanorods (<10 nm) effectively improved the reversible formation/decomposition efficiency of Li_2O_2 film, resulting in a corresponding reduction of overpotential (around 1.23 V, 700 mAh g^{-1}). The experimental results show that the formation and decomposition of the reaction products are effectively controlled by the appropriate surface chemical properties and cathode material structure.

2.1.3 Anode Materials

Lithium metal as an anode of Li–O_2 batteries has many advantages, including high theoretical specific capacity and very low electrochemical negative potential.

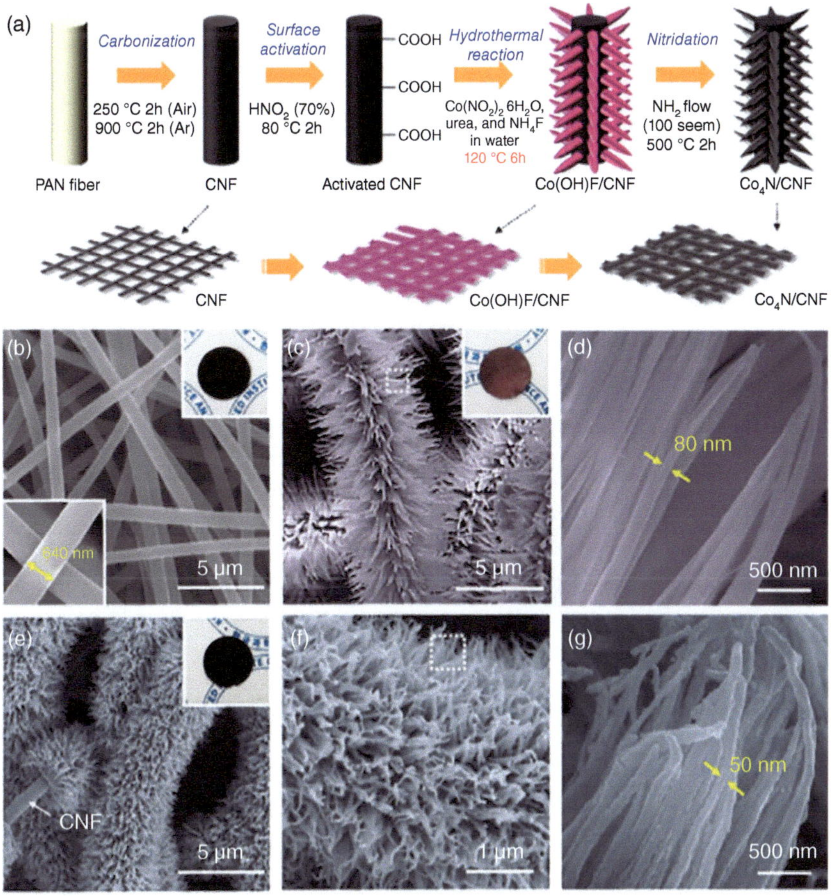

Figure 2.15 (a) Schematic diagrams of Co_4N/CNF synthesis process; SEM and digital images (insets) of (b) CNF, (c, d) Co(OH)F/CNF, and (e–g) Co_4N/CNF with different magnifications. Source: From Yoon et al. [76]. Reproduced with permission of American Chemical Society.

However, uncontrolled lithium dendrite growth and electrolyte decomposition can lead to safety problems and reduce Coulombic efficiency [77–79]. There are many clever ways to protect the lithium foil to increase the safety and stability of Li–O_2 batteries [80, 81]. Zhang and coworkers immersed the lithium electrode plate in a fluoroethylene carbonate solvent to form a double-layer film on the lithium foil anode [82]. The ionic conducting membrane has a dense double-layer structure, with organic components ($ROCO_2Li$ and ROLi) at the top and inorganic components (Li_2CO_3 and LiF) at the bottom. The double-layer interface can prevent the lithium metal anode from being corroded by electrolyte, regulating the homogeneous crystallization of lithium, and realize the lithium metal anode without dendrite (Figure 2.16). Through the surface modification of lithium metal anode, the reasonable construction of double-layer structure interface of safe charging battery was demonstrated.

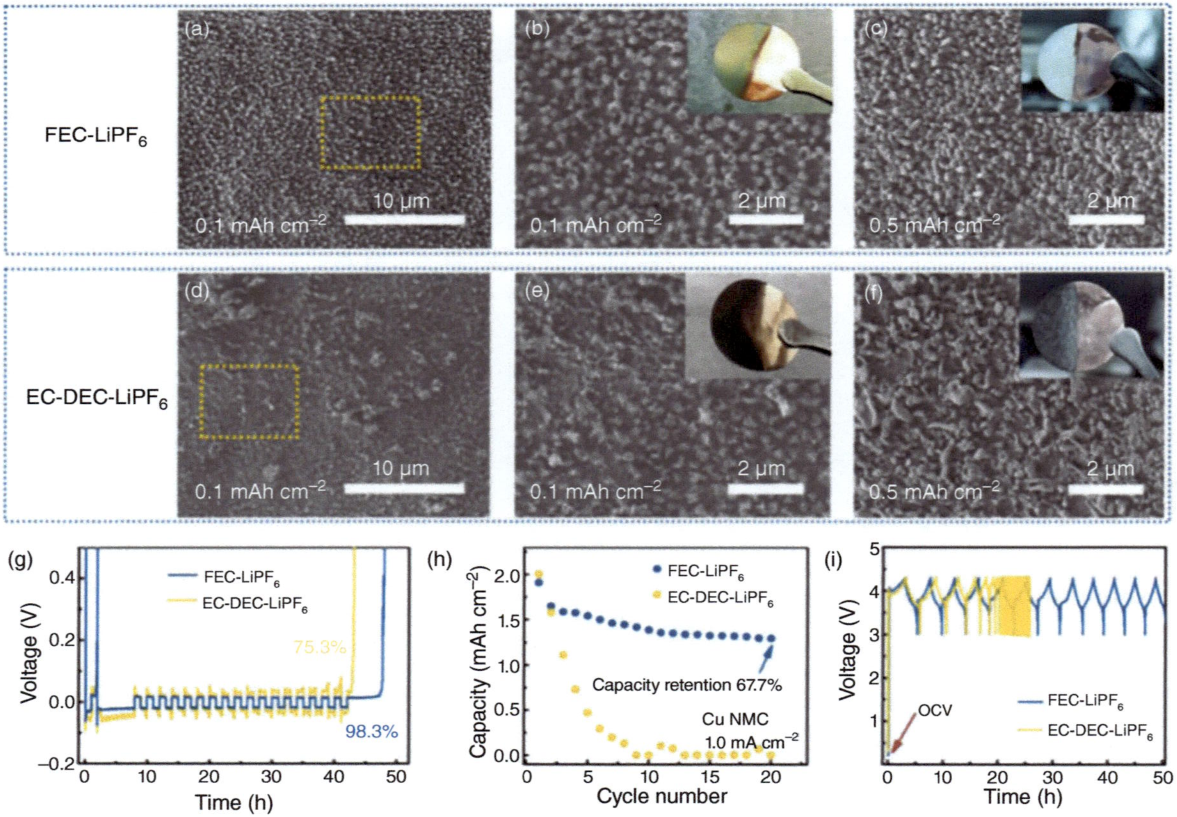

Figure 2.16 SEM images of lithium nucleation and growth in different electrolyte. (a) FEC-$LiPF_6$ can form the uniform nucleation sites on the surface and (c) guide the uniform growth of lithium coating. (d) EC–DEC–$LiPF_6$ electrolyte results in uneven nucleation sites on the surface and. (f) The lithium coating is disordered and uneven. (b) and (e) Enlarged images of the marked areas in (a) and (d). (g) The Columbic efficiency. (h) The cycling performance and (i) the voltage–time curves. Source: From Yan et al. [82]. Reproduced with permission of Wiley VCH.

2.1.4 Electrolyte

To meet the increasing energy density requirements, it is urgent to develop Li–O_2 batteries with ultrahigh theoretical energy density [83]. However, due to the use of organic liquid electrolyte and half-opened air electrode, its safety and cycle life have become the main obstacles to its practical application [84–86]. The electrolyte will decompose seriously in the process of discharge, and the oxygen-free radicals produced by cathode will combine with organic solvent or polymer [87, 88]. Many attempts have been made to find a stable nonaqueous solvent/electrolyte for lithium–oxygen batteries [89, 90]. The carbonate-based electrolyte commonly used in traditional lithium-ion batteries will decompose into Li_2CO_3 during discharge, rather than the ideal discharge product Li_2O_2 [91, 92]. Therefore, researchers have put forward various solutions.

2.1.4.1 Organic Electrolyte

In dimethyl sulfoxide (DMSO), the traditional electrolyte of 1 M bis(trifluoromethanesulfonimide) lithium (LiTFSI) is unstable to lithium metal anode, so it cannot be directly used in practical lithium–oxygen batteries. Zhang and coworkers proved that the high concentration electrolyte based on LiTFSI in DMSO (mole ratio of 1 : 3) can effectively increase the stability of lithium foil anode to DMSO and significantly raise the cycle performance of lithium–oxygen battery [93]. This high concentration electrolyte does not contain free DMSO solvent molecules, only contains $(TFSI^-)_a$-Li^+-$(DMSO)_b$ complexes (where $a + b = 4$), thus enhancing the stability of lithium foil anode (Figure 2.17). In addition, the salt solvent complex has a higher Gibbs activation energy barrier than the free DMSO solvent molecule. The results show that the electrolyte has good stability against superoxide radical anion attack. Therefore, the stability of the high concentration electrolyte on the lithium metal anode and carbon-based air electrode is greatly improved, thus improving the cycle stability of the Li–O_2 battery.

Zhou and coworkers used another idea to optimize the electrolyte [94]. Lithium bis(trifluoromethylsulfonyl)amide (LiTFSA) was dissolved in triglyme and tetraglymer (Gx, $x = 3$ and 4), respectively. A series of nonaqueous electrolytes with different molar ratios were prepared. In the electrolyte, the cycling stability of Li–O_2 battery is strongly dependent on the molar ratio of LiTFSA and Gx. The results show that the molar ratio of 1–5 is very important for the cycling stability of Li–O_2 battery. In LiTFSA-$(Gx)_5$ ($x = 3$ and 4), the high stability of 20 cycles was obtained in 500 mA g^{-1} and oxygen environment. It can be directly detected that the main product of cathode on the first and twentieth discharge electrodes is Li_2O_2, which indicates the excellent rechargeability and feasibility of Li–O_2 battery (Figure 2.18).

2.1.4.2 Quasi-Solid-State Electrolyte

Liquid electrolytes (LES) have the advantages of high ionic conductivity and good wettability at room temperature. However, there are some inherent problems in the use of liquid electrolytes in Li–O_2 batteries [95, 96]. First of all, its flammable properties are inherently dangerous. Second, under the semi-open operating conditions,

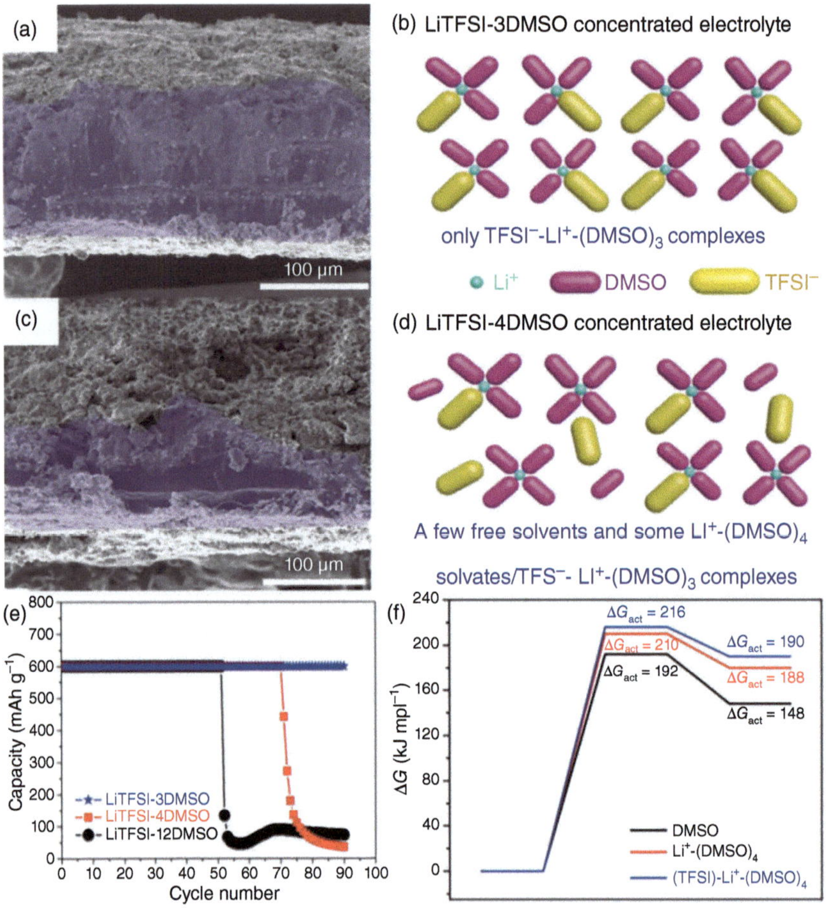

Figure 2.17 The SEM images of lithium foil in charge state using (a) LiTFSI-3DMSO electrolyte and (c) LiTFSI-4DMSO electrolyte after 90 cycles. Schematic diagram of components in (b) LiTFSI-3 DMSO concentrated electrolyte and (d) LiTFSI-4 DMSO high-concentration electrolyte. (e) The cycling performance of the three electrolytes. (f) Simulation of Gibbs activation energy barrier for C—H bond cleavage on CH_3 in different electrolytes. Source: From Liu et al. [93]. Reproduced with permission of Wiley-VCH.

the cycling stability of Li–O_2 battery is seriously deteriorated by the volatilization of LES during the cycle. When operating in ambient air, LES will quickly become wet because H_2O is easy to pass through the open structure of air cathode [97–99]. The rapid corrosion of lithium metal anode leads to battery failure [100, 101]. In addition, the heterogeneous nucleation of lithium metal in LES will produce unwanted dendrimer lithium, which will penetrate the separator and cause serious fire or explosion [102].

The quasi-solid-state electrolyte not only has high ionic conductivity but also can isolate the side effects of oxygen and water on anode lithium [103–107]. Sun and coworkers constructed the quasi-solid-state Li–O_2 battery with gel polymer electrolyte and high-efficiency air electrode, which showed good electrochemical

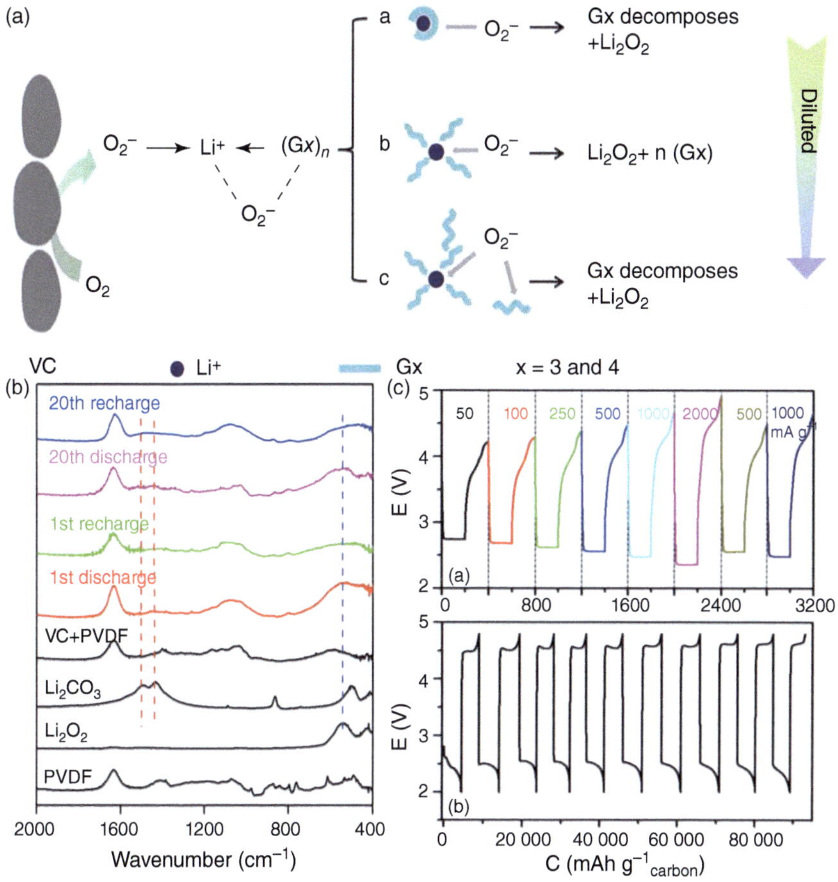

Figure 2.18 (a) The mechanism of O_2 reduction at the carbon black interface and LiTFSI glyme-based electrolyte. (b) The Fourier transform infrared spectroscopy (FTIR) spectra of Li–O_2 battery at the end of the first and 20th charge and discharge. (c) Charge–discharge curves (up) with a cut-off capacity of 200 mAh g^{-1}_{carbon} at various current densities from 25 to 2000 mA g^{-1}_{carbon} and (down) in a voltage range between 2.0 and 4.8 V at 1000 mA g^{-1}_{carbon} with LiTFSA-$(G3)_5$ electrolyte. Source: From Li et al. [94]. Reproduced with permission of Wiley-VCH.

performance [108]. The cycle life of the quasi-solid-state Li–O_2 battery is up to 553 cycles, and the stable operation time is more than 1100 hours (Figure 2.19). Compared with liquid electrolyte, gel polymer electrolyte prevents the diffusion of O_2 to the anode, thus protecting lithium metal and effectively avoiding electrolyte evaporation.

Shao and coworkers prepared a high cross-linking quasi-solid-state electrolyte with high ionic conductivity by using the hydrogen bonding between thermoplastic polyurethane and aerogel SiO_2 in gel polymer [109]. The quasi-solid-state electrolyte has high conductivity, high mechanical flexibility, good flame retardancy, and excellent lithium dendrite impermeability as shown in Figure 2.20a. The prepared

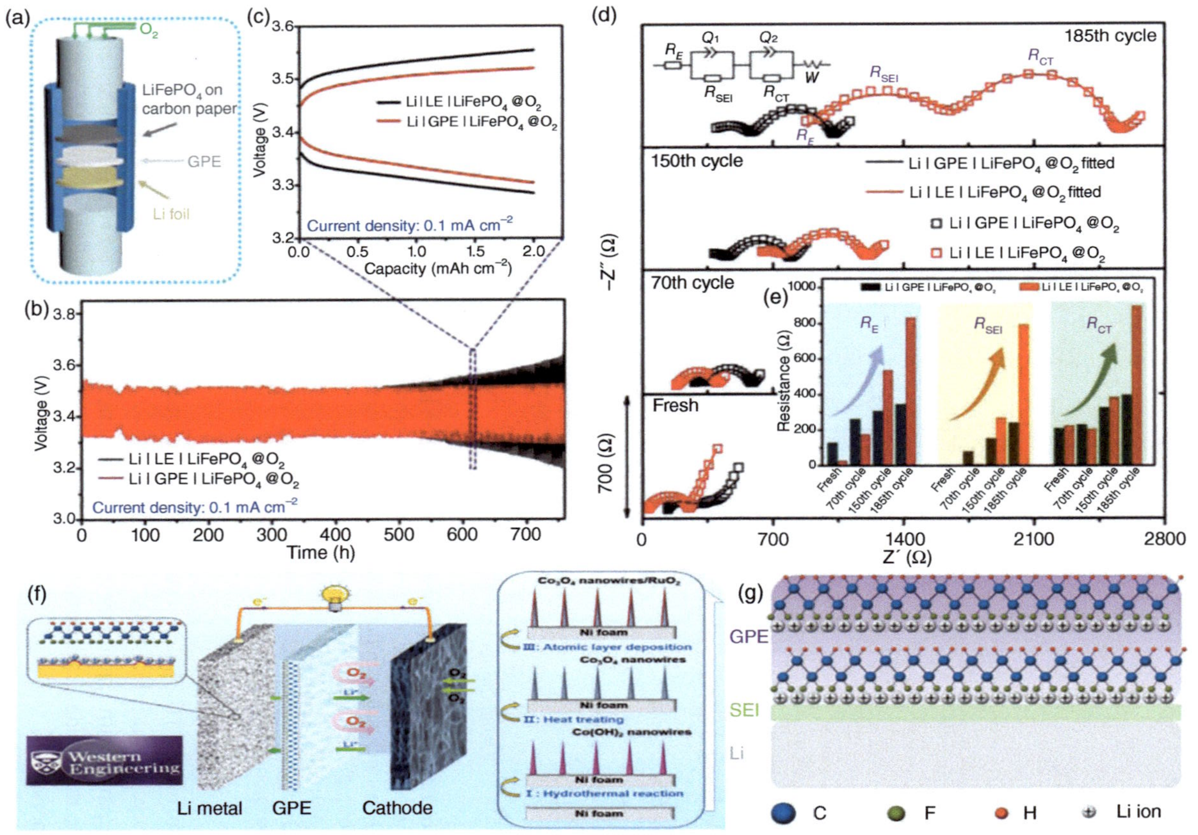

Figure 2.19 (a) Schematic diagram of full battery using gel polymer electrolyte (GPE) in oxygen atmosphere; (b) the corresponding cycling performance and (c) discharge/charge curve of 150th cycle. (d) Nyquist diagrams of LE and GPE of full battery under different cycles in oxygen atmosphere. (e) Histogram of R_E, R_{SEI}, and R_{CT} values of full battery. (f) Schematic diagrams of Li–O_2 battery with GPE and Ni foam@Co_3O_4-50RuO_2 synthesis process. (g) Schematic diagram of lithium deposition behavior with GPE. Source: From Zhao et al. [108]. Reproduced with permission of Wiley-VCH.

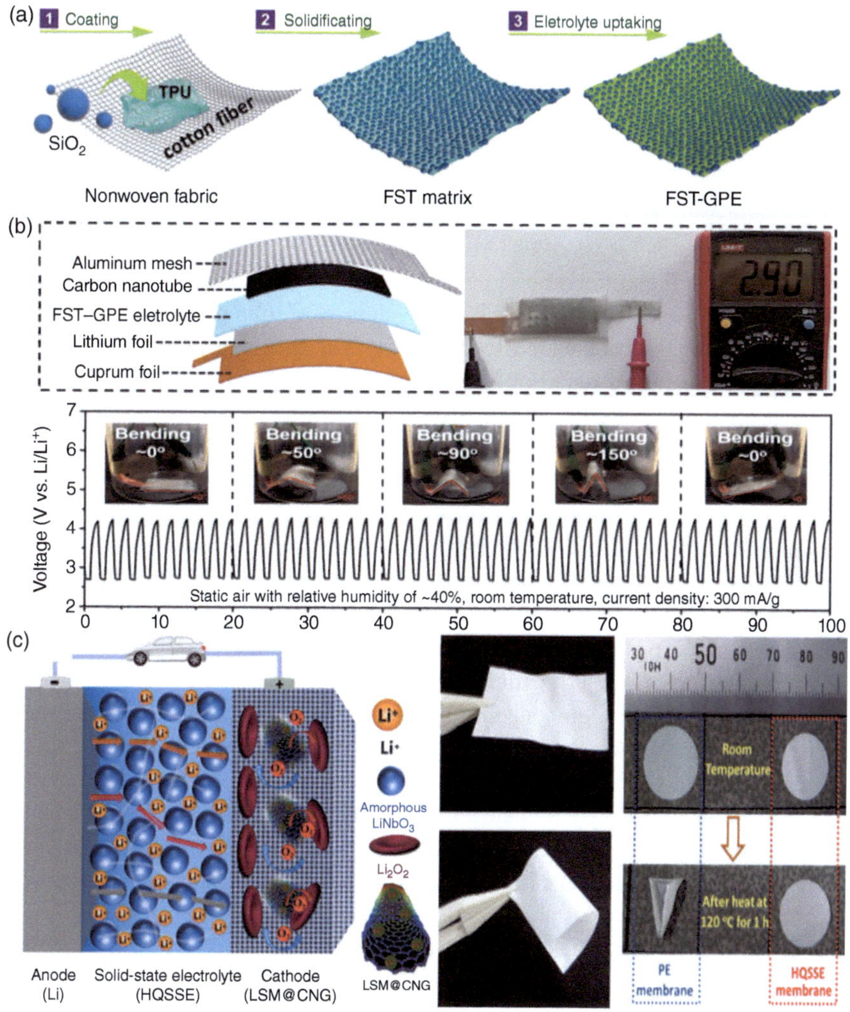

Figure 2.20 (a) Schematic diagram of the FST–GPE manufacturing process of the sample. (b) Test of flexible Li–O_2 battery at different bending angles. Source: From Zou et al. [109]. Reproduced with permission of Wiley-VCH. (c) Li–O_2 battery based on hybrid quasi-solid-state electrolyte (HQSSE) and controllable interface LSM@CNG cathode. Source: From Yi et al. [110]. Reproduced with permission of Wiley-VCH.

quasi-solid-state Li–O_2 battery has high reaction kinetics and stability due to its unique electrode–electrolyte interface and rapid diffusion of oxygen in the cathode. The quasi-solid-state Li–O_2 battery can achieve up to 250 discharge/charging cycles (over 1000 hours) in oxygen. In the air environment, the coin-type quasi-solid-state Li–O_2 battery operating for more than 20 days and working under extreme bending conditions shows excellent performance as shown in Figure 2.20b. On the other hand, Zhou's team designed and explored a mixed quasi-solid-state electrolyte which combines polyelectrolyte and ceramic electrolyte (Figure 2.20c) [110]. The results show that the rechargeable quasi-solid-state Li–O_2 battery has stable cycling performance (>100 cycles). What is more, the quasi-solid-state Li–O_2 battery also shows good safety. The quasi-solid electrolyte can be used as a promising development direction of Li–O_2 batteries. Meanwhile, it also encourages more people to explore other lithium–oxygen mixed solid electrolytes in future applications.

2.1.4.3 Solid-State Electrolyte

Solid-state lithium battery is considered as the ultimate goal of energy storage system due to its high theoretical energy density and improved safety [111–113]. Among those lithium batteries, the theoretical energy density of solid-state Li–O_2 battery is the highest, which has been widely studied [114, 115]. In recent years, Li–air batteries have made a series of remarkable progress. However, there is no suitable solid electrolyte for Li–air batteries, and the design of solid-state air electrodes also faces many challenges. Therefore, the development of all solid-state Li–air batteries is still plagued.

The redox mediators in electrolyte can participate in the charging process at medium voltage [116, 117]. Based on this idea, Janek's group combined high concentration redox mediator with solid-state electrolyte to prepare a new solid-state electrolyte [118]. Janek's group choose TEMPO+/TEMPO redox couple (TEMPO is 2,2,6,6-tetramethyl-1-piperidinyloxy) as redox mediator (Figure 2.21). The capacity and cycle stability of Li–O_2 battery with the solid electrolyte were improved significantly. The use of high-concentration redox mediator significantly improves the discharge capacity because the redox mediator participates in the oxidation and reduction reaction of the battery during the cycle process. The lithium anode is protected by solid electrolyte to realize high-efficiency circulation, which can completely inhibit the adverse deactivation of anodic oxidation species. The solid-state electrolyte can also restrain the harmful side reactions on the catalytic electrode to a great extent and can work stably for a long time below 4.0 V.

Sun and coworkers reported an ionic superconducting halide electrolyte, which can regulate the air–electrode interface and improve the cycle stability and safety of solid-state Li–O_2 batteries [119]. The high ionic conductivity of Li_3InCl_6 and the preparation method of Li_3InCl_6 solution make it work like liquid electrolyte improver on air electrode and adjust the function of interface evenly. The results show that the interface resistance of Li–O_2 battery containing Li_3InCl_6 decreases and the decomposition of discharge products increases (Figure 2.22). This report demonstrates the ability of halide electrolyte to control air–electrode interface and provide battery cycle stability and safety performance in solid-state Li–O_2 battery. The ionic conductivity of Li_3InCl_6 electrolyte is as high as 1.3×10^{-3} S cm^{-1}, and the intermediate product and final product of Li–O_2 battery keep stable in the process of charge and discharge. Li_3InCl_6 can be easily and uniformly distributed by the solution method and keep close contact with nitrogen-doped CNTs and $Li_{1.5}Al_{0.5}Ge_{1.5}(PO_4)_3$. Solid-state Li–$O_2$ batteries were chosen the high ionic conductor as the interface modifier, in which the interface resistance decreases dramatically from 2056 to 569 Ω. This is attributed to the optimization of the ion transport channel. This resistance is equivalent to that of a liquid electrolyte. As

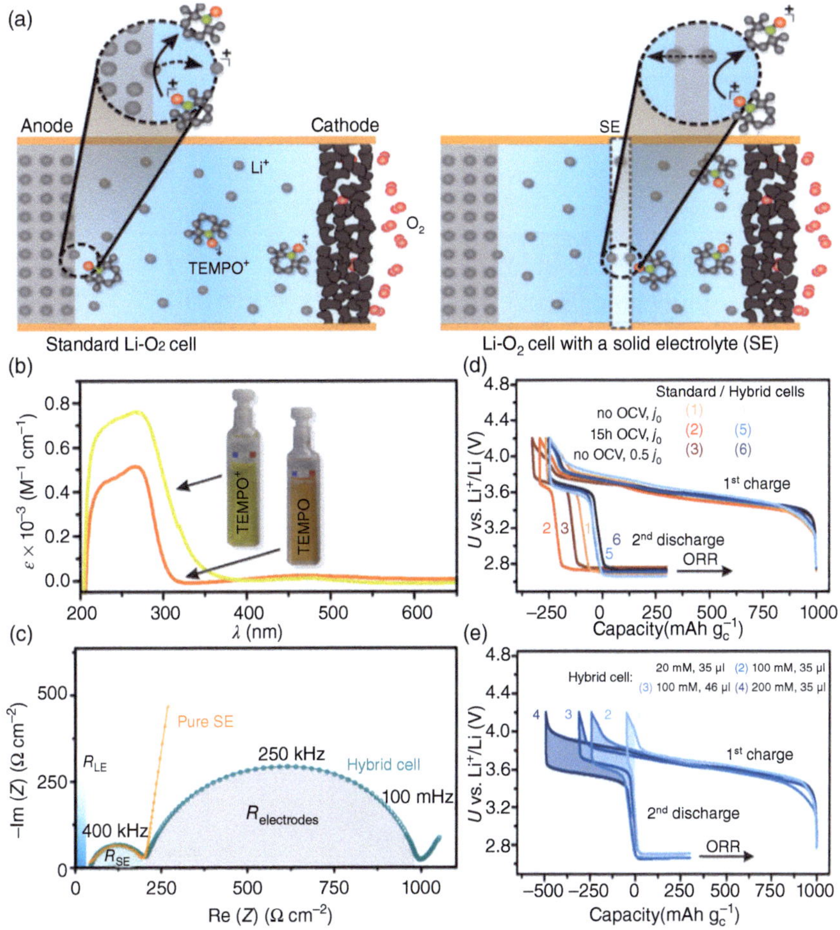

Figure 2.21 (a) The possible reaction of redox mediators in Li–O_2 battery. (b) The UV–vis spectra of the cathode solution before cycling (yellow) and after $K_2S_2O_8$ oxidation (red). Electrochemical performance: (c) impedance spectra of pure SE and hybrid batteries, (d) and (e) charge and discharge curves of different Li–O_2 battery. [118] Source: From Bergner et al. [118]. Reproduced with permission of American Chemical Society.

a modifier, high electron conductor can also effectively adjust the decomposition/formation of discharge products, improves the Coulombic efficiency, and increases the cycling stability.

2.1.5 Separator

With the in-depth study of electrode reaction and various catalysts, the electrochemical performance of air cathode has been significantly improved, however, there are still inevitable problems in the stability of lithium metal interface. Due to the reaction of H_2O and O_2 on the cathode side of air with lithium metal,

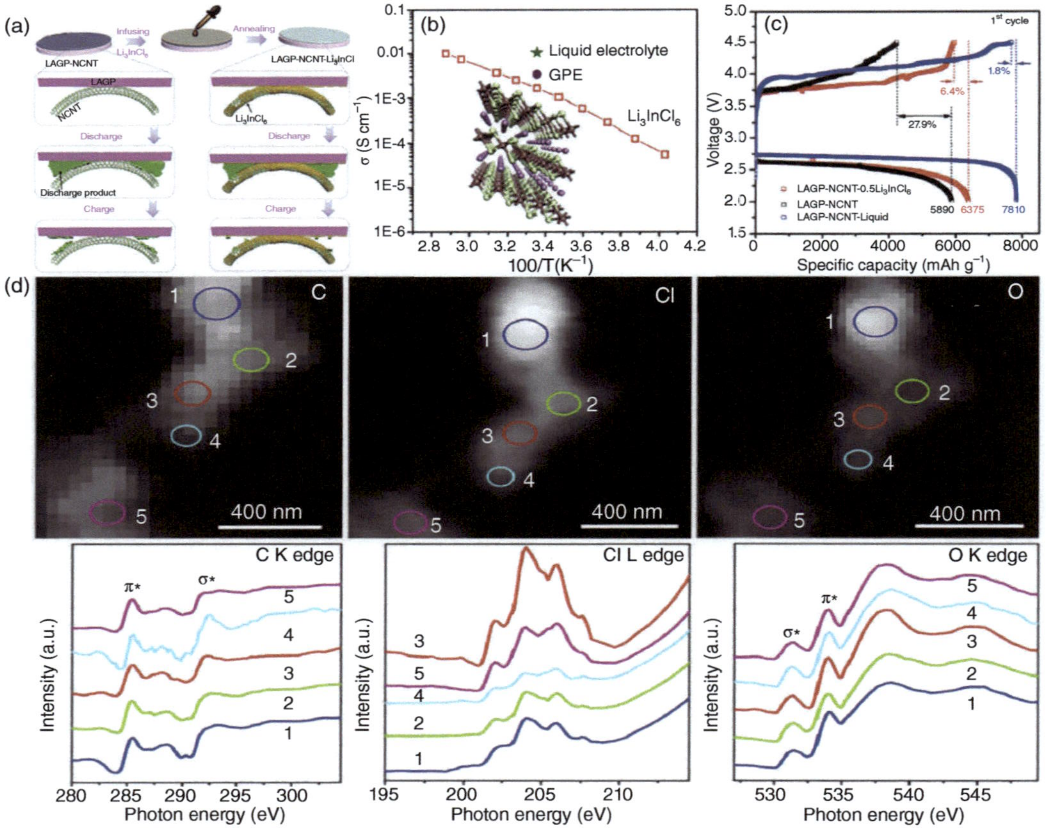

Figure 2.22 (a) Schematic diagrams of the synthesis process of LAGP–NCNT–Li_3InCl_6 air cathode and the influence of Li_3InCl_6 modifier on Li–O_2 battery. (b) The distribution diagram of ionic conductivity of Li_3InCl_6, GPE, and LE. The crystal structure of Li_3InCl_6 is shown in the inset. (c) STXM optical density images of C, Cl, and O distributions in the discharged state and the corresponding XANES spectra. Source: From Zhao et al. [119]. Reproduced with permission of Elsevier.

the instability of lithium–metal interface limits the cycling performance of Li–O_2 battery [120, 121]. The separator with high stability plays a key role here. The development of new moisture- and oxygen-impermeable separator can effectively improve the interface stability of lithium foil and then improve the electrochemical stability of Li–O_2 battery [122–125].

Choi group reported a cheap nonporous polyurethane separator, which can effectively inhibit H_2O and O_2 penetration and allow selective diffusion of lithium ions [126]. Based on the nonporous polyurethane separator, the continuous capacity of Li–O_2 battery can maintain more than 200 cycles at 600 mAh g^{-1}. The separator uses the well-known component poly(ethylene oxide) in gel polymer electrolyte, and its polarity is equivalent to the polarity of the typical battery electrolyte (Figure 2.23). The appropriate polarity will make the electrolyte have good wettability and promote the diffusion of lithium ion in polyurethane film, while the nonporous membrane can effectively prevent the permeation of H_2O and O_2.

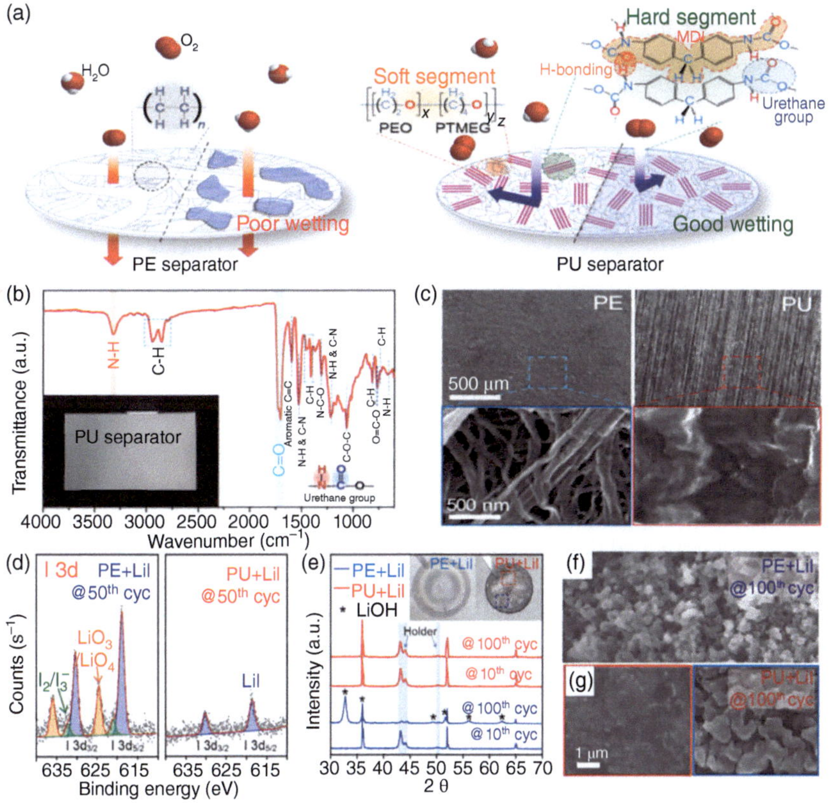

Figure 2.23 (a) Schematic diagrams of the separator influence on electrolyte wetting and gas/water permeation. (b) FT-IR pattern of PU. (c) SEM images of PE and PU separators. (d) X-ray photoelectron spectroscopy (XPS) spectra of the lithium foils in Li–O_2 batteries after 50 cycles. (e) XRD patterns and digital photos of the Li metal. (f) and (g) The corresponding SEM image of lithium foils after 100 cycles. Source: From Kim et al. [126]. Reproduced with permission of Wiley-VCH.

2.1.6 From Li–O_2 Batteries to Li–Air Batteries

At present, most of the reported results are based on pure O_2 atmosphere, so these types of cells are called Li–O_2 batteries. To make Li–O_2 battery have better practicability, the influence of nitrogen, water, and carbon dioxide in ambient air on the performance of Li–O_2 battery needs to be considered comprehensively. Although the concentration of carbon dioxide in the atmosphere is extremely low, it has high solubility in organic solvents and participates in redox reaction [127–130]. Although there are many difficult problems to solve, there are still many reports of Li–CO_2 batteries due to its CO_2 capture capacity and ultrahigh energy density [131–133]. While Li_2CO_3 shows higher chemical stability than that of Li_2O_2, so CO_2 should be paid more attention [134, 135]. Therefore, it has been reported that a series of suitable catalysts have been developed to promote the decomposition of Li_2CO_3 to obtain better cycle stability [136, 137]. Porous MnO [138] was prepared as catalytic cathode, which could decompose Li_2CO_3 below 4 V. Another important factor is the effect of water on the performance of lithium–oxygen batteries. Because of the periodic change of climate, the water content in the air is stable and unstable. Zhou's group reported an ionic liquid gel electrolyte to selectively permeate oxygen and prevent water from soaking in (Figure 2.24) [139]. As the main component of environmental atmosphere, nitrogen also reacts with lithium [140]. The development of lithium–air battery is still in its infancy, and there is still a long way to go to develop practical Li–air battery.

2.1.7 Summary and Perspective

This chapter introduces the basic characteristics of Li–air battery and its mainly aprotic Li–O_2 battery research status. The challenges faced by Li–air batteries are summarized. In recent years, Li–air battery has made great breakthrough in raising electrode catalytic ability, cycle stability, safety, etc. Li–air battery has experienced a high-speed development stage. The super high theoretical energy density of Li–air battery promotes the research and development of practical applications. However, there is still a big gap between the basic theoretical research and industrial application. It cannot be denied that the industrial application of Li–air battery is still in its infancy.

As a new complex energy storage system with high energy density, each part of lithium–air battery seriously affects its energy storage performance. The key challenge of Li–air battery is to improve the electrochemical stability of components in the presence of complex redox products while maintaining good energy storage performance. Air cathodes need highly efficient bifunctional catalytic activity, good conductivity, and stable 3D structure. For Li anode, improving its resistance to oxygen, carbon dioxide, and water is an effective way to ensure its electrochemical stability. In addition, more attention should be paid to the flammability of organic electrolytes and the growth of lithium dendrites. Due to the semi-open structure of Li–air battery, quasi solid electrolyte may be a good choice to solve the problem of low ionic conductivity and poor safety. There are many reasons to believe that high-performance Li–air batteries will be used in practical applications.

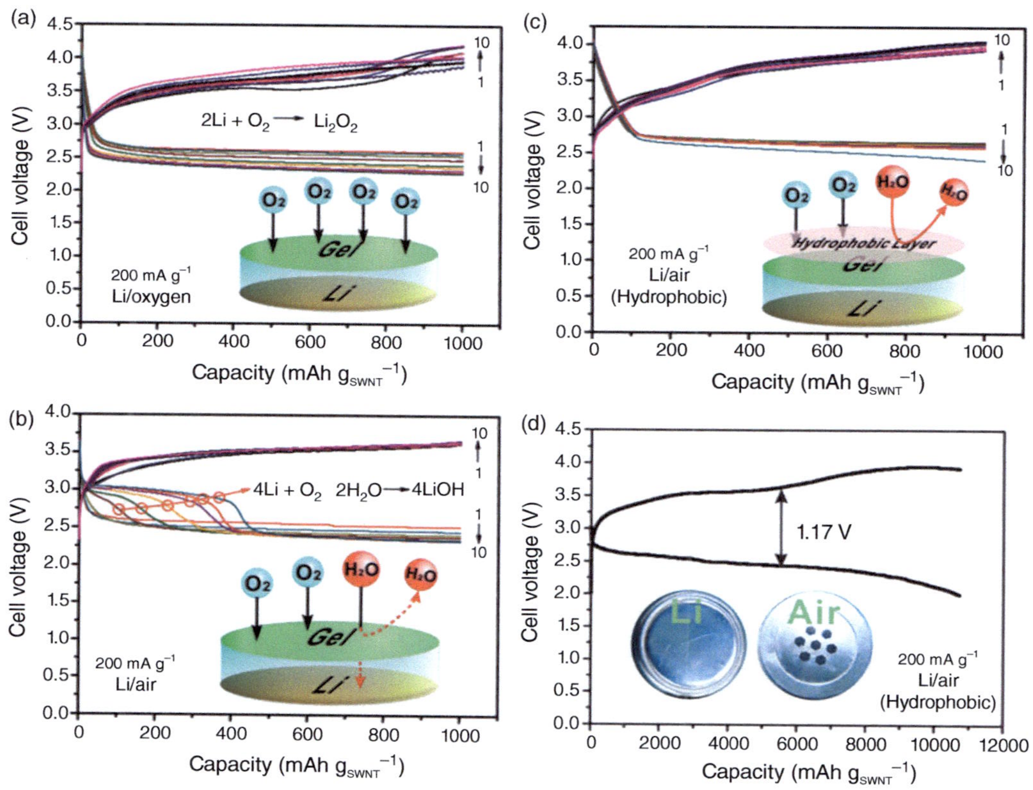

Figure 2.24 (a) and (b) elaborate the influence of H_2O pollution in atmosphere on Li–O_2 battery. (c) and (d) elaborate the effect of hydrophobic layer on the electrochemical behavior of Li–air battery. Source: From Zhang et al. [139]. Reproduced with permission of Wiley-VCH.

Acknowledgments

The authors gratefully acknowledge the financial support for this research which was provided by the National Nature Science Foundation of China (No. 51801136) and the Tianjin Municipal Education Committee scientific research projects (No. 2017KJ075).

References

1 Park, M., Sun, H., Lee, H. et al. (2012). Lithium-air batteries: survey on the current status and perspectives towards automotive applications from a battery industry standpoint. *Adv. Energy Mater.* 2 (7): 780–800.
2 Li, F. and Chen, J. (2017). Mechanistic evolution of aprotic lithium-oxygen batteries. *Adv. Energy Mater.* 7 (24): 1602934.
3 Wang, C., Xie, Z., and Zhou, Z. (2019). Lithium-air batteries: challenges coexist with opportunities. *APL Materials* 7 (4): 040701.
4 Semkow, K.W. and Sammells, A.F. (1987). A lithium oxygen secondary battery. *J. Electrochem. Soc.* 134: 2084.
5 Abraham, K.M. and Jiang, Z. (1996). A polymer electrolyte-based rechargeable lithium/oxygen battery. *J. Electrochem. Soc.* 143 (1): 1.
6 Geng, D., Ding, N., Hor, T.S.A. et al. (2016). From lithium-oxygen to lithium-air batteries: challenges and opportunities. *Adv. Energy Mater.* 6 (9): 1502164.
7 Blurton, K.F. and Sammells, A.F. (1979). Metal/air batteries: their status and potential – a review. *J. Power Sources* 4 (4): 263–279.
8 Lai, J., Xing, Y., Chen, N. et al. (2020). Electrolytes for rechargeable lithium–air batteries. *Angew. Chem. Int. Ed.* 59 (8): 2974–2997.
9 Ma, Z., Yuan, X., Li, L. et al. (2015). A review of cathode materials and structures for rechargeable lithium–air batteries. *Energy Environ. Sci.* 8 (8): 2144–2198.
10 Rahman, M.A., Wang, X., and Wen, C. (2014). A review of high energy density lithium–air battery technology. *J. Appl. Electrochem.* 44 (1): 5–22.
11 Rahman, M.A., Wang, X., and Wen, C. (2013). High energy density metal-air batteries: a review. *J. Electrochem. Soc.* 160 (10): A1759–A1771.
12 Lu, L., Han, X., Li, J. et al. (2013). A review on the key issues for lithium-ion battery management in electric vehicles. *J. Power Sources* 226: 272–288.
13 Shu, C., Wang, J., Long, J. et al. (2019). Understanding the reaction chemistry during charging in aprotic lithium–oxygen batteries: existing problems and solutions. *Adv. Mater.* 31 (15): 1804587.
14 Freunberger, S.A., Chen, Y., Peng, Z. et al. (2011). Reactions in the rechargeable lithium–O_2 battery with alkyl carbonate electrolytes. *J. Am. Chem. Soc.* 133 (20): 8040–8047.
15 Ogasawara, T., Débart, A., Holzapfel, M. et al. (2006). Rechargeable Li_2O_2 electrode for lithium batteries. *J. Am. Chem. Soc.* 128 (4): 1390–1393.

16 McCloskey, B.D., Bethune, D.S., Shelby, R.M. et al. (2011). Solvents' critical role in nonaqueous lithium–oxygen battery electrochemistry. *J. Phys. Chem. Lett.* 2 (10): 1161–1166.

17 Freunberger, S.A., Chen, Y., Drewett, N.E. et al. (2011). The lithium–oxygen battery with ether-based electrolytes. *Angew. Chem. Int. Ed.* 50 (37): 8609–8613.

18 Lu, J., Jung Lee, Y., Luo, X. et al. (2016). A lithium–oxygen battery based on lithium superoxide. *Nature* 529 (7586): 377–382.

19 Read, J., Mutolo, K., Ervin, M. et al. (2003). Oxygen transport properties of organic electrolytes and performance of lithium/oxygen battery. *J. Electrochem. Soc.* 150 (10): A1351.

20 Li, Y., Wang, J., Li, X. et al. (2011). Superior energy capacity of graphene nanosheets for a nonaqueous lithium–oxygen battery. *Chem. Commun.* 47 (33): 9438–9440.

21 Balaish, M., Jung, J.-W., Kim, I.-D., and Ein-Eli, Y. (2020). A critical review on functionalization of air-cathodes for nonaqueous Li–O_2 batteries. *Adv. Funct. Mater.* 30 (18): 1808303.

22 Hu, X., Wang, J., Li, Z. et al. (2017). MCNTs@MnO_2 nanocomposite cathode integrated with soluble O_2-carrier co-salen in electrolyte for high-performance Li–air batteries. *Nano Lett.* 17 (3): 2073–2078.

23 Park, J.O., Kim, M., Kim, J.-H. et al. (2019). A 1000 Wh kg^{-1} Li–air battery: cell design and performance. *J. Power Sources* 419: 112–118.

24 Han, L., Wu, S., Hu, Z. et al. (2020). Hierarchically porous MoS_2–carbon hollow rhomboids for superior performance of the anode of sodium-ion batteries. *ACS Appl. Mater. Interfaces* 12 (9): 10402–10409.

25 Chao, D., Zhou, W., Ye, C. et al. (2019). An electrolytic Zn–MnO_2 battery for high-voltage and scalable energy storage. *Angew. Chem. Int. Ed.* 58 (23): 7823–7828.

26 Peng, H., Yao, B., Wei, X. et al. (2019). Pore and heteroatom engineered carbon foams for supercapacitors. *Adv. Energy Mater.* 9 (19): 1803665.

27 Li, J., Shu, C., Ran, Z. et al. (2019). Heteroatom-induced electronic structure modulation of vertically oriented oxygen vacancy-rich NiFe layered double oxide nanoflakes to boost bifunctional catalytic activity in Li–O_2 battery. *ACS Appl. Mater. Interfaces* 11 (33): 29868–29878.

28 Islam, A.K.M.M., Lee, H.-S., Ro, J.-H. et al. (2019). Application of high-surface-area graphitized carbon black with primary secondary amine as an alternative quick, easy, cheap, effective, rugged, and safe cleanup material for pesticide multi-residue analysis in spinach. *J. Sep. Sci.* 42 (14): 2379–2389.

29 Lin, H., Liu, Z., Mao, Y. et al. (2016). Effect of nitrogen-doped carbon/ketjenblack composite on the morphology of Li_2O_2 for high-energy-density Li–air batteries. *Carbon* 96: 965–971.

30 Samira, S., Deshpande, S., Roberts, C.A. et al. (2019). Nonprecious metal catalysts for tuning discharge product distribution at solid–solid interfaces of aprotic Li–O_2 batteries. *Chem. Mater.* 31 (18): 7300–7310.

31 Zhang, X., Lu, W., Zhou, G., and Li, Q. (2020). Understanding the mechanical and conductive properties of carbon nanotube fibers for smart electronics. *Adv. Mater.* 32 (5): 1902028.

32 Cai, W., Zhang, Y., Jia, Y., and Yan, J. (2020). Flexible heteroatom-doped porous carbon nanofiber cages for electrode scaffolds. *Carbon Energy* 141: 561–567.

33 Ye, J., Zhou, M., Le, Y. et al. (2020). Three-dimensional carbon foam supported MnO_2/Pt for rapid capture and catalytic oxidation of formaldehyde at room temperature. *Appl. Catal., B* 267: 118689.

34 Qin, L., Lv, W., Wei, W. et al. (2019). Oxygen-enriched carbon nanotubes as a bifunctional catalyst promote the oxygen reduction/evolution reactions in Li-O_2 batteries. *Carbon* 141: 561–567.

35 Wang, L., Ara, M., Wadumesthrige, K. et al. (2013). Graphene nanosheet supported bifunctional catalyst for high cycle life Li–air batteries. *J. Power Sources* 234: 8–15.

36 Shi, L. and Zhao, T. (2017). Recent advances in inorganic 2D materials and their applications in lithium and sodium batteries. *J. Mater. Chem. A* 5 (8): 3735–3758.

37 Dong, H., Ji, Y., Hou, T., and Li, Y. (2018). Two-dimensional siligraphenes as cathode catalysts for nonaqueous lithium–oxygen batteries. *Carbon* 126: 580–587.

38 Sun, B., Huang, X., Chen, S. et al. (2014). Porous graphene nanoarchitectures: an efficient catalyst for low charge-overpotential, long life, and high capacity lithium–oxygen batteries. *Nano Lett.* 14 (6): 3145–3152.

39 Shui, J., Lin, Y., Connell, J.W. et al. (2016). Nitrogen-doped holey graphene for high-performance rechargeable Li–O_2 batteries. *ACS Energy Lett.* 1 (1): 260–265.

40 Jiang, J., He, P., Tong, S. et al. (2016). Ruthenium functionalized graphene aerogels with hierarchical and three-dimensional porosity as a free-standing cathode for rechargeable lithium–oxygen batteries. *NPG Asia Mater.* 8 (1): e239–e239.

41 Xie, J., Wang, J., Lee, H.R. et al. (2018). Engineering stable interfaces for three-dimensional lithium metal anodes. *Sci. Adv.* 4 (7): eaat5168.

42 Zhong, X., Papandrea, B., Xu, Y. et al. (2017). Three-dimensional graphene membrane cathode for high energy density rechargeable lithium–air batteries in ambient conditions. *Nano Res.* 10 (2): 472–482.

43 Asadi, M., Sayahpour, B., Abbasi, P. et al. (2018). A lithium–oxygen battery with a long cycle life in an air-like atmosphere. *Nature* 555 (7697): 502–506.

44 He, M., Zhang, P., Liu, L. et al. (2016). Hierarchical porous nitrogen doped three-dimensional graphene as a free-standing cathode for rechargeable lithium–oxygen batteries. *Electrochim. Acta* 191: 90–97.

45 Gong, H., Xue, H., Wang, T. et al. (2016). High-loading nickel cobaltate nanoparticles anchored on three-dimensional N-doped graphene as an efficient bifunctional catalyst for lithium–oxygen batteries. *ACS Appl. Mater. Interfaces* 8 (28): 18060–18068.

46 Xie, J., Yao, X., Cheng, Q. et al. (2015). Three dimensionally ordered mesoporous carbon as a stable, high-performance Li–O_2 battery cathode. *Angew. Chem. Int. Ed.* 54 (14): 4299–4303.

47 Guo, Z., Zhou, D., Dong, X. et al. (2013). Ordered hierarchical mesoporous/macroporous carbon: a high-performance catalyst for rechargeable Li–O_2 batteries. *Adv. Mater.* 25 (39): 5668–5672.
48 Shu, C., Lin, Y., and Su, D. (2016). N-doped onion-like carbon as an efficient oxygen electrode for long-life Li–O_2 battery. *J. Mater. Chem. A* 4 (6): 2128–2136.
49 Sun, B., Chen, S., Liu, H., and Wang, G. (2015). Mesoporous carbon nanocube architecture for high-performance lithium–oxygen batteries. *Adv. Funct. Mater.* 25 (28): 4436–4444.
50 Luo, N., Ji, G.-J., Wang, H.-F. et al. (2020). Process for a free-standing and stable all-metal structure for symmetrical lithium–oxygen batteries. *ACS Nano* 14 (3): 3281–3289.
51 Zhang, T., Zou, B., Bi, X. et al. (2019). Selective growth of a discontinuous subnanometer Pd film on carbon defects for Li–O_2 batteries. *ACS Energy Lett.* 4 (12): 2782–2786.
52 Luo, W.-B., Gao, X.-W., Chou, S.-L. et al. (2015). Porous AgPd–Pd composite nanotubes as highly efficient electrocatalysts for lithium–oxygen batteries. *Adv. Mater.* 27 (43): 6862–6869.
53 Chitturi, V.R., Ara, M., Fawaz, W. et al. (2016). Enhanced lithium–oxygen battery performances with Pt subnanocluster decorated N-doped single-walled carbon nanotube cathodes. *ACS Catal.* 6 (10): 7088–7097.
54 Liu, M., Sun, K., Zhang, Q. et al. (2020). Rationally designed three-dimensional N-doped graphene architecture mounted with Ru nanoclusters as a high-performance air cathode for lithium–oxygen batteries. *ACS Sustainable Chem. Eng.* 8 (15): 6109–6117.
55 Guo, Y., Dai, Z., Lu, J. et al. (2019). Lithiation-induced non-Noble metal nanoparticles for Li–O_2 batteries. *ACS Appl. Mater. Interfaces* 11 (1): 811–818.
56 Tu, Y., Li, H., Deng, D. et al. (2016). Low charge overpotential of lithium-oxygen batteries with metallic Co encapsulated in single-layer graphene shell as the catalyst. *Nano Energy* 30: 877–884.
57 Luo, X., Ge, L., Ma, L. et al. (2018). Effect of componential proportion in bimetallic electrocatalysts on the aprotic lithium-oxygen battery performance. *Adv. Energy Mater.* 8 (20): 1703230.
58 Song, K., Agyeman, D.A., Park, M. et al. (2017). High-energy-density metal–oxygen batteries: lithium–oxygen batteries vs sodium–oxygen batteries. *Adv. Mater.* 29 (48): 1606572.
59 Sevim, M., Francia, C., Amici, J. et al. (2016). Bimetallic MPt (M: Co, Cu, Ni) alloy nanoparticles assembled on reduced graphene oxide as high performance cathode catalysts for rechargable lithium-oxygen batteries. *J. Alloys Compd.* 683: 231–240.
60 Park, M., Liang, C., Lee, T.H. et al. (2020). Regulating the catalytic dynamics through a crystal structure modulation of bimetallic catalyst. *Adv. Energy Mater.* 10 (8): 1903225.
61 Mei, J., Liao, T., Liang, J. et al. (2020). Toward promising cathode catalysts for nonlithium metal–oxygen batteries. *Adv. Energy Mater.* 10 (11): 1901997.

62 Liu, Z., Feng, N., Shen, Z. et al. (2017). Carbon-free O_2 cathode with three-dimensional ultralight nickel foam-supported ruthenium electrocatalysts for Li–O_2 batteries. *ChemSusChem* 10 (13): 2714–2719.

63 Chen, Y., Zhang, Q., Zhang, Z. et al. (2015). Two better than one: cobalt–copper bimetallic yolk–shell nanoparticles supported on graphene as excellent cathode catalysts for Li–O_2 batteries. *J. Mater. Chem. A* 3 (34): 17874–17879.

64 Li, Z., Ganapathy, S., Xu, Y. et al. (2018). Fe_2O_3 nanoparticle seed catalysts enhance cyclability on deep (dis)charge in aprotic Li–O_2 batteries. *Adv. Energy Mater.* 8 (18): 1703513.

65 Shang, C., Zhang, X., Shui, L. et al. (2018). Fe_3O_4@CoO mesospheres with core-shell nanostructure as catalyst for Li-O_2 batteries. *Appl. Surf. Sci.* 457: 804–808.

66 Wang, H., Wang, H., Huang, J. et al. (2019). Hierarchical mesoporous/macroporous co-doped NiO Nanosheet arrays as free-standing electrode materials for rechargeable Li–O_2 batteries. *ACS Appl. Mater. Interfaces* 11 (47): 44556–44565.

67 Kim, D.-S., Lee, G.-H., Lee, S. et al. (2017). Electrocatalytic performance of CuO/graphene nanocomposites for Li–O_2 batteries. *J. Alloys Compd.* 707: 275–280.

68 Ryu, W.-H., Yoon, T.-H., Song, S.H. et al. (2013). Bifunctional composite catalysts using Co_3O_4 nanofibers immobilized on nonoxidized graphene nanoflakes for high-capacity and long-cycle Li–O_2 batteries. *Nano Lett.* 13 (9): 4190–4197.

69 Gao, R., Li, Z., Zhang, X. et al. (2016). Carbon-dotted defective CoO with oxygen vacancies: a synergetic design of bifunctional cathode catalyst for Li–O_2 batteries. *ACS Catal.* 6 (1): 400–406.

70 Gu, T.-H., Agyeman, D.A., Shin, S.-J. et al. (2018). α-MnO_2 nanowire-anchored highly oxidized cluster as a catalyst for Li-O_2 batteries: superior electrocatalytic activity and high functionality. *Angew. Chem. Int. Ed.* 57 (49): 15984–15989.

71 Jeong, M.-G., Kwak, W.-J., Shin, H.-J. et al. (2020). Perpendicularly aligned TiC-coated carbon cloth cathode for high-performance Li-O_2 batteries. *Chem. Eng. J.* 399: 125699.

72 Zhang, S., Huang, Z., Wen, Z. et al. (2017). Local lattice distortion activate metastable metal sulfide as catalyst with stable full discharge–charge capability for Li–O_2 batteries. *Nano Lett.* 17 (6): 3518–3526.

73 Park, J., Jun, Y.-S., Lee, W.-r. et al. (2013). Bimodal mesoporous titanium nitride/carbon microfibers as efficient and stable electrocatalysts for Li–O_2 batteries. *Chem. Mater.* 25 (19): 3779–3781.

74 Kundu, D., Black, R., Adams, B. et al. (2015). Nanostructured metal carbides for aprotic Li–O_2 batteries: new insights into interfacial reactions and cathode stability. *J. Phys. Chem. Lett.* 6 (12): 2252–2258.

75 Song, M., Tan, H., Li, X. et al. (2020). Atomic-layer-deposited amorphous MoS_2 for durable and flexible Li–O_2 batteries. *Small Methods* 4 (6): 1900274.

76 Yoon, K.R., Shin, K., Park, J. et al. (2018). Brush-like cobalt nitride anchored carbon nanofiber membrane: current collector-catalyst integrated cathode for long cycle Li–O_2 batteries. *ACS Nano* 12 (1): 128–139.

77 Li, D., Zhang, S., Zhang, Q. et al. (2020). Pencil-drawing on nitrogen and sulfur co-doped carbon paper: an effective and stable host to pre-store Li for high-performance lithium–air batteries. *Energy Storage Mater.* 26: 593–603.

78 Liang, W., Lian, F., Meng, N. et al. (2020). Adaptive formed dual-phase interface for highly durable lithium metal anode in lithium–air batteries. *Energy Storage Mater.* 28: 350–356.

79 Li, C., Wei, J., Qiu, K., and Wang, Y. (2020). Li–air battery with a superhydrophobic Li-protective layer. *ACS Appl. Mater. Interfaces* 12 (20): 23010–23016.

80 Wang, D., Zhang, W., Zheng, W. et al. (2017). Towards high-safe lithium metal anodes: suppressing lithium dendrites via tuning surface energy. *Adv. Sci.* 4 (1): 1600168.

81 Li, W., Yao, H., Yan, K. et al. (2015). The synergetic effect of lithium polysulfide and lithium nitrate to prevent lithium dendrite growth. *Nat. Commun.* 6 (1): 7436.

82 Yan, C., Cheng, X.-B., Tian, Y. et al. (2018). Dual-layered film protected lithium metal anode to enable dendrite-free lithium deposition. *Adv. Mater.* 30 (25): 1707629.

83 McCloskey, B.D., Speidel, A., Scheffler, R. et al. (2012). Twin problems of interfacial carbonate formation in nonaqueous Li–O_2 batteries. *J. Phys. Chem. Lett.* 3 (8): 997–1001.

84 Black, R., Adams, B., and Nazar, L.F. (2012). Non-aqueous and hybrid Li-O_2 batteries. *Adv. Energy Mater.* 2 (7): 801–815.

85 Nasybulin, E., Xu, W., Engelhard, M.H. et al. (2013). Effects of electrolyte salts on the performance of Li–O_2 batteries. *J. Phys. Chem. C* 117 (6): 2635–2645.

86 Li, Y., Wang, X., Dong, S. et al. (2016). Recent advances in non-aqueous electrolyte for rechargeable Li–O_2 batteries. *Adv. Energy Mater.* 6 (18): 1600751.

87 McCloskey, B.D., Valery, A., Luntz, A.C. et al. (2013). Combining accurate O_2 and Li_2O_2 assays to separate discharge and charge stability limitations in nonaqueous Li–O_2 batteries. *J. Phys. Chem. Lett.* 4 (17): 2989–2993.

88 Liu, T., Leskes, M., Yu, W. et al. (2015). Cycling Li–O_2 batteries via LiOH formation and decomposition. *Science* 350 (6260): 530.

89 Viswanathan, V., Thygesen, K.S., Hummelshøj, J.S. et al. (2011). Electrical conductivity in Li_2O_2 and its role in determining capacity limitations in non-aqueous Li–O_2 batteries. *J. Chem. Phys.* 135 (21): 214704.

90 Gowda, S.R., Brunet, A., Wallraff, G.M., and McCloskey, B.D. (2013). Implications of CO_2 contamination in rechargeable nonaqueous Li–O_2 batteries. *J. Phys. Chem. Lett.* 4 (2): 276–279.

91 Zhang, X., Xie, Z., and Zhou, Z. (2019). Recent progress in protecting lithium anodes for Li–O_2 batteries. *ChemElectroChem* 6 (7): 1969–1977.

92 Guo, H., Hou, G., Dai, L. et al. (2020). Stable lithium anode of Li–O_2 batteries in a wet electrolyte enabled by a high-current treatment. *J. Phys. Chem. Lett.* 11 (1): 172–178.

93 Liu, B., Xu, W., Yan, P. et al. (2017). Stabilization of Li metal anode in DMSO-based electrolytes via optimization of salt–solvent coordination for Li–O_2 batteries. *Adv. Energy Mater.* 7 (14): 1602605.

94 Li, F., Zhang, T., Yamada, Y. et al. (2013). Enhanced cycling performance of Li–O_2 batteries by the optimized electrolyte concentration of LiTFSA in Glymes. *Adv. Energy Mater.* 3 (4): 532–538.

95 Zhang, S.S., Foster, D., and Read, J. (2010). Discharge characteristic of a non-aqueous electrolyte Li/O_2 battery. *J. Power Sources* 195 (4): 1235–1240.

96 Peng, J.-Y., Huang, J., Li, W.-J. et al. (2018). A high-performance rechargeable Li–O_2 battery with quasi-solid-state electrolyte. *Chin. Phys. B* 27 (7): 078201.

97 Peng, Z., Freunberger, S.A., Chen, Y., and Bruce, P.G. (2012). A reversible and higher-rate Li–O_2 battery. *Science* 337 (6094): 563.

98 Zhou, B., Guo, L., Zhang, Y. et al. (2017). A high-performance Li–O_2 battery with a strongly solvating hexamethylphosphoramide electrolyte and a LiPON-protected lithium anode. *Adv. Mater.* 29 (30): 1701568.

99 Younesi, R., Hahlin, M., Treskow, M. et al. (2012). Ether based electrolyte, LiB(CN)4 salt and binder degradation in the Li–O_2 battery studied by hard X-ray photoelectron spectroscopy (HAXPES). *J. Phys. Chem. C* 116 (35): 18597–18604.

100 Qiao, Y., Wang, Q., Mu, X. et al. (2019). Advanced hybrid electrolyte Li–O_2 battery realized by dual superlyophobic membrane. *Joule* 3 (12): 2986–3001.

101 Hou, Z., Feng, S., Hei, P. et al. (2019). Morphology regulation of Li_2O_2 by flower-like $ZnCo_2S_4$ enabling high performance Li–O_2 battery. *J. Power Sources* 441: 227168.

102 Vivek, J.P., Homewood, T., and Garcia-Araez, N. (2019). An unsuitable Li–O_2 battery electrolyte made suitable with the use of redox mediators. *J. Phys. Chem. C* 123 (33): 20241–20250.

103 Kim, H., Kim, T.Y., Roev, V. et al. (2016). Enhanced electrochemical stability of quasi-solid-state electrolyte containing SiO_2 nanoparticles for Li–O_2 battery applications. *ACS Appl. Mater. Interfaces* 8 (2): 1344–1350.

104 Gao, K.-N., Wang, H.-R., He, M.-H. et al. (2020). Interfacial integration and roll forming of quasi-solid-state Li–O_2 battery through solidification and gelation of ionic liquid. *J. Power Sources* 463: 228179.

105 Ren, M., Zhang, J., Zhang, C. et al. (2020). Quasi-solid-state Li–O_2 batteries with laser-induced graphene cathode catalysts. *ACS Appl. Energy Mater.* 3 (2): 1702–1709.

106 Cho, S.M., Shim, J., Cho, S.H. et al. (2018). Quasi-solid-state rechargeable Li–O_2 batteries with high safety and Long cycle life at room temperature. *ACS Appl. Mater. Interfaces* 10 (18): 15634–15641.

107 Liu, K., Sun, H., Dong, S. et al. (2017). A rational design of high-performance sandwich-structured quasisolid state Li–O_2 battery with redox mediator. *Adv. Mater. Interfaces* 4 (23): 1700693.

108 Zhao, C., Liang, J., Sun, Q. et al. (2019). Ultralong-life quasi-solid-state Li–O_2 batteries enabled by coupling advanced air electrode design with Li metal anode protection. *Small Methods* 3 (2): 1800437.

109 Zou, X., Lu, Q., Zhong, Y. et al. (2018). Flexible, flame-resistant, and dendrite-impermeable gel-polymer electrolyte for Li–O_2/air batteries workable under hurdle conditions. *Small* 14 (34): 1801798.

110 Yi, J. and Zhou, H. (2016). A unique hybrid quasi-solid-state electrolyte for Li–O_2 batteries with improved cycle life and safety. *ChemSusChem* 9 (17): 2391–2396.

111 Yi, J., Liu, Y., Qiao, Y. et al. (2017). Boosting the cycle life of Li–O_2 batteries at elevated temperature by employing a hybrid polymer–ceramic solid electrolyte. *ACS Energy Lett.* 2 (6): 1378–1384.

112 Ma, L., Yu, T., Tzoganakis, E. et al. (2018). Fundamental understanding and material challenges in rechargeable nonaqueous Li–O_2 batteries: recent progress and perspective. *Adv. Energy Mater.* 8 (22): 1800348.

113 Mizuno, F., Nakanishi, S., Shirasawa, A. et al. (2011). Design of non-aqueous liquid electrolytes for rechargeable Li–O_2 batteries. *Electrochemistry* 79 (11): 876–881.

114 Xu, W., Hu, J., Engelhard, M.H. et al. (2012). The stability of organic solvents and carbon electrode in nonaqueous Li–O_2 batteries. *J. Power Sources* 215: 240–247.

115 Feng, N., He, P., and Zhou, H. (2016). Critical challenges in rechargeable aprotic Li–O_2 batteries. *Adv. Energy Mater.* 6 (9): 1502303.

116 Matsuda, S., Mori, S., Kubo, Y. et al. (2015). Cobalt phthalocyanine analogs as soluble catalysts that improve the charging performance of Li–O_2 batteries. *Chem. Phys. Lett.* 620: 78–81.

117 Kwak, W.-J., Hirshberg, D., Sharon, D. et al. (2015). Understanding the behavior of Li–oxygen cells containing LiI. *J. Mater. Chem. A* 3 (16): 8855–8864.

118 Bergner, B.J., Busche, M.R., Pinedo, R. et al. (2016). How to improve capacity and cycling stability for next generation Li–O_2 batteries: approach with a solid electrolyte and elevated redox mediator concentrations. *ACS Appl. Mater. Interfaces* 8 (12): 7756–7765.

119 Zhao, C., Liang, J., Li, X. et al. (2020). Halide-based solid-state electrolyte as an interfacial modifier for high performance solid-state Li–O_2 batteries. *Nano Energy* 75: 105036.

120 Woo, J., Sim, E.S., Je, M. et al. (2020). Theoretical dopant screening and processing optimization for vanadium disulfide as cathode material for Li-air batteries: a density functional theory study. *Appl. Surf. Sci.* 508: 145276.

121 Han, C., Li, W., Liu, H.-K. et al. (2019). Design strategies for developing non-precious metal based bi-functional catalysts for alkaline electrolyte based zinc–air batteries. *Mater. Horiz.* 6 (9): 1812–1827.

122 Wu, C., Li, T., Liao, C. et al. (2017). Tea polyphenol-inspired tannic acid-treated polypropylene membrane as a stable separator for lithium–oxygen batteries. *J. Mater. Chem. A* 5 (25): 12782–12786.

123 Wu, S., Qiao, Y., Deng, H., and Zhou, H. (2018). A single ion conducting separator and dual mediator-based electrolyte for high-performance lithium–oxygen batteries with non-carbon cathodes. *J. Mater. Chem. A* 6 (21): 9816–9822.

124 Xu, J.-J., Liu, Q.-C., Yu, Y. et al. (2017). In situ construction of stable tissue-directed/reinforced bifunctional separator/protection film on lithium anode for lithium–oxygen batteries. *Adv. Mater.* 29 (24): 1606552.

125 Yin, Y.-B., Yang, X.-Y., Chang, Z.-W. et al. (2018). A water-/fireproof flexible lithium–oxygen battery achieved by synergy of novel architecture and multifunctional separator. *Adv. Mater.* 30 (1): 1703791.

126 Kim, B.G., Kim, J.-S., Min, J. et al. (2016). A moisture- and oxygen-impermeable separator for aprotic Li–O_2 batteries. *Adv. Funct. Mater.* 26 (11): 1747–1756.

127 Lee, J.-S., Tai Kim, S., Cao, R. et al. (2011). Metal–air batteries with high energy density: Li–air versus Zn–air. *Adv. Energy Mater.* 1 (1): 34–50.

128 Christensen, J., Albertus, P., Sanchez-Carrera, R.S. et al. (2011). A critical review of Li/air batteries. *J. Electrochem. Soc.* 159 (2): R1–R30.

129 Zhang, R.H., Zhao, T.S., Jiang, H.R. et al. (2019). V_2O_5-NiO composite nanowires: a novel and highly efficient carbon-free electrode for non-aqueous Li-air batteries operated in ambient air. *J. Power Sources* 409: 76–85.

130 Ren, M., Zhang, J., Fan, M. et al. (2019). Li-breathing air batteries catalyzed by MnNiFe/laser-induced graphene catalysts. *Adv. Mater. Interfaces* 6 (19): 1901035.

131 Zhou, J., Li, X., Yang, C. et al. (2019). A quasi-solid-state flexible fiber-shaped Li–CO_2 battery with low overpotential and high energy efficiency. *Adv. Mater.* 31 (3): 1804439.

132 Li, C., Guo, Z., Yang, B. et al. (2017). A rechargeable Li–CO_2 battery with a gel polymer electrolyte. *Angew. Chem. Int. Ed.* 56 (31): 9126–9130.

133 Guo, Z., Li, J., Qi, H. et al. (2019). A highly reversible Long-life Li–CO_2 battery with a RuP_2-based catalytic cathode. *Small* 15 (29): 1803246.

134 Xie, M., Huang, Z., Lin, X. et al. (2019). Oxygen selective membrane based on perfluoropolyether for Li-air battery with long cycle life. *Energy Storage Mater.* 20: 307–314.

135 Zou, X., Liao, K., Wang, D. et al. (2020). Water-proof, electrolyte-nonvolatile, and flexible Li-air batteries via O_2-permeable silica-aerogel-reinforced polydimethylsiloxane external membranes. *Energy Storage Mater.* 27: 297–306.

136 Li, M., Bi, X., Wang, R. et al. (2020). Relating catalysis between fuel cell and metal-air batteries. *Matter* 2 (1): 32–49.

137 Kwak, W.-J., Chae, S., Feng, R. et al. (2020). Optimized electrolyte with high electrochemical stability and oxygen solubility for lithium–oxygen and lithium–air batteries. *ACS Energy Lett.* 5 (7): 2182–2190.

138 Lei, X., Lu, S., Ma, W. et al. (2018). Porous MnO as efficient catalyst towards the decomposition of Li_2CO_3 in ambient Li-air batteries. *Electrochim. Acta* 280: 308–314.

139 Zhang, T. and Zhou, H. (2012). From Li–O_2 to Li–air batteries: carbon nanotubes/ionic liquid gels with a tricontinuous passage of electrons, ions, and oxygen. *Angew. Chem. Int. Ed.* 51 (44): 11062–11067.

140 Wang, X.-G., Zhang, Q., Zhang, X. et al. (2019). Promoting nitrogen electroreduction on Mo_2C nanoparticles highly dispersed on N-doped carbon nanosheets toward rechargeable Li–N_2 batteries. *Small Methods* 3 (6): 1800334.

3

Li–Sulfur Battery

Xiaoqun Qi, Fengyi Yang, and Long Qie

Tongji University Shanghai, Institute of New Energy for Vehicles, School of Materials Science and Engineering, 201804, China

3.1 Introduction

Lithium-ion batteries (LiBs) based on intercalation cathodes (Such as $LiFePO_4$, $LiNi_xCo_yMn_{1-x-y}O_2$, $LiCoO_2$, and $LiMn_2O_4$) and graphite anodes experience rapid development in the last two decades and are approaching their theoretical energy density. To meet the ever-increasing demands for the high-energy rechargeable battery systems, more and more research studies are moving their interest into the chemistry beyond lithium-ion technologies, among which lithium–sulfur (Li–S) batteries that use high-capacity sulfur (S, 1675 mAh g^{-1}) cathodes and lithium (Li) metal anodes are the most promising ones (Figure 3.1). Based on the conversion reaction between Li and S ($S_8 + 16Li \leftrightarrow 8Li_2S$), Li–S batteries possess a high theoretical energy density of 2600 Wh kg^{-1}, much higher than those of the conventional Li-ion batteries. What is more, the abundance reserve in the earth's crust and low cost of sulfur additionally make Li–S batteries competitive.

In the last two decades, with extensive investigations on Li–S batteries, impressive progress has been achieved on both the fundamental understandings of electrochemical processes and performance improvement. However, the current Li–S batteries are still far from replacing the Li-ion batteries, which is mainly due to their failure to achieve high overall energy density and long cycling life concurrently in a practical cell. More efforts on such as optimization of electrodes, electrolyte, and cell configuration are still desired to improve the competitiveness of Li–S batteries further.

In this chapter, we will first introduce the fundamental understandings of Li–S batteries in commonly utilized ether electrolytes and then highlight the recent advances of Li–S batteries in cathodes (S and Li_2S cathodes, interlayers, and separator coatings), electrolytes (liquid with new formulas and additives, polymer/solid electrolytes), and anodes (Li metal, carbon, and silicon materials). At last, the challenges and perspectives for Li–S batteries are summarized.

Battery Technologies: Materials and Components, First Edition. Edited by Jianmin Ma.

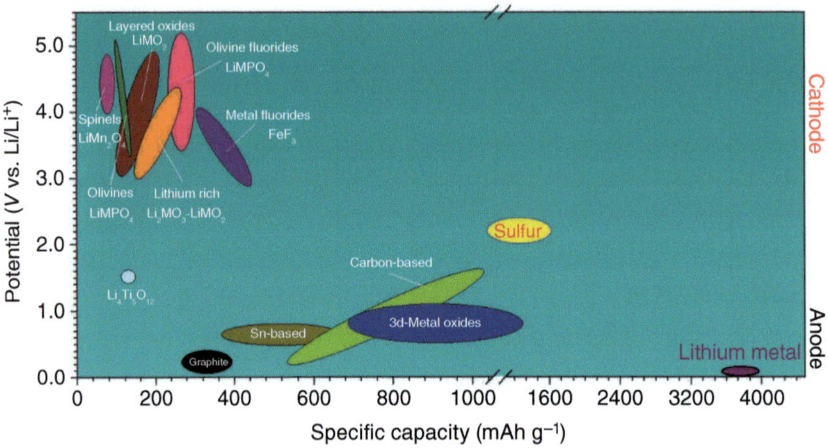

Figure 3.1 The comparison of the specific capacity and potential for the typical cathode and anode materials used in Li-based batteries.

3.2 Fundamentals

A typical Li–S cell (Figure 3.2a) consists of an S@C cathode and a Li metal anode separated by a porous polymeric separator and typically an ether-based electrolyte, e.g. 1 M lithium bis(trifluoromethanesulfonyl)imide (LiTFSI) in 1,3-dioxolane/dimethoxyethane (DOL/DME, 1 : 1 by volume). Due to the insulting nature of S ($\sim 10^{-30}$ S cm^{-1}) and its lithiated product (Li_2S), the S cathode is usually fabricated by dispersing sulfur homogeneously onto conductive matrices, mostly carbon materials, to guarantee the high sulfur utilization and fast conversion kinetics. Currently, ether-based electrolytes are commonly utilized in Li–S batteries, especially the combination of DME and DOL, with the former having high

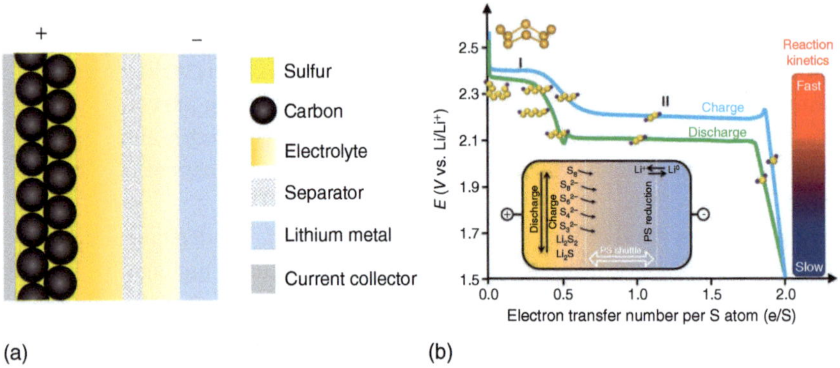

Figure 3.2 (a) A typical configuration of Li–S batteries; (b) the conversion mechanism of Li–S cells in ether electrolytes with the schematic diagram illustrating the "shuttle effect" [1]. Source: Wang et al. [1]. Reproduced with permission of Royal Scoiety of Chemistry.

solubility to long-chain lithium polysulfides (LiPSs, Li_2S_n, $4 \leq n \leq 8$), and the latter stabilizing the solid electrolyte interphase (SEI) formation on Li metal surface [2]. A typical discharge–charge profile of such predominant Li–S batteries is exhibited in Figure 3.2b. In the discharge process, there are two plateaus at around 2.3 and 2.1 V vs. Li^+/Li, corresponding respectively to the conversion of S_8 to dissolved LiPSs and then to the insoluble products (Li_2S_2 and Li_2S). During charging, the cell undergoes an adverse conversion from Li_2S to S_8 with the formation of the LiPS intermediates as well.

Such a solid–liquid–solid reaction pathway ($S_8 \leftrightarrow$ LiPS $\leftrightarrow$ Li_2S_2/Li_2S) of the Li–S batteries in ether electrolyte has a complex and profound impact on the performance of Li–S batteries. First, the generation and dissolution of the LiPS intermediates expose the internal S_8 molecules of S particles to the Li^+ in the electrolyte, which facilities the reduction process of the insulting S particles. On the other hand, the dissolved LiPS brings about the as-known "shuttle effect": in the discharging process, the dissolved LiPSs penetrate through the separator and diffuse to the anode side, subsequently reacting with the Li anode in a chemical rather than an electrochemical way, leading to the precipitation of Li_2S_2 and Li_2S onto the surface of the Li metal and the formation of the LiPS with shorter chain, for example, $Li_2S_8\,(aq) + 6Li\,(s) \rightarrow 4Li_2S\,(s) + Li_2S_4\,(aq)$; in the charging process, the longer-chain LiPSs are chemically reduced to shorter-chain LiPSs or Li_2S_2/Li_2S by Li metal, while shorter-chain LiPSs are electrochemically oxidized to longer-chain LiPSs at the cathode side, over and over again. Such a "shuttle effect" leads to not only the loss of the active materials but also the continuous corrosion of the Li anode and thus accelerates the failure of the Li–S batteries.

In the last decade, extensive strategies have been investigated to suppress the "shuttle effect" and alleviate the adverse impacts induced by dissolved LiPS, e.g. low Coulombic efficiency ($Q_{lithiation}/Q_{delithiuation}$), low sulfur utilization, and cycling stability. In addition, other Li–S systems with conversion mechanisms beyond the solid–liquid–solid reaction pathway have also been developed by adopting new electrolytes and designing new S-based active materials.

3.3 Cathodes

According to the lithiation states, sulfur cathodes could be simply divided into sulfur-based and Li_2S-based ones. Their fabrication and improving methods are discussed here.

3.3.1 S Cathodes

S and its discharge product (Li_2S) are both electric and ionic insulting, to obtain high-efficient S cathodes, it is a prerequisite to improve their conductivity. In this context, dispersing small-sized active materials homogeneously into the conductive matrices is a precondition to guarantee high sulfur utilization. In addition, an advanced S cathode should also be able to reduce the effects of the "shuttle effect."

According to the physicochemical interaction toward LiPS and conversion mechanisms, the current strategies to design advanced S cathodes could be mainly divided into the following four categories: physical confinement, physical blocking, polymeric organosulfur, and chemical adsorption and catalysis.

3.3.1.1 Physical Confinement

Carbonaceous materials with high electrical conductivity are favorable choices to accommodate the insulate sulfur, and their porous morphology physically restrains the LiPS from diffusing to the anode side. In 2009, Nazar and coworkers [3] made use of the low-melting-point (c. 120 °C) character of sulfur to pioneeringly diffuse sulfur into the channels of highly ordered mesoporous carbons (CMK-3) at 155 °C. Such a melting-diffusion method guarantees the homogeneous distribution of S into the conductive carbon matrix (Figure 3.3a), and the as-obtained S@CMK-3 composite displays a high sulfur utilization and stable cycling performance. This simple and effective encapsulating strategy has been widely adopted by researchers to fabricate S@C composition cathodes. A lot of porous carbons such as carbon spheres [6], hollow porous carbon [7], graphene sheets [8], carbon fibers [9], carbon nanotubes [10], and so on have been investigated as the hosts to encapsulate sulfur. The S@C compositions could also be prepared via solution-based methods. For example, dissolved S in CS_2 could be infiltrated into the micropores of the carbon host with a hierarchical structure, after evaporating the solvent, a sulfur-loaded composite could be obtained (Figure 3.3b) [4]. The fabricated S@C composite not only provides high surface area to improve sulfur utilization, but also facilitates ionic transportation and LiPS accommodation. By combining the solution-based and melting-diffusion methods, Zhang and coworkers [5] fabricated porous carbon nanofiber (CNF)–sulfur composites, as Figure 3.3c demonstrates, sulfur incorporated into CNF through a chemical reaction (step 2) followed by the thermal treatment to impel surface-deposited sulfur to diffuse into pores of CNF (step 3). The CNT-S-based electrodes exhibit a high sulfur utilization benefiting from homogeneous sulfur distribution and alleviated shuttle effect. In addition, the efficiency of LiPS confinement could be further improved by using hollow carbon matrices, Cui and coworkers [11] reported cathodes with sulfur constrained into hollow CNFs, in which sulfur could only contact electrolyte at two openings, meanwhile, Li^+ could easily penetrate the thin carbon wall.

3.3.1.2 Physical Blocking

In addition to physical confinement, introducing a porous interlayer between cathode and separator or a coating layer onto the separator has also been demonstrated to be an effective solution to suppress the LiPS shuttling. Such a physical-blocking method was first developed by Manthiram and coworker [12], they inserted a microporous carbon paper (MCP) between the cathode and separator (Figure 3.4a). The introduced interlayer, which serves as "upper current collector" and "polysulfide stockroom", plays bifunctional roles in improving the electrochemical performance, supplying additional electron pathways for S/Li_2S, and trapping LiPS. Following this direction, Cui and coworkers [13] fabricated a functional separator with a thin

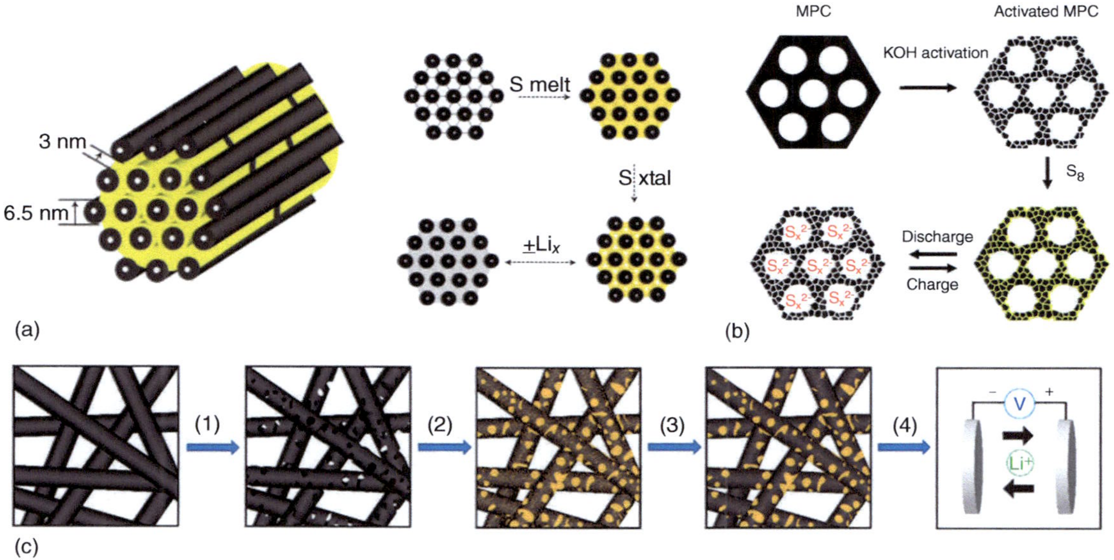

Figure 3.3 (a) S@CMK-3 composite fabricated by a melting-diffusion method [3]. Source: Ji et al. [3]. Reproduced with permission of Springer Nature. (b) The fabrication procedure of a hierarchical S@C composite based on a solution infiltration method and its function of accommodating LiPS during charge/discharge process [4]. Source: Liang et al. [4]. Reproduced with permission of American Chemical Society. (c) Preparation of a CNT-S composite combines the solution-infiltration (step 2) and melting-diffusion (step 3) methods [5]. Source: Ji et al. [5]. Reproduced with permission of Royal Scoiety of Chemistry.

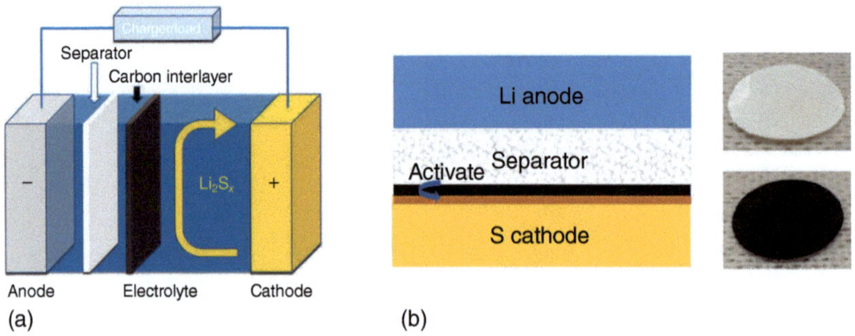

Figure 3.4 Physical-blocking strategies for the suppression of the "shuttle effect": (a) carbon interlayer introduced between the sulfur cathode and the separator [12]. Source: Su et al. [12]. Reproduced with permission of Royal Society of Chemistry. (b) A conductive coating on the separator [13]. Source: Yao et al. [13]. Reproduced with permission of Royal Society of Chemistry.

Super P coating, to activate the utilization of accumulated LiPS in the separator as Figure 3.4b shows. Zhou et al. [14] proposed a graphene–pure sulfur sandwich structure, in which pure sulfur was sandwiched between two graphene membranes. One graphene membrane was coated on sulfur as the current collector, and the other was coated on the separator. These two graphene membranes could not only confine sulfur but also accommodate the volumetric expansion of sulfur. Thanks to the conductivity and flexibility of graphene layers, the sandwich structure significantly improved the electrochemical performance of the cells. The sulfur electrode retained a capacity of 679 mAh g^{-1} after 300 cycles at a current density of 1.5 A g^{-1}. In light of the sandwich structure, a high-areal-capacity sulfur cathode was fabricated by Qie et al. [15] through stacking carbon layers and sulfur layers alternately using porous CNF papers as carbon layers and commercial sulfur powder as S layers. Benefiting from the unique layer-by-layer structure and favorable adsorption effect of layered carbon to LiPS, a high-mass-loading cathode (11.2 mg cm^{-2}) demonstrated an areal capacity as high as 11.3 mAh cm^{-2} and promising cycling stability.

3.3.1.3 Polymeric Organosulfur

Although the abovementioned physical confinement and blocking methods slow down the diffusion of the LiPSs into the anode side, there is still LiPS formation and loss during the electrochemical process. By bonding the S molecules to polymeric compounds chemically, the active materials could be fully retained in the organosulfur cathodes during discharging/charging. One impressive organosulfur composite, which was first fabricated by Wang et al. [16] in 2002, is sulfurized polyacrylonitrile (SPAN). By heating the insulating polyacrylonitrile (PAN) and sulfur at 300 °C, sulfur dehydrogenates generating the SPAN (Figure 3.5a) with acceptable electrical conductivity (10^{-4} S cm^{-1}). The pioneering work achieved high specific capacity (850 mAh g^{-1}) based on the SPAN composite (sulfur content: 53.41 wt%) corresponding to 1591 mAh g^{-1} based on sulfur and stable cycling in a carbonate electrolyte. Making use of the novel electrochemical performance of SPAN, Liu and coworkers [17] designed a three-dimensional polymer framework composed of sulfur–polyaniline nanotubes (SPANI-NT), as Figure 3.5b shows, to alleviate stress and structural deterioration. However, it remains unclear about the exact S-exist state and electrochemical conversion mechanism of SPAN even after intensive investigations. Xiangming He and coworkers [20] speculated the C—S bonding formation by infrared spectra when the heating treatment is higher than 300 °C. Michael R. Buchmeiser and coworkers [21] further proposed a thioamide and poly(sulfide) structures in SPAN. Whereas Lynden A. Archer and coworkers [22] deduced sulfur as S_2/S_3 units covalently attach to dehydrogenated PAN backbone through Raman and FTIR spectra. However, considering the covalent models unable to fully explain the high sulfur utilization (>90%) of SPAN, Wang et al. [23] tend to suppose some complex bonds formed in SPAN, which are stronger than Van der Waals forces, instead of covalent bonds.

Another category of organosulfur compounds, including linear molecules ($R–S_n–R$) and polymers ($–S–R–S_n–$), have been systematically developed since the initial studies toward diphenyl disulfide ($CH_3–S_2–CH_3$) and tetraethylthiuram disulfide ($CH_3–S_3–CH_3$) by Visco and De Jonghe. These organosulfur compounds could be easily tuned at the molecular level by structure design and functionalization, potentially enabling unique Li–S electrochemical performances [24]. However, the inactive segments in organosulfur compounds sacrifice a portion of the theoretical capacity, as calculated by Fu and coworkers [18] (Figure 3.5c), presenting the theoretical capacity increases along with the increase of sulfur content in $R–S_n–R$ linear molecules (namely higher "n" value and lower molecular weight). In this regard, a sulfur-rich organosulfur compound, poly(sulfur-random-1,3-diisopropenylbenzene) (poly[S-r-DIB]), was fabricated by Pyun and coworkers [19] via inverse vulcanization, as Figure 3.5d shows. The poly(S-r-DIB) demonstrates comparable electrochemical performance with elemental sulfur, with a high specific capacity retention of 823 $mAh\,g^{-1}$ after 100 cycles.

3.3.1.4 Chemical Adsorption and Catalysis

Nonpolar carbonaceous materials, which are widely utilized as cathode matrices, interlayers, and coatings, could physically adsorb LiPS, yet insufficient to avoid the "shuttle effect" during long-term cycling. In recent years, more and more polar materials have been introduced into sulfur cathodes to anchor LiPS chemically and even catalyze the conversion between S and Li_2S. The polar materials mainly include (i) polar sulfur hosts, e.g. transition metal oxides, nitrides, sulfides, and carbides; (ii) doped carbonaceous materials, e.g. N-doped CNTs; (iii) functional binders or additives. Zhi Wei Seh et al. [25] designed a sulfur–TiO_2 yolk-shell nanoarchitecture, as Figure 3.6a shows, with void space for accommodating the volume expansion during lithiation process from S to Li_2S. TiO_2 shells well maintained their integrity during cycling and effectively trapped LiPS displaying stable 1000 cycles. Other oxides, such as MnO_2 [30], Nb_2O_5 [31], Fe_3O_4 [32], were also proven to be effective to absorb the LiPS. Zhou et al. [26] systematically investigated the effect of a series of metal sulfides (Ni_3S_2, SnS_2, FeS, CoS_2, VS_2, and TiS_2) in chemically adsorbing LiPS. Among them, VS_2 and TiS_2 demonstrate the most

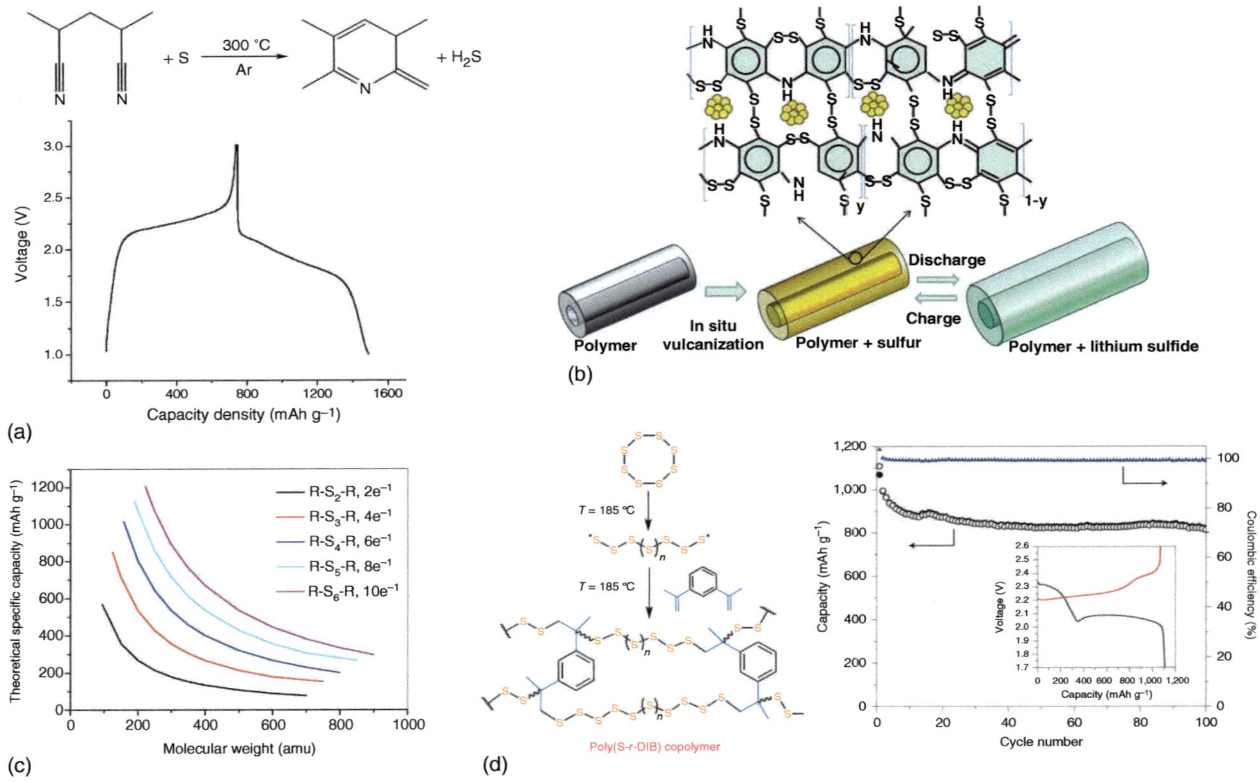

Figure 3.5 (a) The fabrication mechanism of SPAN and its discharge/charge profile in a carbonate electrolyte [16]. Source: Wang et al. [16]. Reproduced with permission of John Wiley & Sons. b) Schematic diagram of the construction and charge/discharge process of the SPANI-NT composite [17]. Source: Xiao et al. [17]. Reproduced with permission of John Wiley & Sons. (c) Calculation on theoretical specific capacity vs. molecular weight of a family of organic compounds [18]. Source: Wang et al. [18]. Reproduced with permission of American Chemical Society. (d) The synthesis procedure and electrochemical performances of the poly(S-r-DIB) material [19]. Source: Chung et al. [19]. Reproduced with permission of Springer Nature.

obvious anchoring effect to Li_2S_6, followed by CoS_2 displaying light yellow in the immersion tests (Figure 3.6b). The chemical adsorption abilities are further verified by first-principle simulations of the binding energy between the metal sulfides and Li_2S_6. As an emerging family of two-dimensional materials, MXenes have also been adopted as the sulfur host by Nazar et al., as Figure 3.6c shows, its chemical affinity to LiPS, along with the unique layered structure and inherent high conductivity, enabled stable cycling with a low capacity decay rate of 0.05% per cycle [27].

Introducing heteroatoms is also demonstrated to be useful to tune the polarization state of the carbonaceous materials and create chemical interaction between the carbon and LiPS. With the characterization of X-ray absorption near-edge structure spectroscopy (XANES) and simulation of density functional theory, Wang and coworkers found that nitrogen (N) doping motivates the LiPS adsorption by oxygen groups in the carbon host [33]. Wei and coworkers [34] also found that nitrogen doping introduced more defects and active sites for the chemical adsorption of LiPS. As a result, impressive capacity retention (95%) within 100 cycles was obtained with the N-doped aligned CNT/graphene hybrid scaffold. In addition, a phosphorous and nitrogen co-doped hierarchical porous carbon was utilized to encapsulate sulfur, and the composites displayed enhanced cycling stability and rate capability [35].

Binder, which is an essential ingredient in electrodes to integrate components and tightly attach them to the current collector, also plays a vital role in LiPS adsorption. However, the commonly utilized polyvinylidene difluoride (PVDF) with the –C–F group displays weak chemical adsorption to LiPS. In contrast, binders containing the carbonyl group (>C = O), e.g. polyvinylpyrrolidone (PVP), have a much higher affinity to LiPS [36]. According to this, Lacey et al. [37] utilized a mixed binder of PVP and poly(ethylene oxide) (PEO) (1 : 4) to improve the capacity retention of Li–S batteries, in which PEO contributes to improving LiPS conversion kinetics.

To promote the conversion between S and Li_2S and avoid the LiPS accumulation, more and more researchers introduced catalytic materials to catalyze the LiPS conversion (Figure 3.6d). Babu et al. [38] investigated the electrocatalysis effects of Pt, Au, and Ni in a carbon-free Li–S system. For Li–S cells with Pt, Au, and Ni deposited (50 nm) current collector (Al foil), the polarization significantly decreased compared with that with pristine Al foil, especially for the Ni-deposited one. Tao et al. [39] investigated the functions of a series of nonconductive metal oxide (MgO, Al_2O_3, CeO_2, La_2O_3, and CaO) nanoparticles-decorated carbon flakes on the behaviors of Li_2S deposition. They found that the LiPS diffusion properties on the nonconductive metal oxides affect the aggregation of Li_2S. A capable candidate should possess strong binding, excellent surface diffusion property, and high surface area for providing sufficient active cites. Metal nitrides such as WN, VN, and Mo_2N are promising catalysts for lithium–sulfur batteries. The WN could form S–W–N bonding between polysulfides and WN, which effectively alleviates the formation of the soluble polysulfides and reduces the loss of active materials [40]. Combining the high affinity of TiO_2 to LiPS and the high conductivity of TiN, Yang and coworkers [29] fabricated a graphene interlayer loaded with TiO_2–TiN heterostructures for anchoring LiPS and accelerating their conversion (Figure 3.6e), which enables stable cycling for 2000 cycles with decent capacity retention.

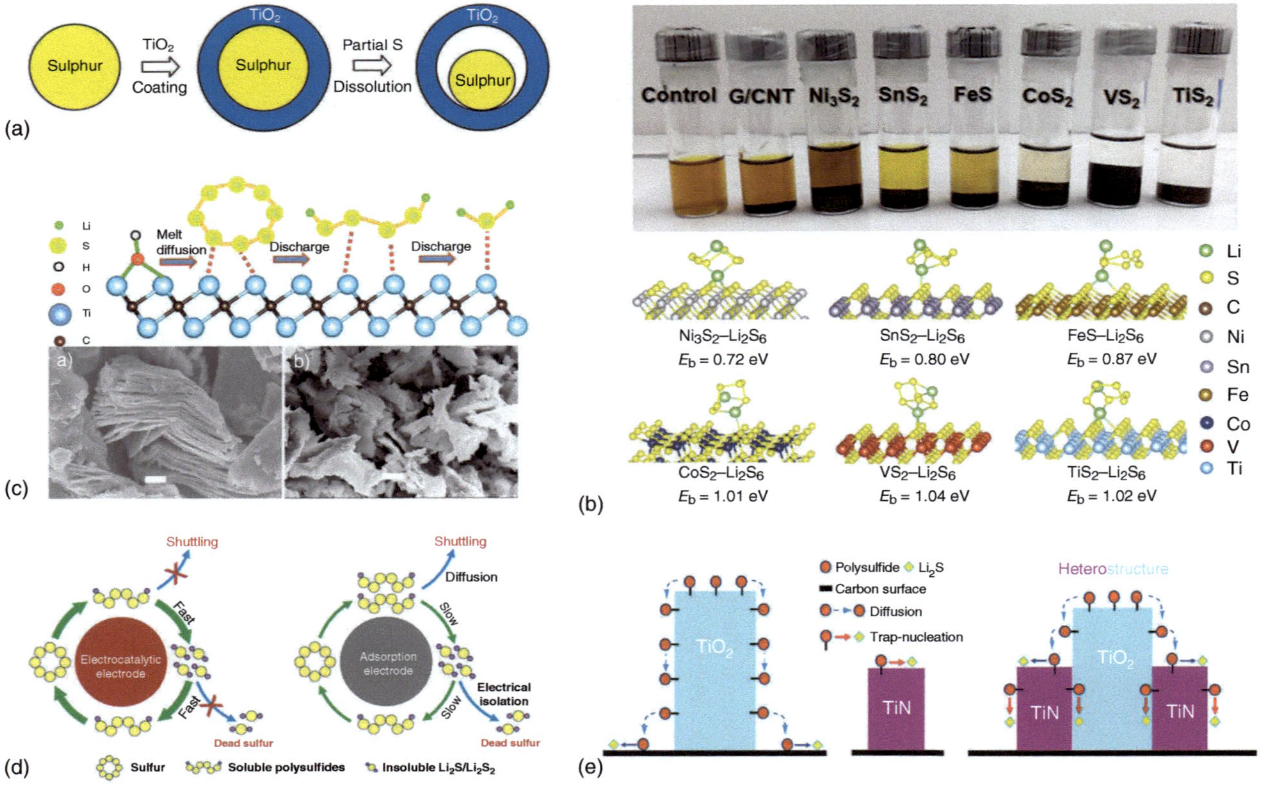

Figure 3.6 Chemical adsorption and catalysis for S cathodes: (a) yolk–shell nanoarchitecture design with TiO_2 as the host accommodating sulfur [25]. Source: Wei Seh et al. [25]. Reproduced with permission of Springer Nature. (b) The comparison of the chemical adsorbing ability of a series of sulfides to LiPS [26]. Source: Zhou et al. [26]. Reproduced with permission of National Academy of Sciences. (c) The interaction between sulfur species (S, LiPS, and Li_2S) and MXene surface [27]. Source: Liang et al. [27]. Reproduced with permission of John Wiley & Sons. (d) Schematics of conversion mechanism in electrocatalytic electrodes and adsorption electrodes [28]. Source: Yu et al. [28]. Reproduced with permission of Elsevier. (e) TiO_2–TiN heterostructures for anchoring LiPSs and accelerating their conversion [29]. Source: Zhou et al. [29]. Reproduced with permission of Royal Society of Chemistry.

3.3.2 Li_2S Cathodes

The traditional cathodes for Li-ion batteries are typically in lithiated states and provide Li^+ ions for the "rocking-chair" reactions. In contrast, for Li–S batteries with S cathodes, the Li^+ ions have to be supplied by the anodes. However, the use of Li metal anode brings about not only safety concerns for potential short-circuit induced by Li dendrites penetrating the separator but also poor cycling stability due to the continuous Li loss and dead Li accumulation. Replacing the Li-metal anodes with non-lithiated anodes such as carbonaceous and silicon-based materials and pairing with the fully lithiated sulfur species (Li_2S) is a strategy for higher safety and enhanced cycling stability. Due to the low electronic and ionic conductivity of Li_2S, a high charging barrier is typically observed during the initial charging of Li_2S. Regarding this phenomenon, Cui and coworkers [41] applied a higher charging cutoff voltage to overcome this barrier and verified LiPS formation by passing over the overpotential peak, which motivates the following redox kinetics (Figure 3.7a).

The melting point of Li_2S is higher than 900 °C, which makes it difficult to be composited with conductive matrices via the melting-diffusion method for S electrodes. Moreover, Li_2S is sensitive to moisture, leading to rigorous demands for fabrication conditions. The good thing is that Li_2S could be dissolved into the ethanol solvent, which makes the preparation of Li_2S cathodes via solution-evaporation method possible. By dropping the Li_2S/ethanol solution (0.5 M) onto a free-standing carbon paper and evaporating the ethanol solvent at 40 °C (Figure 3.7b), Fu and coworkers [42] successfully prepared Li_2S electrodes with uniformly distributed

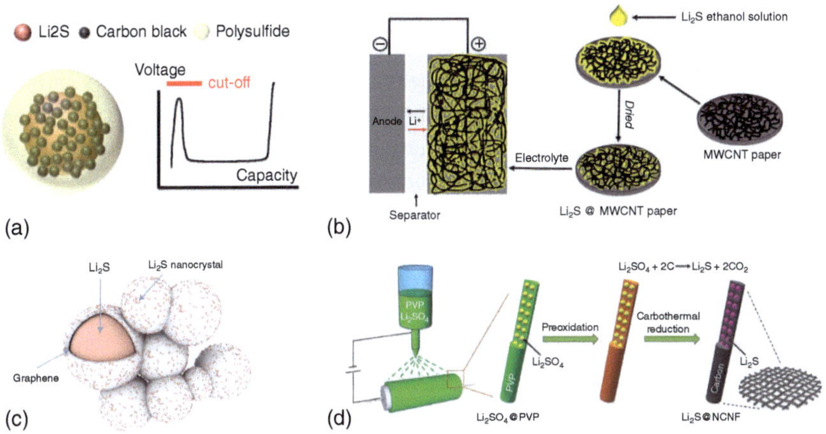

Figure 3.7 (a) Overpotential phenomenon for Li_2S cathodes in the initial charge process [41]. Source: Yang et al. [41]. Reproduced with permission of American Chemical Society. (b) Facile fabrication method for Li_2S@MWCNT paper by a solution method [42]. Source: Wu et al. [42]. Reproduced with permission of American Chemical Society. (c) In situ generated Li_2S@graphene nanocapsules by burning Li in CS_2 [43]. Source: Tan et al. [43]. Reproduced with permission of Springer Nature. (d) Free-standing Li_2S@NCNF paper electrodes fabricated from Li_2SO_4@PVP interconnected nanofibers [44]. Source: Yu et al. [44]. Reproduced with permission of John Wiley & Sons.

Li_2S nanocrystals of ~10 nm. Benefiting from the nano-sized Li_2S and the high-conductivity carbon paper, the as-prepared Li_2S electrode demonstrates a low potential barrier (0.1 V) in the initial charge process. In situ lithiation strategy is also reported for the fabrication of Li_2S/microporous carbon (MC), by the lithiation of the S/MC films with Li metal powder [45]. When paired with a graphite anode, the Li_2S/MC cathode presents stable cycling for 150 cycles in a carbonate electrolyte. Inspired by the reaction of burning Mg in dry ice of CO_2 to form graphene-wrapped MgO nanoparticles, Amine and coworkers [43] fabricated the Li_2S@graphene composite (Figure 3.7c) by burning Li foils in the CS_2 vapor. The as-obtained electrode exhibited an area capacity of 8.1 mAh cm^{-2} at a loading of 10 mg cm^{-2}. Currently, the most popular method for the preparation of Li_2S electrodes is the in situ reduction of Li_2SO_4 by carbon materials ($Li_2SO_4 + 2C \rightarrow Li_2S + 2CO_2$), which has long been utilized in the Leblanc process [46]. Qiu and coworkers [44] prepared a Li_2SO_4@polyvinylpyrrolidone (PVP) nonwoven fabric by electrospinning, which is further carbothermal reduced to the Li_2S@N-doped carbon nanofibers (NCNFs) free-standing electrode (Figure 3.7d). Based on this method, it is facile to obtain high-loading Li_2S cathodes by stacking the Li_2S@NCNF layers. With a high Li_2S loading (9.0 mg cm^{-2}), the cathode delivered an areal specific capacity of 5.76 mAh cm^{-2}.

3.4 Electrolytes

Electrolytes affect the Li–S reactions in directing redox pathways, active materials utilization, SEI and cathode electrolyte interphase (CEI) effectiveness, conversion kinetics, and so on. For Li–S batteries, an ideal electrolyte should meet a variety of criteria: (i) compatible with S and Li, and the reaction intermediates; (ii) appropriate potential window (1–3 V); (iii) high ionic conductivity and low viscosity; (iv) ability to form effective SEI. Various electrolytes have been developed to achieve advanced Li–S batteries, which include ether, carbonate, nitrile, sulfone/sulfoxide, ionic liquid, and polymer/solid-state electrolytes.

3.4.1 Ether Electrolyte

Compared with the commercial carbonate electrolytes used in Li-ion batteries, ethers show lower oxidation resistance at high potential around 4 V vs. Li^+/Li but higher reduction resistance at low potential. As the theoretical lithiation potential of S is 2.275 V vs. Li^+/Li, lower than that of the commercial Li-ion batteries, which makes the ethers capable of Li–S systems. When being used as electrolytes for Li–S batteries, ether solvents such as DOL, DME, tetra(ethylene glycol) dimethyl ether (TEGDME), tetrahydrofuran (THF) are usually mixed to form binary or ternary solvents, such as DOL/DME [47], DOL/TEGDME [48], DOL/DME/diglyme [49], DME/DOL/TEGDME [50], THF/DOL/toluene [51], etc. Among them, the DOL/DME mixture is the most popular selection, in which DOL contributes to generating a protective SEI layer on Li anodes [2], and DME facilitates redox

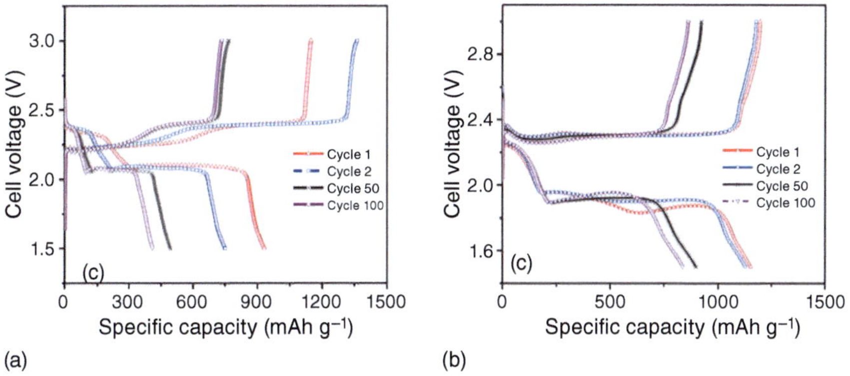

Figure 3.8 (a) Discharge/charge profiles of the Li–S cell in an ether electrolyte and (b) a modified electrolyte with TTE replacing DME [52]. Source: Weng et al. [52]. Reproduced with permission of John Wiley & Sons.

conversion to achieve high sulfur utilization because of its high LiPS solubility. Under this combination, Li–S batteries experience a solid–liquid–solid conversion mechanism, corresponding to S_8-LiPS-Li_2S_2/Li_2S, with the formation of dissoluble LiPS intermediates in the ether electrolyte. However, the dissolution of the LiPS results in serious "shuttle effect" for Li–S batteries, consequently leading to poor Coulombic efficiency (Figure 3.8a). Amine and coworkers [52] replaced DME by a partially fluorinated ether (TTE: 1,1,2,2-tetrafluoroethyl-2,2,3,3-tetrafluoropropyl ether), developed a new electrolyte formulation of DOL/TTE, in which Li–S batteries display high reversibility with Coulombic efficiency to >96% in the first cycle (Figure 3.8b), as contrast, the value is only 70% for the electrolyte of 1.0 M LiTFSI in DOL/DME (1 : 2, v/v). The authors attributed the high Coulombic efficiency to an effective SEI layer on Li anode generated by decomposition of TTE.

The high LiPS solubility in ether electrolytes inevitably brings about severe "shuttle effect," researchers have adopted other liquid electrolytes to decrease LiPS solubility or even avoid LiPS formation through the cooperation of particularly designed sulfur cathodes with carbonate electrolytes. The strategy of lowering LiPS solubility is realized by either selecting a less dissolving-capability solvent to LiPS or increasing Li salt concentration. Comparing to ether electrolytes, Li–S batteries in these LiPS shuttle suppressed electrolytes present higher reversibility and enhanced Coulombic efficiency.

3.4.2 Carbonate-Based

The commercial carbonate solvents in Li-ion batteries, such as ethylene carbonate (EC) and ethylmethyl carbonate (EMC), are generally inapplicable in Li–S batteries, due to their electrophilic nature, leading to a chemical reaction with nucleophilic LiPS through nucleophilic addition or substitution. Taking the EC molecule as an example, as Figure 3.9a shows, Li_2S_n tends to attack the carbonyl-carbon of EC, forming a ring-opened product, which further decomposes into ethylene glycol and

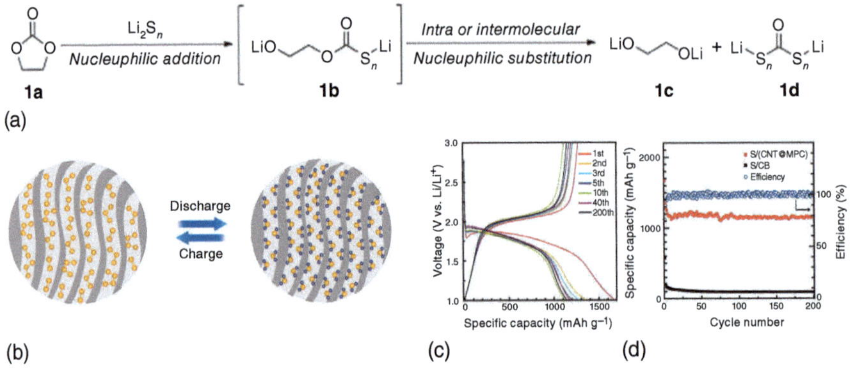

Figure 3.9 (a) Chemical reaction between EC and LiPS species [53]. Source: Yim et al. [53]. Reproduced with permission of Elsevier. (b) Diagram of the sulfur cathode with S_{2-4} confined in the micropores of the carbon matrix and its conversion mechanism. (c) Discharge/charge profiles and (d) cycling performance of Li–S cells utilizing the S_{2-4}-based cathode in a carbonate electrolyte [54]. Source: Xin et al. [54]. Reproduced with permission of American Chemical Society.

thiocarbonate through nucleophilic substitution [53]. As a result, polysulfides are permanently consumed during the first discharge process. Under this circumstance, ester, aldehyde, ketone, and anhydride-type solvents with a higher electrophilic property are also excluded from candidates for Li–S batteries [53].

However, it is reported that carbonate-based electrolytes are suitable for some particular Li–S systems. In 2002, Wang et al. fabricated SPAN by heating PAN and sulfur mixture at 300 °C with the protection of inert gases. The SPAN cathodes work well in the electrolyte of 1 M LiPF6 in EC: DMC (1/1, vol%) with stable cyclability and high sulfur utilization [16]. In addition to the molecular-level mixed SPAN composite, Xin et al. [54] testified smaller sulfur molecules (S_{2-4}) confined in a microporous carbon matrix as well compatible with carbonate-based electrolytes. As Figure 3.9b shows, the pore size of the carbon material is ~0.5 nm, which is only capable of accommodating smaller sulfur molecules instead of cyclo-S_8. In the discharge process, S_{2-4} directly converts into S^{2-} without the polysulfides intermediates generation, corresponding to one discharge plateau at ~1.85 V; in charge process, a reverse conversion from S^{2-} to S_{2-4} occurs under space confinement with a single charge plateau at ~2 V (Figure 3.9c). With the elimination of LiPSs, the S_{2-4}-cathode achieves stable cycling for 200 cycles (Figure 3.9d). Whereas the limited pore space of the microporous carbon matrix restricts sulfur content to a small value, weakens the competitiveness of S_{2-4}@C composite in specific capacity.

3.4.3 Nitrile-Based

Acetonitrile (AN) solution is widely explored considering its high oxidative stability and high ionic conductivity deriving from the pairing of decent dielectric permittivity and a low viscosity [55]. Unfortunately, its poor reductive stability limits the utilization in Li–S batteries, as the severe chemical reaction with Li metal anodes:

$$CH_3CN + 2Li \rightarrow CH_3^- + CN^- + 2Li^+$$

However, the reduction resistance can be enhanced by increasing Li salt concentration. Yamada et al. [56] explored a superconcentrated AN solution with LiTFSI concentration >4 mol l^{-1}. Through the comparison of Raman spectra of a series of LiTFSI/AN solutions (Figure 3.10a), it is found that with increasing Li salt concentration, the Raman signal of free AN molecules (C≡N stretching band) at 2258 cm^{-1} decreases. On the contrary, the signal of Li^+-coordinating AN at 2282 cm^{-1} increases. For the 4.2 M LiTFSI/AN solution, only Li^+-coordinating AN signal is detected, indicating the disappearance of free AN molecule; and the Raman peaks of $TFSI^-$ anion indicate the formation of Li^+-$TFSI^-$ (contact ion pairs, CIP) and ${Li^+}_n$-$TFSI^-$ ($n \geq 2$, aggregates, AGG) coordinating structure, instead of free $TFSI^-$ ions. The graphical representations of dilute and superconcentrated solutions are exhibited in Figure 3.10b. The elimination of free AN molecules in the superconcentrated solution enhances the reduction stability and enables the compatibility with Li metal anodes.

Cuisinier et al. [57] firstly applied the concentrated AN electrolyte into Li–S batteries, which is composed of AN_2-LiTFSI (the molar ratio of AN to LiTFSI is 2 to 1) complex and hydrofluoroether (HFE) as the cosolvent. The addition of HFE is to decrease the viscosity of AN_2-LiTFSI complex. Through operando X-ray absorption spectroscopy and electrochemical methods, it is demonstrated that Li_2S_n is formed in the superconcentrated AN-based electrolytes during cycling, but with no Li_2S_n solvation, as Figure 3.10c displays. By comparing the electrochemical performances of Li–S cells in different electrolytes (Figures 3.10d,e), it exhibits the highest specific capacity and capacity retention after 100 cycles in the [$(AN)_2$LiTFSI]: HFE (1 : 1) electrolyte. Lee et al. [58] investigated Li–S cells in the $(AN)_2$LiTFSI–HFE electrolyte at elevated temperatures (c. 50 °C), which significantly decreases polarization and improves sulfur utilization (~1430 mAh g^{-1} at 55 °C). Moderately rising temperature increases the solubility of lithium polysulfides, thereby facilitating the conversion kinetics.

It is worthy to note that the diluent solvent is essential for the superconcentrated AN electrolytes. Otherwise, the high viscosity leads to sluggish Li^+ conductivity, accounting for high polarization, low utilization, and poor cycle performance. HFE is commonly selected because its highly fluorinated structure is believed barely participating Li^+ solvation. Gewirth and coworkers [59] investigated the effect of HFE on Li^+ solvation in the AN-based electrolytes through Raman spectroscopy and ^{7}Li NMR Spectroscopy. It is found that the addition of HFE slightly increases free AN content in the electrolyte. Moreover, the coordinated AN plays the role of facilitating LiPS formation kinetics.

3.4.4 Sulfones/Sulfoxides-Based

Sulfones are known with high dielectric permittivity, low flammability, and especially high oxidation stability, e.g. the oxidation potentials of isopropyl methyl

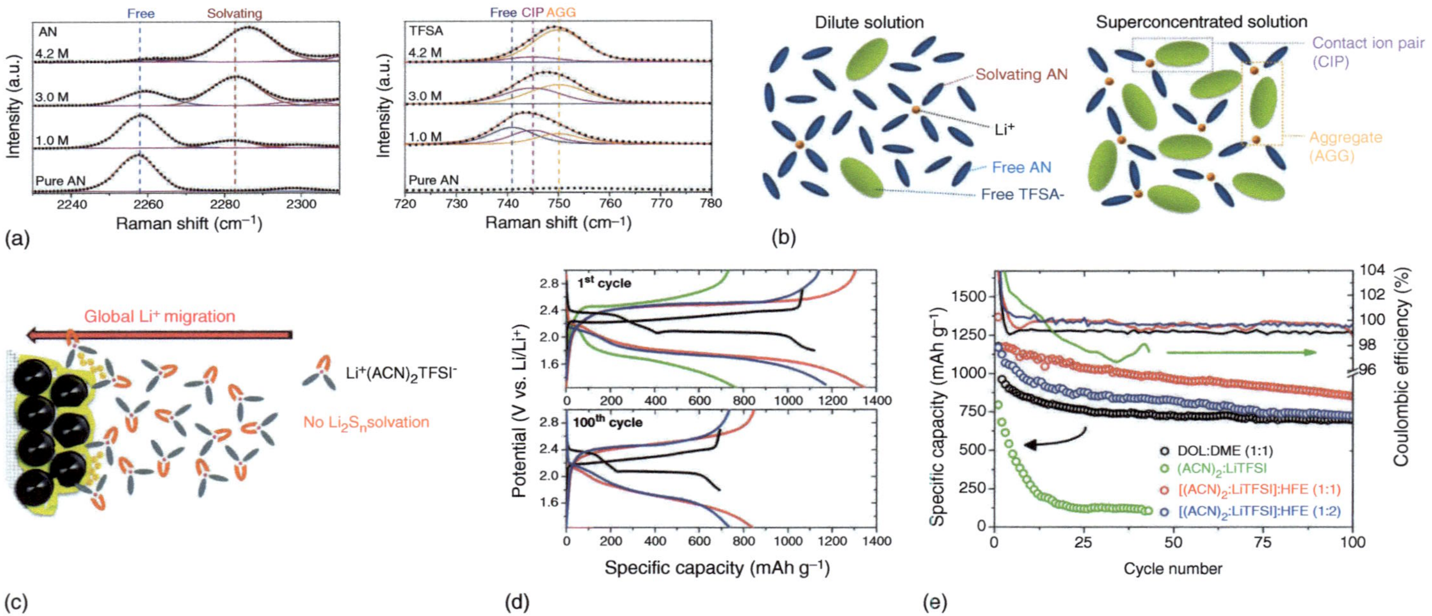

Figure 3.10 (a) Raman spectra of a series of AN solutions with different LiTFSI concentration. (b) Schematic diagram of the coordination structures in dilute and concentrated LiTFSI/AN solutions [56]. Source: Yamada et al. [56]. Reproduced with permission of American Chemical Society. (c) Schematic diagram of the non-solvating effect of LiPS in the concentrated AN electrolyte. (d) and (e) Electrochemical performances of Li–S cells in a series of AN-based concentrated electrolytes with the performance in the DOL: DME electrolyte as the baseline [57]. Source: Cuisinier et al. [57]. Reproduced with permission of Royal Society of Chemistry.

sulfone (IPMS), ethyl isopropyl sulfone (EiPS), dipropyl sulfone (DPS), ethyl methyl sulfone (EMS), tetra methylene sulfone (TMS) are respectively up to 5.6, 5.6, 5.7, 5.7, and 5.1 V vs. Li^+/Li [60]. Whereas the high viscosity of sulfones leads to poor ionic conductivity and wettability to electrodes and separators. In the category of sulfones, TMS and EMS are mostly applied to Li–S systems. However, sulfones usually require a cosolvent to expand liquid-phase temperatures because of their high melting points (e.g. TMS, 27.4 °C; EMS, 34 °C), although Li salts dissolved TMS or EMS solutions could maintain liquid at subzero temperatures [60]. Another function of the cosolvent is decreasing the viscosity of the electrolyte. As for sulfoxides, dimethyl sulfoxide (DMSO) is extensively investigated for Li–S batteries. However, quite different from sulfones, DMSO shows high solubility to not only LiPS but also short-chain sulfides, Li_2S_3 and Li_2S_2. Arising from its high dielectric constant and polarity, the solubility of LiPS in DMSO is even higher than that of the DOL/DME mixed solvent [61].

Comparing with TMS, EMS possesses higher ionic conductivity due to its lower viscosity. In the pioneering work of Nazar and coworkers [3], the electrochemical performance of sulfur cathode fabricated by a melting-diffusion method at 155 °C was tested in the electrolyte of 1 M $LiPF_6$ in EMS. Weller et al. [62] investigated a binary-solvent electrolyte, 1.5 M LiTFSI in TMS/TTE (1 : 1 by volume) (labeled "TT"). Here, TTE is selected as a cosolvent to decrease the viscosity of the electrolyte. LiPS solubility in the TMS-based electrolyte is reduced to 0.033 M and enhanced Coulombic efficiency (c. 98%) comparing with DOL/DME with or without the addition of $LiNO_3$, respectively denoted as DDN or DD (Figure 3.11a). What is more, the TT electrolyte enables Li–S batteries operating under a low E/S ratio (<3 μl mg^{-1}). During cycling, the voltage hysteresis (ΔU) maintains at 0.22 V in the TT electrolyte, while the ΔU of DD electrolyte increases from 0.20 to 0.22 V (Figure 3.11b), indicating that the limited solubility to polysulfides of the TT electrolyte controls the precipitation of Li_2S and S_8 and prevents the polarization from increasing. Pouch cells were assembled utilizing TT and DD electrolytes (Figure 3.11c), after 10 cycles, the cell with the TT electrolyte keeps compact while the one with the DD electrolyte is observed with inflation, demonstrating the advantage of low volatility for the TT electrolyte. Chung and coworkers [60] investigated the physicochemical properties of a series of TMS/DME-based electrolytes and the electrochemical performances when they were applied to Li–S batteries. It was found that Coulombic efficiency displays a negative correlation to the proportion of DME.

Liu and coworkers [61] investigated Li–S redox flow batteries for large-scale energy storage. However, the insoluble short-chain polysulfides (e.g. Li_2S_2) in the DOL/DME electrolyte hinder sulfur utilization and cyclability. Under this circumstance, a DMSO-based electrolyte was proposed for the higher solubility of DMSO to both LiPSs and Li_2S_2 than the DOL/DME mixture (Figure 3.11d). They also found that the solubility to short-chain sulfides in DMSO significantly increases after the addition of Li salts, e.g. LiTFSI and $LiCF_3SO_3$ (LiTf), which is attributed to the interaction force between Li^+ and S_n^{2-}. An optimized formula was proposed with the DMSO to LiTf ratio of 30, in which the solubility of Li_2S_2 is the highest.

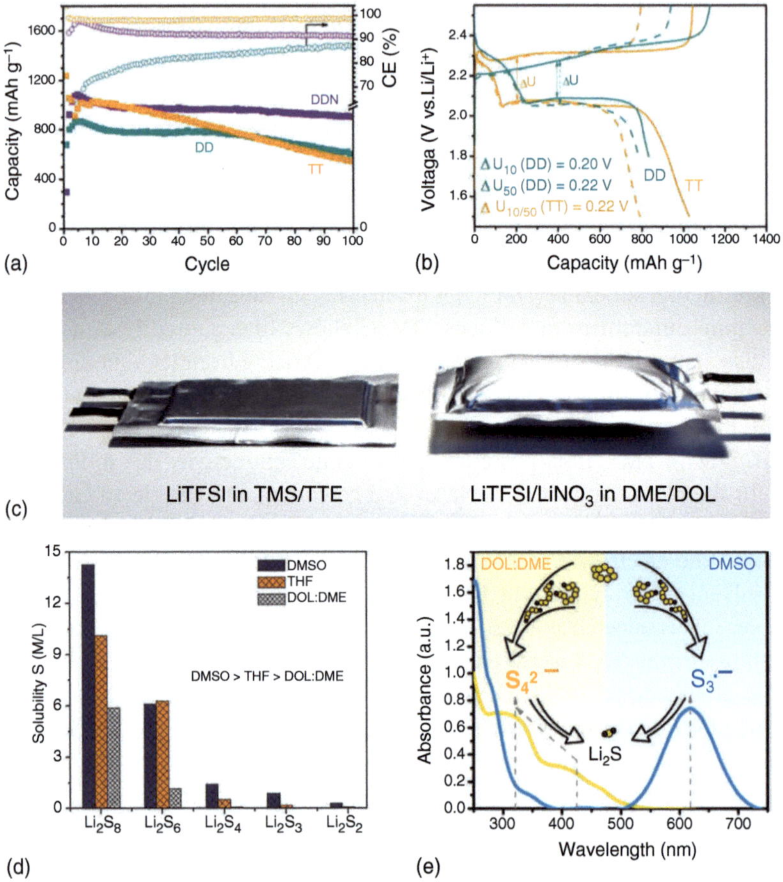

Figure 3.11 Electrochemical comparison between the TMS/TTE and DOL/DME electrolytes in (a) cycling performance and (b) discharge–charge profiles. (c) Inflation tests of pouch cells utilized the TMS/TTE and DOL/DME electrolytes [62]. Source: Weller et al. [62]. Reproduced with permission of IOP Publishing. (d) Solubility comparison of sulfide species in DMSO, THF, and DOL: DME solutions [61]. Source: Pan et al. [61]. Reproduced with permission of John Wiley & Sons. (e) Redox mechanism comparison between Li–S cells in DOL: DME and DMSO electrolytes [63]. Source: Zou and Lu [63]. Reproduced with permission of American Chemical Society.

Due to the incompatibility between DMSO and Li metal anode [64], the Li–S redox flow battery experiences fast capacity decay during cycling in the $DMSO_{30}$: LiTf-based electrolyte with 1 M LiTFSI as the supporting Li salt. After elevating the concentration of LiTFSI to 3 M, the cycling performance is significantly improved, which is attributed to the reduction of side reaction between free DMSO molecules and the Li anode. On the other hand, the higher-concentration strategy also slows down the LiPS shuttle. Making use of operando UV–vis spectroscopy, Lu and coworker [63] revealed that Li–S batteries in DMSO-based electrolyte undergo reaction intermediates including S_8^{2-}, S_6^{2-}, S_4^{2-}, and $S_3{\cdot}^-$, in which $S_3{\cdot}^-$ is the most stable and dominant species, as Figure 3.11e displays, which is different from the process in the lower-donor-number DOL/DME solvent.

3.4.5 Ionic Liquids

The ionic liquids (ILs) entirely consist of cations and anions, maintaining liquid state at room temperature, with broad potential window, high safety arising from non-volatility and low flammability. For Li–S batteries, ILs display inhibited polysulfides dissolution, directing different conversion pathways from ether electrolytes.

Yuan et al. [65] utilized a room-temperature ionic liquid (RTIL) as an electrolyte for Li–S batteries to suppress polysulfides solubility. The synthesized RTIL-based electrolyte is composed of 1 M LiTFSI in *N*-methyl-*N*-butyl-piperidinium bis(trifluoromethanesulfonyl) imide ([PP14][TFSI]), showing a wide potential window in the range from 5.2 to −0.15 V (vs. Li/Li^+) and is compatible with both the sulfur cathode and the Li metal anode. The discharge–charge profiles in the [PP14][TFSI]-based electrolyte are different from those in ether electrolytes, with one single discharge plateau at ~2.1 V, arising from the suppressed polysulfides solubility. A reversible specific capacity of 1055 mAh g^{-1} was achieved in the initial cycle, corresponding to sulfur utilization to 63%. Masayoshi Watanabe and coworkers [66] investigated the performance of Li–S batteries in a series of RTILs, which consist of *N*-methyl-*N*-propyl piperidinium ($[PP13]^+$) pairing with $[TFSI]^-$, or $[BETA]^-$, or $[FSI]^-$. They proposed that the solubility of polysulfides in RTILs is determined by the donor ability, i.e. the weaker the donor ability of the anions in RTILs to Li^+ cations, the lower the Li_2S_n solubility is. However, two discharge plateaus showed up in all the three electrolytes, different from the [PP14][TFSI]-based electrolyte [65]. Among these, [PP13][TFSI]-based electrolyte achieves the highest capacity, most stable cycling, and enhanced Coulombic efficiency, benefiting from its low viscosity.

The above discussed ILs-based electrolytes are binary mixtures of ILs and Li salts, with coexisting ILs cations and Li^+ cations of LiTFSI, decreasing Li^+ transference number and ionic conductivity. A new system of ILs was developed consisting of Li^+ cations (such as $[Li(G3)]^+$, $[Li(G4)]^+$) and anions, named as solvate ILs, with an order of magnitude higher ionic conductivity. Watanabe and coworkers [67] investigated the electrochemical performance of a series of $[Li(G3/G4)_x][TFSI]$ electrolytes. With the x value increasing, solubilities of sulfur species significantly increase, taking the limit solubility of Li_2S_8 as an example, which is ~29.0 mM in $[Li(G3)_1][TFSI]$ and ~5889 mM in $[Li(G3)_4][TFSI]$. Li–S cells in the $[Li(G3)_1][TFSI]$ electrolyte exhibited excellent cycling stability with ~700 mAh g^{-1} retained after 400 cycles and high Coulombic efficiency (>98%).

3.4.6 Polymer/Solid-State Electrolytes

For Li–S batteries, not only the Li metal anode but also the commonly utilized ether-based electrolytes give rise to huge safety concerns because of their low boiling and flash points. In this context, the use of nonflammable polymer/solid-state electrolytes are favorable selections, which also restrain the dissolution and shuttle of polysulfides and suppress Li dendrites growth and dead Li formation.

Inorganic electrolytes include $Li_2S–SiS_2$, $Li_2S–P_2S_5$, Thio-LISICON, $Li_{1.5}Al_{0.5}Ge_{1.5}(PO_4)_3$ (LAGP), $Li_{10}GeP_2S_{12}$, Li_3PS_4, $LiBH_4$, etc. Wang and coworkers [68] integrated a glass solid electrolyte, Li_3PS_4, into Li–S batteries with Li_2S and LiI as start active materials, in which the addition of LiI is to improve the ionic conductivity of Li_2S. The configuration of the all-solid-state battery is graphically demonstrated in Figure 3.12a, the Li_3PS_4 electrolyte is reinforced by Kevlar fiber. The fabricated composite cathode consists of $80Li_2S$-20LiI, the Li_3PS_4 glass electrolyte, and vapor-grown carbon fiber (VGCF), with a mass ratio of 75 : 15 : 10, loading on a stainless steel mesh; and Li metal is directly used as the anode. With a Li_2S loading of 2.54 mg cm^{-2}, the discharge capacity is high to 949.9 mAh g^{-1} (Figure 3.12c), and stable cycling is realized during 100 cycles (at 0.2 C). For an elevated Li_2S loading (7.64 mg cm^{-2}), a high energy density to 370.6 Wh kg^{-1} is achieved at the cell level excluding the current collector.

Xu and coworkers [69] fabricated a rGO@S-$Li_{10}GeP_2S_{12}$-acetylene black (AB) composite as the cathode material, matching with a two-layer solid electrolyte (Figure 3.12c) with a $Li_{10}GeP_2S_{12}$ layer toward the sulfur cathode and a 75%Li_2S-24%P_2S_5−1%P_2O_5 layer contacting the Li anode, to avoid the poor compatibility between $Li_{10}GeP_2S_{12}$ and Li metal [72]. It displays a single discharge plateau at ~2.1 V and one charge plateau at ~2.3 V (Figure 3.12d). At 60 °C, high reversible capacity (1525.6 mAh g^{-1}) is achieved under 0.05 C, and a high capacity of 830 mAh g^{-1} maintains at 1.0 C for 750 cycles.

Hu and coworkers [70] fabricated a bilayer garnet framework, with the dense layer as a solid-state electrolyte and the porous layer as a sulfur host, as Figure 3.12e shows. For the electrolyte–anode interface, a PEO layer with the thickness of 2 μm is coated on the solid-state garnet electrolyte, to achieve better contact and more homogeneous Li^+ cation flux. The designed Li–S batteries display high average Coulombic efficiency to >99% within 35 cycles (in the initial cycle, >99.8%). However, the specific capacities are less than 700 mAh g^{-1}, remaining to be improved.

Polymer electrolytes include PEO, poly(methyl methacrylate) (PMMA), PAN, poly(vinylidene fluoride-*co*-hexafluoropropylene) (PVDF-HFP) electrolytes, of which the PEO electrolyte is the most investigated one. To improve the ionic conductivity and mechanical property of polymer electrolytes, various fillers including Al_2O_3 [73], SiO_2 [74], TiO_2 [73a], ZrO_2 [75], $Li_7La_3Zr_2O_{12}$ (LLZO) [76], and $Li_{3x}La_{(2/3)-x}TiO_3$ (LLTO) [77] are added during the electrolyte preparation. Usually, PEO-based all-solid-state Li–S batteries work at a high temperature around the melting point (c. 63 °C) of PEO to acquire sufficient ionic conductivity (c. 10^{-3} S cm^{-1}). Wan and coworkers [71] investigated the evolution at interfaces of the all-solid-state polymer–ceramic (PEO-$Li_{6.75}La_3Zr_{1.75}Ta_{0.25}O_{12}$) composite electrolyte with both the sulfur cathode and the lithium anode, with the help of real-time optical microscope (OM) imaging, as Figure 3.12g shows. It reveals LiPS dissolves into the composite electrolyte in the discharge process, whereas remaining in the electrolyte in charge process, there is no apparent colorimetric difference in the electrolyte between discharge state to 1.5 V and charge state to

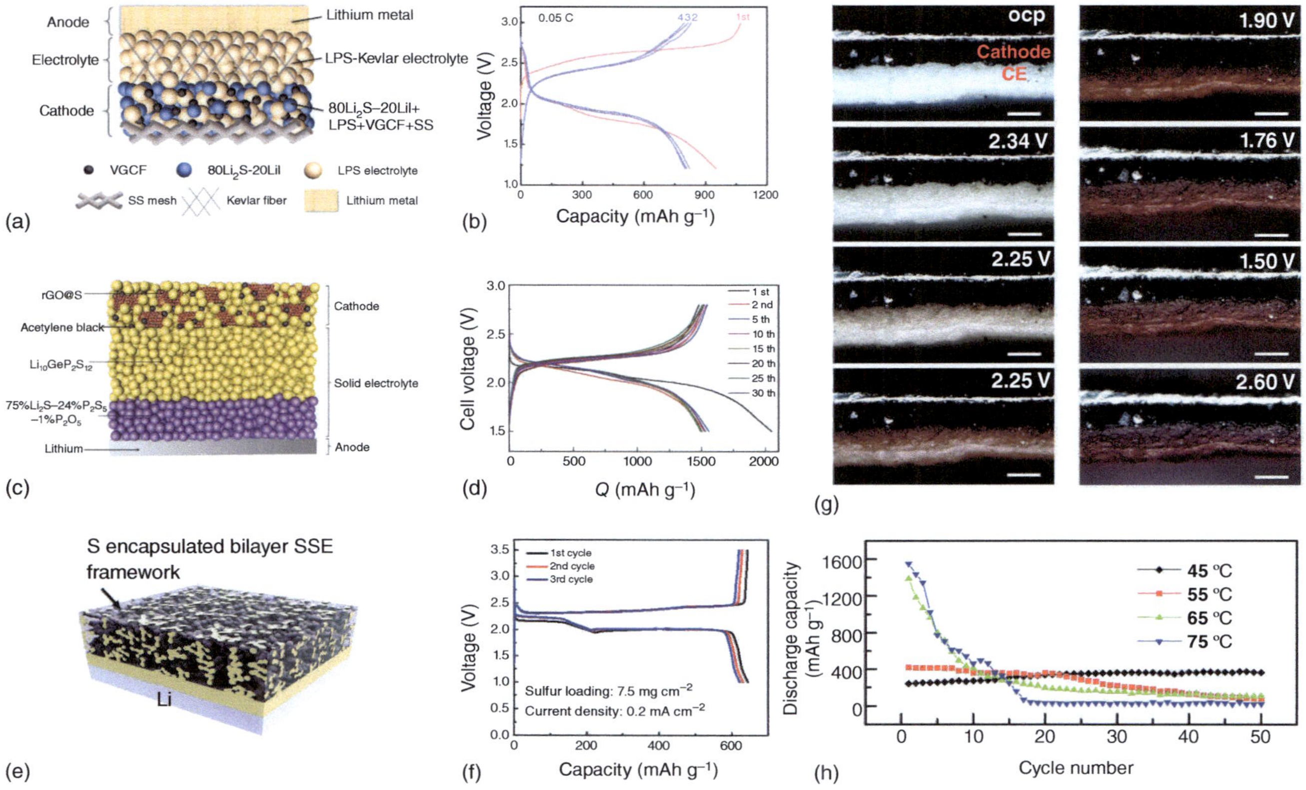

Figure 3.12 Schematic diagrams of solid-state cell configurations and electrochemical performances. (a, b) The Li−Li_2S/LiI battery applied Li_3PS_4 as the solid-state electrolyte [68]. Source: Xu et al. [68]. Reproduced with permission of American Chemical Society. (c, and d) A rGO@S-$Li_{10}GeP_2S_{12}$-acetylene cathode pairs with a Li metal anode with a two-layer solid electrolyte ($Li_{10}GeP_2S_{12}$ and 75%Li_2S-24%P_2S_5-1%P_2O_5) [69]. Source: Yao et al. [69]. Reproduced with permission of John Wiley & Sons. (e, f) A bilayer garnet framework consists of a dense layer as the solid-state electrolyte and a porous layer as the sulfur host [70]. Source: Fu et al. [70]. Reproduced with permission of Royal Society of Chemistry. (g) Real-time optical microscope (OM) imaging recorded the LiPS loss in a PEO-$Li_{6.75}La_3Zr_{1.75}Ta_{0.25}O_{12}$ electrolyte and (h) the cycling performance of Li−S cells under different temperatures [71]. Source: Song et al. [71]. Reproduced with permission of Royal Society of Chemistry.

Table 3.1 Electrolyte additives for Li–S batteries.

Functions	Additives
Li protection	$LiNO_3$ [78], Li_2S_5 [79], trace water [80], toluene [81], P_2S_5 [82], $La(NO_3)_3$ [83], LiI [84], InI_3 [85], biphenyl-4,4′-dithiol (BPD) [86], alphalipoic acid (ALA) [87]
Redox mediator	LiPS [88], Metallocenes (Co, Fe) [89], LiI [84], InI_3 [90]
Cathodes protection	LiI [84], Pyrrole [91], ALA [87]
Flame retardants	Hexafluorocyclotriphosphazene [92], tris(2,2,2-trifluoroethyl) phosphite (TTFP) [93]

2.6 V. Through the characterization of Raman spectroscopy and X-ray photoelectron spectra (XPS), the accumulation of sulfur species is identified on the Li metal anode. Temperature effects on interfaces and electrochemical performance (Figure 3.12h) are also investigated, revealing that increased temperature aggravates polysulfides dissolution and capacity loss.

3.4.7 Additives

Electrolyte additives significantly affect the electrochemical or physical properties of Li–S batteries, which are utilized to achieve the following functions: (i) in situ forming an SEI layer on Li metal surface; (ii) generating a protection layer on sulfur cathode surface to suppress the shuttle of polysulfides; (iii) as redox mediators to promote the conversion kinetics of the sulfur cathode; (iv) as flame retardants to improve safety. Some of these additives are summarized in Table 3.1.

Firstly utilized by Mikhaylik [78], $LiNO_3$ is regarded as one of the most effective additives for Li–S batteries. In ether electrolyte, the reaction between $LiNO_3$ and Li metal during cycling in situ generates a stable SEI layer composed of LiN_xO_y onto the surface of Li-metal anode. With the protection of LiN_xO_y layer, the shuttle effect is remarkably suppressed, bringing enhanced Coulombic efficiency and improved cycling stability. Dissolved polysulfides are also investigated as additives, to form a stable SEI layer on Li anode and facilitate sulfur conversion. The precipitation of insoluble Li_2S and Li_2S_2 on the Li metal surface plays the role of preventing further decomposition of electrolyte solvents and salts. What is more, the formation of polysulfides facilitates the sulfur redox by dragging sulfur into the electrolyte [88], as Figure 3.13a demonstrates, such as the addition of Li_2S_4, performing a process of $S_8 + S_4^{2-} \rightarrow S_6^{2-}$. When $LiNO_3$ and Li_2S_6 are added together, they play a synergetic effect with a bilayer SEI formation (Figure 3.13b), consisting of an inner layer composed of $LiNO_3$ and Li_2S_6 reduced products (LiN_xO_y, Li_2S, Li_2S_2), and an outer layer consisting of oxidized products of Li_2S_6 ($Li_2S_2O_3$, Li_2SO_4) [94]. Cui and coworkers [95] studied the synergetic effect through ex situ SEM and in situ optical microscopy to characterize surface morphology evolution of Li metal and XPS to determine SEI components. In the control experiments with adding $LiNO_3$ and Li_2S_8 alone, either

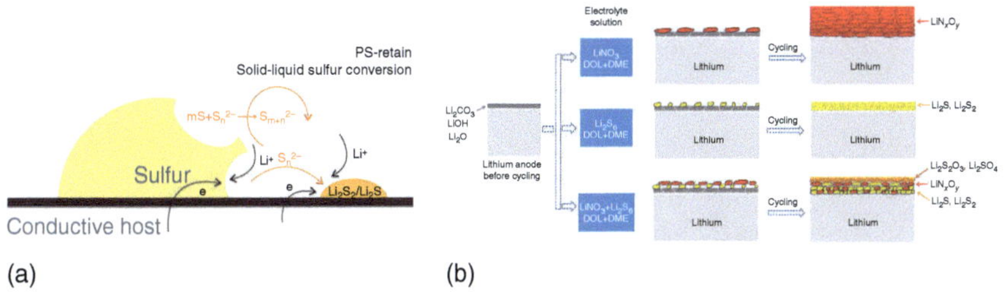

Figure 3.13 (a) The mediation effect of LiPS in the redox conversion at the sulfur cathode [88]. Source: Li et al. [88]. Reproduced with permission of John Wiley & Sons. (b) with the addition of $LiNO_3$, Li_2S_6, and both of them, the components of generated SEI on Li metal anodes [94]. Source: Xiong et al. [94]. Reproduced with permission of Elsevier.

of them fails to form a robust SEI layer enough to prevent electrolyte decomposition. The uneven precipitation layer (Li_2S_2/Li_2S) induces heterogeneous sites which aggravate Li dendrite growth. With the combination of $LiNO_3$ and Li_2S_8, a bottom layer of $LiNO_3$ decomposition product firstly forms on the Li surface, followed by an upper layer of Li_2S_2/Li_2S, thereby protecting the electrolyte from decomposing.

3.5 Anodes

To pair with S cathode, the anode needs to not only supply Li source but also provide a substrate or framework for the reversible Li plating/stripping. Under this circumstance, Li metal is commonly selected considering its lowest electrochemical potential and high specific capacity. However, the use of Li metal brings about issues such as dead Li accumulation and dendrites growth. To comprehend the degradation and failure mechanisms of Li metal anode, extensive and intensive studies were carried out worldwide, obtaining considerable progress. Currently, excess Li is always needed to guarantee the long-term cycling stability of Li–S batteries, which is uneconomical and most importantly, counteracts the advantage of Li–S batteries in terms of energy density. Therefore, a safer and more efficient anode is highly desired for Li–S batteries. When Li_2S is utilized as the start active material in cathode, it makes a Li-free anode applicable as the counterpart, e.g. carbon, silicon, and carbon/silicon composed materials.

3.5.1 Li Anodes

Several decades ago, graphite replaced Li metal as the anode pairing with intercalation-type cathodes, promoting commercialization of Li-ion batteries. However, Li metal with prominent advantages including high theoretical specific capacity ($3860\,mAh\,g^{-1}$) and lowest negative electrochemical potential (3.04 V vs. H_2/H^+) has been attracting extensive investigations all the time. Xiao and coworkers [96] investigated the failure mechanism of Li metal anode. As Figure 3.14a shows, they pointed out that the mossy Li grows inward toward fresh Li bulk under high

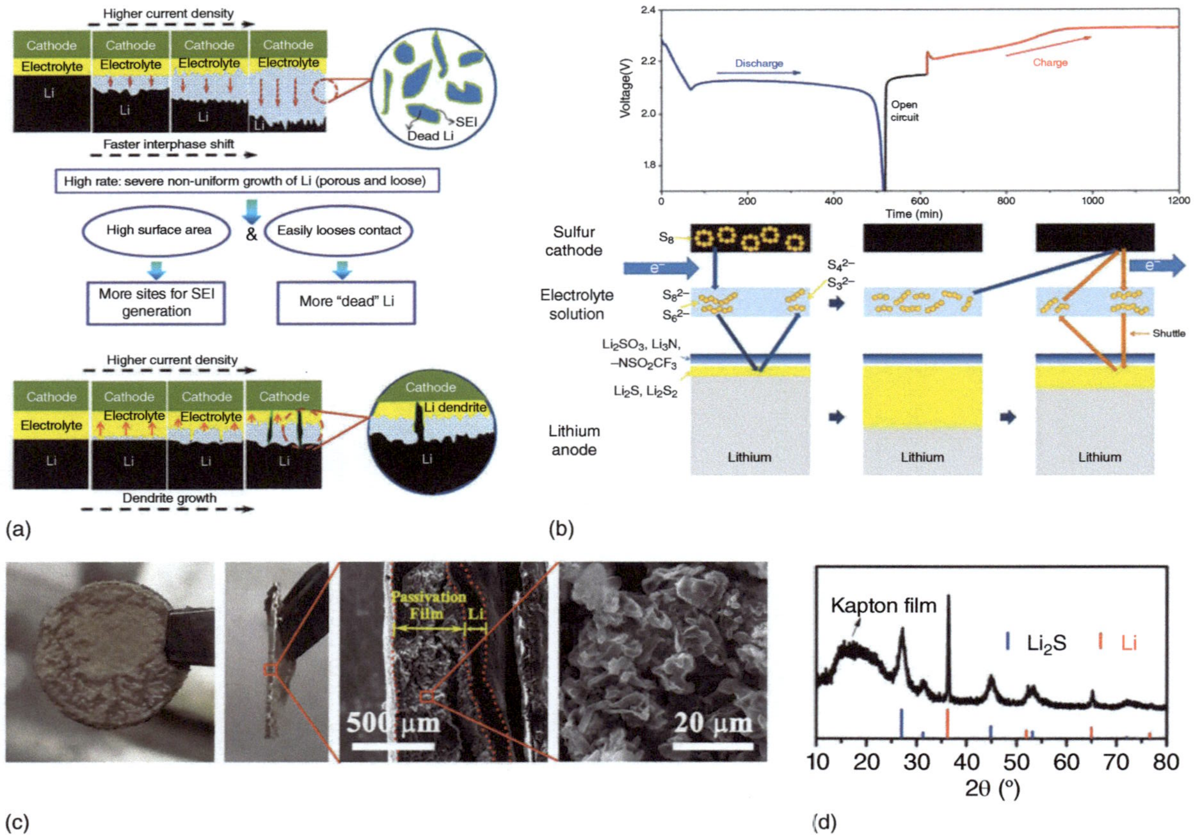

Figure 3.14 Schematic diagrams of (a) dead Li accumulation [96]. Source: Lv et al. [96]. Reproduced with permission of John Wiley & Sons. and (b) The morphology and components evolution on Li metal anode under shuttle effect [97]. Source: Xiong et al. [97]. Reproduced with permission of Elsevier. (c) Morphology detection of the cycled Li anode with porous structure and apparent cracks, and (d) composition determination of the cracks by XRD [98]. Source: Qie et al. [98]. Reproduced with permission of John Wiley & Sons.

current density, and the accumulation of the dead Li accounts for battery failure before Li dendrites penetrating the separator. The circumstance of Li metal anode in Li–S batteries is more complicated with dissolved polysulfides in ether electrolytes. Xiong et al. [97] investigated the evolution of the Li metal anode surface during cycling. As Figure 3.14b shows, in the discharge process, the high-order polysulfides (S_6^{2-} and S_8^{2-}) migrate to the anode side, penetrating line cracks in the top layer composed of electrolyte decomposition, generating low-order polysulfides (S_4^{2-} and S_3^{2-}) and insoluble products (Li_2S_2 and Li_2S) precipitated at the bottom layer. When the cell is recharged to 2.3 V, the high-order polysulfides dissolve into the electrolyte and react with insoluble Li_2S_2 and Li_2S, the as-generated low-order polysulfides migrate back to the sulfur cathode side electrochemically converted to high-order polysulfides again. The bottom layer is thick enough to prevent LiTFSI from further decomposition, thus improves stability. In the beginning, the impacts of Li metal on electrochemical performance degradation of Li–S batteries were usually neglected under immense excess Li supply. When a high loading (18.1 mg cm^{-2}) sulfur cathode is applied, Qie et al. [98] found that the Li–S cell experiences a sudden death after dozens of cycles. After disassembling the dead cell, a thick gray passivation layer (~440 μm) was found on the cycled Li metal anode, exhibiting a porous structure with obvious cracks (Figure 3.14c), which is identified as Li_2S through X-ray diffraction (Figure 3.14d). By reassembling a cell with the cycled sulfur cathode and a pristine Li, and addition of electrolyte, the cell resurrected, further confirming that the sudden death was caused by the failure of Li metal anode and the exhaustion of electrolytes.

To achieve long-cycle Li–S batteries with low Li excess and high sulfur loading, effectively protected and optimized Li anodes are necessary. Generally, the strategies for Li anode protection include: (i) in situ SEI layers generation on Li anode surface with electrolyte additives; (ii) artificial SEI layers formation through gas or liquid treatments and coating; (iii) Li anode architecture design. Due to the high reductive activity of Li metal, liquid electrolyte components spontaneously decompose on the surface of Li metal, forming an SEI layer that is electron-insulated but Li^+-conductive. The SEI layer, with a thickness of 1–100 nm, plays an important role in preventing further consumption of liquid electrolytes and Li anodes protection. It provides a universal method to manipulate SEI components through electrolytes formula adjustment, comprising solvents combination, Li salts selection, and effective additives addition.

On the other hand, before assembling batteries, pretreating Li anode to form an artificial SEI layer is another strategy. In 2003, Lee et al. [99] introduced a polymer protection layer on Li anode through a cross-linking reaction to suppress the attack by LiPSs; consequently, the cyclability was enhanced. Wen and coworkers [100] introduced N_2 treatment to generate a LiN_3 layer, a high Li^+ conductive substance (~10^{-3} S cm^{-1}), onto the surface of Li anode, which enables the stable Li–S cycling for over 500 cycles with an average Coulombic efficiency of 92.3% even without $LiNO_3$ addition in the electrolyte. A fluorination process of F_2 treatment was adopted by Cui and coworkers [101], generating a dense and homogeneous LiF layer on the surface of Li metal or lithiated silicon anodes, to protect anodes from corrosion and enhance cycling stability. In addition to the SEI strategy, architectural design is another way to obtain safer and more efficient Li anodes. Comparing to a planar substrate, the three-dimensional anodes with high surface area straightforwardly decrease the current density, thereby suppressing dendrite growth. Lu et al. [102] fabricated a free-standing Cu nanowire (CuNW) as the current collector to accommodate Li metal. The designed anode presents a low overpotential (~0.04 V) and a high average Coulombic efficiency (~98.6%) during 200 cycles. Liu et al. [103] applied a porous structured $Li_{2.6}BMg_{0.05}$ alloy pairing with sulfur cathodes, showing reduced voltage polarization to the Li–S cell.

3.5.2 Carbon Anodes

The graphite anodes have dominated the Li-ion batteries, in contrast, the Li-metal anodes confront barriers to meet the mass-market application. Replacing Li metal with carbon anodes is a strategy to improve the safety and cycling stability of Li–S batteries. To pair the Li-free carbon anodes, the starting active materials in cathode have to be the lithiated product, Li_2S. Graphite, with a theoretical specific capacity of 372 mAh g^{-1}, is utilized as anode for commercial Li-ion batteries with carbonate-based electrolytes. However, graphite anodes are not well compatible with most noncarbonate solvents, including the most utilized ether electrolytes in Li–S batteries, with the solvent molecules cointercalating with Li^+ into the graphite, destroying its structure [104]. In a graphite||Li_2S cell with an N/P ratio of 1 (no excess Li supply), the SEI layer formation on the graphite surface consumes a portion of Li source in the initial cycle, thereby leading to a considerable irreversible capacity loss in the discharge process. Wang and coworkers [45] fabricated a Li_2S@microporous carbon (Li_2S@MC) cathode through an in situ lithiation method through spraying stabilized lithium metal powder onto S@MC composite, followed by a compression process. For the precursor, S@MC composite, sulfur is mainly in the state of short-chain S_2 molecules. Similar to the small-sulfur cathode, there is no LiPS intermediates formation during redox conversion, thus the assembled graphite||Li_2S full cells operate stably in the carbonate electrolyte of 1 M $LiPF_6$ in EC/DEC (1 : 1, v/v) for 900 cycles, with high capacity retention and Coulombic efficiency of ~100%. What is more, there is extra Li in the in situ fabricated Li_2S@MC cathode besides the active material Li_2S, which could compensate Li loss in the first charge process and achieve enhanced initial Coulombic efficiency.

Except for the particular adoption of carbonate electrolyte, the compatibility between graphite and noncarbonate solvents can be realized by the highly concentrated electrolytes and sometimes with the assistance of additives, e.g. $LiNO_3$. In this respect, highly concentrated AN [56], ether [105], and DMSO-based [106] electrolytes have been adopted to coexist with graphite anodes. Li and coworkers [105] pointed out that the increase of Li salt concentration and the addition of $LiNO_3$ effectively lower the interaction strength between Li^+ and solvent molecules, thus impeding the cointercalation process. An optimized ether-based electrolyte, 2.5 M LiTFSI and 0.4 M $LiNO_3$ dissolved in DME: DOL = 1 : 1 (v/v), demonstrates high compatibility to graphite anode in Li–S full batteries

[105]. Amine and coworkers [43] studied the electrochemical performances of graphite||Li_2S@graphene full cells in an ether-based electrolyte,1 M LiTFSI in partially fluorinated 1,1,2,2-tetrafluoro-3-(1,1,2,2-tetrafluoroethoxy)-propane (D2) and DOL (v/v, 2 : 1), in which the full cells exhibit stable cyclability indicating its compatibility to graphite anodes. However, this formula could also be categorized as the high concentrated electrolyte, considering the barely solvating ability of D2 to Li salts because of its fluorinated structure. The inevitable Li loss in the first charge process makes discharge capacity unsatisfactory.

Carbon-based composite anodes have also been developed. Scrosati and coworker [107] fabricated an Sn/C (1 : 1, mass ratio) composite anode as an alternative for Li metal anode, pairing with Li_2S cathode. It was reported that Sn/C shows high stability to sulfide ions without chemical reaction [108]. They applied a PEO-based gel electrolyte into the Sn/C||Li_2S full cell, exhibiting improved cycling stability compared with that in a conventional liquid electrolyte.

3.5.3 Silicon Anodes

Silicon has a high theoretical specific capacity and is compatible with ether electrolytes. The electrochemical potential of silicon is around 0.4 V vs. Li/Li^+, and correspondingly the Si||Li_2S full cell presents lower discharge voltage. Cui and coworkers [109] fabricated a silicon nanowire anode, pairing with a Li_2S@CMK-3-based cathode, presenting stable cycling in a typical DOL/DME electrolyte. Similar to graphite, comparing to half-cells, the full cell displays a high irreversible capacity in the initial cycle, arising from insufficient available Li source.

3.6 Challenges and Perspectives

In the last decade, impressive progress has been achieved on both the fundamental understandings of electrochemical processes and performance improvement. However, up to date, the practical applications of Li–S batteries are still challenging. As Figure 3.2a shows, a Li–S cell consists of a sulfur@carbon cathode pairing with a Li metal anode and an ether electrolyte (1 M LiTFSI in DOL: DME = 1 : 1, v/v) with the addition of $LiNO_3$. On the cathode side, the inherent poor electric conductivity of S and Li_2S leads to insufficient S utilization and slow redox kinetics. The S particles also experience a significant volume change (~80%) when S_8 fully converts to Li_2S due to their density difference, resulting in the structural collapse during cycling. Usually, sulfur cathodes in most of the literatures are with low sulfur content and/or loading, which fail to meet the demands for practical application. For the SPAN cathodes, although high sulfur utilization can be achieved, the limited sulfur content (<60 wt%) brings down the overall energy density.

For ether electrolyte, the solid–liquid–solid conversion mechanism improves sulfur utilization, meanwhile, brings about the shuttle of solvated polysulfides, leading to a series of problems: (i) lower the Coulombic efficiency because of the chemical reaction between polysulfides and lithium metal instead of electrochemical reaction;

(ii) a passivation layer (insoluble Li_2S and Li_2S_2) generates on the anode surface increasing the interface impedance; (iii) persistent consumption of lithium and sulfur due to the precipitation of Li_2S and Li_2S_2 on the Li anode surface; (iv) a high electrolyte to sulfur ratio (namely E/S ratio) is needed to dissolve the polysulfides, usually higher than 5 μl mg^{-1}, significantly reduces overall energy density when a complete battery is evaluated; (v) the dissolution strategy in ether-based electrolytes degrades the efforts in nano-sulfur fabrication, because of the sulfur redistribution during cycling; (vi) severe self-discharge phenomenon is caused by the dissolution of sulfur and polysulfides in the electrolyte, the relationship of the high plateau capacity (Q_H), shuttle constant (k_s), and storage time (t_s) is quantized by Mikhaylik et al. [110] as the following:

$$\frac{d \ln Q_H}{dt_s} = -k_s$$

It demonstrates the negative correlation between Q_H and k_s, that is to say, the more severe the shuttle effect is, the higher the capacity loss is under the same storage time.

Other electrolyte systems, including liquid and polymer/solid-state electrolytes, exhibit impressive performance in specific aspects, but still face barriers in some way. For the concentrated AN-based electrolyte, highly coordinated AN_2-LiTFSI complex eliminates free AN solvent, consequently, preventing the shuttle of polysulfides and expanding its potential window, especially the reductive stability. What is more, the diluting effect of HFE, decreasing viscosity of the AN-based electrolyte, enabling decent ionic conductivity and acceptable rate performance. However, the compatibility between the AN electrolyte and Li metal anode is insufficient during prolonged cycling. For the carbonate-based electrolyte, its application is restricted within particular cases, such as SPAN composite and smaller sulfur cathodes, in which the sulfur content is urgent but challenging to increase excessing 60 wt%. For the sulfone-based electrolytes, their high viscosity, low ionic conductivity, and low solubility to LiPS lead to unsatisfactory electrochemical performance. DMSO, a sulfoxide solvent with a higher solubility to LiPS than that of DOL/DME mixture, however, its poor compatibility with Li metal impedes the long-term cyclability. As for solid-state electrolytes, several issues should be concerned about: (i) the solubility of LiPS in polymers (e.g. PEO-based) leads to continuous sulfur loss; (ii) large volume change during sulfur conversion deteriorates the contact between inorganic solid electrolytes and cathodes; (iii) sluggish conversion kinetics result in a low S utilization and poor rate performance.

On the Li metal anode side, except for the attack by polysulfides, Li–S batteries are faced with severe safety concerns about potential short circuit caused by Li dendrites penetrating separators. The continuous consumption of lithium needs excess Li supply to support long-cycle performance, decreasing the overall energy density, which is uneconomical and less competitive. Thus, it is challenging and urgent to realize effective and stable SEI layers on the Li anode surface.

No doubt, though inspiring progress of Li–S batteries has been achieved at the lab level, more efforts are necessary for fundamental understanding, material development, electrolyte formulation optimization, and configuration innovation. Above all, the future investigation should concentrate on developing practical Li–S batteries with competitive energy density both gravimetrically and volumetrically.

At present, electrolyte takes up the largest weight fraction of the Li–S battery. Usually, a flooded dosage is added with the electrolyte to sulfur (E/S) ratio > 10 μl mg^{-1}, which restricts the theoretical energy density to lower than 300 Wh kg^{-1} (based on the density of electrolyte of 1.1 g ml^{-1}). A practical E/S ratio should be <5 μl mg^{-1}, ideally lower than 2 μl mg^{-1}. [111] Based on the representative formula, 1 M LiTFSI in DOL/DME with $LiNO_3$ addition, dissolved LiPS lead to locally concentrated zone in the cathode, greatly hindering Li^+ transportation and impeding following conversion of the active material. In this regard, two strategies are promising for reducing the E/S ratio: (i) develop non-solvating or sparingly solvating electrolytes; (ii) exploit higher solvating electrolyte system. Gallagher and coworkers [112] put forward that it should shift from the fully solvating electrolyte to the sparingly solvating electrolyte to fulfill lean electrolyte operation with E/S ratio < 1 μl mg^{-1}. In the non-solvating or sparingly solvating electrolytes, due to the eliminated or limited LiPS solvation, Li–S cells may perform different conversion mechanisms, as presented in the concentrated AN electrolyte investigated by Nazar and coworkers [57]. And a low E/S ratio of 2 μl mg^{-1} for the AN electrolyte was applied to Li–Se batteries by our group [113], where Se is a congener of S, thus the electrochemical behaviors of the Li–Se cell have much in common with those of the Li–S system. Under the lean electrolyte condition, the Li–Se cell demonstrates stable three cycles along with a high specific capacity (>600 mAh g^{-1}), approaching its theoretical value (675 mAh g^{-1}). An opposite strategy is applying highly solvating electrolytes, constituted of solvents with high donor number (DN), such as *N*,*N*-dimethylacetamide (DMA), DMSO, and 1-methylimidazole (MeIm). These solvents display higher LiPS dissolution limitation comparing to DME; therefore, a lower limited E/S ratio is promising to support Li–S cells operating well. However, the compatibility between the high DN electrolytes and Li metal anodes demands to be improved.

For cathodes, high content and high loading of sulfur are needed to counteract the inactive components in configuration, such as current collector, separator, and packaging materials. Disappointingly, literature works generally ignore these parameters to pursue high capacity and better cyclability, by applying low-content (50 wt%) and/or low-loading (2 mg cm^{-2}) cathodes. In contrast, excess Li is always adopted in the anode side to support the long-cycling requirement. The negative to positive capacity (N/P) ratio is a parameter for assessing the capacity of the anode to the cathode. Usually, this N/P ratio is higher than 40, meaning much excess Li supply in Li metal anode, significantly degrading the energy density and the low-price merit of the Li–S system.

To overcome the gap between literatures (Figure 3.15a) and practical demands (Figure 3.15b), the following parameters are necessary (Figure 3.15c): (i) sulfur content >70 wt%; (ii) sulfur loading >5 mg cm^{-2}; (iii) E/S ratio < 5 μl mg^{-1}; (iv) lithium excess <4 (the lithium to sulfur capacity ratio < 5). When it takes account of the volumetric energy density, a compact cathode with low porosity is required as well.

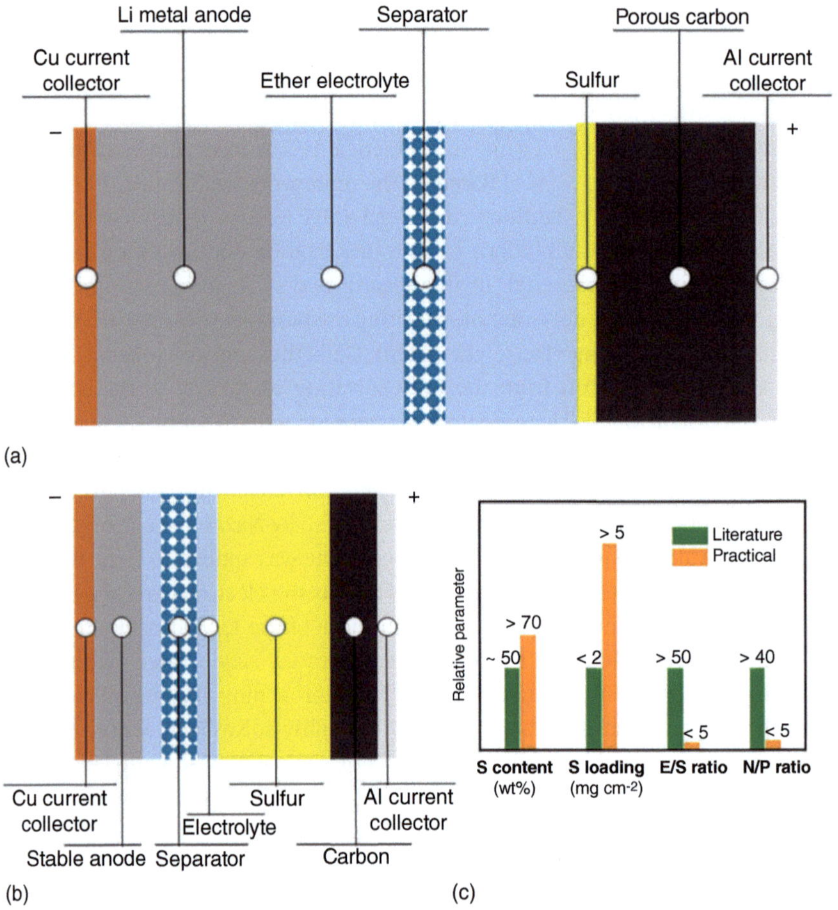

Figure 3.15 The Li–S battery configurations (a) in literature and (b) of practical demands, and (c) comparisons between literature and practical parameters.

References

1 Wang, D.-W., Zeng, Q., Zhou, G. et al. (2013). Carbon–sulfur composites for Li–S batteries: status and prospects. *J. Mater. Chem. A* 1: 9382–9394.

2 Aurbach, D., Pollak, E., Elazari, R. et al. (2009). On the surface chemical aspects of very high energy density, rechargeable Li-sulfur batteries. *J. Electrochem. Soc.* 156: A694–A702.

3 Ji, X., Lee, K.T., and Nazar, L.F. (2009). A highly ordered nanostructured carbon-sulphur cathode for lithium-sulphur batteries. *Nat. Mater.* 8: 500–506.

4 Liang, C., Dudney, N.J., and Howe, J.Y. (2009). Hierarchically structured sulfur/carbon nanocomposite material for high-energy lithium battery. *Chem. Mater.* 21: 4724–4730.

5 Ji, L., Rao, M., Aloni, S. et al. (2011). Porous carbon nanofiber-sulfur composite electrodes for lithium/sulfur cells. *Energy Environ. Sci.* 4: 5053–5059.

6 Zhang, B., Qin, X., Li, G.R., and Gao, X.P. (2010). Enhancement of long stability of sulfur cathode by encapsulating sulfur into micropores of carbon spheres. *Energy Environ. Sci.* 3: 1531–1537.

7 Jayaprakash, N., Shen, J., Moganty, S.S. et al. (2011). Porous hollow carbon@sulfur composites for high-power lithium-sulfur batteries. *Angew. Chem. Int. Ed.* 50: 5904–5908.

8 Cao, Y., Li, X., Aksay, I.A. et al. (2011). Sandwich-type functionalized graphene sheet-sulfur nanocomposite for rechargeable lithium batteries. *Phys. Chem. Chem. Phys.* 13: 7660–7665.

9 Elazari, R., Salitra, G., Garsuch, A. et al. (2011). Sulfur-impregnated activated carbon fiber cloth as a binder-free cathode for rechargeable Li-S batteries. *Adv. Mater.* 23: 5641.

10 Zheng, G., Zhang, Q., Cha, J.J. et al. (2013). Amphiphilic surface modification of hollow carbon nanofibers for improved cycle life of lithium sulfur batteries. *Nano Lett.* 13: 1265–1270.

11 Zheng, G., Yang, Y., Cha, J.J. et al. (2011). Hollow carbon nanofiber-encapsulated sulfur cathodes for high specific capacity rechargeable lithium batteries. *Nano Lett.* 11: 4462–4467.

12 Su, Y.S. and Manthiram, A. (2012). Lithium-sulphur batteries with a microporous carbon paper as a bifunctional interlayer. *Nat. Commun.* 3: 1166.

13 Yao, H., Yan, K., Li, W. et al. (2014). Improved lithium–sulfur batteries with a conductive coating on the separator to prevent the accumulation of inactive S-related species at the cathode–separator interface. *Energy Environ. Sci.* 7: 3381–3390.

14 Zhou, G., Pei, S., Li, L. et al. (2014). A graphene-pure-sulfur sandwich structure for ultrafast, long-life lithium-sulfur batteries. *Adv. Mater.* 26: 625–631.

15 Qie, L. and Manthiram, A. (2015). A facile layer-by-layer approach for high-areal-capacity sulfur cathodes. *Adv. Mater.* 27: 1694–1700.

16 Wang, J.L., Yang, J., Xie, J.Y., and Xu, N.X. (2002). A novel conductive polymer-sulfur composite cathode material for rechargeable lithium batteries. *Adv. Mater.* 14: 963–965.

17 Xiao, L., Cao, Y., Xiao, J. et al. (2012). A soft approach to encapsulate sulfur: polyaniline nanotubes for lithium-sulfur batteries with long cycle life. *Adv. Mater.* 24: 1176–1181.

18 Wang, D.Y., Guo, W., and Fu, Y. (2019). Organosulfides: an emerging class of cathode materials for rechargeable lithium batteries. *Acc. Chem. Res.* 52: 2290–2300.

19 Chung, W.J., Griebel, J.J., Kim, E.T. et al. (2013). The use of elemental sulfur as an alternative feedstock for polymeric materials. *Nat. Chem.* 5: 518–524.

20 Wang, L., He, X., Li, J. et al. (2012). Analysis of the synthesis process of sulphur–poly(acrylonitrile)-based cathode materials for lithium batteries. *J. Mater. Chem.* 22: 22077–22081.

21 Fanous, J., Wegner, M., Grimminger, J. et al. (2011). Structure-related electrochemistry of sulfur-poly(acrylonitrile) composite cathode materials for rechargeable lithium batteries. *Chem. Mater.* 23: 5024–5028.

22 Wei, S., Ma, L., Hendrickson, K.E. et al. (2015). Metal-sulfur battery cathodes based on PAN sulfur composites. *J. Am. Chem. Soc.* 137: 12 143–12 152.

23 Wang, J., He, Y.-S., and Yang, J. (2015). Sulfur-based composite cathode materials for high-energy rechargeable lithium batteries. *Adv. Mater.* 27: 569–575.

24 Liang, Y., Tao, Z., and Chen, J. (2012). Organic electrode materials for rechargeable lithium batteries. *Adv. Energy Mater.* 2: 742–769.

25 Wei, S.Z., Li, W., Cha, J.J. et al. (2013). Sulphur-TiO_2 yolk-shell nanoarchitecture with internal void space for long-cycle lithium-sulphur batteries. *Nat. Commun.* 4: 1331–1331.

26 Zhou, G., Tian, H., Jin, Y. et al. (2017). Catalytic oxidation of Li_2S on the surface of metal sulfides for Li-S batteries. *Proc. Natl. Acad. Sci. U.S.A.* 114: 840–845.

27 Liang, X., Garsuch, A., and Nazar, L.F. (2015). Sulfur cathodes based on conductive MXene nanosheets for high-performance lithium-sulfur batteries. *Angew. Chem. Int. Ed.* 54: 3907–3911.

28 Yu, M., Zhou, S., Wang, Z. et al. (2019). Accelerating polysulfide redox conversion on bifunctional electrocatalytic electrode for stable Li-S batteries. *Energy Storoge Mater.* 20: 98–107.

29 Zhou, T., Lv, W., Li, J. et al. (2017). Twinborn TiO_2–TiN heterostructures enabling smooth trapping–diffusion–conversion of polysulfides towards ultralong life lithium–sulfur batteries. *Energy Environ. Sci.* 10: 1694–1703.

30 Liang, X. and Nazar, L.F. (2016). In situ reactive assembly of scalable core-shell sulfur-MnO_2 composite cathodes. *Acs Nano* 10: 4192–4198.

31 Tao, Y., Wei, Y., Liu, Y. et al. (2016). Kinetically-enhanced polysulfide redox reactions by Nb_2O_5 nanocrystals for high-rate lithium-sulfur battery. *Energy Environ. Sci.* 9: 3230–3239.

32 He, J., Luo, L., Chen, Y., and Manthiram, A. (2017). Yolk-shelled C@Fe_3O_4 nanoboxes as efficient sulfur hosts for high-performance lithium-sulfur batteries. *Adv. Mater.* 29: 1702707.

33 Song, J., Xu, T., Gordin, M.L. et al. (2014). Nitrogen-doped mesoporous carbon promoted chemical adsorption of sulfur and fabrication of high-areal-capacity sulfur cathode with exceptional cycling stability for lithium-sulfur batteries. *Adv. Funct. Mater.* 24: 1243–1250.

34 Tang, C., Zhang, Q., Zhao, M.Q. et al. (2014). Nitrogen-doped aligned carbon nanotube/graphene sandwiches: facile catalytic growth on bifunctional natural catalysts and their applications as scaffolds for high-rate lithium-sulfur batteries. *Adv. Mater.* 26: 6100–6105.

35 Cai, J., Wu, C., Zhu, Y. et al. (2017). Sulfur impregnated N, P co-doped hierarchical porous carbon as cathode for high performance Li-S batteries. *J. Power Sources* 341: 165–174.

36 Seh, Z.W., Zhang, Q., Li, W. et al. (2013). Stable cycling of lithium sulfide cathodes through strong affinity with a bifunctional binder. *Chem. Sci.* 4: 3673–3677.

37 Lacey, M.J., Jeschull, F., Edström, K., and Brandell, D. (2014). Functional, water-soluble binders for improved capacity and stability of lithium–sulfur batteries. *J. Power Sources* 264: 8–14.

38 Babu, G., Ababtain, K., Ng, K.Y., and Arava, L.M. (2015). Electrocatalysis of lithium polysulfides: current collectors as electrodes in Li/S battery configuration. *Sci. Rep.* 5: 8763.

39 Tao, X., Wang, J., Liu, C. et al. (2016). Balancing surface adsorption and diffusion of lithium-polysulfides on nonconductive oxides for lithium-sulfur battery design. *Nat. Commun.* 7 (11): 203.

40 Mosavati, N., Salley, S.O., and Ng, K.Y.S. (2017). Characterization and electrochemical activities of nanostructured transition metal nitrides as cathode materials for lithium sulfur batteries. *J. Power Sources* 340: 210–216.

41 Yang, Y., Zheng, G., Misra, S. et al. (2012). High-capacity micrometer-sized Li_2S particles as cathode materials for advanced rechargeable lithium-ion batteries. *J. Am. Chem. Soc.* 134: 15 387–15 394.

42 Wu, M., Cui, Y., and Fu, Y. (2015). Li2S nanocrystals confined in free-standing carbon paper for high performance lithium-sulfur batteries. *ACS Appl. Mater. Interfaces* 7: 21 479–21 486.

43 Tan, G., Xu, R., Xing, Z. et al. (2017). Burning lithium in CS_2 for high-performing compact Li_2S-graphene nanocapsules for Li-S batteries. *Nat. Energy* 2 (17): 090.

44 Yu, M., Wang, Z., Wang, Y. et al. (2017). Freestanding flexible Li_2S paper electrode with high mass and capacity loading for high-energy Li-S batteries. *Adv. Energy Mater.* 7: 1700018.

45 Shiyou, Z.Y.C., Xu, Y., Yi, F. et al. (2013). In situ formed lithium sulfide/microporous carbon cathodes for lithium-ion batteries. *Acs Nano* 7: 10995–11003.

46 Yang, Z., Guo, J., Das, S.K. et al. (2013). In situ synthesis of lithium sulfide-carbon composites as cathode materials for rechargeable lithium batteries. *J. Mater. Chem. A* 1: 1433–1440.

47 Choi, J.-W., Kim, J.-K., Cheruvally, G. et al. (2007). Rechargeable lithium/sulfur battery with suitable mixed liquid electrolytes. *Electrochim. Acta* 52: 2075–2082.

48 Ryu, H.-S., Ahn, H.-J., Kim, K.-W. et al. (2006). Discharge behavior of lithium/sulfur cell with TEGDME based electrolyte at low temperature. *J. Power Sources* 163: 201–206.

49 Kim, S., Jung, Y., and Lim, H.S. (2004). The effect of solvent component on the discharge performance of lithium–sulfur cell containing various organic electrolytes. *Electrochim. Acta* 50: 889–892.

50 Kim, T.J., Jeong, B.O., Koh, J.Y. et al. (2014). Influence of electrolyte composition on electrochemical performance of Li-S cells. *B Kor. Chem. Soc.* 35: 1299–1304.

51 Peled, E., Sternberg, Y., Gorenshtein, A., and Lavi, Y. (1989). Lithium-sulfur battery: evaluation of dioxolane-based electrolytes. *J. Electrochem. Soc.* **136**: 1621.

52 Weng, W., Pol, V.G., and Amine, K. (2013). Ultrasound assisted design of sulfur/carbon cathodes with partially fluorinated ether electrolytes for highly efficient Li/S batteries. *Adv. Mater.* 25: 1608–1615.

53 Yim, T., Park, M.-S., Yu, J.-S. et al. (2013). Effect of chemical reactivity of polysulfide toward carbonate-based electrolyte on the electrochemical performance of Li–S batteries. *Electrochim. Acta* 107: 454–460.

54 Xin, S., Gu, L., Zhao, N.-H. et al. (2012). Smaller sulfur molecules promise better lithium-sulfur batteries. *J. Am. Chem. Soc.* 134: 18510–18513.

55 Xu, K. (2014). Electrolytes and interphases in Li-ion batteries and beyond. *Chem. Rev.* 114: 11503–11618.

56 Yamada, Y., Furukawa, K., Sodeyama, K. et al. (2014). Unusual stability of acetonitrile-based Superconcentrated electrolytes for fast-charging lithium-ion batteries. *J. Am. Chem. Soc.* 136: 5039–5046.

57 Cuisinier, M., Cabelguen, P.E., Adams, B.D. et al. (2014). Unique behaviour of nonsolvents for polysulphides in lithium–sulphur batteries. *Energy Environ. Sci.* 7: 2697–2705.

58 Lee, C.W., Pang, Q., Ha, S. et al. (2017). Directing the lithium-sulfur reaction pathway via sparingly solvating electrolytes for high energy density batteries. *ACS Cent. Sci.* 3: 605–613.

59 (a) See, K.A., Wu, H.L., Lau, K.C. et al. (2016). Effect of hydrofluoroether cosolvent addition on Li solvation in acetonitrile-based solvate electrolytes and its influence on S reduction in a Li-S battery. *ACS Appl. Mater. Interfaces* 8: 34360–34371; (b)Shin, M., Wu, H.L., Narayanan, B. et al. (2017). Effect of the hydrofluoroether cosolvent structure in acetonitrile-based solvate electrolytes on the Li(+) solvation structure and Li-S battery performance. *ACS Appl. Mater. Interfaces* 9: 39357–39370.

60 Yoon, S., Lee, Y.-H., Shin, K.-H. et al. (2014). Binary sulfone/ether-based electrolytes for rechargeable lithium-sulfur batteries. *Electrochim. Acta* 145: 170–176.

61 Pan, H., Wei, X., Henderson, W.A. et al. (2015). On the way toward understanding solution chemistry of lithium polysulfides for high energy Li-S redox flow batteries. *Adv. Energy Mater.* 5: 1500113.

62 Weller, C., Thieme, S., Härtel, P. et al. (2017). Intrinsic shuttle suppression in lithium-sulfur batteries for pouch cell application. *J. Electrochem. Soc.* 164: A3766–A3771.

63 Zou, Q. and Lu, Y.-C. (2016). Solvent-dictated lithium sulfur redox reactions: an operando UV–vis spectroscopic study. *J. Phys. Chem. Lett.* 7: 1518–1525.

64 Peng, Z., Freunberger, S.A., Chen, Y., and Bruce, P.G. (2012). A reversible and higher-rate Li-O_2 battery. *Science* 337: 563–566.

65 Yuan, L.X., Feng, J.K., Ai, X.P. et al. (2006). Improved dischargeability and reversibility of sulfur cathode in a novel ionic liquid electrolyte. *Electrochem. Commun.* 8: 610–614.

66 Park, J.-W., Ueno, K., Tachikawa, N. et al. (2013). Ionic liquid electrolytes for lithium-sulfur batteries. *J. Phys. Chem. C* 117: 20531–20541.

67 Dokko, K., Tachikawa, N., Yamauchi, K. et al. (2013). Solvate ionic liquid electrolyte for Li–S batteries. *J. Electrochem. Soc.* 160: A1304–A1310.

68 Xu, R., Yue, J., Liu, S. et al. (2019). Cathode-supported all-solid-state lithium–sulfur batteries with high cell-level energy density. *ACS Energy Lett.* 4: 1073–1079.

69 Yao, X., Huang, N., Han, F. et al. (2017). High-performance all-solid-state lithium-sulfur batteries enabled by amorphous sulfur-coated reduced graphene oxide cathodes. *Adv. Energy Mater.* 7: 1602923.

70 Fu, K., Gong, Y.H., Hitz, G.T. et al. (2017). Three-dimensional bilayer garnet solid electrolyte based high energy density lithium metal-sulfur batteries. *Energy Environ. Sci.* 10: 1568–1575.

71 Song, Y.-X., Shi, Y., Wan, J. et al. (2019). Direct tracking of the polysulfide shuttling and interfacial evolution in all-solid-state lithium–sulfur batteries: a degradation mechanism study. *Energy Environ. Sci.* 12: 2496–2506.

72 (a) Mo, Y., Ong, S.P., and Ceder, G. (2012). First principles study of the $Li_{10}GeP_2S_{12}$ lithium super ionic conductor material. *Chem. Mater.* 24: 15–17; (b) Tao, Y., Chen, S., Liu, D. et al. (2016). Lithium superionic conducting oxysulfide solid electrolyte with excellent stability against lithium metal for all-solid-state cells. *J. Electrochem. Soc.* 163: A96–A101.

73 (a) Croce, F., Appetecchi, G.B., Persi, L., and Scrosati, B. (1998). Nanocomposite polymer electrolytes for lithium batteries. *Nature* 394: 456–458; (b) Jeong, S.S., Lim, Y., Choi, Y.J. et al. (2007). Electrochemical properties of lithium sulfur cells using PEO polymer electrolytes prepared under three different mixing conditions. *J. Power Sources* 174: 745–750.

74 Lin, D., Liu, W., Liu, Y. et al. (2016). High ionic conductivity of composite solid polymer electrolyte via in situ synthesis of monodispersed SiO_2 nanospheres in poly(ethylene oxide). *Nano Lett.* 16: 459–465.

75 Hassoun, J. and Scrosati, B. (2010). Moving to a solid-state configuration: a valid approach to making lithium-sulfur batteries viable for practical applications. *Adv. Mater.* 22: 5198.

76 Tao, X., Liu, Y., Liu, W. et al. (2017). Solid-state lithium sulfur batteries operated at 37 degrees C with composites of nanostructured $Li_7La_3Zr_2O_{12}$/carbon foam and polymer. *Nano Lett.* 17: 2967–2972.

77 Liu, W., Liu, N., Sun, J. et al. (2015). Ionic conductivity enhancement of polymer electrolytes with ceramic nanowire fillers. *Nano Lett.* 15: 2740–2745.

78 Mikhaylik Y. (2008) *US Patent, 7,352,680.*

79 Cheng, X.-B., Yan, C., Chen, X. et al. (2017). Implantable solid electrolyte interphase in lithium-metal batteries. *Chem* 2: 258–270.

80 Wu, H.-L., Haasch, R.T., Perdue, B.R. et al. (2017). The effect of water-containing electrolyte on lithium-sulfur batteries. *J. Power Sources* 369: 50–56.

81 Choi, J.-W., Cheruvally, G., Kim, D.-S. et al. (2008). Rechargeable lithium/sulfur battery with liquid electrolytes containing toluene as additive. *J. Power Sources* 183: 441–445.

82 Lin, Z., Liu, Z., Fu, W. et al. (2013). Phosphorous Pentasulfide as a novel additive for high-performance lithium-sulfur batteries. *Adv. Funct. Mater.* 23: 1064–1069.

83 Liu, S., Li, G.-R., and Gao, X.-P. (2016). Lanthanum nitrate as electrolyte additive to stabilize the surface morphology of lithium anode for lithium-sulfur battery. *ACS Appl. Mater. Interfaces* 8: 7783–7789.

84 Wu, F., Lee, J.T., Nitta, N. et al. (2015). Lithium iodide as a promising electrolyte additive for lithium-sulfur batteries: mechanisms of performance enhancement. *Adv. Mater.* 27: 101–108.

85 Ren, Y.X., Zhao, T.S., Liu, M. et al. (2017). A self-cleaning Li-S battery enabled by a bifunctional redox mediator. *J. Power Sources* 361: 203–210.

86 Wu, H.-L., Shin, M., Liu, Y.-M. et al. (2017). Thiol-based electrolyte additives for high-performance lithium-sulfur batteries. *Nano Energy* 32: 50–58.

87 Song, J., Noh, H., Lee, H. et al. (2015). Polysulfide rejection layer from alpha-lipoic acid for high performance lithium-sulfur battery. *J. Mater. Chem. A* 3: 323–330.

88 Li, G., Wang, S., Zhang, Y. et al. (2018). Revisiting the role of polysulfides in lithium-sulfur batteries. *Adv. Mater.* 30: 1705590.

89 Meini, S., Elazari, R., Rosenman, A. et al. (2014). The use of redox mediators for enhancing utilization of Li_2S cathodes for advanced Li-S battery systems. *J. Phys. Chem. Lett.* 5: 915–918.

90 Liu, M., Ren, Y.X., Jiang, H.R. et al. (2017). An efficient Li_2S-based lithium-ion sulfur battery realized by a bifunctional electrolyte additive. *Nano Energy* 40: 240–247.

91 Yang, W., Yang, W., Song, A. et al. (2017). Pyrrole as a promising electrolyte additive to trap polysulfides for lithium-sulfur batteries. *J. Power Sources* 348: 175–182.

92 Fei, H., An, Y., Feng, J. et al. (2016). Enhancing the safety and electrochemical performance of ether based lithium sulfur batteries by introducing an efficient flame retarding additive. *RSC Adv.* 6: 53560–53565.

93 Yang, H., Guo, C., Chen, J. et al. (2019). An intrinsic flame-retardant organic electrolyte for safe lithium-sulfur batteries. *Angew. Chem. Int. Ed.* 58: 791–795.

94 Xiong, S., Xie, K., Diao, Y., and Hong, X. (2014). Characterization of the solid electrolyte interphase on lithium anode for preventing the shuttle mechanism in lithium–sulfur batteries. *J. Power Sources* 246: 840–845.

95 Li, W., Yao, H., Yan, K. et al. (2015). The synergetic effect of lithium polysulfide and lithium nitrate to prevent lithium dendrite growth. *Nat. Commun.* 6: 7436.

96 Lv, D., Shao, Y., Lozano, T. et al. (2015). Failure mechanism for fast-charged lithium metal batteries with liquid electrolytes. *Adv. Energy Mater.* 5: 1400993.

97 Xiong, S., Xie, K., Diao, Y., and Hong, X. (2013). On the role of polysulfides for a stable solid electrolyte interphase on the lithium anode cycled in lithium–sulfur batteries. *J. Power Sources* 236: 181–187.

98 Qie, L., Zu, C., and Manthiram, A. (2016). A high energy lithium-sulfur battery with ultrahigh-loading lithium polysulfide cathode and its failure mechanism. *Adv. Energy Mater.* 6: 1502459.

99 Lee, Y.M., Choi, N.S., Park, J.H., and Park, J.K. (2003). Electrochemical performance of lithium/sulfur batteries with protected Li anodes. *J. Power Sources* 119: 964–972.

100 Ma, G., Wen, Z., Wu, M. et al. (2014). A lithium anode protection guided highly-stable lithium-sulfur battery. *Chem. Commun.* 50: 14209–14212.

101 Zhao, J., Liao, L., Shi, F. et al. (2017). Surface fluorination of reactive battery anode materials for enhanced stability. *J. Am. Chem. Soc.* 139: 11550–11558.

102 Lu, L.L., Ge, J., Yang, J.N. et al. (2016). Free-standing copper nanowire network current collector for improving lithium anode performance. *Nano Lett.* 16: 4431–4437.

103 Liu, S., Yang, J., Yin, L. et al. (2011). Lithium-rich $Li_{2.6}BMg_{0.05}$ alloy as an alternative anode to metallic lithium for rechargeable lithium batteries. *Electrochim. Acta* 56: 8900–8905.

104 Aurbach, D., Markovsky, B., Weissman, I. et al. (1999). On the correlation between surface chemistry and performance of graphite negative electrodes for Li ion batteries. *Electrochim. Acta* 45: 67–86.

105 Ming, J., Cao, Z., Wahyudi, W. et al. (2018). New insights on graphite anode stability in rechargeable batteries: Li ion coordination structures prevail over solid electrolyte interphases. *ACS Energy Lett.* 3: 335–340.

106 (a) Yamada, Y., Usui, K., Chiang, C.H. et al. (2014). General observation of lithium intercalation into graphite in ethylene-carbonate-free superconcentrated slectrolytes. *ACS Appl. Mater. Interfaces* 6: 10892–10899; (b) Liu, X.-R., Wang, L., Wan, L.-J., and Wang, D. (2015). In situ observation of electrolyte-concentration-dependent solid electrolyte interphase on graphite in dimethyl sulfoxide. *ACS Appl. Mater. Interfaces* 7: 9573–9580.

107 Hassoun, J. and Scrosati, B. (2010). A high-performance polymer tin sulfur lithium ion battery. *Angew. Chem. Int. Ed.* 49: 2371–2374.

108 Hassoun, J., Derrien, G., Panero, S., and Scrosati, B. (2008). A nanostructured Sn-C composite lithium battery electrode with unique stability and high electrochemical performance. *Adv. Mater.* 20: 3169–3175.

109 Yang, Y., McDowell, M.T., Jackson, A. et al. (2010). New nanostructured Li_2S/silicon rechargeable battery with high specific energy. *Nano Lett.* 10: 1486–1491.

110 Mikhaylik, Y.V. and Akridge, J.R. (2004). Polysulfide shuttle study in the Li/S battery system. *J. Electrochem. Soc.* 151: A1969–A1976.

111 Gupta, A., Bhargav, A., and Manthiram, A. (2019). Highly solvating electrolytes for lithium-sulfur batteries. *Adv. Energy Mater.* 9: 1803096.

112 Cheng, L., Curtiss, L.A., Zavadil, K.R. et al. (2016). Sparingly solvating electrolytes for high energy density lithium-sulfur batteries. *ACS Energy Lett.* 1: 503–509.

113 Qi, X., Yang, Y., Jin, Q. et al. (2020). Two-plateau Li-se chemistry for high volumetric capacity se cathodes. *Angew. Chem. Int. Ed.* 59: 2–9.

4

Na-Ion Battery

Xiaochuan Duan[1], *Lei Wang*[2], *and Jianmin Ma*[3]

[1]*College of Materials Science and Engineering, Taiyuan University of Technology, No.79 West Street Yingze, Taiyuan 030024, P. R. China*
[2]*Hunan University, State Key Laboratory for Chemo/Biosensing and Chemometrics, and College of Chemistry and Chemical Engineering, Hunan Key Laboratory of Two-Dimensional Materials, Chengdu 410082, P. R. China*
[3]*University of Electronic Science and Technology of China, School of Materials and Energy, Changsha 611731, P. R. China*

4.1 Introduction

4.1.1 History of Sodium-Ion Batteries

Lithium-ion batteries (LIBs) are widely used in portable electronic devices, wearable electronic devices, electric vehicles, and other fields due to their high energy density and power density, which profoundly affected people's daily lives. However, the rapid development of mobile electronic devices and new energy vehicles requires higher performance of LIBs, such as higher energy density and longer cycle life. Especially with the development of new energy vehicle companies such as Tesla, the field of electrochemical energy storage is experiencing a real revolution. Not only that, the further promotion of LIBs has also accelerated the consumption of global lithium resources. Up to now, the content of lithium in the earth's crust is only about two hundred thousandths, and its global resource distribution is uneven, mainly distributed in South America. According to the consumption rate and reserves of existing lithium sources, lithium sources can only last less than 70 years. The rapid consumption of lithium resources inevitably makes the manufacturing cost of LIBs rise accordingly. These factors will undoubtedly hinder the large-scale energy storage of LIBs application. Therefore, the development of new secondary batteries that can replace LIBs with a wide range of raw material resources and low prices to meet the needs of large-scale energy storage has become a hot spot in the academic and industrial circles. In recent years, sodium-ion batteries (SIBs) have received increasing attention from researchers due to their abundant sodium resource reserves, suitable electrochemical platforms, low prices, and similar energy storage mechanisms to LIBs [1]. Therefore, SIBs are considered to be a promising alternative to LIBs as the next generation of energy storage system [2].

Battery Technologies: Materials and Components, First Edition. Edited by Jianmin Ma.

The research on SIBs can be traced back to the 1970s, even earlier than the research on LIBs. However, the larger ion radius and molar mass of sodium ions lead to slow ion diffusion kinetics and limited cycle life, which hindered the development of SIBs [3]. On one hand, the heavier mass and more negative potential sodium atoms make it difficult to find suitable sodium storage materials with higher sodium storage capacity and suitable working potential; on the other hand, the radius of sodium ion (1.02 Å) is larger than that of lithium ion (0.76 Å), which makes sodium ions exhibit slow insertion and insertion kinetics in the main lattice of the electrode material, thereby reducing the rate performance of the battery. To find more suitable electrodes for SIBs, researchers believed that sodium ions and lithium ions have similar physical and chemical properties, and many lithium storage materials should have sodium storage properties. However, as the study found, the electrode materials used in LIBs usually exhibit different storage mechanisms and ion dynamics when applied to SIBs, which in turn affects the overall performance of the battery. For example, Ceder and coworkers used first-principles calculations to find that the diffusion impedance of sodium ion migration in layered oxides is lower than that of lithium ion migration, although sodium ions have a larger ion radius and mass [4]. In addition, they also found that the electrochemical performance of the sodium ion intercalation compound is very sensitive to the phase structure of the material, and the performance of the battery can be improved by adjusting the chemical composition and lattice structure of the material. To date, the development of SIBs is still in its infancy, and finding suitable sodium ion battery electrode materials is the key to promoting the practical application of SIBs. At the same time, it involves the in-depth study of the mechanism of action of sodium ions and electrode materials, development of sodium ion intercalation, conversion, alloying, and other electrochemical theories in order to provide the basic understanding and theoretical guidance for the design of new sodium ion electrode materials. The challenge for the commercialization of SIBs is to explore suitable electrode materials with high capacity, long cycle stability, suitable voltage platform, and high safety, while optimizing the composition of the electrolyte [5].

4.1.2 Composition and Working Mechanism of SIBs

Generally, the cell structure of SIBs is similar to LIBs, which is mainly composed of the cathode materials, anode materials, separator, electrolyte, packaging, and other parts (Figure 4.1) [6]. In particular, the cathode and anode materials are critical for SIBs that directly determine the performance of SIBs. The working mechanism of SIBs is also similar as LIBs that based on the rocking chair battery model proposed by M. Armand. The working principle involves that sodium ions shuttle back and forth between the cathode and anode materials through the electrolyte during charging/discharging cycles. During the charging process, sodium ions are deintercalated from the cathode material under the external electric field, then enter the electrolyte and pass through the separator, and electrochemically react with the anode materials while the electrons provided by the power supply. At this time, the anode material is in a sodium-rich state. During the discharging process, the anode material releases sodium ions and electrons, and the sodium ions are combined with the

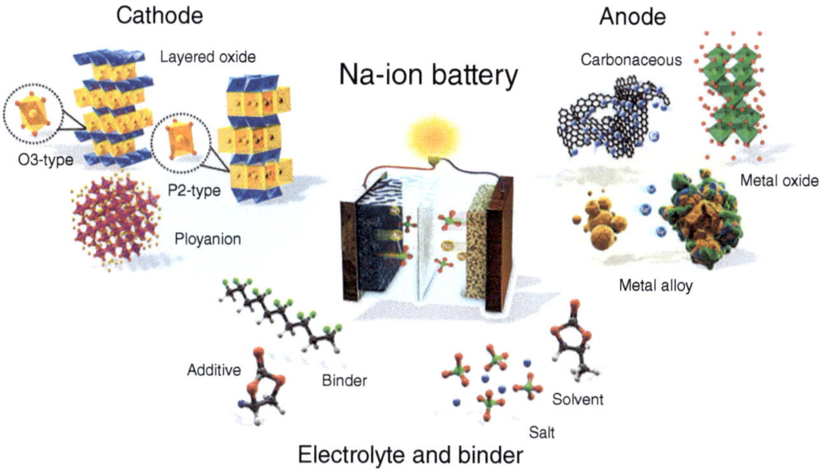

Figure 4.1 Representative illustration of SIBs. Source: Hwang et al. [6]. Licenced under CC BY 3.0.

cathode material through the electrolyte, and thus stored again in the cathode material, thereby realizing an electrochemical cycle.

Compared with LIBs, the advantages of SIBs generally have the following aspects:

- The low cost of SIBs makes it more suitable for large-scale energy storage systems. On one hand, sodium reserves are abundant, evenly distributed, and low cost; on the other hand, the current collectors for both cathode and anode materials in SIBs can use cheap aluminum foil since aluminum does not react with sodium at low potential.
- The manufacturing process of SIBs is compatible with current commercial LIBs, since the working principle of SIBs is similar to that of LIBs.
- The solvation energy of sodium ions is lower than that of lithium ions, that is, it has a better interfacial ion diffusion ability.
- The Stokes diameter of sodium ions is smaller than that of lithium ions, and the solvation energy of sodium ions is lower than that of lithium ions, that is, it has better interfacial ion diffusion ability.
- The internal resistance of SIBs is slightly higher than that of LIBs, resulting in less heat and lower temperature rise in safety tests such as short circuits, thus achieving good safety performance.

Overall, the challenge for the commercialization of SIBs is to explore positive and negative electrode materials with high capacity, long cycle stability, suitable voltage platform, and high safety, while optimizing the composition of the electrolyte.

4.2 Cathode Materials for SIBs

Cathode material, is one of the main components of SIBs, serves as a host accommodating sodium ions and a source of active Na^+ for the entire battery system. The reversible intercalation/deintercalation behavior of Na^+ ions and corresponding electron transfer is the operation mode of the cathodes, which constitutes a complete electrochemical sodium storage process together with the functions of anodes. Thus, the cathode materials determine the performance of the full cell to a large extent. Superior cathode materials ought to be qualified with high capacity, remarkable rate properties, and excellent durability, etc.

The theoretical specific capacity (Q) of cathode materials can be calculated by the following equation:

$$Q = \frac{nF}{3.6M_w}$$

where F is the Faraday constant (96 485 C mol^{-1}), n is the number of electrons transferred, and M_w is the relative molecular mass. Q is directly proportional to n and inversely proportional to M_w. Smaller molecular weight and larger electron transfer number are the prerequisites for obtaining high specific capacity.

In addition, the cycle stability and rate performance of the cathodes always depend on its electrochemical stability and charge (electrons and ions) transport capability. These are the basis for the selection and design of cathode materials. In the following section, we classify the four types of layered transition metal oxide, phosphates/fluorophosphate, hexacyanoferrate, and organic compounds, detailing the characteristics of various cathodes and the involved electrochemical behaviors.

4.2.1 Layered Transition Metal Oxide

Layered sodium transition metal oxides became the most common type of cathode for SIBs with characteristics of simple structure, large capacity, easy synthesis [7–10]. Their general formula (Na_xMO_2) was first established by C. Delmas and coworkers [11], where M usually represents a single or binary or ternary or more transition metal elements, such as Mn, Ni, Co, and Cr. Typically, the edge-shared MO_6 octahedra are stacked to form MO_2 layers with a 2D channel for accommodating Na^+ ions. Based on the coordination environment of sodium and the number of MO_2 layers in a single stacking unit, layered metal oxides exhibit multiple crystal structures, such as O3, O2, P2, and P3 phase.

Among these, P2 phase and O3 phase are the most common, where alkali sodium ions occupy prismatic site between two MO_2 layers and octahedral site between two and three MO_2 layers, respectively (Figure 4.2) [1]. In the O3-type phase, always at high Na content, MO_2 layers stack through ABCABC pattern, while Na geometries share one edge. In the P2-type phase, always at low Na content, MO_2 layers stack through ABAB pattern, while Na geometries share one face.

Different phases indicate significantly special electrochemistry due to the variable pristine amount of sodium and the distinct sodium occupation between MO_2 layers. Due to higher pristine contents of sodium, O3 phase usually can deliver a higher initial capacity. However, the migration of Na^+ ions in P2 phase possesses lower barrier by passing shared face directly to the next prismatic site, unlike passing through the adjacent octahedron in O3 phase. Moreover, P2 phase exhibits stronger electrochemical stability, considering that the O3 phase always undergoes a series of slab gliding during the de/intercalation processes.

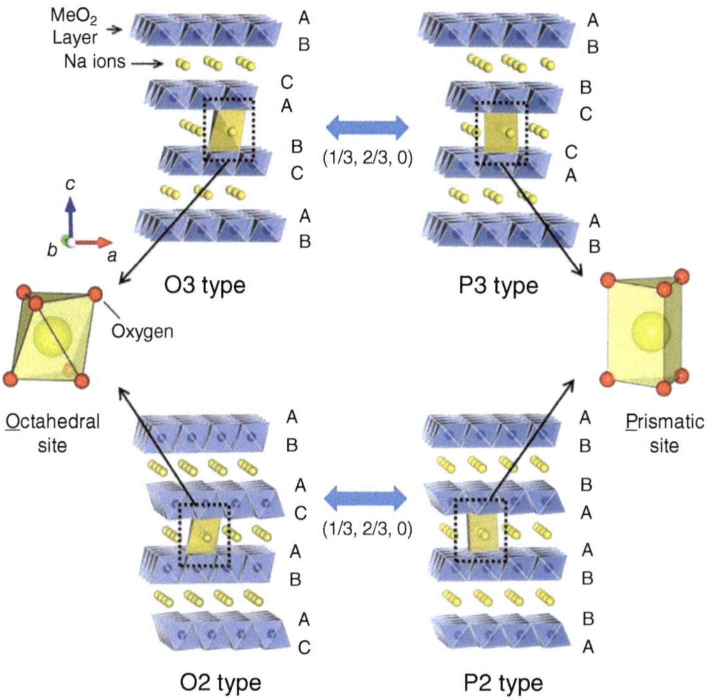

Figure 4.2 Crystal structure of O3-, O2-, P2-, and P3-type phase-layered sodium transition metal oxides. Source: Yabuuchi et al. [1]. Reproduced with permission of American Chemical Society.

Apart from the type of phase, transition metal elements also play a vital role in electrochemical performance due to the differences in valency, ionic radius, and chemical bonds with oxygen. In addition, from the individual differences of a single transition metal element to the synergistic effect of binary, ternary, or more transition metal elements, it also affects the quality of electrochemical performance. Here, we discuss in detail the application of layered sodium transition metal oxides in SIBs cathodes through some specific examples.

Sodium cobalt oxide (Na_xCoO_2, $0 \leq x \leq 1$) and sodium manganese oxide (Na_xMnO_2, $0 \leq x \leq 1$) were studied very early which can be traced back to the research in the 1980s [11, 12]. During the de-/intercalation of Na^+, a reversible phase transition of O3 → O′3 → P′3 was found in O3-$NaCoO_2$, while no obvious phase change occurred in the P2-Na_xCoO_2 [13, 14]. However, the capacity of sodium cobalt oxide tends to be less than 100 mAh g^{-1}. For the Na–Mn–O system, Mn^{3+}-induced Jahn–Teller effect has largely caused severe capacity degradation, although Na_xMnO_2 exhibited a high capacity benefit from the excellent electrochemical reaction of Mn^{3+}/Mn^{4+}. Otherwise, O3 or P2 phase Na_xFeO_2, Na_xNiO_2, Na_xCrO_2, and other layered oxides have also received many attentions.

Combined with the unique advantages of single-metal oxides, such as the high ion diffusion coefficient of Na–Co–O system, the high capacity of Na–Mn–O system, and the high redox potential of Na–Ni–O system, binary and ternary transition metal oxides further improve the electrochemical behaviors of the cathode

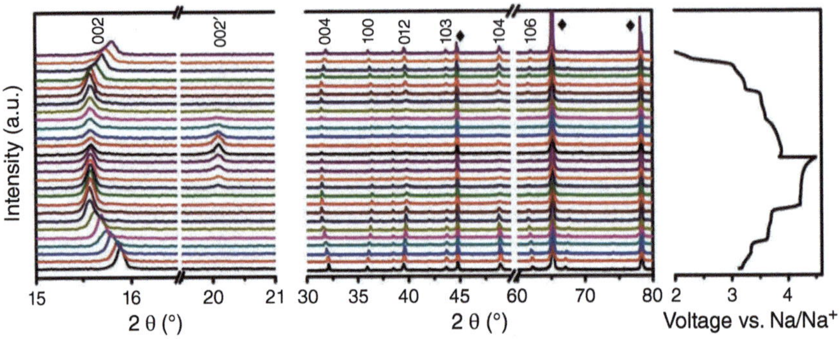

Figure 4.3 The transition from P2 phase to O2 phase of binary Na–Ni/Mn–O system based on in situ XRD analysis. Source: Wang et al. [17]. Reproduced with permission of Royal Society of Chemistry.

materials, thanks to the synergistic effect between multiple metals. Among them, the combinations of Ni, Fe, Mn, and Co are the most extensively studied. Komaba et al. reported that P2-$Na_{2/3}Fe_{1/2}Mn_{1/2}O_2$ could deliver a reversible capacity of 190 mAh g^{-1} in the sodium cells based on the electrochemically active Fe^{3+}/Fe^{4+} redox [15]. P2-phase $Na_xNi_{1/3}Mn_{2/3}O_2$ is considered to be a layered oxide that being the most widely studied, due to superior the $Ni^{2+}/Ni^{3+}/Ni^{4+}$ couple (>3.0 V vs. Na/Na^+) with high redox potential and high specific capacity [16]. However, a detrimental phase transition (P2–O2) occurred during deep charging (>4.2 V), as shown in Figure 4.3 [17]. Sodium ternary transition metal oxides were also proposed, such as O3-phase $NaNi_{1/3}Fe_{1/3}Mn_{1/3}O_2$ [18, 19] and P2-phase $Na_{0.63}Ni_{0.22}Co_{0.11}Mn_{0.66}O_2$ [20]. But the practical performance of these cathode materials is always limited by the instability in air, dissolution of transition metal elements, Jahn–Teller effect, irreversible phase transition, and ordered Na^+/vacancies.

High-valence metal ions have ultra-high oxidation, being easy to react with O_2 and/or H_2O in the air. Moreover, when a large amount of sodium is extracted from the metal oxide layer, the structural damaging is induced together with phase transition and Na^+/vacancies order, which eventually leads to irreversible capacity and attenuation. Doping or substitution in MO_2 layer is regarded as an effective way to suppress irreversible phase transition. The electrochemically inactive elements, such as Ti, Mg, and Al, are introduced into transition metal site as factors to stabilize the structure. The concentration of inactive metal elements needs to be precisely regulated to avoid excessive capacity loss. It is worth mentioning that the participation of certain active transition metal elements also plays a certain role in inhibiting the above-described issues.

4.2.2 Polyanionic Compounds

Polyanionic compounds possess a framework with high structural stability which is composed by strong P—O covalent bonds [21]. During the de/-intercalation of sodium ions, the stable framework ensures almost no volume expansion and phase change. So far, polyanionic compounds such as transition metal phosphates,

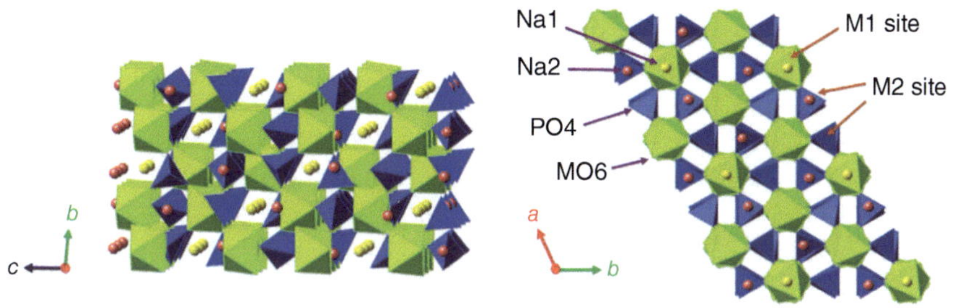

Figure 4.4 The crystal structure of NASICON-type $Na_3M_2(PO_4)_3$, M = V, Fe, Ti, etc. Source: Fang et al. [24]. Licenced under CC BY 4.0.

fluorophosphates, and polyphosphates, have been investigated as a host of sodium ions. For example, olivine-type $NaFePO_4$ and NASICON-type $Na_3V_2(PO_4)_3$ have gained more investigations among the single-transition metal phosphates. The specific electrochemical behavior is discussed below.

Lithium iron phosphate ($LiFePO_4$), as a typical phosphate, attracted much attention once being discovered. Because of its low cost, excellent thermal stability, security, and high specific capacity, it has won considerable recognition in the commercial market. Sodium iron phosphate ($NaFePO_4$), as an analog of $LiFePO_4$, is also beginning to be studied. Olivine-type $NaFePO_4$ can deliver a high theoretical capacity (154 $mAh\,g^{-1}$) with a moderate operating potential (2.9 V vs. Na^+/Na), depending on the Fe^{3+}/Fe^{2+} redox [22, 23]. An intermediate $Na_{2/3}FePO_4$ phase is generated during the Na^+ ions de-intercalation, leading to two potential plateaus in charge process.

NASICON was named by Na super ionic conductor, which has high Na ionic conductivity (~10^{-3} $S\,cm^{-1}$). For NASICON-type $Na_3V_2(PO_4)_3$, three PO_4 tetrahedra and VO_6 octahedra form a 3D $[V_2(PO_4)_3]^-$ framework by sharing corners, where Na^+ ion being at two different interstitial sites, as shown in Figure 4.4 [24]. After $Na_3V_2(PO_4)_3$ being firstly synthesized by Delmas et al. the Na storage properties of $Na_3V_2(PO_4)_3$ were first investigated by Yamaki and coworkers [25]. Similar to $Na_3V_2(PO_4)_3$, other NASICON-type compounds, such as $Na_3Fe_2(PO_4)_3$ and $NaTi_2(PO_4)_3$, were also reported as cathodes for SIBs [26–28]. However, the overall electrochemical properties of NASICON-type electrodes are limited by the intrinsic low electronic conductivity due to the presence of phosphate skeleton. Carbon coating and metal element doping or substitution have proven to be effective strategies to improve electrical conductivity, ultimately improving the performance of this kind of cathodes.

The combination of fluoride ion and phosphate creates another type of polyanion compound (fluorophosphates), being as positive electrode for SIBs. The research of fluorophosphates is mainly focused on Na_2FePO_4F and $Na_3V_2(PO_4)_2F_3$ [29–32]. As shown in Figure 4.5, $FePO_4F$ layers are composed of FeO_4F_2 octahedra and PO_4 tetrahedra, while Na^+ ions occupy in a 2-D interlayer space [1]. The electrochemical storage sodium properties of Na_2FePO_4F were evaluated for the first time by Tarascon and coworkers. Although composite with carbon is considered

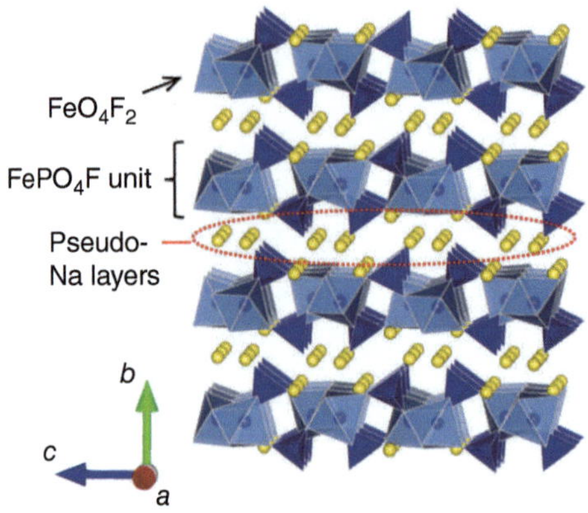

Figure 4.5 The layered structure of Na_2FePO_4F. Source: Yabuuchi et al. [1]. Reproduced with permission of American Chemical Society.

to be effective in further improving the performance, the capacity still remained at about 100 mAh g^{-1}. $Na_3V_2(PO_4)_2F_3$ (Na vanadium fluorophosphates) is originated from fluoride-substituted NASICON-type $Na_3V_2(PO_4)_3$, equipping with three-dimensional sodium ion accommodating space. The $Na_3V_2(PO_4)_2F_3$ cathode can deliver a high operating potential (~3.9 V vs. Na/Na^+) based on two charging voltage plateaus at 3.7 and 4.2 V, which is the basis for obtaining ultra-high energy density. However, the V^{3+}/V^{4+} redox between 2.0 and 4.5 V only offers a relatively low theoretical capacity of 128 mAh g^{-1}, which is consistent with that two-thirds of Na^+ ions can be reversibly extracted from the host materials. Modification by oxygen substitution at the site of fluorine has been proved to be an effective method. $Na_3(VO_x)_2(PO_4)_2F_{3-2x}$ ($0 \leq x \leq 1$) with the addition of V^{4+} exhibits more excellent sodium storage properties based on the mixed $V^{3+}/V^{4+}/V^{5+}$ redox couples. For instance, $Na_3(VO_{0.8})_2(PO_4)_2F_{1.4}$ was proposed to achieve 1.2-electron transfer, resulting a high energy density [33].

Pyrophosphate is also a class of polyanionic compounds used in the positive electrodes of SIBs. The general formula of pyrophosphates can be written as $Na_2MP_2O_7$ (M = transition metal), where the transition metal oxide layers MO_6 and P_2O_7 units form a 3D space to accommodate Na^+ ions. $Na_2FeP_2O_7$ was first reported by Yamada and coworkers as cathode for SIBs, delivering a reversible capacity of 82 mAh g^{-1} based on the $Fe^{3+/2+}$ redox mainly at 2.5–3 V vs. Na/Na^+ [34]. For each molecule of $Na_2FeP_2O_7$, only one-half of sodium being extracted results in a theoretical capacity of only 97 mAh g^{-1}, which is the reason for the low specific capacity. $Na_2MnP_2O_7$, $Na_2CoP_2O_7$, and $NaVP_2O_7$ also were proposed, but their capacities are all limited to less than 100 mAh g^{-1} [35–37]. It should be noted that these cathodes tend to have higher average voltage platform (>3.8 V).

4.2.3 Hexacyanoferrates

Sodium transition metal hexacyanoferrates, as a type of Prussian blue analogs (PBAs), possesses a general chemical formula of $Na_xM[Fe(CN)_6]_{1-y} \cdot nH_2O$ ($0 < x \leq 2$, $0 \leq y < 1$), where y represents the amount of $[Fe(CN)_6]$ vacancy [38, 39]. All of Na^+ ions occupy 8c sites and are transferred by a wide 3D diffusion channel in the framework. In the electrochemical process, $Na_xM[Fe(CN)_6]_{1-y}\cdot nH_2O$ can theoretically involve two electron transfer or two sodium ion de-/intercalation, delivering a theoretical specific capacity of 170 mAh g^{-1}.

However, the actual capacities are much lower than the theoretical value, due to the primary sodium content, the presence of vacancies, and H_2O molecules, which is related to the common co-precipitation synthesis. Reasonably controlling the content of vacancies and coordinated water is the focus of current research. Among them, doping with electrochemically inactive metals proved to be effective. In addition, the different transition metal elements and their controlling compositions can effectively improve the electron transfer and ion transport of sodium transition metal hexacyanoferrates, ultimately improving the performance of electrochemical sodium storage.

4.2.4 Organic Compounds

Organic compounds, as a class of metal-free electrode materials, have received much attention [40, 41]. Compared with traditional inorganic materials, organic materials are composed of naturally abundant elements (C, H, O, N, S, etc.), and the preparation process is simpler. In addition, organic materials have structural diversity, flexible molecular structure, which facilitates obtaining fast reaction kinetics, even for Na^+ ions with a large ionic radius. More importantly, the low molecular weights and involved multi-electron reactions are the guarantee of high specific capacity and power/energy density. In general, the charge storage mechanism of organic compounds is divided into two categories: cations de-/intercalation reaction and anions de-/intercalation mechanism. Disodium rhodizonate ($Na_2C_6O_6$) is a typical Na^+-deintercalation-type cathode, having high capacities, because no less than 2 sodium per molecule participates in the electrochemical reaction [42, 43]. Meanwhile, many polymers and copolymers with high work potentials (>3.0 V vs. Na/Na^+) were proposed as anions-de-/intercalation-type cathodes.

4.3 Anode Materials for SIBs

At present, it is still a challenge to find a suitable anode material for SIBs. Owing to the success of graphite anode for commercial LIBs, people first explored graphite anode as ideal anode material for SIBs. However, the experiment found that sodium ions are difficult to reversibly insert and extract in ordinary graphite anode materials, because the sodium ion diameter (2.04 Å) is larger than the graphite interlayer channel size (1.86 Å); in addition, previous studies also showed that the metal sodium plating layer is much easier to form than sodium ions embedded in the graphite layer, which means that ordinary graphite does not have considerable sodium storage performance. Although sodium metal has an electrode potential similar to that of lithium metal (−2.714 V, relative to the standard hydrogen electrode), its theoretical energy density is low, and when sodium metal is used as the anode in the full battery system, the battery's cycle performance is low. Moreover, the high reactivity of metallic sodium in organic electrolytes makes sodium dendrites form during the deposition of sodium metal, thereby causing battery safety issues. Therefore, relative to the research on stable anode materials for SIBs, the development of anode materials with high capacity, high structural stability, and suitable working potential is particularly important for the commercialization of SIBs. Up to now, a variety of anode materials including carbon-based materials and metal oxygen/sulfur/phosphorus/fluoride have been studied. According to their sodium storage mechanism, they are mainly divided into the following three categories: (i) insertion anode materials; (ii) conversion anode materials; and (iii) alloying anode materials.

4.3.1 Insertion Anode Materials

4.3.1.1 Carbon Materials

Graphite. Although graphite with interlayer distance of 0.34 nm is not suitable for the storage of sodium ions, the graphite structure can be modified to realize the application of graphite and graphite-like structures in SIBs. Wen et al. increased the interlayer distance of graphite to 0.43 nm by means of oxidation and subsequent heat treatment (Figure 4.6) [44]. This expanded graphite still maintains a long-range order in structure and can maintain a stable structure during electrochemical charge and discharge processes. As a result, the expanded graphite can achieve a high specific capacity of 284 $mAh\,g^{-1}$ at a high current density of 20 $mA\,g^{-1}$, and can maintain 74% capacity after 2000 cycles. The excellent sodium storage performance is due to the mechanism of sodium ion insertion and extraction reactions with small volume changes, and also the oxygen-containing functional groups and defect sites in the expanded graphite that contribute to a certain sodium storage capacity. In addition, the high diffusion energy barrier of sodium ions in the graphite anode and the high formation energy of the sodium–graphite intercalation compound make the intercalation process of sodium in graphite anode far from thermodynamic equilibrium process, so sodium ion cannot be effectively intercalated in graphite anode materials [45].

Hard carbon. The carbon crystallites in hard carbon have less carbon sheet stacking in the *c*-axis direction and exhibit a random orientation, so there are more voids

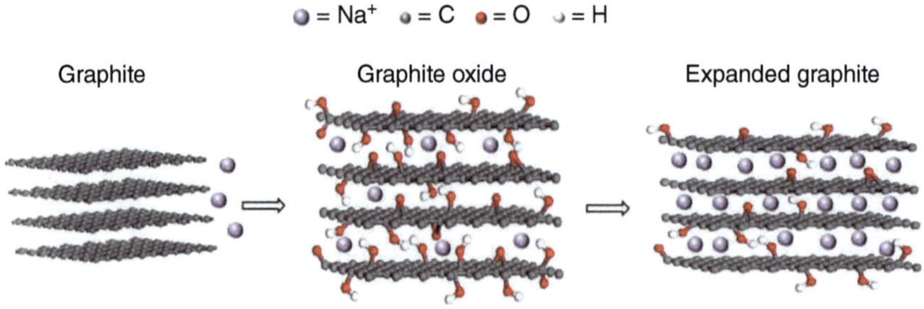

Figure 4.6 Schematic illustration of sodium storage in graphite-based materials with enlarged interlayer spacing. Source: Wen et al. [44]. Reproduced with permission of Springer Nature.

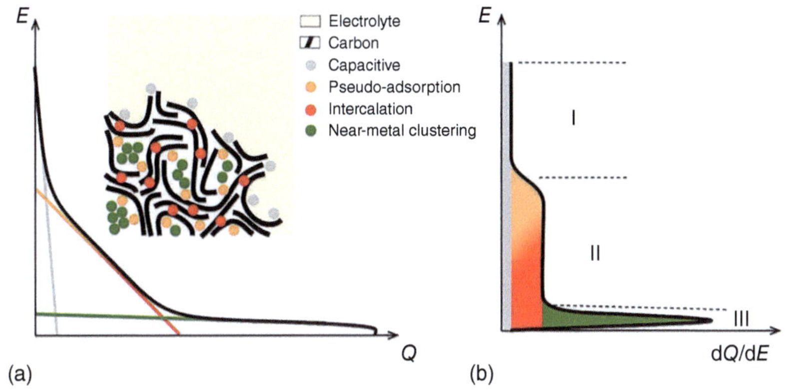

Figure 4.7 Schematic illustration of typical constant current chronopotentiogram of (a) hard carbon and (b) its derivative curve, with the three main voltage regions highlighted: initial drop (I), sloping voltage (II), and low voltage plateau (III). Source: Saurel et al. [46]. Reproduced with permission of John Wiley & Sons.

inside for electrochemical storage of sodium ions. Hard carbon as anode material for SIBs has the advantages of high sodium storage capacity and low sodium storage potential (has a higher sodium storage capacity in the 0.1 V platform area), which is considered to be an ideal anode material for SIBs. Based on the previous studies, there are three mechanisms of sodium storage in hard carbon anodes: sodium ion adsorption mechanism of surface active sites, sodium pore filling/adsorption mechanism in nanopores, and interlayer sodium storage mechanism. At present, there are different understandings of the sodium storage method and the corresponding area attribution in the specific capacity–voltage curve. Generally, the part with a larger slope in the charge–discharge curve corresponds to the insertion/extraction process of sodium ions in the nano-graphite stacking layer, while the part with a flat charge–discharge curve corresponds to the filling, adsorption, and inverse process of sodium ions in the amorphous graphite region. Saurel et al. systematically summarized the current main sodium storage process and mechanism (Figure 4.7) [46]: (i) sodium ions are stored in the place where the electrolyte on the surface of the material that can be wetted by capacitive adsorption; (ii) sodium ions are stored in the material near the surface by means of pseudocapacitance; (iii) sodium ions are stored by intercalation reaction; (iv) sodium ions form atomic clusters in closed pores. Since the mechanism of sodium storage proposed is mainly based on structural information obtained using ex situ characterization techniques (such as ex situ XRD (X-ray diffraction), HRTEM (high resolution transmission electron microscope), etc.), it is difficult to ensure the sodium storage status in hard carbon during the ex situ characterization test. Therefore, the structural information obtained may not be consistent with the actual situation. For the mechanism of sodium storage in hard carbon, there are still needed more studies related advanced characterization techniques to establish a more complete and accurate sodium storage model.

Soft carbon. The carbon platelets of carbon microcrystals inside the soft carbon exhibit the characteristics of short-range order and long-range disorder, so it is a

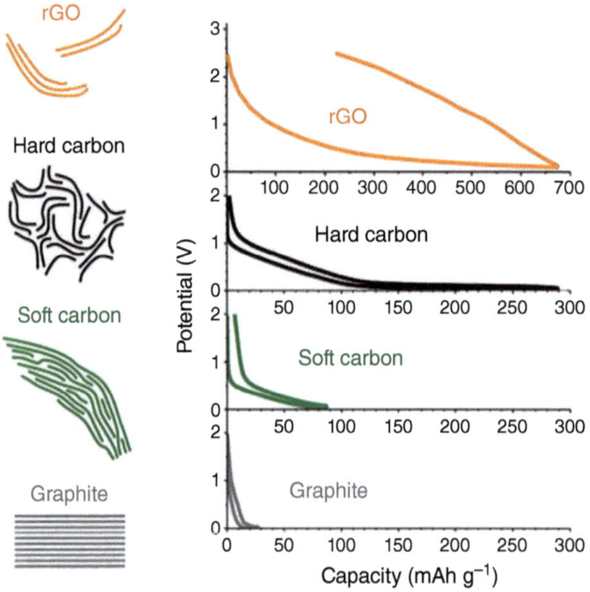

Figure 4.8 Graphical representation of the fine structure for the carbon anode materials, and their corresponding voltage charge profiles at second cycle in sodium half-cell. Source: Saurel et al. [46]. Reproduced with permission of John Wiley & Sons.

turbostratic structure. The structure of the carbon layer of this structure is better than hard carbon, so it has higher conductivity. In addition, soft carbon can be converted into high graphitization carbon after high-temperature heat treatment (above 2000 °C), so soft carbon is also called graphitizable carbon. Sodium storage behavior of soft charcoal exhibits a slope characteristic that the specific capacity gradually changes with voltage (Figure 4.8) [46]. The main sodium storage capacity of soft carbon anodes is concentrated above 0.2 V, which can effectively avoid the formation of sodium dendrites. In addition, when sodium ions are intercalated in the chaotic layer structure, it will cause expansion between the layers and part of the sodium ions will be trapped in the intercalation position. The slope of soft carbon storage area is more reversible than hard carbon materials. Soft carbon materials have better cycle performance and rate retention when storing sodium ions, and the use of soft carbon's high electrical conductivity to improve hard carbon materials to develop power sodium ion battery anode materials is the key development direction in the future.

Graphene. As a super star of the carbon material family, graphene is widely used in electrochemical energy storage due to its high specific surface area, excellent electrical conductivity, and good flexibility. Graphene can be used to efficiently construct the conductive network inside the energy storage material to improve the rate performance. The flexibility of graphene can buffer the volume expansion of the active material during the electrochemical reaction to improve the cycle performance. The basic sodium storage for graphene is the chemical adsorption of sodium ions at the functional groups or edges and defects on the surface. In order to further increase the storage capacity of graphene sodium, the introduction of heteroatom doping is a practical way. Various nitrogen-doped graphene, sulfur-doped graphene, and phosphorus-doped graphene have been reported in succession, confirming that heteroatom doping can increase capacity, interface wetting, and conductivity, and promote charge transfer and electrode–electrolyte reaction. Although graphene-based materials have considerable sodium storage performance, their lower density and first-time Coulomb efficiency limit graphene to be the main material for sodium storage in practical applications. At present, graphene is mainly used to construct composite materials or three-dimensional conductive networks and flexible frameworks in electrodes to achieve good performance anode materials for SIBs.

4.3.1.2 Titanium-Based Oxide

Titanium-based materials have become a very attractive type of insertion anode materials for SIBs because of their suitable sodium storage voltage, good cycle stability, high safety performance, high-volume energy density, and low environmental protection cost. Generally, titanium-based anode materials should be divided into three categories, including titanium dioxide (TiO_2), sodium titanate ($Na_2O{\cdot}nTiO_2$), and layered Na/Ti-based oxides.

Titanium dioxide. The crystalline structure of TiO_2 mainly includes anatase, rutile, brookite, and $TiO_2(B)$. The sodium storage properties of TiO_2 with different crystal types show obvious differences due to their different structures. According to the density functional theory (DFT), the energy barrier of anatase TiO_2 during sodium insertion is significantly lower than that of rutile TiO_2, and other studies have also shown that the anatase TiO_2 has the best cycle performance. Therefore, anatase TiO_2 can be used as a promising anode material for SIBs. The structure of anatase TiO_2 belongs to the tetragonal system and is based on octahedron TiO_6 (Figure 4.9). Each TiO_6 is connected with the surrounding eight TiO_6 (four of TiO_6 co-edge and four of TiO_6 co-angle). The continuous vacancies allow the TiO_2 unit cell to form a bidirectional void channel in the *a*-axis and *b*-axis directions, which can store more Na^+ and serve as an anode material for SIBs, thereby exhibiting higher capacity [47]. However, when Na^+ is inserted in these channels, no effective electric field is formed, so that Na^+ cannot be completely removed. Therefore, its first Coulombic efficiency and actual reversible capacity are low. At present, various effective strategies have been developed to improve its electrochemical performances, such as design a special morphology with a large specific surface area, compound it with a material with a high specific capacity and high conductivity, and doping with heteroatoms.

Sodium titanate. Generally, sodium titanate ($Na_xTi_yO_z$) can be divided into multiple crystal phases according to the sodium–titanium ratio, mainly Na_2TiO_3, $Na_2Ti_3O_7$, $Na_2Ti_6O_{13}$, $Na_4Ti_5O_{12}$, etc., which show obvious differences in chemical properties. Among them, the layered structure of $Na_2Ti_3O_7$ has the characteristics of high reversible capacity and low voltage platform (0.3 V vs. Na/Na^+), which becomes an ideal anode material for SIBs. Its crystal structure is formed by TiO_6 octahedral co-edges and common vertices forming a transition metal layer, in which every two Ti—O bonds are shared by three TiO_6 octahedrons and will gradually

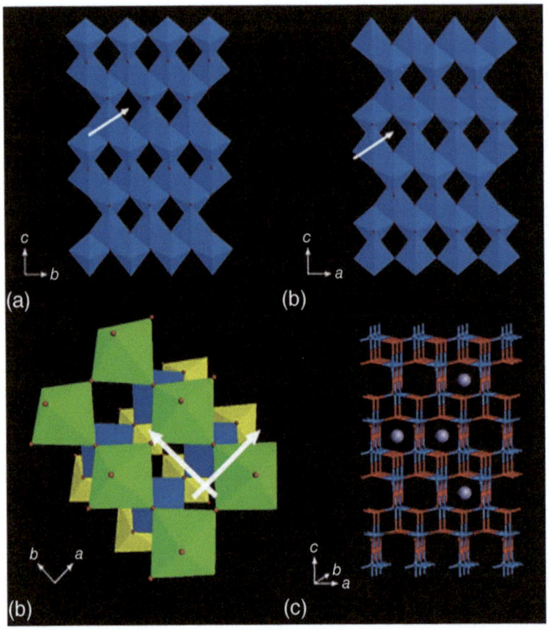

Figure 4.9 Crystal structures of anatase TiO_2 projected along (a) (100) and (b) (010), showing Na ions diffusion path; (c) perspective view projected along (001) (each layer of TiO_6 octahedra is shown in different colors), showing the interconnection between the diffusion paths along (100) and (010); (d) supercell structures in ball-stick mode, showing the possible filling sites of Na ions. Source: Xu et al. [47]. Reproduced with permission of Royal Society of Chemistry.

slide to form parallel axes, eventually forming a zigzag $(Ti_3O_7)^{2-}$ transition metal layer, the structure of which is stabilized by covalent bonds (Figure 4.10) [48]. The sodium layer is between the transition metal layers, forming two different embedding sites (Na1 and Na2), sodium ions can be reversibly deintercalated, and can form ionic bonds with the TiO_6 octahedron in the inner layer to make the material have a higher stability. During the first sodium extraction process, a clear voltage platform is generated at 0.3–0.4 V. The process is as follows:

$$Na_2Ti_3O_7 + (x-2)Na^+ + (x-2)e \leftrightarrow Na_xTi_3O_7$$

Pristine $Na_2Ti_3O_7$ has the problem of poor cycle performance. As an insertion-type anode material, although $Na_2Ti_3O_7$ did not break the chemical bond during sodium intercalation, the position of sodium intercalation changed significantly from the beginning 7 coordination and 9 coordination become 6 coordination. Therefore, the crystal structure of $Na_2Ti_3O_7$ is destroyed during the electrochemical reaction, resulting in poor cycle performance. In addition, this material has the common defect of poor conductivity of titanium-based materials, making its rate performance poorly satisfactory. At present, the improvement of the electrochemical performance of $Na_2Ti_3O_7$ mainly includes the following points: (i) design and preparation of structured nanomaterials with special morphology; (ii) compounding with highly conductive substances (carbon materials and polymer materials); and (iii) bulk doping or co-doping with anions and cations.

4.3.2 Alloyed Anode Materials

Up to now, the more widely studied alloying materials mainly include the group IVA elements (Ge and Sn) and the group VA elements (P, Sb, and Bi). Metal materials based on alloying reaction can exhibit a higher theoretical specific capacity.

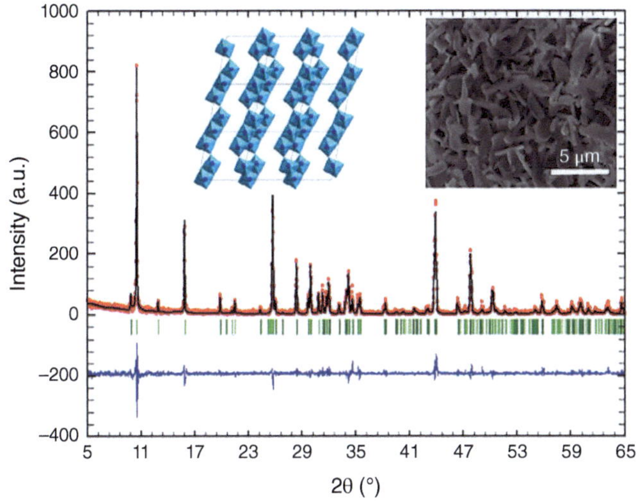

Figure 4.10 Rietveld refinement of the XRD pattern and crystal structure of pristine $Na_2Ti_3O_7$. Source: Senguttuvan et al. [48]. Reproduced with permission of American Chemical Society.

However, the problem of capacity decay has always existed in practical applications of alloyed materials, mainly due to the dramatic volume change during alloying and dealloying with sodium that causes the aggregation and powdering of active materials, resulting in loss of electrical contact. In addition, the solid interface electrolyte (SEI) film is continuously formed on the surface of the newly exposed active material during the cycle, and the thick and uneven SEI film formed hinders charge transfer and causes capacity decay. At present, strategies to improve the electrochemical performance of alloyed materials mainly include the design of nanostructures, the introduction of conductive carbon substrates, and the use of suitable binders.

The more widely studied group IVA elements include Ge and Sn. Among them, the metal Ge can react with Na to form NaGe. Compared with other alloyed materials, the theoretical capacity is relatively low (396 mAh g^{-1}). Besides, Na^+ has a high diffusion barrier in the Ge lattice, so the crystalline Ge has poor sodium storage activity. However, the amorphous Ge film can provide a reversible capacity of 350 mAh g^{-1} (close to its theoretical specific capacity). Unlike Ge, Sn can form $Na_{15}Sn_4$ alloy with Na, with a theoretical specific capacity of 847 mAh g^{-1}. Unfortunately, it will cause a volume expansion of 420%, resulting in severe powdering of electrode materials. At present, the problem of volume expansion can be solved by mainly preparing Sn nanostructures and introducing carbon substrate.

Some elements in the group VA (M = P, Sb, and Bi) can be alloyed with Na to form Na_3M, but the theoretical capacity differs greatly, where the theoretical capacity of P can reach 2596 mAh g^{-1}. P has a lower redox potential (0.4 V vs. Na^+/Na), low cost, and environmental protection advantages. Red phosphorus and black phosphorus are two widely studied allotropes of P. Among them, black phosphorus has a layered orthorhombic crystal structure with high conductivity (300 S m^{-1}), but the volume changes during charging and the air stability is poor. Black phosphorus reacts with water and oxygen in the air to cause its rapid oxidation, resulting in deterioration of electrochemical performance. Red phosphorus is a polymer chain structure with better stability, but lower conductivity (10^{-12} S m^{-1}) and large volume expansion (~300%) limit its application as a negative electrode material. Metallic Sb has a high theoretical capacity (660 mAh g^{-1}) and a suitable oxidation/reduction potential (0.5–0.8 V vs. Na^+/Na). The volume expansion of Sb during sodium insertion (~300%) is relatively large. More researched solutions include controlling morphology or compounding with conductive carbon. Metallic Bi is a layered metal material with a large interlayer distance (d(003) = 3.95 Å), which has a theoretical capacity of 384 mAh g^{-1} with high-volume expansion rate (~250%). In general, sodium storage materials based on alloying reaction show extremely high reversible capacity, which is a very promising anode material for SIBs, but how to solve the huge volume expansion problem in the cycling process is still a great challenge.

4.3.3 Conversion-Type Anode Materials

Conversion anode materials including transition metal oxides, sulfides, nitrides, and phosphides have played an important role in SIBs. Compared with alloyed electrode materials, these materials tend to have a larger theoretical sodium storage capacity. The conversion reaction material undergoes phase transition during sodium ion insertion and extraction and is accompanied by bond breakage and formation. The conversion reaction can be expressed by the following equation:

$$M_aX_b + (b \cdot z)Na \leftrightarrow aM + bNa_zX$$

Among them, M stands for metal element and X represents nonmetal element. For a typical conversion reaction, M is a transition metal element, such as Fe, Co, Ni, Cu, and Mn, while X includes O, N, F, S, Se, P, and H. In addition, when M is Si, Ge, Sn, Sb, and Bi, after the first conversion reaction occurs, sodium can also alloy with these metals or semi-metals. For this type of material, the capacity comes from both the conversion reaction and the alloying reaction, so it can provide a higher capacity than the pure metal M.

Metal oxides. The transition metal oxides (M_xO_y) are widely used anode materials with low cost and high specific capacity. They can be divided into two types according to the different sodium storage mechanisms: (i) when M is an electrochemically inactive element with Na (such as Fe, Co, Ni, and Cu), it reacts with sodium to form M and Na_2O: $M_xO_y + 2yNa^+ + 2ye^- \rightarrow xM + yNa_2O$; (ii) when M is an electrochemically active element (such as Sn and Sb) with Na, the conversion reaction takes place first to generate elemental metals M and Na_2O, and then the metal M is alloyed with Na to generate Na_nM: $M + nNa^+ + ne^- \rightarrow Na_nM$. Iron-based oxides and copper-based oxides have the advantages of wide distribution, stable chemical properties, high capacity (theoretical capacities of Fe_2O_3 and CuO are 1005 and 674 mAh g^{-1}, respectively), and low cost, but their conductivity is low, and a large volume expansion occurs during the charge and discharge cycle. These unfavorable factors make the electrode cracked and pulverized, causing the electrode conductivity to be destroyed, and the final result is capacity attenuation. Moreover, due to the larger radius of sodium ions, the slow migration ability of sodium ions is still a challenge to fully exert its theoretical capacity. Since each conversion cycle exposes a fresh electrode surface, this will accelerate the decomposition of the electrolyte. The results of the study confirmed that the materials that construct the nanostructures can effectively release the stress caused by the volume change. In addition, nanomaterials can provide more electrode/electrolyte contact and shorten the diffusion distance of sodium ions in the material, resulting in higher rate performance. However, its side effect is to accelerate the decomposition of the electrolyte. The large voltage lag during the conversion reaction can also be reduced by optimizing the material microstructure and electrode structure, which can improve the energy efficiency of the sodium ion battery based on the conversion reaction.

Unlike inactive metal oxides, SnO_2 with a theoretical specific capacity of 1378 mAh g^{-1} has two different reactions in the sodium storage process: conversion reaction and alloying reaction:

$$\text{conversion process}: SnO_2 + 4Na^+ + 4e^- \rightarrow Sn + 2Na_2O, 711\ \text{mAh g}^{-1}.$$

$$\text{alloying process}: Sn + 3.75Na^+ + 3.75e^- \rightarrow Na_{3.75}Sn\ (Na_{15}Sn_4), 667\ \text{mAh g}^{-1}.$$

Due to the poor kinetics of Na_2O formation in the first conversion reaction, it is difficult to achieve a complete reverse reaction in this step. Therefore, SnO_2 anode materials have poor performance at large rates, and how to overcome the problem of slow conversion reaction kinetics becomes the key. Another problem with SnO_2 is still the huge volume change during the cycle, resulting in poor cycle performance. At present, the main modification methods are (i) preparation of nanomaterials with loose structure and (ii) compounding with carbon-based or other matrix materials to alleviate volume changes.

Metal sulfides. As a typical class of anode materials for SIBs, metal sulfide (MS_x) has also attracted great attention from researchers. Such materials mainly include layered disulfides and non-layered sulfides. MS_x undergoes a conversion reaction similar to that of metal oxides during the electrochemical cycle. The M—S bond in the metal sulfide is weaker than the corresponding M—O bond in the metal oxide, which is kinetically beneficial for the conversion reaction with sodium ions. Among them, the layered disulfide usually undergoes Na^+ deintercalation reaction at a high potential first, and then a conversion reaction at a low potential to generate elemental metals M and Na_2S. Some of these materials, such as SnS_2, also undergo alloying reactions at lower potentials. Similarly, similar to metal oxide materials, metal sulfide will have a large volume change when the conversion reaction occurs, affecting the material's cycle stability.

Metal selenides. Compared with metal oxides, metal selenides (MSe_x) have relatively small polarization and relatively high electrical conductivity; compared with metal sulfides that have polysulfide ion dissolution problems during electrochemical cycling, the metal selenides can effectively avoid the above reaction during the cycle. At the same time, the selenium atom has a larger diameter and stronger metallity than the sulfur atom. The metal selenides have a larger layer and higher conductivity than the metal sulfides. However, the electrical conductivity of metal selenides still cannot meet the needs of large rate charge and discharge. There is still a certain volume change during the insertion/extraction of sodium ions, which leads to changes in the structure of the electrode material and the instability of the electrolyte intermediate phase, which limited their electrochemical performances. At present, various measures adopted to improve their sodium storage performance: (i) to alleviate the volume change during charge and discharge through the nanostructure of the material; (ii) to compound with materials with high electronic conductivity, while protecting the integrity of the electrode material; (iii) to select the best system of liquid electrolyte to improve the electrochemical performance of the material; and (iv) properly adjust the charge and discharge cutoff voltage window, etc.

Metal selenides can be divided into layered selenides and non-layered selenides according to structure type. Most of the layered metal selenides have M–Se–M sandwich-layered structure. Typical layered metal selenides are $MoSe_2$ and WSe_2. There are two kinds of coordination methods between selenium and metal: trigonal cylinder coordination and octahedral coordination. During the charging/discharging process, sodium ions break the interlayer van der Waals force and insert in it, and the larger interlayer distance is conducive to the insertion and extraction of sodium ions. Non-layered metal selenides mainly include $FeSe_2$, $CoSe_2$, $NiSe_2$, $CuSe_2$, and $ZnSe_2$. These selenides can be obtained from natural ores which have a higher specific capacity and a multi-electron reaction mechanism but will be generated due to the insertion and extraction of sodium ions, and the volume change causes the electrode to shatter and collapse.

Metal phosphides. Phosphorus as anode material for SIBs has a theoretical specific capacity of up to 2596 mAh g^{-1}. However, its own insulating properties and the huge volume expansion during the reaction make the phosphorus anode materials exhibit poor electrochemical performance. The electrochemical reaction process of transition metal phosphide (MP_x) is similar to that of MO_x and MS_x. During the charging/discharging process, MP_x undergoes a conversion reaction with Na metals to produce M and Na_3P. The elemental metal M is evenly dispersed in the Na_3P phase, which can accelerate the kinetic process of the Na_3P oxidation reaction, thus making MP_x exhibit excellent electrochemical performance.

4.4 Electrolytes for SIBs

The electrolyte plays a crucial role in the performance, cycle life, and safety of SIBs. The interaction between the electrolyte and the electrode material greatly affects the formation of the SEI layer and the internal structure of the electrode material. As shown in Figure 4.11, the electrolyte can be divided into three types: water system, organic liquid system, and solid electrolyte [49].

The SIBs based on the aqueous electrolyte have the advantages of being green and environmentally friendly and has a high safety factor. However, its average operating voltage is concentrated between 1.0 and 1.5 V, and the corresponding battery energy density is usually less than 100 Wh kg^{-1}, which greatly restricted their

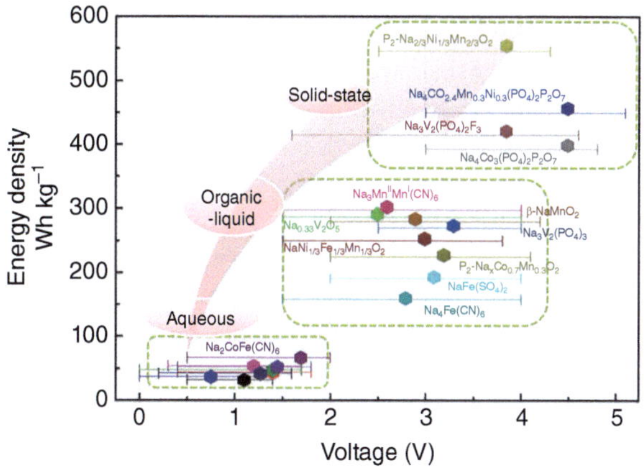

Figure 4.11 The energy density of various SIBs based on different electrolyte systems. Source: Che et al. [49]. Reproduced with permission of Royal Society of Chemistry.

practical application. SIBs based on organic electrolytes have an operating voltage window that can be extended to 1.5–4.25 V, which makes their energy density up to 150–300 Wh kg^{-1}. Moreover, there are many types of organic electrolytes, and they usually have a higher ionic conductivity than the existing ones. Therefore, they are currently the most widely studied electrolyte systems. Compared with organic liquid electrolytes, solid electrolytes have a wider electrochemical window and higher stability and safety, making them suitable for high-voltage cathode materials and metal sodium anodes. The energy density of SIBs based on solid electrolytes can reach 500 Wh kg^{-1}, but at this stage it is also facing the problems of poor ionic conductivity and electrode electrolyte contact interface.

The composition of the electrolyte solution mainly includes salts, solvents, and additives. Among them, the salts and solvents play a major role in the properties of the electrolyte solution. At present, commonly used salts are sodium hexafluorophosphate ($NaPF_6$), sodium perchlorate ($NaClO_4$), sodium bis(trifluoromethanesulfonyl)imide (NaTFSI), sodium fluorosulfonyl (trifluoromethanesulfonyl)imide (NaFTFSI), sodium bis(fluorosulfonyl)imide (NaFSI), etc. Among the most widely studied organic electrolyte systems, they can be divided into two types, such as ester electrolytes and ether electrolytes, according to different organic solvents. Ester electrolytes have a wide electrochemical window, high ionic conductivity, and good thermal stability. They are widely concerned in the research of SIBs. Organic solvents mainly include ethylene carbonate (EC) and propylene carbonate (PC). In addition, in order to build a stable SEI membrane on the electrode surface and reduce the occurrence of side reactions in the electrolyte, fluoroethylene carbonate (FEC) is usually used as a functional additive. Compared with ester electrolytes, ether electrolytes have a narrow voltage window and are not suitable for high-voltage SIBs, but they play an important role in graphite anode materials, organic cathode materials, and sulfur-containing compounds, which can significantly improve the above electrochemical performance of similar materials. The most studied organic solvents mainly include diglyme (DGM) and ethylene glycol dimethyl ether (DME).

4.4.1 Aqueous Electrolytes

The aqueous SIBs have a promising application prospect due to their characteristics of better safety, environmental friendliness, low price, and less corrosiveness. At present, Na_2SO_4 solution is a more widely studied aqueous electrolyte in aqueous SIBs. As a strong electrolyte, Na_2SO_4 can provide Na^+ in aqueous solutions. Its mechanism is similar to that of aqueous LIBs. The "rocking chair" aqueous secondary battery is constructed by the ion intercalation reaction between the two electrodes. Up to now, the research on the aqueous electrolyte of SIBs is still in its infancy, and there are many problems, such as the narrow electrochemical window of the aqueous solution and the occurrence of side reactions of hydrogen precipitation, which have a certain impact on the performance of the battery. In addition, considering the hydrogen evolution reaction and other issues, the insertion potential of the negative electrode material is high, resulting in a low operating voltage and low energy density of the aqueous SIBs. Therefore, the development of other stable electrolyte systems and matching with different cathode and anode materials to obtain more excellent electrochemical performance is the development direction of aqueous electrolytes for SIBs.

4.4.2 Organic Electrolytes

Generally, the viscosity, ionic conductivity, and solubility of salts in solvents of the liquid electrolyte greatly affect the performance of SIBs. Therefore, most studies are devoted to reducing the viscosity of electrolyte solutions, increasing the number of Na^+ migration, and increasing the solubility of salts in solvents, in order to improve the performance of electrolyte solutions. Komaba et al. systematically studied the storage behavior of hard carbon anode materials in an electrolyte solution composed of 1 $mol\,l^{-1}$ $NaClO_4$ and a series of ester organic solvents. The research shows that the electrolyte solution composed of 1 $mol\,l^{-1}$ $NaClO_4$ and PC, EC, and EC/DEC shows better electrochemical performance. Ponrouch et al. tested the room temperature conductivity of the commonly used sodium salts $NaClO_4$ and $NaPF_6$ in PC solvents. The results show that $NaPF_6$ has the maximum ionic conductivity ($0.78 \times 10^{-3}\,S\,cm^{-1}$), because the PF_6^- group has a small polarizability that can promote the decomposition of salt; the ionic conductivity of $NaClO_4$ is slightly lower ($0.62 \times 10^{-3} \sim 0.78 \times 10^{-3}\,S\,cm^{-1}$), but its thermal stability is better. Although the liquid electrolyte of SIBs has high conductivity and good wettability of the electrode surface, it is similar to the liquid electrolyte of the lithium ion battery, and it is also easy to leak, flammable, and easy to consume and form unstable SEI film during the cycle, and low mechanical strength, which greatly limits their development. At present, $NaClO_4$ and $NaPF_6$ salts are often used in academic research with electrolytes such as EC:PC and EC:DEC. In addition, since organic electrolytes can easily corrode sodium electrodes and affect the electrochemical performance of batteries, film-forming additives are usually added to improve this situation. FEC, vinyl sulfite (ES), vinylene carbonate (VC), etc. are very effective film-forming additives in LIBs. They can form a passive film on the surface of the negative electrode to prevent the electrolyte from reacting with the negative electrode. However, in SIBs, only the addition of FEC has been found to have a better effect.

4.4.3 Solid-State Electrolytes

In order to avoid the potential safety hazards of liquid electrolyte leakage and combustion, the development of solid electrolytes with high safety and high energy density has become a new way to improve the above problems. The solid electrolyte is light in volume and low in cost, and it can effectively avoid leakage of the electrolyte and can adapt to the change of the volume of the electrode material during the charging and discharging process, so it has been paid more and more attention. Sodium ion solid electrolyte materials can be divided into solid polymer electrolytes and inorganic solid electrolytes.

4.4.3.1 Solid Polymer Electrolytes

The solid polymer electrolytes are considered to be the most promising electrolytes due to their high safety, flexibility, and easy to process properties. The solid polymer electrolytes are mainly composed of polymer substrates and electrolyte salts. The common polymer matrices include polyoxyethylene (PEO), polyvinylpyrrolidone (PVP), polyacrylonitrile (PAN), polyvinyl alcohol (PVA), and polymethylmethacrylate (PMMA). The migration of sodium ions in the solid polymer electrolyte is mainly based on the coordination of the sodium salt in the polymer backbone. The thermal motion of the polymer backbone is used to transport ions under an external electric field. Therefore, they usually show low conductivity at room temperature. Taking the most extensively studied PEO-based solid polymer electrolyte as an example, its ion conduction mechanism is mainly through the continuous complexation and dissociation reaction between sodium ions and oxygen atoms of etheroxy groups in the PEO polymer chain. With the movement of the PEO segment, ion migration is achieved. However, due to the high crystallinity of the PEO polymer at room temperature, its ionic conductivity is very low, only $10^{-7}\ S\ cm^{-1}$. Therefore, increasing the proportion of the PEO amorphous region is one of the effective measures to improve its electrical conductivity. At present, it is mainly modified by adding inorganic fillers (such as SiO_2, Al_2O_3, TiO_2) through crosslinking or copolymerization. Among them, the addition of inorganic fillers on the one hand can disrupt the arrangement of PEO chains to reduce the crystallinity of the matrix; on the other hand, the surface of the filler can also provide more ion transmission channels, thereby improving the conductivity of the polymer.

In addition, the introduction of solvent molecules into the solid polymer electrolyte provides strong molecular polarization for the ions to form a gel polymer electrolyte, which can also effectively improve its ionic conductivity. Thus, gel polymer electrolytes have also received extensive attention. The intermolecular force between the skeleton molecule and the solvent molecule is used to obtain a swelling system, which improves the room temperature ionic conductivity and solves the problem of electrode–electrolyte interface infiltration.

4.4.3.2 Inorganic Solid Electrolytes

Inorganic solid electrolytes are also very promising electrolytes in SIBs because of their high ionic conductivity and strong thermal stability. The inorganic solid electrolytes currently used in SIBs mainly include Na–β/β″–Al_2O_3 solid electrolytes, NASICON solid electrolyte materials, sulfide solid electrolytes, and boron–hydrogen compound solid electrolytes.

The traditional Na–β/β″–Al_2O_3 materials are mainly used in sodium–sulfur batteries and solid SIBs. The spinel layers of β–Al_2O_3 and Na–O layers are alternately stacked, and the stacking of Na–O layers is loose, while β″–Al_2O_3 contains more Na^+ ions and has higher ionic conductivity (2×10^{-3} S cm^{-1} at room temperature, 0.2 ~ 0.4 S cm^{-1} at 300 °C), which is an important electrolyte material for solid sodium batteries. Unfortunately, the synthesis of Na–β″–Al_2O_3 requires high-temperature sintering conditions of 1200–1600 °C, and its ionic conductivity is better under high-temperature conditions, which is bound to limit its large-scale application.

NASICON solid electrolyte materials have attracted the attention of researchers because of their advantages such as higher conductivity, lower expansion coefficient, and suitability for working at high temperatures. Goodenough et al. first reported the NASICON ion transport material $Na_{1+x}Zr_2P_{3-x}Si_xO_{12}$. In this structure system, when $x = 2$ (the structure is $Na_3Zr_2Si_2PO_{12}$), the material has the highest ion conductivity, and its ion conductivity at room temperature is 6.7×10^{-4} S cm^{-1}. Element doping or substitution methods can further improve the room temperature conductivity. For example, when Ce^{4+} and La^{3+} replace part of Zr^{4+}, the room temperature ion conductivity of the sample can reach 9.0×10^{-4} S cm^{-1}, 3.4×10^{-3} S cm^{-1}, respectively. Although there are many reports of NASICON solid electrolytes, the solid–solid contact between the electrode and the electrolyte in the solid sodium battery composed of it has a large impedance, resulting in poor battery performance. There are two main strategies: first, the size of the electrode material is reduced to the nanometer level, which can effectively release the internal stress generated by the electrode during the cycle; second, the electrode surface is coated with carbon material or other amorphous phase buffer layer with high ion conductivity, and also it can alleviate the occurrence of surface cracks or powdering caused by volume changes and stress concentration. In addition, the drop of ionic liquid on the positive electrode side can effectively reduce the solid–solid interface impedance and improve the ionic conductivity to improve the interface dynamics.

The sulfide solid electrolyte has higher ion conductivity and lower grain boundary impedance, which attracts much attention. The sulfide sodium ion conductor materials are mainly divided into crystalline, glass, and glass–ceramics. In general, the glass–ceramic of Na–sulfide solid electrolyte is higher in conductivity than its corresponding crystalline or glassy material, such as the conductivity of Na_3PS_4 is 1 to 2 orders of magnitude higher than its crystalline and glassy state. Because the temperature of the synthetic sulfide is not too high, it can greatly reduce the grain boundary resistance, making it more suitable for use in room temperature solid sodium batteries. However, crystalline sulfides containing alkali metals are easy to absorb water and emit toxic hydrogen sulfide gas. Therefore, its preparation process needs to be carried out under the protection of inert gas. At present, the stability of the material to moisture is mainly improved by doping metal elements in the electrolyte. For example, $Na_3P_{0.62}As_{0.38}S_4$ is obtained by doping Na_3PS_4 with As, after doping the ionic conductivity of the solid electrolyte is significantly improved (1.46×10^{-3} S cm^{-1} at room temperature) and the stability to moisture is significantly improved. This is because As has weak expansion and has a strong interaction with S, thus the Na—S bond is weakened. Therefore, improving the synthesis conditions of the sulfide solid electrolyte and improving its stability in air are the key points for popularization and application.

4.5 Separators for SIBs

The separator, as one of the key internal components of the battery, is a thin film that separates the positive and negative electrodes to prevent direct reaction on the basis of allowing the ions in the electrolyte to pass through [50, 51]. The performance of separator determines the interface structure and internal resistance of the cell and directly affects the battery's capacity, cycle, and safety performance. The separator should have good chemical stability in the organic electrolyte system, high tensile strength and puncture strength to meet the requirements of assembly, high porosity and uniform pore size distribution to avoid local overheating caused by uneven current density, good electrolyte wettability and high liquid absorption rate to improve ion conductivity, and wide operating temperature range ensures the safe use of the battery.

4.5.1 Glass Fiber Separator

At present, glass fiber (GF) is selected as separator by most SIBs due to its excellent electrolyte wettability, high porosity, high electrolyte absorption rate, high ion conductivity, and other advantages. However, the large-scale commercial use of GF separator on SIBs is still facing many challenges. The presence of micron-sized macropores may cause internal micro short circuits between the electrodes. The thickness of GF separators is always hundreds of microns, an order of magnitude thicker than commercial separators, thereby reducing the volume and weight energy density. The fragile nature of the GF separator will also be detrimental to the commercial battery assembly process.

4.5.2 Modified Polyolefin Separator

The traditional commercial separator materials for LIBs are mainly polyolefin separators, such as polyethylene (PE) and polypropylene (PP). However, these conventional polyolefin separators are not suitable for SIBs due to the extremely poor wettability of carbonate-based electrolytes commonly used for SIBs. In addition, conventional separators also have problems such as high thermal shrinkage and low ionic conductivity.

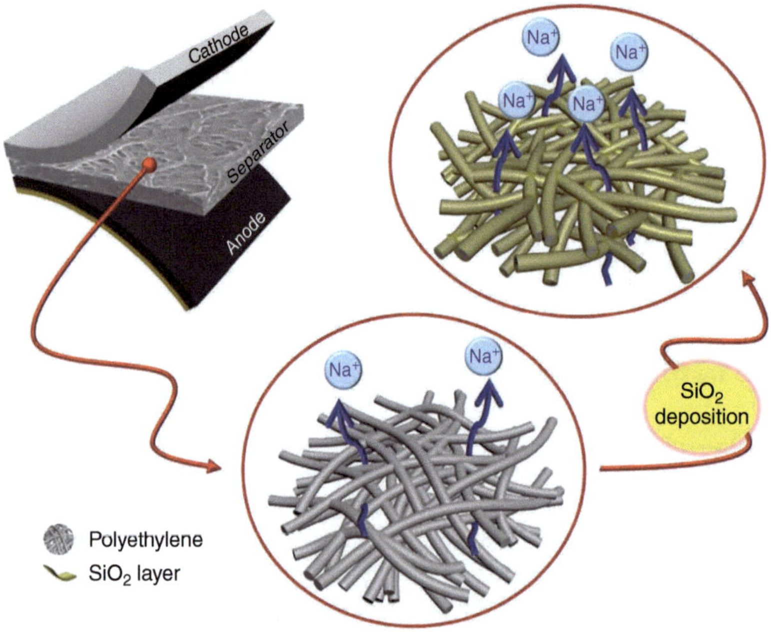

Figure 4.12 SiO_2 layer coating on a PE separator. Source: Kim et al. [54]. Reproduced with permission of The Electrochemical Society.

In order to improve the thermal stability and the electrolyte wettability of conventional polyolefin separators, the strategies of surface modification or coating are feasible. The principle of this type of strategy is to make full use of the thermal stability and polarity of the coating layer without increasing the thickness of the diaphragm and the ionic conductivity. Usually, ceramics, metal oxides, polymers, and their composites are coated or in situ grown on the separators surface by various methods [52–55]. SiO_2 with high polarity and thermal stability are suitable in modifying separators. As shown in Figure 4.11, the coating of silicon oxide increased the electrolyte wettability of the separator and reduced its thermal shrinkage rate, promoting the transmission of sodium ions and enhancing the high-temperature electrochemical stability, respectively. Other inorganic compounds such as ZrO_2, and organic compounds such as polyvinylidene fluoride (PVDF), also have similar functions (Figure 4.12).

4.5.3 Other Separator

Apart from GF separator and modified polyolefin separator, porous PVDF membrane [56], cellulose acetate film [57], and chitin nanofiber membrane [58] also have been tried as robust separators for SIBs. In addition, composite separators have been also rationally designed through the combination of inorganic materials and polymers, such as SiO_2-PAN nonwoven [59], cellulose–PAN–Al_2O_3 composite [60], SiO_2–PVDF/hexafluoropropylene (PVDF/HFP) fibers [61], and

$BaTiO_3$–PVDF/HFP–poly(butyl methacrylate) blend [62]. During preparation, combined with other surface treatment solutions, the electrospinning technology with advantages in fiber manufacturing is often used. These separators all exhibited high chemical stability with small shrinkage rate and electrolyte wettability with small contact angles. These are ultimately conducive to obtaining high-performance SIBs. Unfortunately, research on the development of new type separators is still limited in response to the faced dilemma for traditional commercial separator materials.

References

1 Yabuuchi, N., Kubota, K., Dahbi, M., and Komaba, S. (2014). Research development on sodium-ion batteries. *Chem. Rev.* 114 (23): 11636–11682.

2 Slater, M.D., Kim, D., Lee, E., and Johnson, C.S. (2013). Sodium-ion batteries. *Adv. Funct. Mater.* 23 (8): 947–958.

3 Choi, J.W. and Aurbach, D. (2016). Promise and reality of post-lithium-ion batteries with high energy densities. *Nat. Rev. Mater.* 1 (4): 16013.

4 Ong, S.P., Chevrier, V.L., Hautier, G. et al. (2011). Voltage, stability and diffusion barrier differences between sodium-ion and lithium-ion intercalation materials. *Energy Environ. Sci.* 4 (9): 3680–3688.

5 Kundu, D., Talaie, E., Duffort, V., and Nazar, L.F. (2015). The emerging chemistry of sodium ion batteries for electrochemical energy storage. *Angew. Chem. Int. Ed.* 54 (11): 3431–3448.

6 Hwang, J.-Y., Myung, S.-T., and Sun, Y.-K. (2017). Sodium-ion batteries: present and future. *Chem. Soc. Rev.* 46 (12): 3529–3614.

7 Browne, M.P., Sofer, Z., and Pumera, M. (2019). Layered and two dimensional metal oxides for electrochemical energy conversion. *Energy Environ. Sci.* 12 (1): 41–58.

8 Wang, P.-F., You, Y., Yin, Y.-X., and Guo, Y.-G. (2018). Layered oxide cathodes for sodium-ion batteries: phase transition, air stability, and performance. *Adv. Energy Mater.* 8 (8): 1701912.

9 Han, M.H., Gonzalo, E., Singh, G., and Rojo, T. (2015). A comprehensive review of sodium layered oxides: powerful cathodes for Na-ion batteries. *Energy Environ. Sci.* 8 (1): 81–102.

10 Liu, Q., Hu, Z., Chen, M. et al. (2019). Recent progress of layered transition metal oxide cathodes for sodium-ion batteries. *Small* 15 (32): 1805381.

11 Mendiboure, A., Delmas, C., and Hagenmuller, P. (1985). Electrochemical intercalation and deintercalation of Na_xMnO_2 bronzes. *J. Solid State Chem.* 57 (3): 323–331.

12 Braconnier, J.-J., Delmas, C., Fouassier, C., and Hagenmuller, P. (1980). Comportement electrochimique des phases Na_xCoO_2. *Mater. Res. Bull.* 15 (12): 1797–1804.

13 Berthelot, R., Carlier, D., and Delmas, C. (2011). Electrochemical investigation of the P2-Na_xCoO_2 phase diagram. *Nat. Mater.* 10 (1): 74–80.

14 Delmas, C., Braconnier, J.-J., Fouassier, C., and Hagenmuller, P. (1981). Electrochemical intercalation of sodium in Na_xCoO_2 bronzes. *Solid State Ionics* 3: 165–169.

15 Yabuuchi, N., Kajiyama, M., Iwatate, J. et al. (2012). P2-type $Na_x[Fe_{1/2}Mn_{1/2}]O_2$ made from earth-abundant elements for rechargeable Na batteries. *Nat. Mater.* 11 (6): 512–517.

16 Lu, Z. and Dahn, J.R. (2001). In situ X-ray diffraction study of P2-$Na_{2/3}[Ni_{1/3}Mn_{2/3}]O_2$. *J. Electrochem. Soc.* 148 (11): A1225.

17 Wang, L., Sun, Y.-G., Hu, L.-L. et al. (2017). Copper-substituted $Na_{0.67}Ni_{0.3-x}Cu_xMn_{0.7}O_2$ cathode materials for sodium-ion batteries with suppressed P2–O2 phase transition. *J. Mater. Chem. A* 5 (18): 8752–8761.

18 Xie, Y., Wang, H., Xu, G. et al. (2016). In operando XRD and TXM study on the metastable structure change of $NaNi_{1/3}Fe_{1/3}Mn_{1/3}O_2$ under electrochemical sodium-ion intercalation. *Adv. Energy Mater.* 6 (24): 1601306.

19 Jeong, M., Lee, H., Yoon, J., and Yoon, W.-S. (2019). O3-type $NaNi_{1/3}Fe_{1/3}Mn_{1/3}O_2$ layered cathode for Na-ion batteries: structural evolution and redox mechanism upon Na (de) intercalation. *J. Power Sources* 439: 227064.

20 Buchholz, D., Moretti, A., Kloepsch, R. et al. (2013). Toward Na-ion batteries—synthesis and characterization of a novel high capacity Na ion intercalation material. *Chem. Mater.* 25 (2): 142–148.

21 Ni, Q., Bai, Y., Wu, F., and Wu, C. (2017). Polyanion-type electrode materials for sodium-ion batteries. *Adv. Sci.* 4 (3): 1600275.

22 Moreau, P., Guyomard, D., Gaubicher, J., and Boucher, F. (2010). Structure and stability of sodium intercalated phases in olivine $FePO_4$. *Chem. Mater.* 22 (14): 4126–4128.

23 Oh, S.-M., Myung, S.-T., Hassoun, J. et al. (2012). Reversible $NaFePO_4$ electrode for sodium secondary batteries. *Electrochem. Commun.* 22: 149–152.

24 Fang, Y., Zhang, J., Xiao, L. et al. (2017). Phosphate framework electrode materials for sodium ion batteries. *Adv. Sci.* 4 (5): 1600392.

25 Y. Uebou, T. Kiyabu, S. Okada and J.-I. Yamaki. (2002) Report of the Institute of Functional Material Science, Kyushu University. 16: 1–5.

26 Rajagopalan, R., Chen, B., Zhang, Z. et al. (2017). Improved reversibility of Fe^{3+}/Fe^{4+} redox couple in sodium super ion conductor type $Na_3Fe_2(PO_4)_3$ for sodium-ion batteries. *Adv. Mater.* 29 (12): 1605694.

27 Zhou, W., Xue, L., Lü, X. et al. (2016). $Na_xMV(PO_4)_3$ (M = Mn, Fe, Ni) structure and properties for sodium extraction. *Nano Lett.* 16 (12): 7836–7841.

28 Wu, C., Kopold, P., Ding, Y.-L. et al. (2015). Synthesizing porous $NaTi_2(PO_4)_3$ nanoparticles embedded in 3D graphene networks for high-rate and long cycle-life sodium electrodes. *ACS Nano* 9 (6): 6610–6618.

29 Li, Q., Liu, Z., Zheng, F. et al. (2018). Identifying the structural evolution of the sodium ion battery Na_2FePO_4F cathode. *Angew. Chem.* 130 (37): 12094–12099.

30 Kawabe, Y., Yabuuchi, N., Kajiyama, M. et al. (2011). Synthesis and electrode performance of carbon coated Na_2FePO_4F for rechargeable Na batteries. *Electrochem. Commun.* 13 (11): 1225–1228.

31 Cai, Y., Cao, X., Luo, Z. et al. (2018). Caging $Na_3V_2(PO_4)_2F_3$ microcubes in cross-linked graphene enabling ultrafast sodium storage and long-term cycling. *Adv. Sci.* 5 (9): 1800680.

32 Li, L., Xu, Y., Sun, X. et al. (2018). Fluorophosphates from solid-state synthesis and electrochemical ion exchange: $NaVPO_4F$ or $Na_3V_2(PO_4)_2F_3$? *Adv. Energy Mater.* 8 (24): 1801064.

33 Park, Y.-U., Seo, D.-H., Kwon, H.-S. et al. (2013). A new high-energy cathode for a Na-ion battery with ultrahigh stability. *J. Am. Chem. Soc.* 135 (37): 13870–13878.

34 Barpanda, P., Ye, T., Nishimura, S.-i. et al. (2012). Sodium iron pyrophosphate: a novel 3.0V iron-based cathode for sodium-ion batteries. *Electrochem. Commun.* 24: 116–119.

35 Li, H., Chen, X., Jin, T. et al. (2019). Robust graphene layer modified $Na_2MnP_2O_7$ as a durable high-rate and high energy cathode for Na-ion batteries. *Energy Storage Mater.* 16: 383–390.

36 Kim, H., Park, C.S., Choi, J.W., and Jung, Y. (2016). Defect-controlled formation of triclinic $Na_2CoP_2O_7$ for 4 V sodium-ion batteries. *Angew. Chem. Int. Ed.* 55 (23): 6662–6666.

37 Drozhzhin, O.A., Tertov, I.V., Alekseeva, A.M. et al. (2019). β-$NaVP_2O_7$ as a superior electrode material for Na-ion batteries. *Chem. Mater.* 31 (18): 7463–7469.

38 Hurlbutt, K., Wheeler, S., Capone, I., and Pasta, M. (2018). Prussian blue analogs as battery materials. *Joule* 2 (10): 1950–1960.

39 Ma, F., Li, Q., Wang, T. et al. (2017). Energy storage materials derived from Prussian blue analogues. *Sci. Bull.* 62 (5): 358–368.

40 Xu, Y., Zhou, M., and Lei, Y. (2018). Organic materials for rechargeable sodium-ion batteries. *Mater. Today* 21 (1): 60–78.

41 Rajagopalan, R., Tang, Y., Jia, C. et al. (2020). Understanding the sodium storage mechanisms of organic electrodes in sodium ion batteries: issues and solutions. *Energy Environ. Sci.* 13: 1568–1592.

42 Wang, Y., Ding, Y., Pan, L. et al. (2016). Understanding the size-dependent sodium storage properties of $Na_2C_6O_6$-based organic electrodes for sodium-ion batteries. *Nano Lett.* 16 (5): 3329–3334.

43 Chihara, K., Chujo, N., Kitajou, A., and Okada, S. (2013). Cathode properties of $Na_2C_6O_6$ for sodium-ion batteries. *Electrochim. Acta* 110: 240–246.

44 Wen, Y., He, K., Zhu, Y. et al. (2014). Expanded graphite as superior anode for sodium-ion batteries. *Nat. Commun.* 5 (1): 4033.

45 Jache, B. and Adelhelm, P. (2014). Use of graphite as a highly reversible electrode with superior cycle life for sodium-ion batteries by making use of co-intercalation phenomena. *Angew. Chem. Int. Ed.* 53 (38): 10169–10173.

46 Saurel, D., Orayech, B., Xiao, B. et al. (2018). From charge storage mechanism to performance: a roadmap toward high specific energy sodium-ion batteries through carbon anode optimization. *Adv. Energy Mater.* 8 (17): 1703268.

47 Xu, Y., Memarzadeh Lotfabad, E., Wang, H. et al. (2013). Nanocrystalline anatase TiO_2: a new anode material for rechargeable sodium ion batteries. *Chem. Commun.* 49 (79): 8973–8975.

48 Senguttuvan, P., Rousse, G., Seznec, V. et al. (2011). $Na_2Ti_3O_7$: lowest voltage ever reported oxide insertion electrode for sodium ion batteries. *Chem. Mater.* 23 (18): 4109–4111.

49 Che, H., Chen, S., Xie, Y. et al. (2017). Electrolyte design strategies and research progress for room-temperature sodium-ion batteries. *Energy Environ. Sci.* 10 (5): 1075–1101.

50 Arora, P. and Zhang, Z. (2004). Battery separators. *Chem. Rev.* 104 (10): 4419–4462.

51 Pan, Y., Chou, S., Liu, H.K., and Dou, S.X. (2017). Functional membrane separators for next-generation high-energy rechargeable batteries. *Nat. Sci. Rev.* 4 (6): 917–933.

52 Suharto, Y., Lee, Y., Yu, J.-S. et al. (2018). Microporous ceramic coated separators with superior wettability for enhancing the electrochemical performance of sodium-ion batteries. *J. Power Sources* 376: 184–190.

53 Mun, J., Yim, T., Gap Kwon, Y., and Jae Kim, K. Self-assembled nano-silica-embedded polyethylene separator with outstanding physicochemical and thermal properties for advanced sodium ion batteries. *Chem. Eng. J.* 2020: 125844.

54 Kim, J.I., Heo, J., and Park, J.H. (2017). Tailored metal oxide thin film on polyethylene separators for sodium-ion batteries. *J. Electrochem. Soc.* 164 (9): A1965–A1969.

55 Janakiraman, S., Khalifa, M., Biswal, R. et al. (2020). High performance electrospun nanofiber coated polypropylene membrane as a separator for sodium ion batteries. *J. Power Sources* 460: 228060.

56 Janakiraman, S., Surendran, A., Ghosh, S. et al. (2016). Electroactive poly(vinylidene fluoride) fluoride separator for sodium ion battery with high coulombic efficiency. *Solid State Ionics* 292: 130–135.

57 Chen, W., Zhang, L., Liu, C. et al. (2018). Electrospun flexible cellulose acetate-based separators for sodium-ion batteries with ultralong cycle stability and excellent wettability: the role of Interface chemical groups. *ACS Appl. Mater. Interfaces* 10 (28): 23883–23890.

58 Zhang, T.-W., Shen, B., Yao, H.-B. et al. (2017). Prawn Shell derived chitin nanofiber membranes as advanced sustainable separators for Li/Na-ion batteries. *Nano Lett.* 17 (8): 4894–4901.

59 Zhang, L., Feng, G., Li, X. et al. (2019). Synergism of surface group transfer and in-situ growth of silica-aerogel induced high-performance modified polyacrylonitrile separator for lithium/sodium-ion batteries. *J. Membr. Sci.* 577: 137–144.

60 Jo, J.H., Jo, C.-H., Qiu, Z. et al. (2020). Nature-derived cellulose-based composite separator for sodium-ion batteries. *Front. Chem.* 8 (153).

61 Coustan, L., Tarascon, J.-M., and Laberty-Robert, C. (2019). Thin fiber-based separators for high-rate sodium ion batteries. *ACS Appl. Energy Mater.* 2 (12): 8369–8375.

62 R, A., Vijaya Kumar Saroja, A.P., and Sundara, R. (2019). Barium Titanate-based porous ceramic flexible membrane as a separator for room-temperature sodium-ion battery. *ACS Appl. Mater. Interfaces* 11 (4): 3889–3896.

5

$Na{-}O_2$ Battery

Haiying Lu[1], Xianghong Chen[1], Yu Lei[1], Feng Xiao[1], Weiyin Gao[2], Jiakui Zhang[1], Sai Zhao[1], Min Yan[1], Chenxin Ran[2], and Jiantie Xu[1]

[1]*South China University of Technology, School of Environment and Energy, National Engineering Laboratory for VOCs Pollution Control Technology and Equipment, Guangdong Provincial Key Laboratory of Solid Wastes Pollution Control and Recycling, 382 Outer Ring Road, Guangzhou University City, Panyu District, Guangzhou 510640, China*
[2]*Northwestern Polytechnical University, Xi'an Institute of Flexible Electronics (IFE), Frontiers Science Center for Flexible Electronics, 127 West Youyi Road, Xi'an 710072, China*

5.1 Introduction

Due to the gradual depletion of traditional fossil fuels and increasingly serious environmental pollution, many efforts have been devoted to developing the utilization of renewable energy (e.g. solar, wind, and tidal energy). However, the developed renewable energies still face two major challenges, such as the intermittency and geographical constraints. As one of the most important energy storage systems, rechargeable lithium-ion batteries (LIBs) is a key to the utilization of renewable energy. Despite their great success in the market of portable and smart electronic devices, limited specific energy density (350–400 Wh kg^{-1}) of LIBs severely limits their further application in long-range electric vehicles (EVs) and large-scale energy storage systems. Compared to traditional LIBs, metal–O_2 cells are expected to be promising alternatives due to their high theoretical energy densities. Among various metal–O_2 batteries, Li–O_2 batteries have attracted great attention due to their super high energy density of 3500 Wh kg^{-1}. Nevertheless, the reported Li–O_2 batteries still suffer from large overpotential (>1 V), poor cycling performance (<100 cycles), and severe parasitic reactions [1–6]. In addition, the natural scarcity and high price of lithium resources are critical issues for large-scale applications of LIBs. As a result, sodium-ion batteries (SIBs) are emerging as ideal alternatives to LIBs owing to the abundant and low cost of sodium (Na) resources [7–9]. Based on the formation of Na_2O_2 and NaO_2, the theoretical energy densities of Na–O_2 batteries are calculated to be 1602 and 1105 Wh kg^{-1}, respectively, which are nearly six and nine times higher than that of LIBs [10]. In 2010, Peled et al. demonstrated the first example of Na–O_2 battery by using liquid Na to replace metallic lithium anode. By taking advantage of deposition–dissolution process of Na in polymer electrolytes at 105 °C,

Battery Technologies: Materials and Components, First Edition. Edited by Jianmin Ma.

the authors presented the feasibility of operating a liquid-sodium–oxygen cell with polymer electrolytes at above 100 °C [11]. Subsequently, Sun et al. demonstrated the first rechargeable non-aqueous Na–O_2 battery at room temperature in 2012 [12]. Thereafter, Na–O_2 battery has received great concerns and inspired extensive research in the academic community.

Nevertheless, Na–O_2 battery still faces great challenges in the main components. For the anode, the growth of Na dendrites and the continuous formation of solid electrolyte interface (SEI) layer could result in the constant depletion of active sodium and electrolyte, which ultimately leads to the low Coulomb efficiency and poor cycle life of Na–O_2 batteries [13, 14]. Moreover, the growing Na dendrites may pierce through the separator, resulting in serious safety concerns [15]. For the cathode, the deposition of insoluble discharge products on the surface of the cathode could block oxygen diffusion channels and increase the charge overpotential of the battery, leading to the sluggish kinetics of oxygen reduction reaction (ORR) and oxygen evolution reaction (OER) processes. Ultimately, the battery will suffer from poor cycling stability, low rate performance, and low round-trip efficiency [16, 17]. For the electrolyte, the decomposition of organic electrolytes could lead to the generation of insulating carbonate-based parasitic products, which ultimately increases the charge overpotential of the battery. For the discharge products in Na–O_2 battery, their formation mechanisms and chemical compositions are still controversial. Unlike Li–O_2 battery, where Li_2O_2 has been identified as the major discharge product [18–20], various discharge products of Na–O_2 battery, such as NaO_2 [21–23], Na_2O_2 [24, 25], $Na_2O_2{\cdot}2H_2O$ [26, 27], and Na_2CO_3 [28, 29], have been reported by different groups. The chemical composition of discharge product could largely affect the overpotential of Na–O_2 battery. For instance, charge overpotential could be as low as 0.2 V with cubic NaO_2 as the major discharge product [21], while it could be as high as 1.5 V with Na_2O_2 as the major discharge product [25]. Therefore, despite its great development prospect, Na–O_2 battery is currently confronted with great challenges.

In this chapter, we review the advanced progress in the field of Na–O_2 battery. We begin with the introduction of fundamental principles of Na–O_2 battery. Sequentially, current development on the design of Na–O_2 battery is systematically summarized, including cathode design (i.e. carbon materials, metals, transition metal oxides, and their complexes), anode design (metallic Na protection and Na dendrite inhibition) and electrolyte design (key parameters that affect electrolyte stability and cell performance). Also, the development of in situ techniques for the mechanism study of electrochemistry in Na–O_2 battery is discussed in details. Finally, emerging challenges and some perspectives on the development of Na–O_2 battery are also concluded.

5.2 Fundamental Principles

A typical Na–O_2 battery is composed of a Na-containing anode, air cathode, electrolyte, and a separator between cathode and anode, as shown in Figure 5.1. During the discharge process, solid Na on anode will lose one electron to form Na^+ cation,

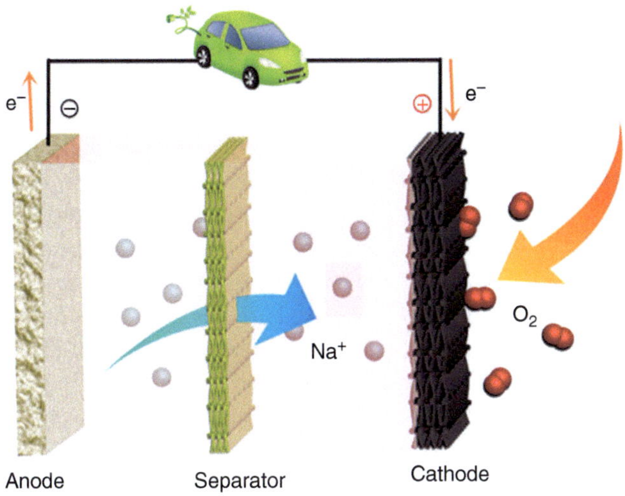

Figure 5.1 Working principle of a nonaqueous Na–O_2 battery. Source: Lin et al. [10]. Reproduced with permission of John Wiley & Sons.

which further moves to cathode through the internal circuit channels (i.e. electrolyte and separator). Meanwhile, the lost electrons will transfer to cathode through the external circuit, generating external current. At the cathode side, O_2 diffused from air will react with the electrons transferred from external circuit and then is reduced into oxide anions O_2^- or O_2^{2-}, which simultaneously react with Na^+ diffused from internal circuit to form solid discharge products NaO_2 or Na_2O_2 that deposit on the cathode (called as oxygen reduction reaction, or ORR). During the charging process, the discharge products will be converted back into O_2 and Na^+ (called as oxygen evolution reaction, or OER) and Na^+ will move back to anode through the internal circuit channels. Meanwhile, the collected Na^+ on the anode will be reduced into solid Na. The electrochemical reactions and potentials of the reactions in typical Na–O_2 battery could be described as follows [30–32]:

$$\text{At anode electrode: } \mathrm{Na(s)} \rightarrow \mathrm{Na^+ + e^-}$$

$$\text{At cathode electrode: } \mathrm{O_2 + e^- = O_2^-}$$

$$\text{Or } \mathrm{O_2 + 2e^- = O_2^{2-}}$$

$$\text{Overall: } \mathrm{Na^+ + O_2 = NaO_2}\ \ \mathrm{E_0 = 2.27v}$$

$$\text{Or } 2Na^+ + O_2 = Na_2O_2\ \ E_0 = 2.33v$$

5.3 Cathode Materials

The air cathode electrode of Na–O_2 battery is a place for ORR and OER occurring, and the efficiencies of these two reactions play important roles in determining the performance of Na–O_2 battery. Commonly, the poor electrochemical performances of Na–O_2 batteries (e.g. short cycle life, poor rate performance, and low round-trip efficiency) are closely related to poor ORR and OER activities at cathode. Over the past several years, carbon materials (e.g. graphene [25], carbon nanotubes [CNTs] [27], and porous carbon [33]) have been developed as efficient ORR and OER electrocatalysts for Na–O_2 batteries. Due to the high specific surface area, high porosity, and electrical conductivity of carbon materials, carbon-based cathode can accommodate more discharge products and promote fast transportation of O_2 molecules and electrons [31, 34, 35]. In addition to carbon materials, metals, transition metal oxides, and their complexes have also been developed as efficient cathode catalysts for Na–O_2 batteries. In this section, state-of-the-art cathode materials for Na–O_2 batteries are comprehensively discussed.

5.3.1 Carbon Materials

In 2012, Sun et al. reported rechargeable Na–O_2 battery at room temperature for the first time using diamond-like carbon (DLC) thin film as the cathode electrode [12]. The Na–O_2 battery delivers discharge capacities of 1884 mAh g^{-1} (0.56 mAh cm^{-2}) at 1/10 C and 3600 mAh g^{-1} at 1/60 C. In this Na–O_2 battery, the discharge products are found to be crystallized Na_2O_2 and amorphous carbonate salts. Since then, much attention has been paid on the development of carbon materials as cathode for Na–O_2 batteries. Various carbon materials, such as carbon black [29], carbon–fiber gas diffusion layer (GDL) [21], graphene nanosheets (GNSs) [25], as cathode for Na–O_2 batteries have been reported by different groups. These carbon materials are found to show significant differences in discharge capacity, composition of discharge products, and morphology of discharge products. Bender et al. studied the discharge products and capacities of Na–O_2 batteries using various carbon materials with different structures as cathodes [23]. It was found that NaO_2 was the main discharge product no matter which carbon materials were used as cathode. Moreover, the discharge capacities using different carbon materials varied between 300 and 4000 mAh g^{-1}. Such result suggests that the discharge capacity of Na–O_2 battery is largely determined by the morphology and/or specific surface area of carbon material. Yadegari et al. investigated the effect of morphology and specific surface area of carbon material on the performance of Na–O_2 battery. A series of carbon materials with controllable specific surface area and porous structure were specially designed through heat treatment (Figure 5.2a) and used as cathode electrodes for Na–O_2 batteries [34]. It was found that the discharge capacity of Na–O_2 battery was linearly correlated with the surface area of carbon material (Figure 5.2b). In addition, the morphology of the discharge product also strongly depended on the surface area as well as pore size of carbon material. It can be seen from Figure 5.2c,d, low surface area carbon material with limit nucleation sites could lead the formation of rod-shaped particles, while carbon material with high specific surface area could provide more nucleation sites for Na deposition, leading to the dominant production with a film-like morphology.

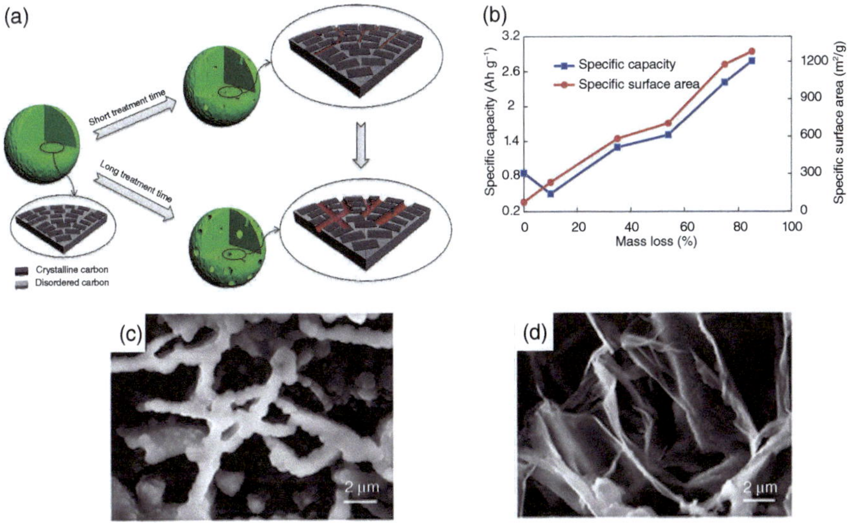

Figure 5.2 (a) Schematic diagram of the preparation of carbon black electrode with controllable morphology during the heat-treatment process. (b) Plot of specific capacities and specific surface area of a NH_3-treated carbon material as the cathode electrode as a function of mass loss. scanning electron microscopy (SEM) micrographs of discharge products for the carbon material under various durations of NH_3 treatment with a mass loss of (c) 13% and (d) 85%. Source: Yadegari et al. [34]. Reproduced with permission of Royal Society of Chemistry.

Sun et al. developed a hierarchical porous carbon sphere (PCS) material with controllable porosity as cathode for Na–O_2 battery (Figure 5.3a) [33]. The PCSs possess unique porous structure (Figure 5.3b), which could effectively facilitate electrolyte impregnation and oxygen diffusion to the inner part of the cathode electrode. As a result, Na–O_2 battery using this PCS-based cathode achieves a much higher discharge capacity of 16 500 mAh g^{-1} than that using carbon-black-based cathode (7790 mAh g^{-1}) at 500 mA g^{-1} (Figure 5.3c). Most importantly, PCS-based Na–O_2 battery achieves superior reversibility owing to the formation of easily decomposed NaO_2 discharge product with conformal film-like structure. Kwak et al. investigated the key role of surface structure of carbon cathode in determining the electrochemical performance of Na–O_2 battery [28]. An ordered mesoporous carbon (OMC) with high specific surface area (1544 m^2 g^{-1}) and narrow pore size (2.7 nm) (OCM-2.7, Figure 5.3d) was synthesized and used as cathode for Na–O_2 battery. The Na–O_2 battery using OCM-2.7 as a cathode delivers a high discharge capacity of 7987 mAh g^{-1} at 100 mA g^{-1}, while that using super P as cathode only shows a discharge capacity of 4864 mAh g^{-1} under the same condition (Figure 5.3e).

In addition to the structural design of conventional carbon materials, low-dimensional carbon materials (e.g. CNTs and graphene) have also been developed as the cathode material for Na–O_2 battery. In 2014, Jian et al. reported a three-dimensional (3D) interpenetrating CNT paper as a binder-free cathode for Na–O_2 battery for the first time [27]. The cathode made from 3D interpenetrating binder-free CNT paper could be thinner and possess more porous, and thus leading

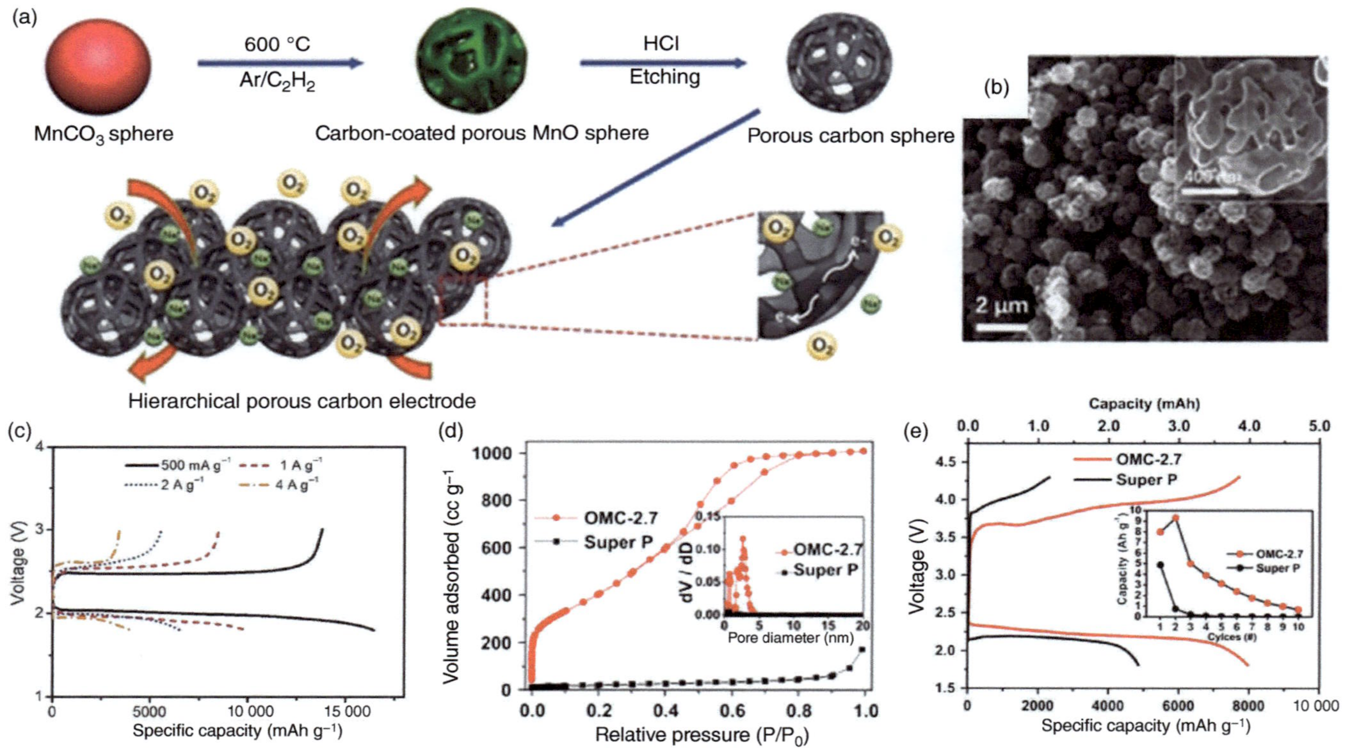

Figure 5.3 (a) Schematic illustration of the synthesis processes of PCSs and the hierarchical porous oxygen cathode made of PCSs. (b) SEM images of as-prepared PCSs, the inset shows magnified SEM image of PCSs. (c) Charge/discharge profiles of PCSs as cathodes. Source: Sun et al. [33]. Reproduced with permission of Wiley-VCH. (d) Nitrogen adsorption–desorption isotherm and pore size distribution plots (inset) of OMC-2.7 and super P. (e) Initial charge–discharge curves of OMC-2.7 (red) and Super P (black) as cathode, the inset shows corresponding discharge capacities versus cycle number. Source: Kwak et al. [28]. Reproduced with permission of Elsevier.

to improved electronic conductivity and expanded void spaces to accommodate the discharge products. This binder-free cathode exhibits high discharge and charge capacities of 7530 and 3300 mAh g^{-1} at 500 mA g^{-1}, respectively. Zhao et al. reported a vertically aligned carbon nanotubes (VACNTs) grown on stainless steel (SS) networks as cathode for Na–O_2 batteries [36]. It was found that cuboid-shaped NaO_2 particles were the main discharge product in static Ar/O_2 (80/20 vol%) (Figure 5.4a), whereas $Na_2O_2{\cdot}2H_2O$ became the main discharge product in a flowing O_2 atmosphere (Figure 5.4b). The quick formation of $Na_2O_2{\cdot}2H_2O$ during the first discharge is likely due to the introduction of substantial humidity in a flowing atmosphere, leading to the hydration of NaO_2. In addition, the electrolyte decomposes into H_2O, CO_2, and other by-products in the presence of O_2, and the degree of side reactions is greater in a flowing atmosphere. The Na–O_2 battery with VACNTs as a cathode achieves a cycle life of 130 cycles with a high cut-off capacity of 750 mAh g^{-1} and high discharge capacity of 4200 mAh g^{-1} in static Ar/O_2 (80/20 vol%) (Figure 5.4c). The good performance of the Na–O_2 battery was attributed to the abundant nucleation sites of VACNTs for NaO_2 and the optimal atmospheric conditions (static Ar/O_2) that minimized the side effects on cathode. Recently, Lin et al. demonstrated a "O_2 breathable" cathode for Na–O_2 batteries, where the cathode was composed of 3D printing reduced graphene oxide (rGO) with hierarchical porous structure [37]. This unique cathode with small open pores (diameter of ~350 μm, called 3DP-SP) was proposed to facilitate sufficient O_2 access even in the inner cathode and provide sufficient space for electrolyte permeation and NaO_2 accommodation. Moreover, the packed rGO sheets ensure sufficient electronic conductivity owing to the high degree of exposure of active sites on 3DP-SP cathode for ORR and OER. Consequently, Na–O_2 batteries using these "O_2 breathable" cathode electrodes achieve a high capacity of 13 484.6 mAh g^{-1} at 0.2 A g^{-1} and a stable cycling performance over 100 cycles with a cut-off capacity of 500 mAh g^{-1} at 1.0 A g^{-1} (Figure 5.4d,e). Most importantly, the formation and decomposition of discharge product, small cubic NaO_2 ($\leq$ 5 μm), were found to be highly reversible on the 3DP-SP cathode electrode (Figure 5.4f).

Compared to those for heteroatom/defect-doped carbon, the catalytic activities of un-doped carbon materials are usually low. Doping foreign atoms (e.g. N, B, and O) into carbon materials is an effective strategy that can introduce more electrocatalytic active sites, as well as improved electron/ionic conductivities to facilitate ORR and/or OER of the carbon materials [24, 38]. Li et al. reported the use of N-doped graphene nanosheets (N-GNSs) as cathode electrode for Na–O_2 batteries [24]. Owing to the additional active sites induced by nitrogen (N) doping, the N-GNS-based Na–O_2 batteries display a discharge capacity of 8600 mAh g^{-1} at 75 mA g^{-1}, which is about two times higher than pristine GNSs counterpart (4350 mAh g^{-1}) (Figure 5.5a). It is worth mentioning that the discharge product, Na_2O_2, was found to be particle in smaller size and disperse more uniformly on the surface of N-GNSs cathode than that on GNSs one (Figure 5.5b), leading to the lower overpotential and electrode polarization. Zhang et al. synthesized free-standing 3D N-doped graphene aerogels on Ni foam (3D N-GA@Ni) as cathode for Na–O_2 batteries (Figure 5.5c–f) [39]. The presence of structural nitrogen in

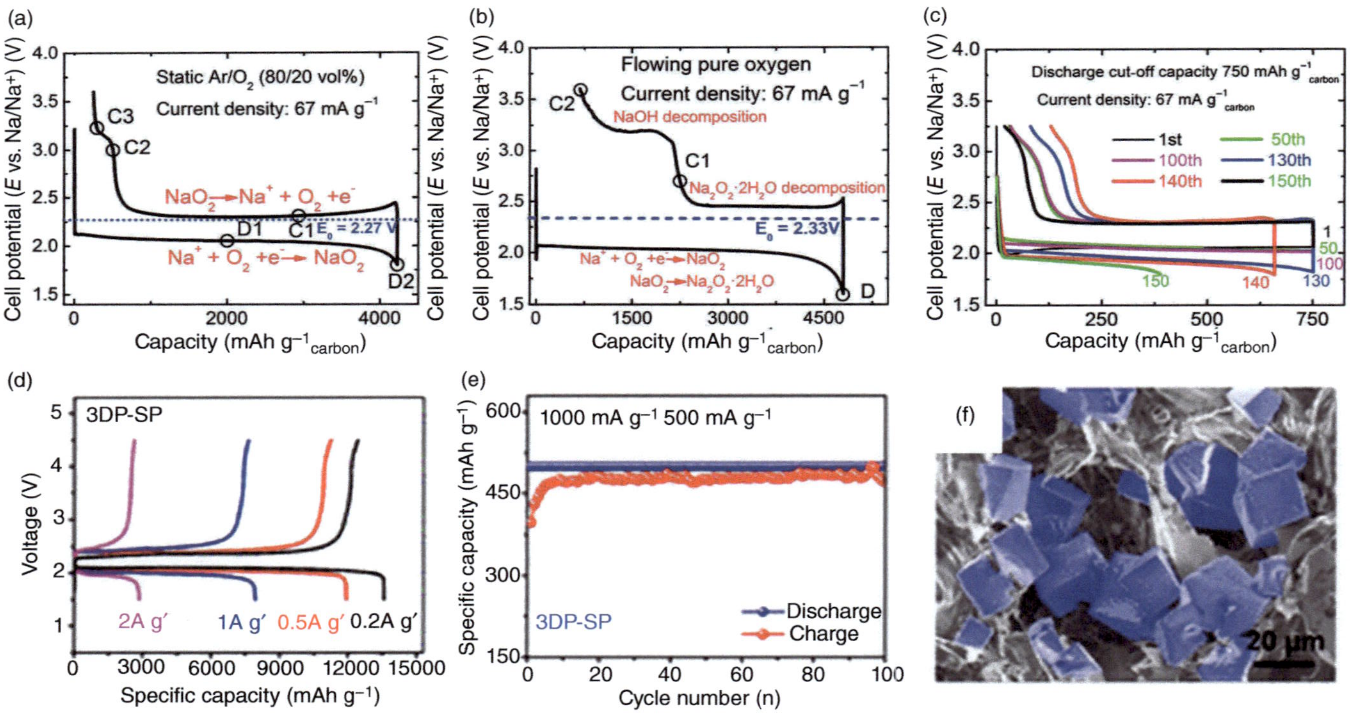

Figure 5.4 Discharge/charge profiles of VACNT cathode operated in (a) static Ar/O_2 (80/20 vol%) and (b) flowing pure O_2. (c) Cycling performance of $Na-O_2$ batteries in static Ar/O_2 (80/20 vol%). Source: Zhao et al. [36]. Reproduced with permission of Royal Society of Chemistry. (d) Initial discharge/charge profiles of 3DP-SP cathode at different current densities. (e) Cycling performance of 3DP-SP cathode with a cut-off capacity of 500 mAh g^{-1} at 1.0 A g^{-1}. (f) SEM image of the partially discharged 3DP-SP cathode at 0.2 A g^{-1}. Source: Lin et al. [37]. Reproduced with permission of American Chemical Society.

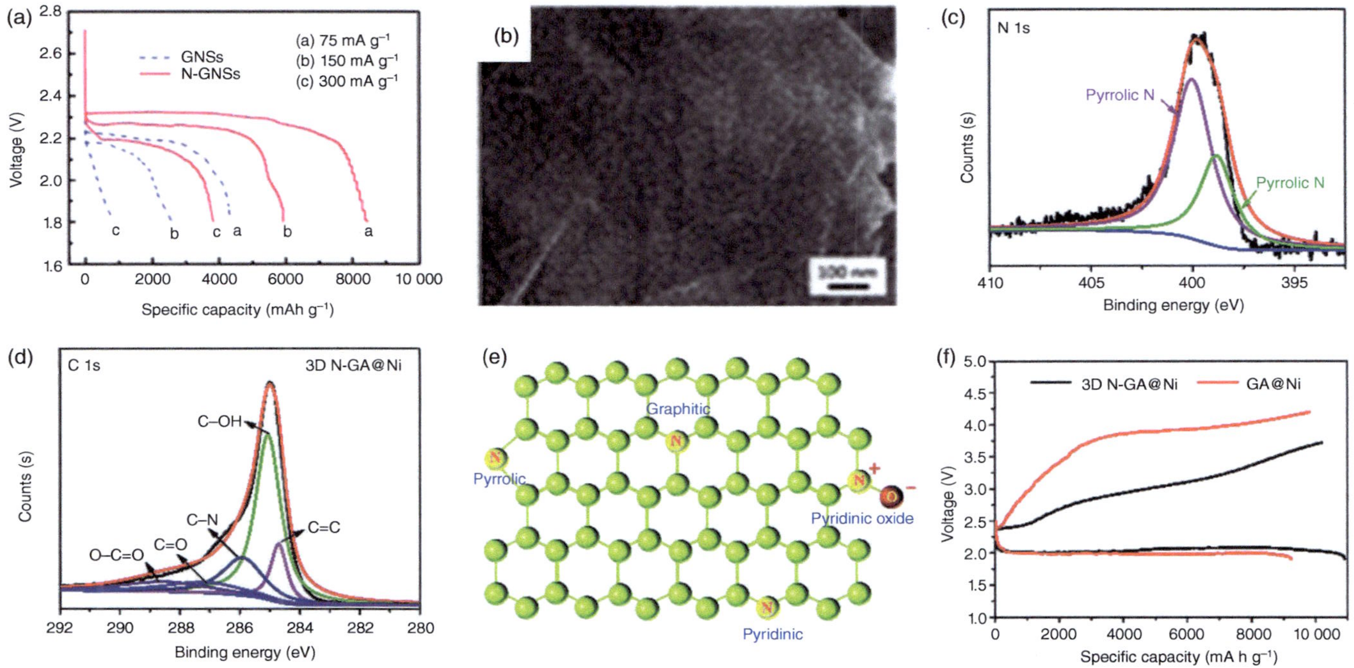

Figure 5.5 (a) Discharge curves of GNS and N-GNS electrodes. (b) SEM image of the N-GNS electrodes discharged for 12 hours. Source: Li et al. [24]. Reproduced with permission of Royal Society of Chemistry. High-resolution XPS spectra of (c) N, (d) C. (e) Sketch of N-doped sites on 3D N-GA@Ni. (f) Discharge/charge voltage profile of $Na-O_2$ batteries based on 3D N-GA@Ni and GA@Ni at 100 mA g^{-1}. Source: Zhang et al. [39]. Reproduced with permission of Royal Society of Chemistry.

the 3D N-GA@Ni was determined by X-ray photoelectron spectroscopy (XPS) measurement. It was found that the active N-groups in 3D N-GA@Ni could control the uniform deposition of discharge products at the nanoscale and provide active sites for decreasing overpotential, which enabled the 3D N-GA@Ni with a large discharge capacity of 10 905 mAh g^{-1} at 100 mA g^{-1}, long cycle life over 100 cycles at 100 mA g^{-1} with cut-off capacity of 500 mAh g^{-1}, and high rate performance over 50 cycles at 300 mA g^{-1}. Sun et al. reported a self-stacked N-doped carbon nanotubes (NCNTs) as cathode for Na–O_2 batteries [40]. Due to the introduction of active sites, Na–O_2 batteries based on NCNT show increased rate performance and prolonged cycle life compared to those based on pristine CNT counterpart. Notably, NCNTs were also found to show the morphology tenability of the discharge products by controlling the current density, where core–shell and nanorod-like morphology could be obtained under low and high current densities, respectively. Yadegari et al. developed a binder-free 3D vertically grown N-doped CNTs on carbon paper (NCNT-CP) for Na–O_2 batteries [41]. The NCNT-CP electrode exhibits a specific capacity that is 17 times higher than bare CP cathode. Ma and Zhang prepared N-doped porous carbon cathode by sol–gel method [42]. Through controlling the heat treatment temperature, N-doped carbon materials with different N contents, species, and hierarchical macropore/mesopore structures are formed for Na–O_2 battery. The high surface area and optimized N distribution enable the result N-doped porous carbon cathode with improved discharge capacity of 6905 mAh g^{-1}. What is more, the morphology of the discharge products could also be regulated.

In addition to N-doping, boron (B)-doping of carbon materials has also been developed for Na–O_2 batteries. Shu et al. synthesized mesoporous B-doped onion-like carbon (B-OLC) microspheres as cathode for Na–O_2 batteries (Figure 5.6a–f) [43]. The introduction of boron produces extra structural defects compared to un-doped carbon. The mesoporous structure of B-OLC cathode was proposed to facilitate the electrolyte impregnation, oxygen diffusion, and provide a plenty of room to accommodate solid products. Meanwhile, the active sites in B-OLC formed by B-doping enhance the capability of O_2 adsorption, resulting in improved ORR kinetics. Consequently, the B-OLCs as cathode for Na–O_2 batteries deliver a high specific capacity of ~10 200 mAh g^{-1} at 0.15 mA g^{-1}, and good cycling stability over 125 cycles at 0.3 mA cm^{-2}. Similar to N-doping, B-doping of carbon materials was also found to have a significant influence on the morphology of discharge products, where thick sheet-shaped structure could transform into an ultrathin sheet-shaped structure.

Additionally, Liang et al. reported that multielement doping of carbon materials could achieve better catalytic activity than single-element doping due to the synergistic effect of different elements [44]. The authors prepared N and S dual-doped graphene (N–S–G) as an electrochemical catalyst for ORR. The N–S–G catalyst shows much better ORR performance than graphene catalysts doped by solely S atoms (S-G) or N atoms (N–G), comparable to the commercial Pt/C catalyst. Moreover, density functional theory (DFT) calculation revealed that the ORR performance enhancement originated from the redistribution of spin density and charge density of graphene induced by the dual doping of S and N atoms, which resulted in the formation of a large number of active sites on adjacent carbon atoms.

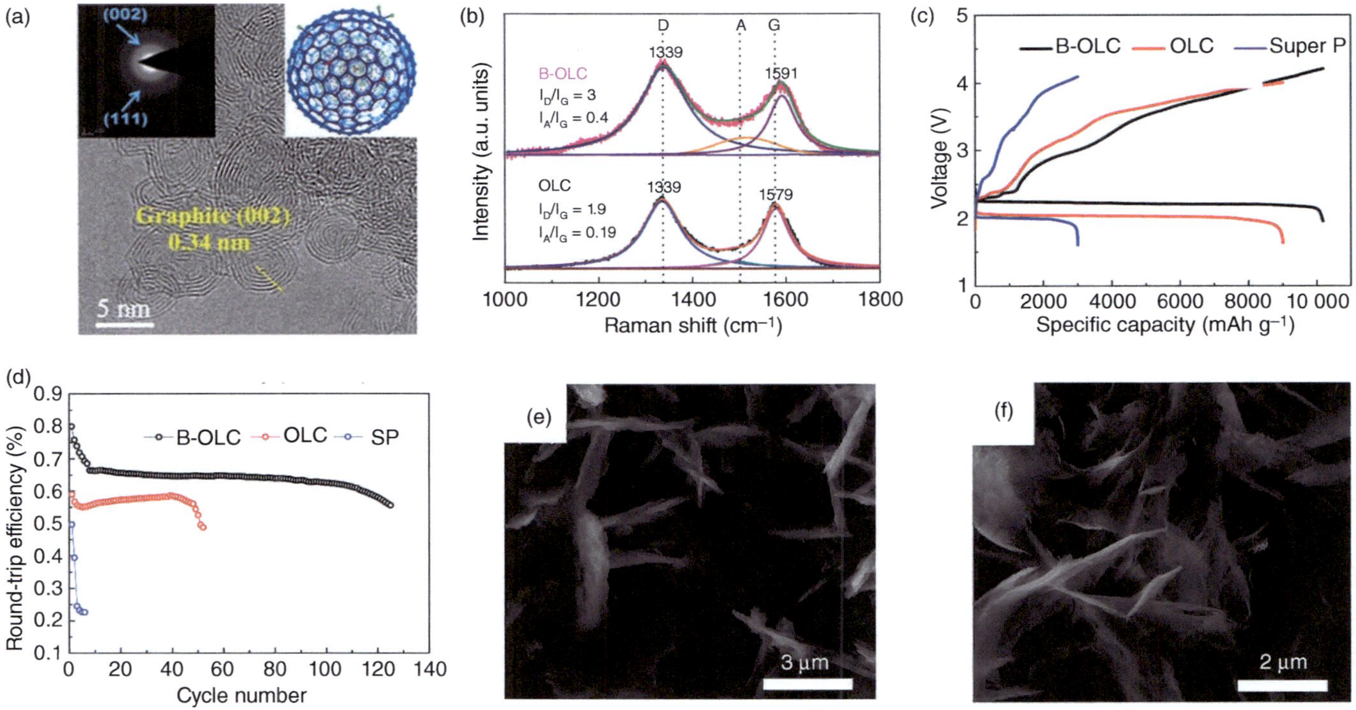

Figure 5.6 (a) High-resolution TEM image and selected area electron diffraction pattern (inset) of B-OLC. (b) Raman spectra of B-OLC and OLC. (c) Discharge/charge curves of B-OLC, OLC, and super P electrodes for $Na-O_2$ at 0.15 mA cm^{-2}. (d) Cycle performance of B-OLC, OLC, and super P electrodes at 0.3 mA cm^{-2}. SEM images of the discharge products of (e) OLC and (f) B-OLC. Source: Shu et al. [43]. Reproduced with permission of Royal Society of Chemistry.

5.3.2 Metals and Their Oxides

5.3.2.1 Noble Metals and Their Oxides

Non-carbon-based materials (e.g. metals and metal oxides) have been shown to exhibit high catalytic activities in wide applications, such as fuel cells [45–47] and Li–O_2 cells [48–50]. Similarly, metals and metal oxides have also been investigated as efficient cathodes for Na–O_2 batteries. In 2015, Yang and coworkers successfully prepared graphene nanosheets loaded with highly dispersed platinum nanoparticles (Pt@GNSs) and studied the electrochemical performance of Pt@GNSs as cathode of Na–O_2 batteries for the first time [51]. The presence of Pt nanoparticles in the graphene not only provides more oxygen adsorption sites to increase discharge capacity but also facilitates the uniform deposition of discharge products. Most importantly, the Pt@GNS structure effectively alleviates the blockage of discharge product aggregation to the oxygen diffusion channel. As a result, Na–O_2 battery based on the Pt@GNS cathode shows a high discharge capacity of 7574 mA h g^{-1} at 0.1 mA cm^{-2} and good cycling performance with a limited discharge capacity of 1000 mAh g^{-1}. Kumar et al. reported the use of Ag nanoparticle-decorated rGO (Ag-rGO) as a bifunctional catalyst for ORR and OER for Na–O_2 batteries [52]. The 10th cycle discharge capacity reaches 566 mAh g^{-1} at 0.1 mA cm^{-2} with Coulombic efficiency higher than 80%. Besides, NaO_2, Na_2O_2, and Na_2O are identified to be the stable discharged products. Lutz et al. investigated electrochemical oxygen redox behaviors of gas diffusion layer (GDL) cathode and gold coated GDL (Au-GDL) cathode for Na–O_2 batteries [53]. The results indicate that the surface properties of cathode materials play a key role in the growth of discharge product and ultimately determine the capacity. Specifically, Au-GDL surface displays a high surface interaction with O_2 due to the high surface energy and large superoxide affinity of Au (Figure 5.7a). Compared to GDL, Au-GDL shows increased nucleation rate and altered growth process of Na_2O discharge product. Consequently, large cubes (10 μm length) of NaO_2 and thin (3 μm thickness) flakes of NaO_2 are formed on the surface of GDL and Au-GDL, respectively (Figure 5.7b,c).

Kang et al. reported the dual catalytic effect of CNTs decorated with ruthenium nanoparticles (CNT/Ru) cathode for Na–O_2 batteries for the first time [26]. The CNT/Ru cathode shows a high catalytic efficiency of OER, leading to a reduced overpotential. Meanwhile, the relatively strong interaction between adsorbed oxygen and Ru surface further promotes the ORR process, which results in the formation of discharge product of $Na_{2-x}O_2$ rather than NaO_2. The formation of $Na_{2-x}O_2$ instead of NaO_2 for Na–O_2 enhances the stability and reversibility of Na–O_2 battery. As a result, the CNT/Ru cathode exhibits stable cycling performance over 100 cycles and the cycling efficiency of battery before failure (due to the Na anode) approach 100%. Apart from Ru metal particle, Wu et al. prepared a micrometer-sized RuO_2 coated on B-doped rGO (m-RuO_2-B-rGO) as cathode for Na–O_2 batteries [54]. By comparing with Na–O_2 batteries using rGO and B-rGO cathodes, different morphologies of discharge products and electrochemical performance are shown in that using m-RuO_2-B-rGO cathode. Specifically, micrometer-sized cubic NaO_2, film-like NaO_2, and nano-sized amorphous spherical $Na_{2-x}O_2$ are characterized as

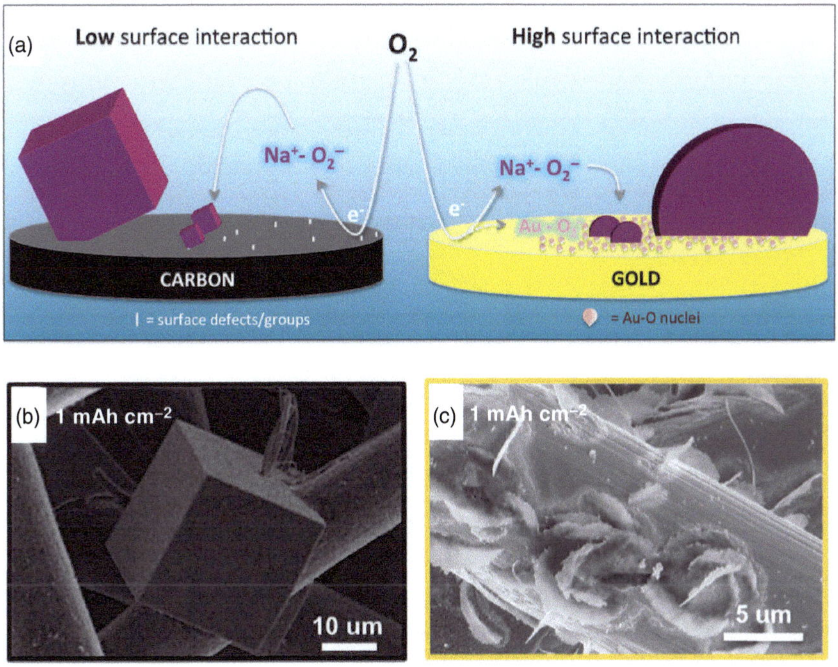

Figure 5.7 (a) Scheme of NaO_2 crystallization pathways on carbon and Au surfaces. SEM images of fully discharged (b) GDL and (c) Au-GDL cathode. Source: Lutz et al. [53]. Reproduced with permission of Wiley-VCH.

the main morphology for rGO, B-rGO, and m-RuO_2-B-rGO cathodes, respectively (Figure 5.8a–c). Besides, the m-RuO_2-B-rGO cathode exhibits longer cycle life over 100 cycles than other two cathodes. The better cycling performance of m-RuO_2-B-rGO is mainly attributed to the high affinity of RuO_2 to oxygen and good electrical contact between amorphous $Na_{2-x}O_2$ discharge product and RuO_2 particles. Meanwhile, the micrometer-sized RuO_2 particles provide sufficient active sites and space for ORR, as well as the minimize side reactions between discharge products and carbon defects. It suggests that the affinity of the surface structure of cathode to oxygen (or superoxide) could affect the growth process of discharge products (Figure 5.8d), resulting in the change of the morphology of discharge products, and thus affecting the final performance of the battery. Although noble metals and their oxides catalysts have made considerable progress in enhancing the performance of Na–O_2 batteries, the high cost of noble metal still limits their large-scale application.

5.3.2.2 Non-noble Metals and Their Oxides

To replace expensive noble metal electrocatalysts, abundant and relatively low-cost transition metal oxides materials have been extensively investigated as promising candidates for Na–O_2 batteries ORR and OER catalysts. In 2014, Liu et al. synthesized $NiCo_2O_4$ nanosheets supported on Ni foam as a carbon-free and binder-free cathode for Na–O_2 batteries [55], which exhibit an initial discharge capacity of

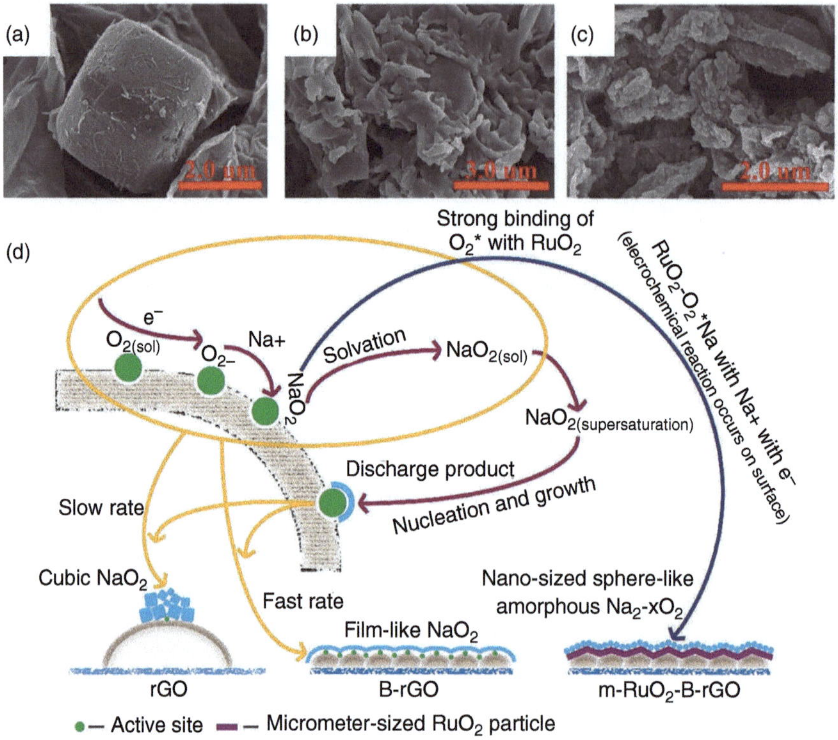

Figure 5.8 SEM images of discharge products formed on (a) rGO, (b) B-rGO, and (c) m-RuO_2-B-rGO cathodes. (d) Schematic illustration of the proposed mechanism accounting for the different morphology of discharge products on different cathodes. Source: From Wu et al. [54]. Reproduced with permission of Wiley-VCH.

1762 mAh g^{-1} with a low polarization of 0.96 V at 20 mA g^{-1}. Na_2O_2 with nanosheet morphology is observed as the discharged product (Figure 5.9a,b). In 2017, Sun et al. prepared CNT decorated by Co_3O_4 material (CNT@Co_3O_4) via atomic layer deposition (ALD) as a non-noble metal oxide catalyst for Na–O_2 batteries [57]. Well-dispersed Co_3O_4 on CNTs serves as functionalized active sites, which enables rapid electron exchange and high oxygen reduction/evolution activities, promotes the electrochemical decomposition of sodium peroxide, superoxide, and carbonates, and further reduces the overpotential of battery. Furthermore, dissolved Co^{2+}/Co^{3+} ions was proposed to affect reaction intermediate and thus control the morphology of the discharge products. Mai and coworkers designed and synthesized a high-spin Co_3O_4 as cathode for Na–O_2 batteries [56]. Interestingly, discharge product of Na_2O with nanowire (NW) shape is observed nucleating on the surface layer of catalyst particles and growing through the space between them (Figure 5.9c). Benefiting from the discharge product of Na_2O nanowires, Na–O_2 batteries using this high-spin Co_3O_4 cathode can run more than 400 cycles with a fixed capacity of 1000 mAh g^{-1} at current densities of 100 and 200 mA g^{-1} and achieve a very low charge overpotential (~60 mV) and near-zero discharge overpotential. This strategy provides a unique

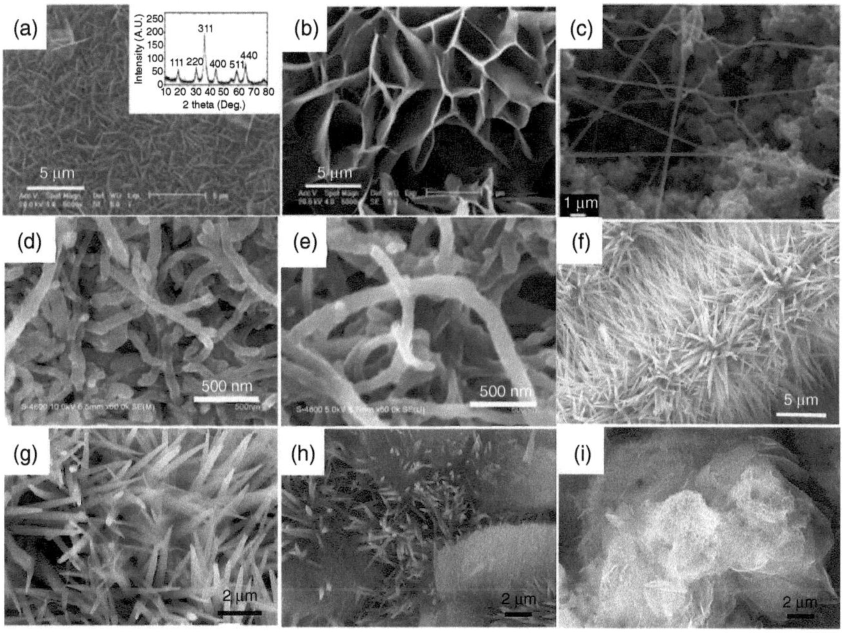

Figure 5.9 (a) SEM image of $NiCo_2O_4$ nanosheets on Ni foam, and the inset shows XRD pattern of the nanosheets. (b) SEM image of the discharge product of $NiCo_2O_4$ nanosheets cathode. Source: Liu et al. [55]. Reproduced with permission of Elsevier. (c) SEM image of the discharge product of high-spin Co_3O_4 cathode. Source: Khajehbashi et al. [56]. Reproduced with permission of American Chemical Society. SEM images of the discharge product of ALD CNT@Co_3O_4 at (d) 150 and (e) 300 mA g^{-1}. Source: Sun et al. [57]. Reproduced with permission of Wiley-VCH. (f) SEM image of pristine C@$NiCo_2O_4$-NAs. SEM images of C@$NiCo_2O_4$-NAs after discharging to (g) 800 mAh g^{-1}, (h) 3000 mAh g^{-1}, and (i) 6500 mAh g^{-1}. Source: Liu et al. [58]. Reproduced with permission of Elsevier.

way to control the morphology of discharge product to achieve high-performance Na–O_2 batteries. Recently, Liu et al. designed a vertically self-standing C@$NiCo_2O_4$ nanoneedle arrays (C@$NiCo_2O_4$-NAs) as a binder-free cathode for Na–O_2 batteries [58]. The score-shell structures of C@$NiCo_2O_4$-NA are favorable electron transfer and O_2 diffusion path. Such high ORR and OER activities enable C@$NiCo_2O_4$-NAs with high initial specific capacity of 6500 mAh g^{-1} and low overpotential of 0.6 V at 50 mA g^{-1}. The formation mechanism and the morphological evolution of discharge products on C@$NiCo_2O_4$-NAs were studied under different depths of discharge. The results indicate that the homogeneity of discharge products is greatly influenced by the discharging depth (Figure 5.9f–i). Zhang et al. developed a binder-free cathode based on Co in situ embedded and coated on N-doped carbon fibers (Co-ECNCFs) for Na–O_2 batteries [8]. The Co-ECNCFs for Na–O_2 battery display improved electrochemical performance, including low charge overpotential (reduced by 200 mV), high discharge capacity (increased by ~57%), and long cycle life (up to 112 cycles). The improved performance is attributed to fiber shape and binder-free Co-ECNCFs with porous structure, accelerating the transportation of mass and electron within the cathode. Most importantly, the synergistic effect between

N-doping and Co encapsulating/coating in Co-ECNCFs leads to the enhanced and stabilized kinetics of ORR and OER, as well as the uniform formation of film-like discharge products. Recently, CuO nanowires [59] and MnO_2 nanowires [60] have been reported as cathode for solid-state Na–O_2 batteries. In these works, real-time imaging technique using in situ aberration-corrected environmental transmission electron microscope (ETEM) was utilized to reveal the catalytic mechanism of the catalyst for the ORR process.

Other non-noble metal-based materials have also been developed as efficient catalysts for Na–O_2 batteries. For example, Hu et al. reported the porous $CaMnO_3$ microspheres as an efficient electrocatalyst for Na–O_2 batteries [61]. The good performance of $CaMnO_3$/C is attributed to the porous structure of $CaMnO_3$ microspheres that provide abundant active sites for bifunctional ORR/OER and effective space for the accommodation of discharge product. Li et al. developed pyrochlore-type oxides, $La_2Co_2O_7$, as cathode catalysts for Na–O_2 batteries [62]. Due to metallic character and the increased specific surface area, the $La_2Co_2O_7$ cathode displays high capacity up to 20 184.2 mAh g^{-1} and good cycle stability up to 167 cycles. Besides, it was also reported that porous cobalt boride (CoB) nanosheets for Na–O_2 possess high catalytic activity and electronic conductivity, which make the material suitable to be used as a cathode catalyst for Na–O_2 batteries [63]. Recently, Tang et al. reported that the heterostructures constructed by transition-metal dichalcogenides (TMDs) with MXenes (M = early transition metal, X = C, N, or both, ene = surface modification group such as O, OH, or F) have great application potential for Na–O_2 batteries [64]. Based on adsorption energy and Bader charge analysis, it was predicted that VS_2/Ti_2CO_2 with O facet of the heterostructure could be a promising candidate for Na–O_2 batteries. Although considerable progress has been made in the study of non-noble metals and their oxides as catalysts for cathode for Na–O_2 batteries, the achieved catalytic capacity is still far from satisfactory.

5.3.2.3 Dual Functional Composites

Recently, the design of nanomaterials with dual function catalytic effect for Na–O_2 batteries has aroused wide interest. Sun and coworkers developed a 3D-structured carbonaceous cathode composed by graphene, NCNTs, Mn_3O_4, and Pd (Gr/NCNT/MnO/Pd). The bifunctional role was mainly realized by the mesoporous Mn_3O_4 as an ORR catalyst and Pd nanoclusters as OER catalysts [65]. The enhanced ORR catalytic activity is mainly attributed to the adsorption of molecular oxygen on the catalyst surface that reduces the activation energy. The improved OER catalytic activity is mainly due to the ability of the catalyst to stabilize the superoxide intermediate. In addition, Pd nanoclusters and mesoporous Mn_3O_4 in Gr/NCNT/MnO/Pd composite display a synergetic effect for OER, which is correlated with the electron exchange between their metallic d orbitals. Amine et al. developed a composite of Pd nanoparticles on the ZnO-passivated porous carbon (Pd/ZnO/C) as cathode for Na–O_2 batteries [66]. The bond between Pd and ZnO allows the stable anchoring of Pd catalysts on the cathode and improves the battery cycle life. Interestingly, the composite shows a low overpotential (~0.5 V) due to the formation of an oxygen-rich superoxide-like product in the presence of

Pd nanoparticles. $Na_2O_2{\cdot}2H_2O$ with porous structure was observed as the main discharge product, which is different from the main discharge product of Na_2CO_3 in the reported Pt@GNSs cathode [51]. Wang et al. synthesized CoO/CoP nanosheets with heterogeneous structure and interpenetration interface between P and O atoms and used the composite as a bifunctional catalyst for Na–O_2 batteries [67]. The synthesized CoO/CoP heterostructured nanosheets inherit both the high OER activity of CoP and the high ORR activity of CoO. Owing to the formation of the O–P interpenetration interface, OER and ORR activities of CoO/CoP heterostructure are simultaneously enhanced and even exceed that of the parent CoO or CoP as bifunctional catalysts for Na–O_2 batteries. As a result, the initial capacity, rate capacity, and cycling performance of Na–O_2 battery based on CoO/CoP heterostructured nanosheet cathode are significantly improved.

Recently, Tovini et al. prepared a 3D RuO_2/Mn_2O_3/carbon nanofiber (CNF) composite by a facile two-step microwave synthesis and applied the composite as a bifunctional electrocatalyst for Na–O_2 batteries [68]. Due to the synergistic effect of RuO_2 and Mn_2O_3, the RuO_2/Mn_2O_3/CNF composite exhibits much higher specific capacity of 9352 mAh g^{-1} than CNF (1395 mAh g^{-1}), Mn_2O_3/CNF (3108 mAh g^{-1}), and RuO_2/CNF (4859 mAh g^{-1}) for Na–O_2 batteries. More recently, Liu et al. reported hierarchical Co_3O_4@$MnCo_2O_{4.5}$ nanocubes (h-Co_3O_4@$MnCo_2O_{4.5}$ Ns) composite with high specific surface area of 130.4 m^2 g^{-1}, hierarchical macro- and mesoporous structure, and synergistic yolk-shelled active sites [69], delivering enhanced electrocatalytic activities toward both ORR and OER. The h-Co_3O_4@$MnCo_2O_{4.5}$ Ns as cathode for Na–O_2 batteries displays a high initial discharge capacity of 8400 mAh g^{-1}, charge–discharge overpotential of 0.45 V, and long cycling life over 135 cycles.

5.4 Anode Materials

The anode material also plays an important role in the overall performance of Na–O_2 batteries. Na metal is considered as an attractive anode candidate for Na–O_2 battery due to its high theoretical specific capacity of 1166 mAh g^{-1} and low electrochemical potential of −2.71 V (vs. standard hydrogen electrode) [70–72]. However, metallic Na suffers from several disadvantages during electrochemical cycling, which hinders its further application in Na–O_2 batteries. For example, (i) the inhomogeneous electrochemical deposition of Na^+ on the surface of Na metal anode during the charging process could result in the growth of sodium dendrites, leading to constant loss of active Na, reduced Coulomb efficiency, and the potential safety hazard of short circuit inside the battery [73]. (ii) The formation of unstable SEI layer could lead to the depletion of electrolyte and electroactive Na and thus shortens the cycle life of the battery [13, 14]. (iii) The hostless nature of the Na metal anode results in infinite volume change during the electrochemical plating/stripping process [74–76]. Therefore, various strategies have been proposed to address these problems, such as constructing artificial protective layers, forming in situ stable SEI layer, introducing solid electrolyte to protect metallic Na, guiding Na deposition behavior by carbon materials modification, and developing alternative anode materials. In this section, the recent progress in dealing with the challenges of Na metal anode for Na–O_2 batteries will be reviewed.

5.4.1 Modification of Na Metal Anode

Surface modification of Na metal anode with an interfacial or protective layer is an effective strategy to protect Na metal anode. Previous studies have shown that modifying separator as a blocking interlayer is an attractive method to inhibit the growth of Na dendrite. This is because it can be well compatible with previous battery fabrication techniques. The modified separator could not only play the same function as a pristine separator but also show high mechanical strength to prevent Na dendrite from penetrating the separator. In 2015, Bi et al. fabricated a Nafion-Na^+ membrane into glass fiber separator to physically prevent the penetration of Na dendrite [15], leading to the enhanced cycling stability up to 120 cycles. Subsequently, Ma et al. reported a fibrillar polyvinylidene fluoride (f-PVDF) film with nonthrough pore as a multifunctional blocking interlayer [77] (Figure 5.10a). No Na dendrites were observed on the f-PVDF interlayer after 3 mAh deposition (Figure 5.10b), superior to its counterparts (e.g. compact-PVDF, porous-PVDF, polyethylene oxide [PEO] film, polytetrafluoroethylene [PTFE] film). This result indicates that the f-PVDF interlayer with uniform fiber structure could effectively suppress the growth of Na dendrite, leading to the higher rate capacity and longer cycle life (up to 87 cycles) than other counterparts (Figure 5.10c,d). This unique structure with high ion conductivity, strong polar C–F function groups, and fine affinity with electrolyte is favorable for electrolyte uptake, Na ion flux, and homogeneous Na deposition. Inspired by an inorganic filler (SiO_2) sandwiched separator that can restrain the growth of Li dendrite [79], a new polypropylene–TiO_2–polypropylene (PP–TiO_2–PP)-sandwiched separator was also designed for Na—O_2 battery as shown in Figure 5.10e [78]. TiO_2 nanoparticles filled between two layers of PP separators could react with Na to slow down the growth of Na dendrites. PP–TiO_2–PP-sandwiched separator for Na—O_2 battery increases the cycle life from 82 to 137 cycles with a cut-off capacity of 1000 mAh g^{-1} at 200 mA g^{-1} (Figure 5.10f). Ansari et al. prepared a mechanically reinforced glass microfiber (RGMF) by a simple infusion of high modulus inorganic particles and investigated the use of RGMF as a separator for Na–O_2 battery [80]. This separator could not only inhibit dendrite growth but also efficiently restrict oxygen crossover toward the anode. Na–O_2 batteries based on RGMF separator realize the reversible cycle of more than 400 cycles at the current density of 0.28 mA m^{-2} to the capacity of 1.0 mAh cm^{-2}.

Apart from separator modification, artificial SEI coating on the surface of Na metal anode is a promising approach to protect metallic Na anode. For example, Choudhury et al. reported that uniformly coating NaBr on a Na metal by direct chemical reaction not only restricted dendritic formation but also prevented unwanted side reactions between the electrode and electrolyte (Figure 5.11a) [81]. Recently, a stable high-performance metal alloying Na/Bi anode was fabricated by

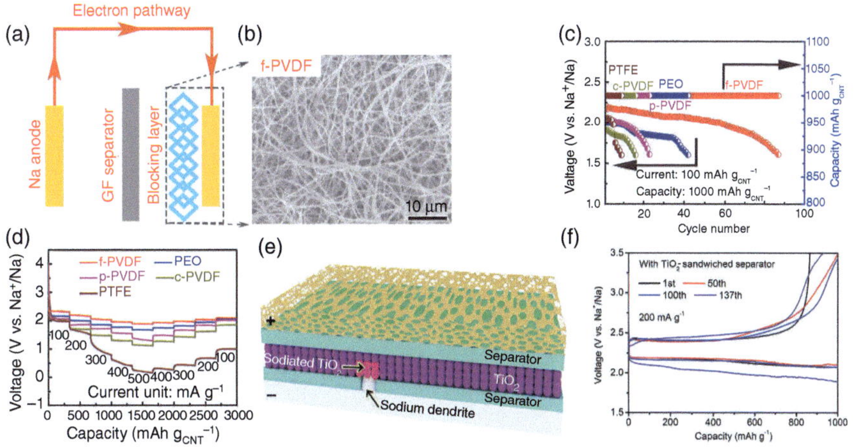

Figure 5.10 (a) Schematics of the symmetrical battery design. (b) SEM image of f-PVDF. (c) Cycling performance and (d) rate capability of Na–O_2 batteries with the five blocking interlayers. Source: Ma et al. [77]. Reproduced with permission of Wiley-VCH. (e) Scheme of suppressed growth of Na dendrite by the PP–TiO_2–PP sandwiched separator. (f) Cycling performance of Na–O_2 batteries with the PP–TiO_2–PP sandwiched separator. Source: Yang et al. [78]. Reproduced with permission of Royal Society of Chemistry.

in situ deposition of Bi layer on the surface of Na metal via a simple replacement reaction [82]. The resulting Bi layer is about 10 μm thick and uniformly distributed on Na metal, which effectively prevents the reaction of Na with electrolyte and suppressed the formation of Na dendrite (Figure 5.11b–f). The Na/Bi composite anode could run for 50 cycles without the formation of dendrite and exhibit good high-rate performance (Figure 5.11g,h). Previous works have revealed that gas phase deposition methods, such as ALD and molecular layer deposition (MLD), are effective tools to deposit artificial SEI layers due to their exclusive advantages of extremely uniform and controllable deposition ability under relatively low temperatures [71]. For example, ALD Al_2O_3 coatings and MLD alucone (Al-ethylene glycol) have been reported as protective layers for Na metal anodes [83–85]. However, their performance for applying in Na–O_2 batteries still needs further verification.

In addition to inhibiting Na dendrite growth by constructing an artificial protective layer, the in situ formation of a stable SEI layer by optimizing the electrolyte formulation is also an effective strategy to stabilize Na metal. The electrolyte composition, such as organic solvents, Na salts, and electrolyte additives, has been shown to significantly affect the stability of the electrolyte. $NaPF_6$ is found to be able to form a uniform and compact SEI layer composed by Na_2O and NaF, which is highly impermeable to the electrolyte solvent and conducive to nondendritic growth [86]. Lutz et al. also reported that $NaPF_6$ salt in ether solvents could form a stable SEI composed of NaF, achieving the high efficiency (>98%) and performance for Na–O_2 battery [87]. Several studies indicate that NaF is a crucial component to realize highly stable SEI layer. Fluoroethylene carbonate (FEC) is regarded as an effective electrolyte additive for producing stable NaF-based SEI layer and improving Coulombic efficiency of Na–O_2 battery [88, 89]. Wu et al. reported a stable Na–O_2 battery with

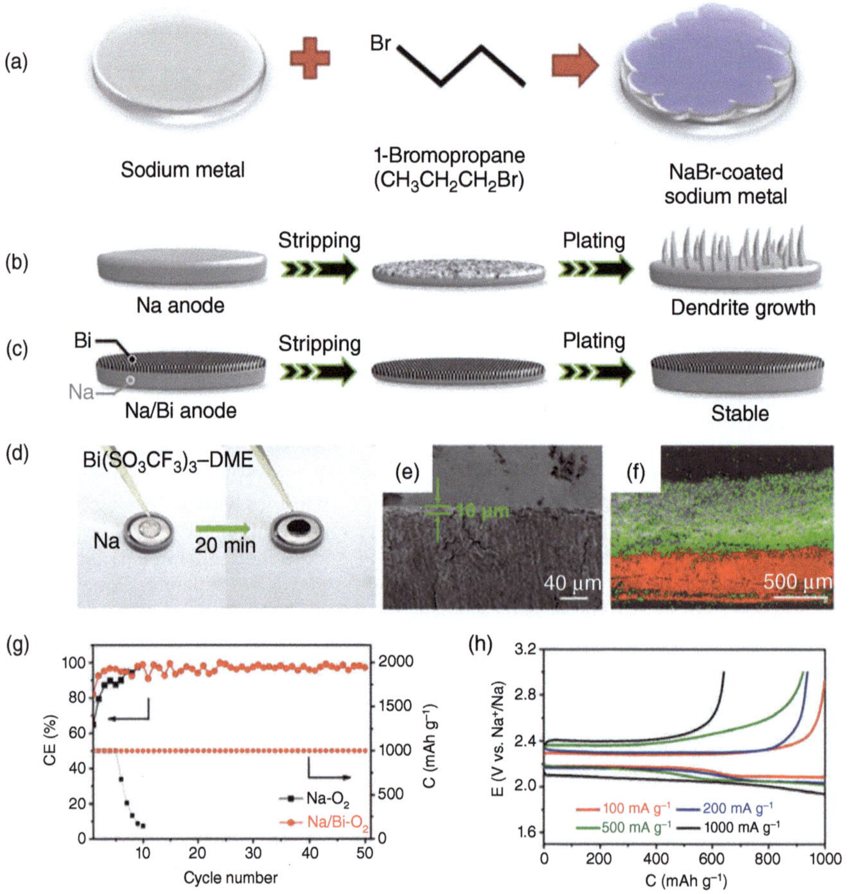

Figure 5.11 (a) Scheme of the procedure used to coat Na metal with NaBr. Source: Choudhury et al. [81]. Reproduced with permission of Springer Nature. Models of Na dendrite growth on (b) Na and (c) Na/Bi composite anodes. (d) Photographs of the synthesis process of Na/Bi anode. (e) Cross-sectional SEM image and (f) energy dispersive X-ray (EDS) mappings of Bi (green) and Na (red) of the Na/Bi anode. (g) Coulombic efficiency and discharge capacity of Na/Bi–O_2 battery and Na–O_2 battery during charging/discharging process at 500 mA g^{-1}. (h) Rate performance of Na/Bi–O_2 battery at different current rates. Source: Ma et al. [82]. Reproduced with permission of Wiley-VCH.

FEC as an electrolyte additive [90]. FEC was found to be preferentially reduced to form NaF passivation film on the surface of Na metal anode, namely tailored Na anode (TNa). TNa effectively prevents electrolyte decomposition and suppresses O_2 crossover, which often occurs on pristine Na anode (PNa) (Figure 5.12a,b). The TNa displays a discharge capacity of 1000 mAh g^{-1} with high Coulomb efficiency over 50 cycles (Figure 5.12c–e). As shown in Figure 5.12f–h, the TNa also shows stable Na electroplating and deposition ability and no dendrite formation was observed in the characterization of metal anode after 30 minutes cycling.

Recently, Shi et al. reported that potassium bis(trifluoromethylsulfonyl)imide (KTFSI) could be used as a bifunctional electrolyte additive to stabilize Na metal

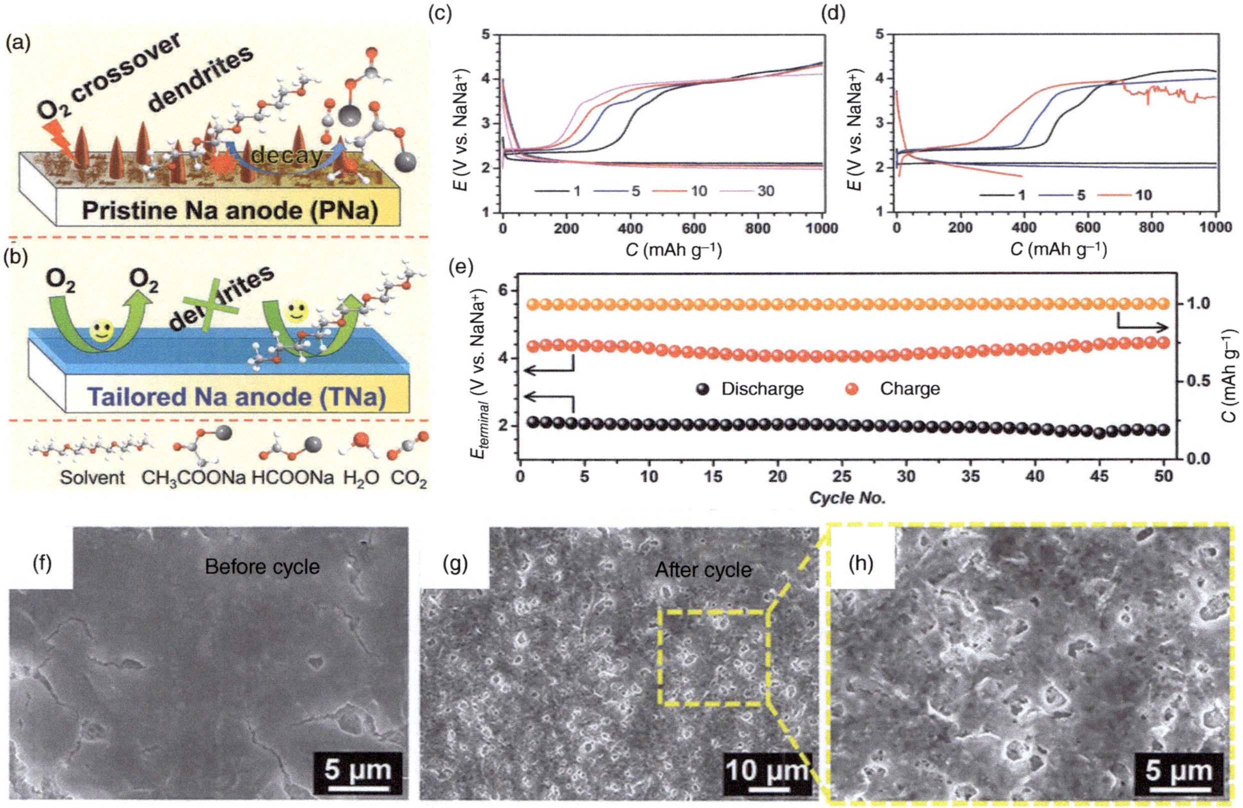

Figure 5.12 (a) Illustration of negative effects (O_2 crossover, dendrite formation, and electrolyte decay) using PNa for $Na-O_2$ batteries. (b) Illustration of the suppression of the negative effects using TNa with NaF passivation film for $Na-O_2$ batteries. Discharge/charge profiles of $Na-O_2$ batteries (c) with and (d) without FEC in the electrolyte. (e) Cycling performance of $Na-O_2$ battery with FEC in the electrolyte. SEM images of (f) PNa anode and (g) TNa anode after stripping/plating cycles. (h) Magnified SEM image in (g). Source: Wu et al. [90]. Reproduced with permission of Wiley-VCH.

anode [91]. Specifically, $TFSI^-$ of KTFSI assists the formation of a desirable SEI layer containing sodium nitride (Na_3N) and sodium oxynitrides (NaN_xO_y), while K^+ of KTFSI preferentially adsorbs onto Na protrusions and provides electrostatic shielding for homogeneous Na^+ flux. By using this KTFSI additive in electrolyte, Na metal anode achieves a stable cycling capacity of 10 mAh cm^{-2} for hundreds of hours. Recently, it was demonstrated that highly concentrated electrolytes can inhibit dendrite growth, which is beneficial to stabilize SEI layer. For example, Ruiz-Martínez et al. [92] and Zheng et al. [93] proposed that, highly concentrated $NaY{\cdot}xNH_3$ inorganic electrolyte and localized high-concentration electrolyte, respectively, can lead to highly reversible plating–stripping of Na and the dendrite-free Na deposition. These studies provide a new concept for the development of dendrite-free and high-performance Na–O_2 batteries.

Although Na dendrite growth on Na metal anode could be inhibited by artificial protective layer or in situ formed stable SEI, the inevitable decomposition of organic electrolyte still exists because of the liquid nature of organic electrolyte. With this in mind, the development of solid-state electrolytes (SSEs), such as ceramic-ion conductors and solid polymer electrolytes (SPEs), could be of great significance. SSEs are expected to greatly enhance the battery safety and electrochemical stability due to the avoidance of the oxidative decomposition of organic solvents that protects Na metal anode. In 2018, Gao et al. designed an all-solid-state sodium battery with a NASICON $Na_3Zr_2Si_2PO_{12}$ solid electrolyte [94]. It was proved that the interfacial parasitic reactions and the growth of Na dendrite can be effectively suppressed in the all-solid-state sodium battery [94]. Liu et al. constructed an all-solid-state Na–O_2 battery with in situ formed Na_2O on Na metal surface as the solid electrolyte for Na^+ conduction, and no carbonate formation was observed after cycling [59]. This is due to the absence of carbon source in the whole battery setup. Kwak et al. also revealed the reaction pathways and morphological evolution of an all-solid-state Na–O_2 nanobattery using in situ transmission electron microscopy (TEM) [95]. These results confirm that solid electrolytes show significant advantage in limiting the side reactions of Na metal anode.

5.4.2 Carbon Materials Modified Na Anode

Although the above-mentioned strategies have been shown to be effective in inhibiting Na dendrite growth, they only deal with the already formed Na dendrites rather than the initial behavior of Na deposition. In other words, those strategies do not address the fundamental problem of Na dendrite growth. Theoretically, the nonuniform distribution of current density on Na metal anode is the primary reason for the growth of Na dendrite [96]. Therefore, controlling the distribution of current collection on Na metal anode to achieve uniform initial nucleation behavior of Na is an effective strategy to fundamentally solve the problem of Na dendrite growth. To realize it, the introduction of a collector or host with high specific surface area on anode could increase the surface area, dissipate local current density, promote uniform nucleation, and inhibit the growth of Na dendrite on Na metal anode [71, 97]. For example, Liu et al. reported that porous Al foils, serving as the plating substrate, could significantly increase the available surface for Na nucleation and decrease the Na^+ flux distribution, leading to the homogeneous plating of Na with suppressed Na dendrite growth [98]. Na–O_2 battery with porous Al foil substrate can maintain the stable capacity over 200 cycles, which is 10 times higher than that using planar Al substrate with fast decay after only 20 cycles. Notably, carbon materials are ideal candidates with high surface area, superior conductivity, and multiple functions. Therefore, the modification of metallic Na with carbon materials not only effectively reduces the local current density and facilitates uniform nucleation on Na metal anode but also provides sufficient space for the deposited Na to alleviate the infinite volume changes.

Currently, carbon fiber paper (CFP) [99], carbonized wood [100], and rGO [101] etc. have been reported as current collecting matrix/host for metallic Na. In 2017, Wang et al. prepared the Na@rGO composite anode by absorbing melted Na into the space between rGO sheets (Figure 5.13a) [101]. It was found that Na@r-GO composite film could be molded to various shapes with controllable size by adjusting the thickness and shape of the densely stacked GO film. Moreover, rGO sheets were found to flatten the surface of the Na@rGO composite, leading to a more evenly distributed Na^+ flux. Besides, rGO can stabilize SEI with homogeneous Na plating because it can avoid the direct contact between metallic Na and electrolyte. Compared with Na metal anode, the plating/stripping cycling of Na@rGO composite anode is significantly extended in both ether and carbonate electrolytes with depressed dendrite formation. More importantly, Na@rGO-based symmetric cell could maintain a stable plating/stripping more than 600 hours (Figure 5.13b), lower charge/discharge overpotential, and better cycling stability. Luo et al. reported the preparation of a stable Na carbonized wood (Na–wood) composite as an anode (Figure 5.13c,d) [100]. The channels of carbonized wood function as a high-surface-area, conductive, mechanically stable skeleton, which lowers the effective current density, ensure a uniform Na nucleation and restrict the volume change on the anode over cycles. Although the application of Na–wood composite anode to Na–O_2 battery has not been investigated, it provides some guidance for the reasonable design of carbon materials modified Na metal anode for Na–O_2 batteries.

Furthermore, decorating the carbon materials with sodiophilic functional groups was found to increase the affinity of substrate to Na, which could guide the uniform nucleation of Na and fundamentally suppress the formation of Na dendrite. A N and S co-doped carbon nanotube (NSCNT) paper was developed as the interlayer on the Na metal anode for Na–O_2 batteries [97]. The N- and S-containing functional groups on CNTs induce the high "sodiophilic" nature of NSCNTs, which could guide the initial Na nucleation and direct Na to distribute uniformly on the NSCNT paper. The Na/NSCNTs show a dendrite-free morphology and excellent cycling stability during repeated Na plating and stripping. Recently, Ye et al. synthesized an oxygen-functionalized 3D CNT network (O_f-CNT) [102] with sodiophilic and robust sodiophilic interphases. This interphase shows a strong interaction between Na metal and the functional groups and facilitated a homogeneous Na nucleation. The Na@O_f-CNT anode exhibits a long cycle life up to 3000 cycles and a capacity

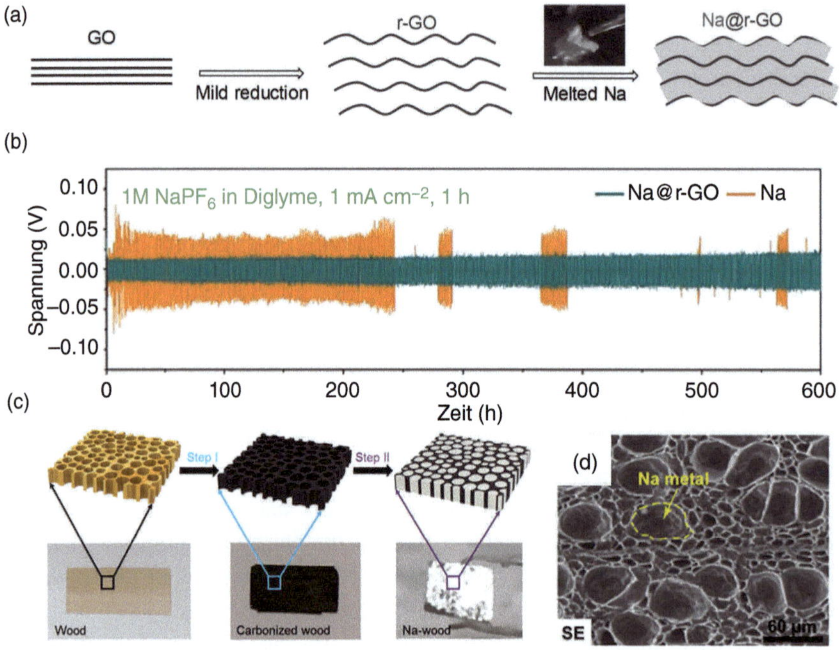

Figure 5.13 (a) Scheme of the preparation process of Na@rGO composite. (b) Electrochemical performance of symmetric cells using Na@rGO anode and pure Na metal anode at 1.0 mA cm^{-2} with a capacity of 10 mA h cm^{-2}. Source: Wang et al. [101]. Reproduced with permission of Wiley-VCH. (c) Scheme of the preparation process of Na-wood composite electrode. (d) SEM image of carbonized wood after melt infusion of metallic Na. Source: Luo et al. [100]. Reproduced with permission of American Chemical Society.

of 1078 mAh g^{-1} with an area capacity of 10 mAh cm^{-2} which is five times higher than that with a bare Na metal anode. More recently, He et al. designed a fibrous hydroxylated MXene/CNT (h-Ti_3C_2/CNT) composite as a scaffold for dendrite-free Na metal anode [103]. The h-Ti_3C_2/CNTs containing abundant sodiophilic O and F functional groups significantly induce homogeneous nucleation of Na. Besides, CNTs in the composite provide high tensile strength and ease film-forming property, and thus h-Ti_3C_2/CNT scaffold exhibits a high average Coulombic efficiency of 99.2% without Na dendrite formation after 1000 cycles. Moreover, the h-Ti_3C_2/CNTs/Na anode exhibits a low potential gap of 0.11 V after 70 cycles.

In addition to modifying Na metal, the development of alternative materials for metallic Na is an effective strategy to achieve high stability of Na–O_2 batteries because the use of metallic Na-free anode can avoid the negative effects associated with Na metals, especially sodium dendrite growth and sodium corrosion. For example, a sodiated carbon electrodes based on a GDL as anode for Na–O_2 batteries was reported by Bender et al. [104]. Using sodiated carbon instead of Na metal effectively prevents dendrite formation and avoids the short circuiting of the cell. Although replacing Na metal with sodiated carbon reduces the available capacity, it still has significant advantages in terms of device safety and stability. It is believed that through continuous research efforts and development of novel Na metal substitutes, stable Na–O_2 battery with high specific capacity could be realized in the future.

5.4.3 Metal Alloys/Composites/Hybrids

Dilimon et al. reported the use of pre-sodiated antimony (Sb) instead of Na metal as anode in dimethyl sulfoxide (DMSO)-based Na–O_2 battery to prevent DMSO from reacting with the metal anode [105]. It provides possibility novel concept to realize the good reversibility and performance of Na–O_2 battery. Recently, a bimetal Li–Na alloy anode for Na–O_2 battery was reported by Ma et al. [106]. As shown in Figure 5.14a, the Li–Na alloy is found to be well tolerant to oxygen and electrolyte. In addition, the addition of Li metal into Na could effectively suppress the growth of Na dendrite due to the strong electrostatic shield effect of Li^+ during the plating process. Furthermore, the introduction of 1,3-dioxolane (DOL) additive into the electrolyte could react with alkali metal anode to form a solid and highly elastic SEI layer on the interface of anode. It not only blocks the direct reaction between the alloy anode and the electrolyte but also effectively buffers the volume expansion effect. The Li–Na alloy anode achieves a long cycling stability of 137 cycles, which is four times longer than that based on pure Na metal anode (31 cycles) (Figure 5.14b).

Oxygen crossover from cathode to the anode in Na–O_2 battery has adverse effects on electrolyte and anode, thus limiting the performance of the battery [108, 109]. As expected, blocking O_2/O_2^- crossover is of primary importance for enhancing the electrochemical performance of Na–O_2 battery. Recently, Lin et al. successfully developed a hybrid solid-state (HSS) Na–O_2 battery based on NASICON-type $Na_{3.25}Zr_2Si_{2.25}P_{0.75}O_{12}$ (NZSPO) as SSE and CP modified Na as an anode (Figure 5.14c) [107]. The dense structure of NZSPO SSE is Na^+ conductive but impermeable to O_2/O_2^-, which effectively suppresses the O_2/O_2^- crossover, thus avoiding Na metal corrosion and improving the cell reversibility. Meanwhile, the introduction of CP between SSE and Na metal anode could facilitate uniform Na deposition. Owing to the synergistic advantages of SSE and CP modification, the developed HSS Na–O_2 battery delivers high initial discharge capacities of 5.15 and 4.21 mAh cm^{-2}, high Coulomb efficiency of 97.6% and 93.5%, at 0.1 and 0.2 mA cm^{-2}, respectively (Figure 5.14d), and stable cycling for over 160 cycles at 0.2 mA cm^{-2} (Figure 5.14e). Furthermore, the depth profile analysis through XPS shows that the SSE shielding of O_2/O_2^- is attributed to the formation of NaF-rich inorganic SEI protective layer.

The above-discussed results indicate that anode plays a crucial role in determining the overall Na–O_2 battery performance. Rational anode design could greatly promote the development of advanced Na–O_2 batteries with improved cycling performance. Although recent advances have partially resolved the problems faced by Na metal anode, further development of effective strategies to realize highly reversible Na metal anode remains a big challenge in the field of Na–O_2 battery toward large-scale application.

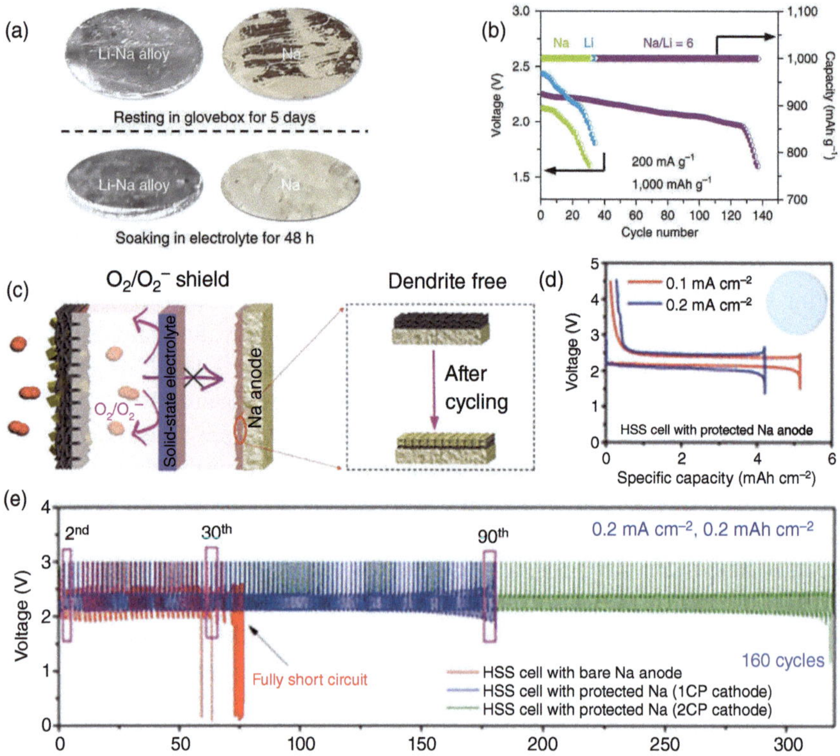

Figure 5.14 (a) Optical photographs of the Li–Na alloy anode and Na anode after undergoing different treatments. (b) Cycling performance of Na–O_2 batteries based on different anodes at 200 mA g^{-1}. Source: Ma et al. [106]. Reproduced with permission of Spinger Nature. (c) Schematic illustration of the effects of O_2/O_2^- crossover prevention by SSE. (d) The discharge/charge curves of HSS cells using NZSPO SSE and CP-protected Na anode at 0.1 and 0.2 mA cm^{-2}, the inset shows optical image of the SSE collected after fully recharged at 0.2 mA cm^{-2}. (e) Cycling performance of HSS Na–O_2 batteries with or without CP protection on the Na anode at 0.2 mA cm^{-2} with a limited capacity of 0.2 mAh cm^{-2}. Source: Lin et al. [107]. Reproduced with permission of American Chemical Society.

5.5 Electrolytes

Instability of electrolyte is one of the major challenges for the development of stable Na–O_2 batteries. The nature characteristics of organic electrolytes and their decomposition to insulation side products during the discharge/charge process greatly affect the electrochemical performance of Na–O_2 battery. To obtain high-performance battery, an ideal electrolyte should have the advantages as follows: high toleration of the highly oxidative environment, high solubility of electrolyte salts and O_2, high ionic conductivity, and high compatibility with electrodes [32, 110]. So far, electrolytes for Na–O_2 batteries are adopted from that in well-studied Li–O_2 batteries, such as organic electrolytes, DMSO, acetonitrile (ACN), and ionic liquids (ILs). In this section, the application of these electrolytes for Na–O_2 batteries will be introduced.

5.5.1 Carbonate-Based Electrolyte

In nonaqueous Li–O_2 batteries, carbonate-based electrolytes are commonly used as solvents. Similarly, early studies of Na–O_2 batteries were also conducted using carbonate-based electrolytes. In 2012, Sun et al. reported Na–O_2 battery for the first time using $NaPF_6$ dissolved in ethylene carbonate (EC) and dimethyl carbonate (DMC) as electrolyte at room temperature [12]. In addition to crystallized Na_2O_2 as major discharge product, Na_2CO_3 or NaOCORO was also detected from the discharge product, which was proposed to be the result of carbonate solvents being attacked by the reduction product of O_2. However, the complex reaction routes of carbonate solvents remain unclear. In 2013, Kim et al. investigated the electrochemical reaction of Na–O_2 batteries based on propylene carbonate (PC)-based electrolyte [29] and it was found that Na_2CO_3 was the main discharge product in the carbonate-based Na–O_2 battery. As shown in Figure 5.15, the formation and decomposition of Na_2CO_3 are responsible for the cyclic process. During the discharging process (Figure 5.15a), the reaction starts from O_2^- attacking the ethereal C atoms of the CH_2 group in PC by S_N2 substitution, which finally forms Na_2CO_3 with the evolution of O_2 gas through a series of triggering reactions. During the charging process (Figure 5.15b), the resulting Na_2CO_3 decomposes into H_2O and CO_2 again.

In addition, other research groups also studied Na–O_2 batteries with carbonate electrolyte and Na_2O_2 or Na_2CO_3 was reported as the main discharge product [28, 51]. However, Na_2CO_3 and Na_2O_2 with poor electronic conductivity cannot be completely decomposed during the charging process. Moreover, these discharge products will accumulate on the cathode electrode during the deep cycles, leading to the poor electrochemical performance.

5.5.2 Ether-Based Electrolyte

Compared to carbon-based electrolytes subjecting to be unstable and vulnerable to superoxide ion attacking, ether-based electrolytes with better stability have been widely utilized for Na–O_2 batteries [12]. Zhao and Guo revealed that the acid dissociation constant (pK_a) of the electrolyte solvent was closely related to the stability of the electrolyte [111]. For example, tetraethylene glycol dimethyl ether (TEGDME)-based electrolyte with relatively large value of pK_a shows the good stability with the predominant yield of NaO_2 upon discharge. The battery also shows the largest electrical energy efficiency approaching 90% (Figure 5.16a). Aldous et al. analyzed the chemical reactions of a series of solvents for Na–O_2 batteries [112]. They revealed that the Gutmann donor number (D_N) of the solvent is correlated with the ability of solvent to form ion pairs with O_2^- intermediates. Notably, solvents with higher D_N could lead to the initial formation of an ion pair, which is removed from the surface, and then aggregate and precipitate out in the form of NaO_2 during the discharge process (Figure 5.16b). Lutz et al. compared the influence of electrolytes (dimethyl ether [DME], diethylene glycol dimethyl etherv[DGME], and tetraethylene glycol dimethyl ether [TGME]) with different ether-chain lengths on the formation mechanism of discharge product for Na–O_2 batteries [113]. The results indicate that long-chain ethers with strong

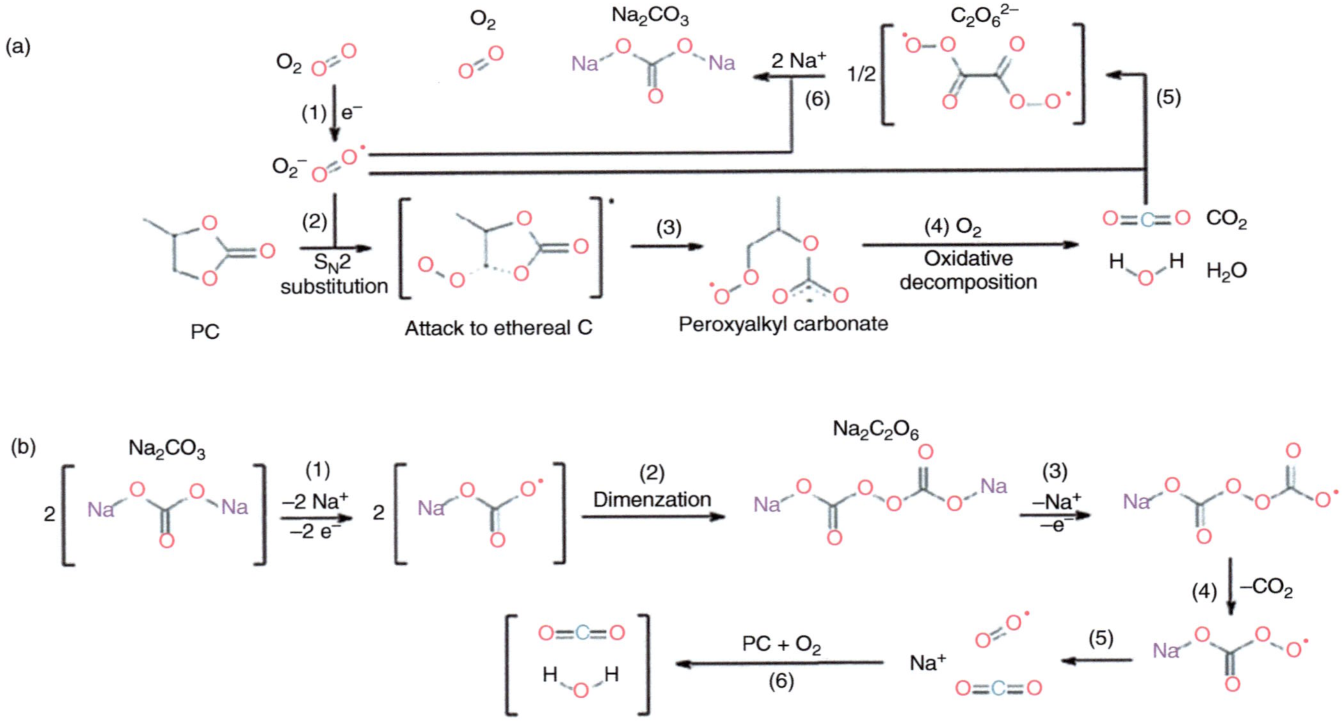

Figure 5.15 Schematic diagram of the proposed reaction mechanisms for the (a) discharge and (b) charge processes with PC-based electrolyte. Source: Kim et al. [29]. Reproduced with permission of Royal Society of Chemistry.

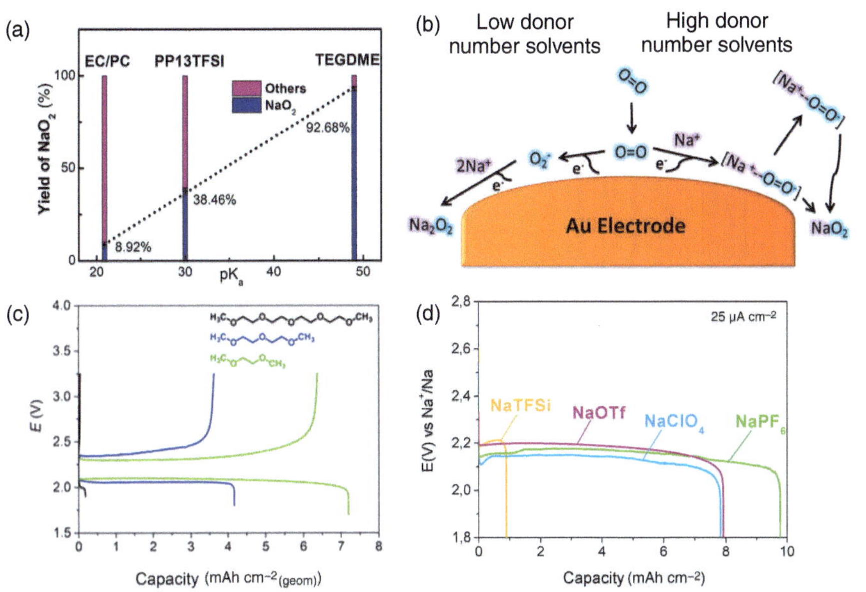

Figure 5.16 (a) Correlation between NaO_2 yield and acidity of solvent. Source: Zhao and Guo [111]. Reproduced with permission of American Chemical Society. (b) Scheme of oxygen reduction mechanism in different non-aqueous solvents. Source: Aldous and Hardwick [112]. Licenced under CC BY 4.0. (c) Discharge/charge curves of Na–O_2 batteries at 25 μA cm^{-2} using 0.5 M NaOTf in DME, DGME, and TGME. Source: Lutz et al. [113]. Reproduced with permission of American Chemical Society. (d) Discharge curve of Na–O_2 batteries at current densities of 25 μA cm^{-2} using DME with 0.5 M of $NaPF_6$, NaOTF, $NaCl_4$, NaTFSI. Source: Lutz et al. [87]. Reproduced with permission of American Chemical Society.

solvent–solute interactions shift the formation of NaO_2 toward a surface process, resulting in submicrometric crystallites and low capacities. Whereas the short-chain ethers conduct the solution-mediated growth mechanism, leading to the formation of large cubic NaO_2 crystals (c.10 μm) with high capacities (c.7.5 mAh cm^{-2}) (Figure 5.16c). The study highlights the importance of moderating solvent–solute interaction, which could allow for the formation of contact-ion pair on the growth of NaO_2 and achieve high discharge capacity.

The effects of electrolyte salt and concentration of ether-based electrolyte on the electrochemical performance of Na–O_2 batteries have been intensively investigated. For example, Lutz et al. investigated the effect of DME with four different anions (ClO_4^-, PF_6^-, OTf^-, and $TFSi^-$) on the discharge capacity of Na–O_2 batteries [87]. It was found that no correlation was identified between the strength of electron donation from anion to cation and the discharge capacity (Figure 5.16d). Liu et al. revealed the relationship between domain size and morphology of discharge products and D_N of electrolyte salt [7]. OTf^- is found to possess a six to seven times higher D_N than PF_6^-, as well as its stronger strength of the Na^+–solvent complex. As expected, a higher amount of Na–O_2 complex with slow nucleation could be stabilized in the electrolyte, resulting in the formation of a larger number of cubic-shaped NaO_2 particles, as shown in Figure 5.17. Recently, Aetukuri

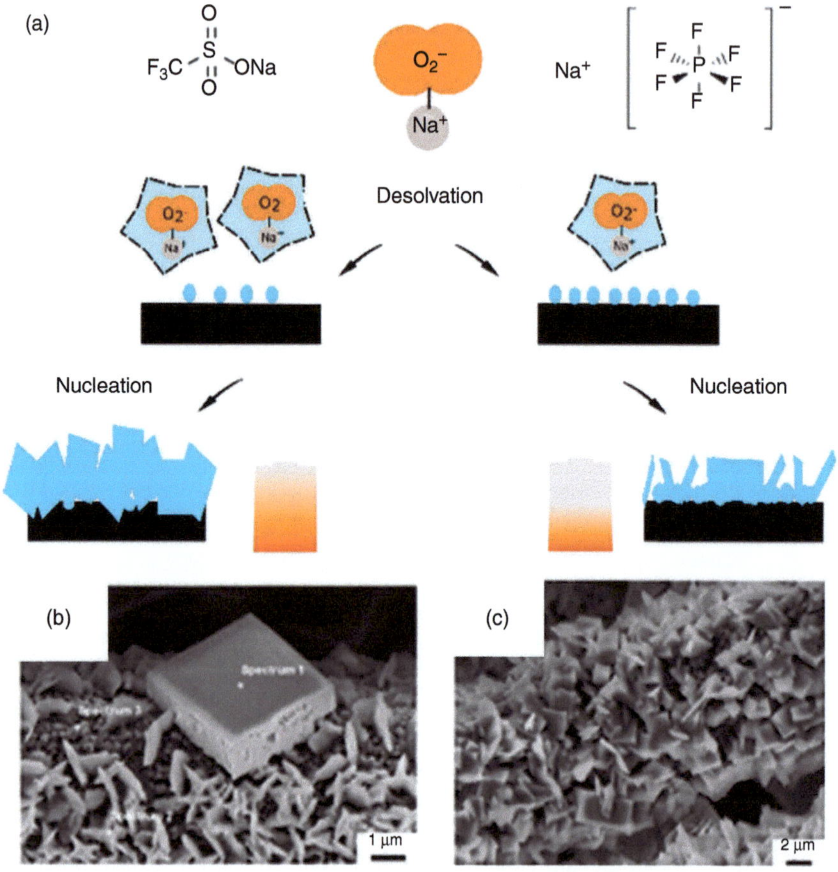

Figure 5.17 (a) Scheme of growth of discharge products under different electrolyte salt features. Morphology of discharge products of Na–O$_2$ batteries based on (b) NaPF$_6$ and (c) NaOTf salt. Source: Liu et al. [7]. Reproduced with permission of American Chemical Society.

et al. reported that using electrolyte with low salt concentration could result in the weakened cation–anion (Na^+–$O_2^{\cdot-}$) interaction, strong growth kinetics, and accelerated NaO_2 crystal growth [114]. As a confirmatory experiment, the addition of cation-coordinating crown molecules to NaOTf/DME as an electrolyte is shown to weaken the ion pairing, leading to the improved discharge capacity. However, Tatara et al. argued that with NaTFSI concentration increased, Na^+ activity and free DME activity increased and decreased, respectively [115]. The ability of Na^+ to stabilize NaO_2 was found to be stronger than that of free DME, and thus the solubility of NaO_2 increased and maximized at medium concentration (NaTFSI:DME = 1 : 8). The increased solubility of NaO_2 leads to a volcano-like morphology, which yields the maximum discharge capacity and cubic shape of NaO_2 with size of ~20 μm.

In addition to electrolyte salt and concentration, electrolyte additive also has critical effects on the performance of Na–O$_2$ batteries. Electrolyte additive can assist the formation of stable SEI to protect Na metal anode, thus improving the cycling performance as discussed in the previous sections. Another function of electrolyte additive is to serve as effective catalysts for ORR/OER. In 2015, Yin et al. investigated the electrochemical cycling performance of Na–O_2 batteries using NaI as additive in $NaClO_4$/DME electrolyte for Na–O_2 batteries [116]. The battery could run up to 150 cycles due to the oxidation reaction of active iodine anions. However, the volatilization of iodine during the cycling process could result in the deterioration of the electrochemical performance of the two-electrode cell. Following this work, they further developed a new nonvolatile and soluble catalyst ferrocene ($Fe[C_5H_5]_2$) in $NaClO_4$/DME as an electrolyte for Na–O_2 batteries [117]. It was found that the soluble $Fe(C_5H_5)_2$ exhibited superior OER electrocatalytic activity and reduced charging voltage, resulting in a distinct improvement in the cyclic performance. Frith et al. proposed that introducing ethyl viologen in NaOTf/diglyme electrolyte could enhance the performance of Na–O_2 batteries [118]. The ethyl viologen was demonstrated to facilitate ORR process, promote the solution mechanism, and avoid the fast saturation of NaO_2 in the electrolyte nearby the oxygen electrode, resulting in increased discharge capacity by 50% and low overpotential of Na–O_2 battery.

Ether-based electrolytes are generally considered to be more stable than carbonate-based electrolytes. Nevertheless, Black et al. suggested that ether-based electrolytes are also unstable due to the strong nucleophilic character of the discharge product NaO_2 and the reactivity of the O_2^-/HO_2 radical [119]. They found that the decomposition of diglyme electrolyte mainly occurs in the discharge process of Na–O_2 battery, while solvated O_2^- and NaO_2 react with both the carbon cathode and the diglyme electrolyte to form a variety of Na–carboxylate decomposition products. Yadegari et al. reported that the reaction between oxygen-rich phase of discharge product (e.g. NaO_2) and ether-based electrolyte results in the production of carbonate parasitic side products [41]. Lutz et al. suggested that radical species produced by the chemical reduction of the solvent could lead to the polymerization of the solvent and the decomposition of the salt anions [87]. They also investigated the SEI formation on Na metal using different sodium salts (e.g. $NaPF_6$, $NaClO_4$, NaTFSI, and NaOTf) in DME. By analyzing the chemical components of SEI on the surface of Na metal using XPS technique, the presence of organic oligomers, Na–halides, and anion decomposition fragments was observed. Recently, Yadegari et al. investigated the decomposition mechanism of Na_2O in ether-based electrolyte by in situ Raman imaging. It was found that Na_2O decomposition rate is fast, resulting in the formation oxalate-based side products [120]. It was proposed that ether-based electrolytes might promote Na_2O decomposition.

5.5.3 DMSO- and ACN-Based Electrolytes

Apart from ether- and carbon-based electrolytes, DMSO- [105, 121] and ACN- [87] based electrolytes for Na–O_2 batteries have also been studied. In 2017, Lutz et al. revealed the environment of Na^+ in different solvents (e.g. DME, ACN, and DMSO) by means of ^{23}Na-NMR [87]. As shown in Figure 5.18, in the weakly solvating DME with relative low D_N and acceptor number (A_N), the anion and cation are in close

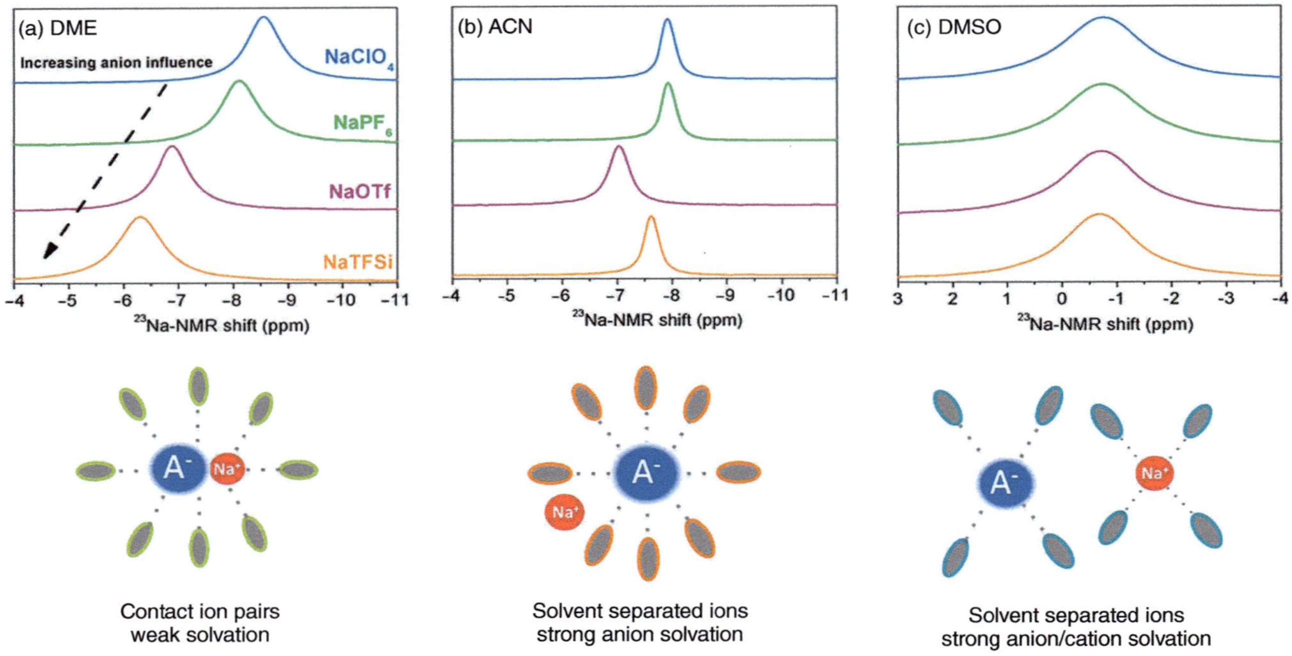

Figure 5.18 ^{23}Na-NMR of ClO_4^- (blue line), PF_6^- (green line), OTf^- (pink line), and $TFSI^-$ (orange line) anion on the Na^+ chemical shift in (a) DME, (b) ACN, and (c) DMSO. Source: Lutz et al. [87]. Reproduced with permission of American Chemical Society.

contact with each other to form contact-ion pair, in which the anions show sizable effect on the chemical shift of Na^+. It was claimed that ACN with high A_N ($A_N = 18.9$) but low D_N ($D_N = 14.1$) allows a stronger solvation of the anion by electron acceptance of the ACN solvent, which ultimately restricts the impact of the salt anion on Na^+, leading to the separation of salt anions and cations. In addition, anion is found to show little effect on the chemical shift of Na^+, and only a slight downfield shift is observed for OTf^-. While for DMSO with high D_N and A_N ($D_N = 29.8$, $A_N = 19.3$), both the cation and the anion are found to be stabilized in solution, therefore solvent-separated ion pairs can be formed in solution, which screens the Na^+ cation from the anion, leading to an identical chemical Na^+ shift for all sodium salts.

Nevertheless, the application of DMSO in Na–O_2 batteries is limited owing to the reaction of DMSO with metallic Na [122]. In 2016, He et al. reported the enhanced stability of Na in DMSO solution containing concentrated NaTFSI salts (>3 mol kg^{-1}) [121]. This is due to the solvation of Na^+ in concentrated solutions with majority DMSO molecules, reducing the available free DMSO molecules. This process renders the decomposition of $TFSI^-$ to form compact surface layer. The reversible Na deposition/dissolution in 3.2 mol kg^{-1} NaTFSI/DMSO solution is observed (Figure 5.19a). Such electrolyte enables the batteries delivering long cycle life over 150 cycles with a Coulombic efficiency (up to 90%) (Figure 5.19b). As expected, it provides a feasible solution for DMSO application in Na–O_2 batteries. Besides, Dilimon et al. reported that high D_N pairs of anion/solvent encouraged the superoxide stability and the reversibility of Na–O_2 batteries [105]. In this work, Na–O_2 batteries were constructed by using pre-sodiated Sb (Sb:Na) as anode and $CF_3SO_3^-$/DMSO as high-D_N-pair electrolyte. Based on Sb:Na anode and $CF_3SO_3^-$/DMSO electrolyte, the formed NaO_2 product in the first step is stable, whereas in the PF_6^-/DMSO electrolyte, the formed NaO_2 product immediately disproportionates to the surface adsorbed Na_2O_2. Furthermore, the high D_N pair ($CF_3SO_3^-$/DMSO) shows higher reversibility than the Hartmann's anion/solvent pair ($CF_3SO_3^-$/DEGDME) and the pair with low-D_N anion (PF_6^-/DMSO) (Figure 5.19c,d). These results indicate that high D_N anion is more likely to inhibit the chemical disproportionation or electrochemical transformation of superoxide to peroxide. Unfortunately, alloying reactions of Na/Sb in high D_N solvents is not as reversible as in conventional organic electrolytes for rechargeable battery, though high D_N solvents contribute more to superoxide stability. Therefore, it is of great significance to explore high D_N solvents that do not react with Na metal or novel anode materials that can replace Na metal to realize good reversibility and performance of Na–O_2 batteries.

5.5.4 Ionic Liquid-Based Electrolyte

The use of conventional organic electrolytes for Na–O_2 batteries is prone to electrolyte decomposition due to the high reduction environment of the alkali metal anode. To solve this problem, ionic liquids (ILs), which are entirely composed of anions and cations, are developed as an alternative electrolyte candidate. ILs have aroused great research interest due to their nature characteristics of low reactivity,

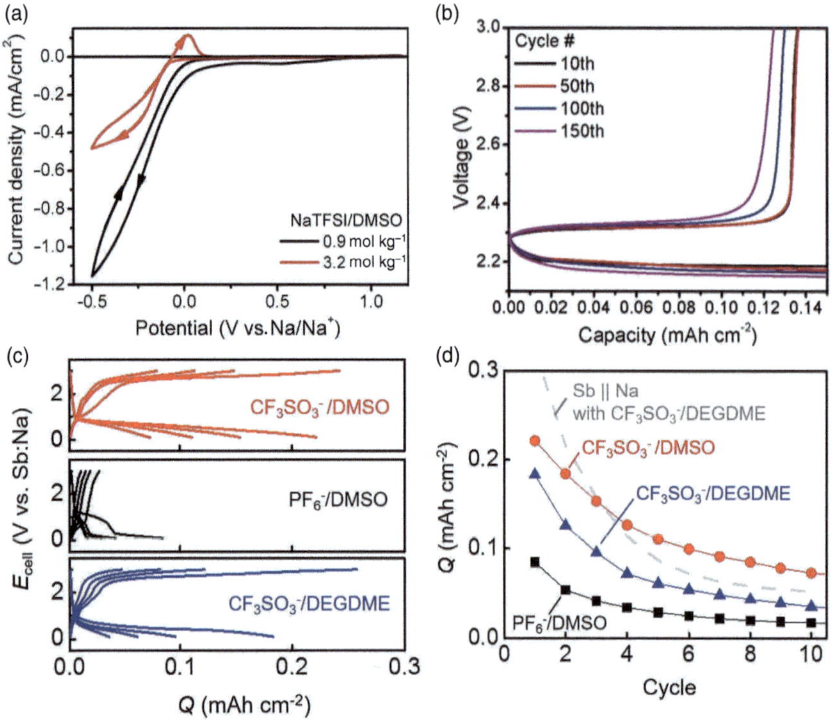

Figure 5.19 (a) Cyclic Voltammograms of Na plating/striping on Cu foils in 0.9 and 3.2 mol/kg NaTFSI/DMSO solutions at 10 mV s^{-1}. (b) Cycle performance of Na–O_2 battery using 3.2 mol/kg NaTFSI/DMSO. Source: He et al. [121].. Reproduced with permission of John Wiley & Sons. (c) Discharge/charge profiles of Na–O_2 batteries using Sb:Na anode at the 1st, 3rd, 5th, and 10th cycle. (d) Capacity retention (Q) as a function of different cycles. Source: Dilimon et al. [105]. Licenced under CC BY 4.0.

negligible vapor pressure, excellent electrochemical and thermal stability, and high ionic conductivity. These unique features endow ILs a wide operational potential range and operating safety of batteries [123]. So far, a great number of ILs (e.g. phosphonium- [124], piperidinium- [111] and pyrrolidinium-based [125–127] ILs) have been reported as potential electrolytes for metal–air batteries. Zhao and Guo for the first time reported the application of *N*-methyl-*N*-propylpiperidinium bis(trifluoromethansulfonyl) imide (PP13TFSI) IL as electrolyte in Na–O_2 batteries [111]. However, PP13TFSI-based electrolyte severely decomposes during the discharge process to form a large amount of side products. Moreover, imidazolium-based ILs with a positive charge in the carbon atom are susceptible to be attacked by nucleophilic species [128]. Unlike imidazolium-based ILs, pyrrolidinium-based (e.g. $[C4mpyr]^+$) ILs without positively charged carbon atoms are stable to superoxide, which shows great potentials for application in Na–O_2 batteries. For example, in neat [C4mpyr][TFSI] IL, the mechanism of ORR has been proved to be the one-electron reversible reaction of O_2/O_2^- redox couple [128]. Plating and stripping of Na in [C4mpyr][TFSI] IL have also been investigated [129].

In 2012, Abraham and coworkers provided experimental evidence to support the ORR mechanism in ILs, which is revealed by the hard soft acid base (HSAB) [130] theory[131]. Similarly, in 2017, Azaceta et al. studied the influence of the metal cation in NaTFSI/[C4mpyr][TFSI] electrolyte on ORR process of Na–O_2 batteries. It was found that the ORR process is governed by the Lewis acidity of the metal cation [127]. Systems with soft and intermediate Mn^+ acidity could facilitate oxygen reduction and metal oxide formation, whereas oxygen reduction is hampered by hard acid cations, such as Na^+ and Li^+. Furthermore, Na^+ concentration of electrolyte is shown to control the electrochemical pathway of the discharge product growth through either solution precipitation or surface deposition. Pozo-Gonzalo et al. investigated the ORR mechanism in ([C_4mpyr][TFSI]) IL at different Na^+ concentrations [125]. Upon increasing the Na^+ concentration, the ORR process becomes more efficient. Since Na^+ has a higher charge density (the ionic radius of Na^+ is 0.96 Å [132] vs. 3.30 Å for [C4mpyr]$^+$ [133]), the solvation between superoxide and Na^+ is more stronger compared to that between superoxide and [C4mpyr]$^+$, resulting in a shift of equilibrium potential toward more positive value. In another report [126], they confirmed that the presence of Na^+ had a significant effect on the electrochemistry of the ORR. In the absence of Na^+, the superoxide is coordinated by four [C4mpyr]$^+$, while both Na^+ and [C_4mpyr]$^+$ could coordinate with superoxide at low Na^+ concentration (c. 1.13 mol%). However, by increasing Na^+ concentration, the superoxide preferentially coordinated with Na^+, compared to that with less available to coordinate with [C_4mpyr]$^+$. Therefore, the large aggregate of the simplified formula of [$O_2^{\cdot-}$][C_4mpyr]$^{n+}$[Na^+]^m could be generated, and the composition varied with the variation of Na^+ concentration in [C_4mpyr][TFSI] [126]. These theoretical studies provide important guidance for the application of IL electrolyte in Na–O_2 batteries.

In 2018, Zhang et al. studied NaTFSI/[C4mpyr][TFSI] as electrolyte for Na–O_2 batteries using an in-house designed three electrode pipette cell [134]. With the increase in concentration of NaTFSI salt in IL electrolyte, the discharge capacity and cycling performance significantly increase (Figure 5.20a,b). This is due to the solvation structure between $O_2^{\cdot-}$, Na^+, and [C_4mpyr]$^+$, leading to the different nucleation and growth mechanism of discharge product. The higher Na^+ concentration causes the interaction of superoxide anion ($O_2^{\cdot-}$) with Na^+ rather than with [C_4mpyr]$^+$. This interaction induces a lower desolvation energy for deposition of NaO_2 from the electrolyte and triggers the formation of the discharge products on the electrode surface. Recently, Ortiz-Vitoriano et al. reported that a synergistic effect of diglyme and IL solvents could increase the stability of $O_2^{\cdot-}$ and limit the formation of side reaction products during the discharge process [135]. ORR kinetics of two concentrations of NaTFSI (16.6 and 35 mol%), in a mixture of diethylene glycol dimethyl ether (G2) and [C_4mpyr][TFSI] IL, was investigated. It was found that both electrolytes show a low overpotential of 170 and 310 mV and a considerable discharge capacity of 4.5 vs 2.8 mAh cm^{-2} for 16.6 and 35 mol%, respectively (Figure 5.20c). Discharge products with cube size NaO_2 are observed in both electrolytes (Figure 5.20d,e). However, the 16.6 mol% IL/G2 mixture shows larger and more uniform particle coverage. This is because the 16.6 mol% IL/G2 presents a more positive chemical shift

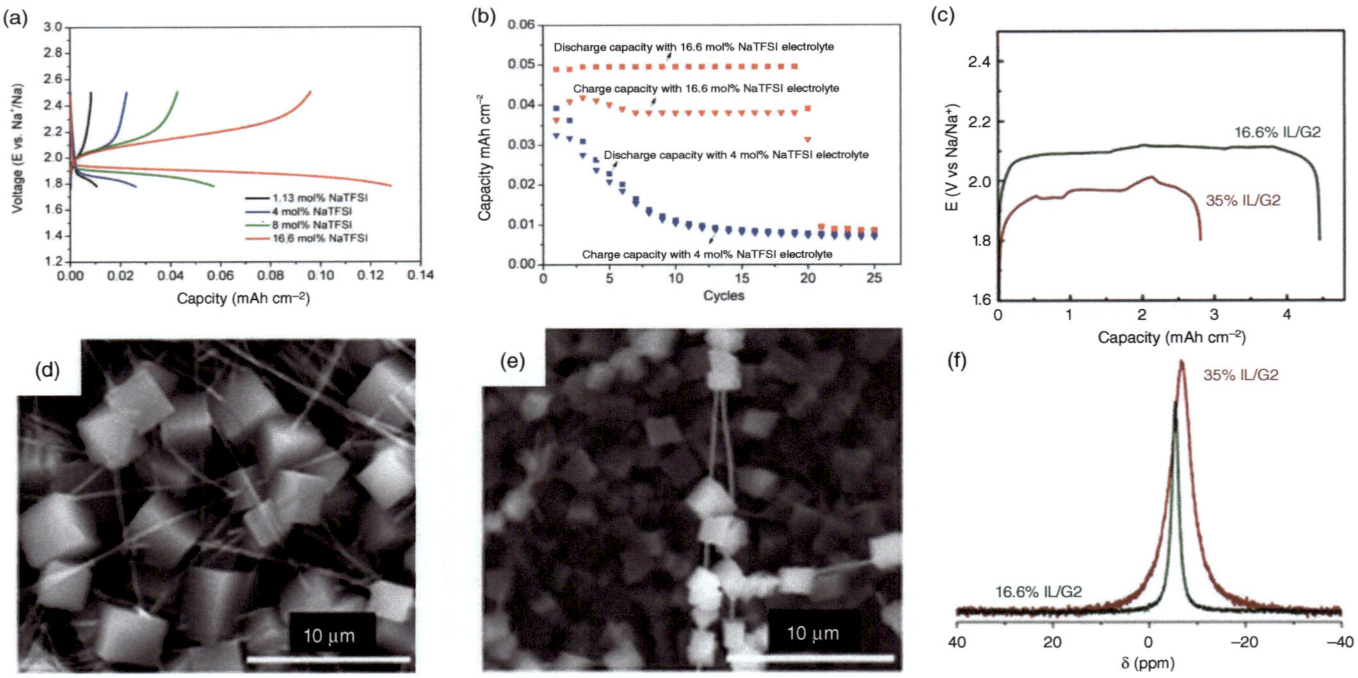

Figure 5.20 (a) Initial discharge/charge profiles of $Na-O_2$ batteries using [C_4mpyr][TFSI] IL-based electrolyte mixtures. (b) Cycling performance of $Na-O_2$ batteries using 4 and 16.6 mol% NaTFSI/[C_4mpyr][TFSI] IL electrolyte mixtures. Source: Zhang et al. [134]. Reproduced with permission of American Chemical Society. (c) Discharge curves of two different NaTFSI concentrations in IL/G2. SEM images of the discharged cathodes using (d) 16.6 mol% and (e) 35 mol% NaTFSI in IL/G2. (f) ^{23}Na chemical shift in the 16.6 and 35 mol% NaTFSI-based electrolyte mixtures (IL/G2) at 20 °C. Source: Ortiz-Vitoriano et al. [135]. Reproduced with permission of American Chemical Society.

than the 35 mol% IL/G2 analog (e.g. −5.4 vs. −6.7 ppm, respectively) (Figure 5.20f), exemplifying a stronger coordination ability in the former electrolyte mixture, and thus showing slower nucleation and growth rate.

5.6 Mechanism Studies

So far, the development of Na–O_2 batteries is still at early stage. Apart from achieving improved electrochemical performance through the optimization of electrodes and electrolytes, deep understanding on the charging and discharging mechanism as well as key parameters that affect the battery performance is highly desirable. In this respect, the development and application of in situ technologies have shown to provide effective means for the mechanism study of Na–O_2 batteries.

To evaluate the performance of Na–O_2 batteries, it is of great significance to quantitatively measure the amount of main formed/decomposed reaction products and degree of parasitic reactions. In situ or in operando X-ray diffraction (XRD) provides a powerful technical support. Liu et al. used operando synchrotron radiation powder X-ray diffraction (SR-PXD) to quantitatively track the formation of NaO_2 [7]. The results indicate that the growth of NaO_2 could be divided into two time stages. The efficiency of the discharge reaction in the second stage is high (>90%). Meanwhile, the influence of electrolyte salt on discharge capacity and the formation of NaO_2 was also revealed. The Na^+–solvent complex in the high D_N salt anion-based electrolyte shows strong interaction. It could stabilize large amounts of Na–O_2 complex by slowing the nucleation rate to form a larger number of cubic-shaped NaO_2 particles, leading to the improved discharge capacity.

Apart from SR-PXD, Banis et al. developed an in situ soft X-ray absorption spectra (XAS) technique to study the formation and decomposition of the discharge products during the cycling [136]. As shown in Figure 5.21, all spectra collected from Na–O_2 batteries based on the cathode show a broad feature at ~536 eV. When the nucleation and deposition of NaO_2 occur during the discharge process, a peak gradually emerged at 532 eV. This is due to the transition of O_2^- anions from O 1s to π^*. During the charge process, the behavior of the π^* transition is reversed. However, a small broad feature remaining at ~532 to 534 eV is observed. It is correlated with the decomposition reaction between the discharge product and the cell electrolyte. By in situ Raman imaging technique, they further revealed the chemical mechanism of the decomposition reaction of NaO_2 with diglyme-based electrolyte [120]. As shown in Figure 5.22a, NaO_2 is decomposed in the presence of diglyme-based electrolyte and no superoxide could be detected after 150 hours. Meanwhile, the chemical maps at 1136 cm^{-1} show an upward trend in the first 100 hours and then a downward trend. This is due to the presence of NaO_2 phase in the (Na/H)OH environment (Figure 5.22b). Besides, the chemical diagram obtained at 860 cm^{-1} indicates that the continuous formed oxalate-based side products completely cover the surface of NaO_2 after 170 hours.

Among the in situ techniques, in situ TEM technology is also a powerful analytical tool that can directly observe the reaction pathway and dynamic evolution of the

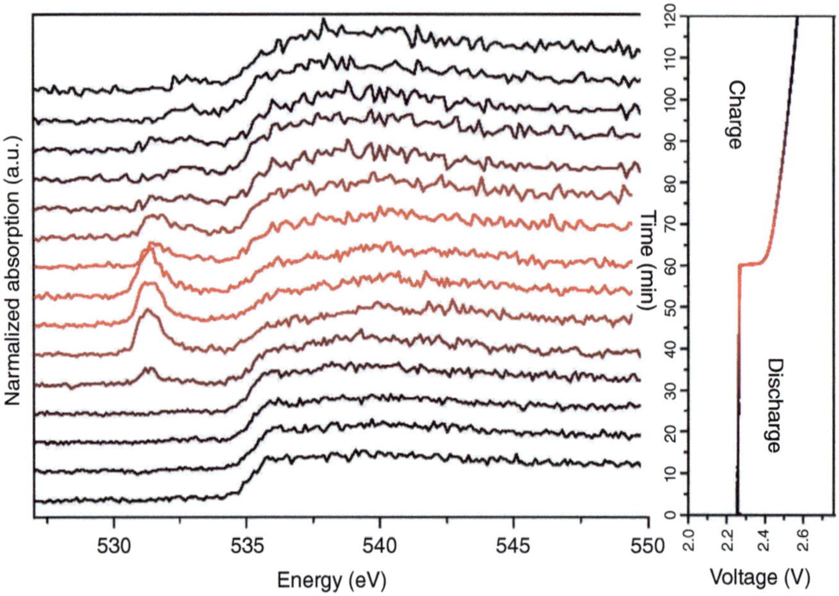

Figure 5.21 In-situ XAS measurements of Na–O_2 battery. Source: Banis et al. [136]. Reproduced with permission of Royal Society of Chemistry.

discharge–charge products. In 2018, Lutz et al. used liquid-electrochemical TEM to monitor the operation of the solution-mediated discharge and charge process in Na–O_2 batteries (Figure 5.23a) [137]. During the discharge and charge processes, the formation and dissolution of NaO_2 cubes can be governed by the equilibrium between $NaO_{2(solv)} \leftrightarrow NaO_{2(solid)}$. Besides, magnified TEM images obtained at the end of discharge show that Na_2O cubes are surrounded by the shell of the parasitic reaction products. This mainly originates from the instability of the glyme electrolyte contacting highly oxidizing NaO_2. However, part of parasitic products remains on the electrode surface at the end of charge and accumulates continuously in the subsequent cycling. The existence of parasitic products tends to block crucial O_2 redox and further nucleation of NaO_2, leading to low efficiency and poor cycling stability.

Kwak et al. employed in situ TEM (Figure 5.23b) technique together with solid electrolyte instead of liquid electrolyte to reveal the reaction mechanism and structural evolution of the discharge products [95]. During discharging, the size of the discharge products will increase in the discharge process and the agglomeration of small particles of discharge products also leads to the formation of large NaO_2 particles. During charging, larger cubic NaO_2 particles reversibly decompose. It should be noted that NaO_2 and Na_2O_2 layers on the surface of CNT cathode are not completely removed, leading to the enlarged overpotential. Moreover, when the reaction was suspended and then kept in a vacuum environment for eight hours, no side products were formed in either the discharge products with large cubic particles or the conformal coating layer. Only part of small particles experienced phase transition on the surface of the cubic particle. It is suggested that the chemical

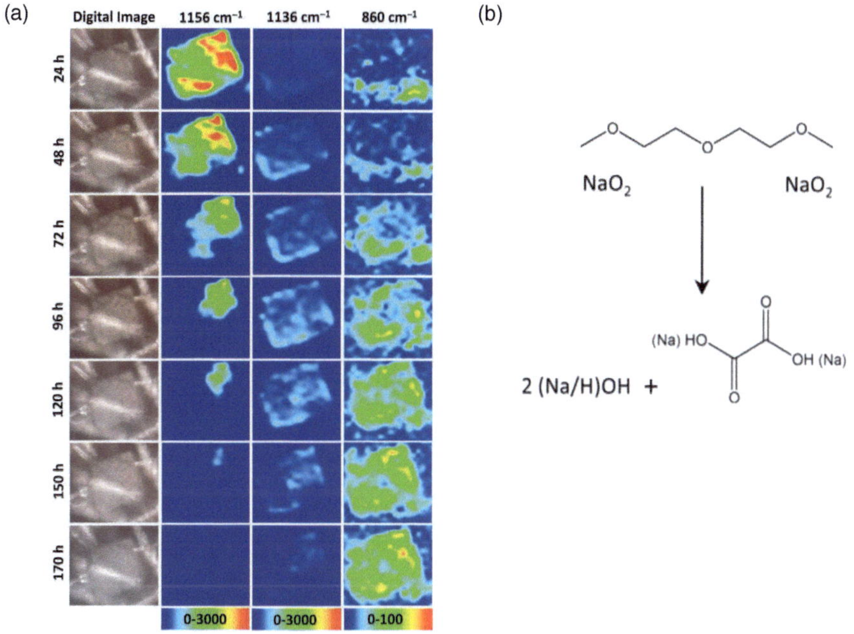

Figure 5.22 (a) Chemical maps at various frequencies reproduced from Raman spectra recorded on a NaO_2 cube air electrode at different time intervals. (b) The chemical reaction of the formation of oxalate-based side products. Source: Yadegari et al. [120]. Reproduced with permission of American Chemical Society.

side reaction between NaO_2 and electrolyte is absent when the solid electrolyte is used. Such result provides a new insight for the design and optimization of Na–O_2 batteries. Besides, Liu et al. demonstrated real-time imaging of the ORR in all-solid-state Na–O_2 batteries with CuO nanowires (NWs) as cathode. The experiment was operated in an aberration-corrected environmental TEM under an oxygen environment [59]. At the initial stage, CuO is converted to Cu_2O and then to Cu. Then, the freshly formed ultrafine-grained Cu catalyzes the following ORR process to produce NaO_2, followed by being disproportionate to form Na_2O_2 and O_2. Remarkably, no carbonate formation is observed during the battery cycle. These results provide fundamental understanding on the oxygen chemistry in the carbon-free cathode of Na–O_2 batteries.

More recently, Han et al. directly observed the ORR and OER processes of Na–O_2 batteries with CuS as cathode by in situ environmental TEM [138]. During the ORR process, the Na_2S extrudes the formation of Cu after the sodiation of CuS crystal. After releasing O_2 into the chamber, the Na_2S begins to transform to NaS, NaO_2, and Na_2O_2. The instability of NaO_2 phase leads to the formation of mesoporous particle-shaped Na_2O_2 with the release of O_2 gas. As the ORR product, the Na_2O_2 uniformly covers the whole wire-shape cathode. In the following OER process, mesoporous Na_2O_2 transforms to NaO_2, leading to the volume expansion, and then reversibly decomposes into Na^+ and O_2 gas. Compared to CuO-based cathode, CuS-based one is shown to be highly stable with negligible volume expansion

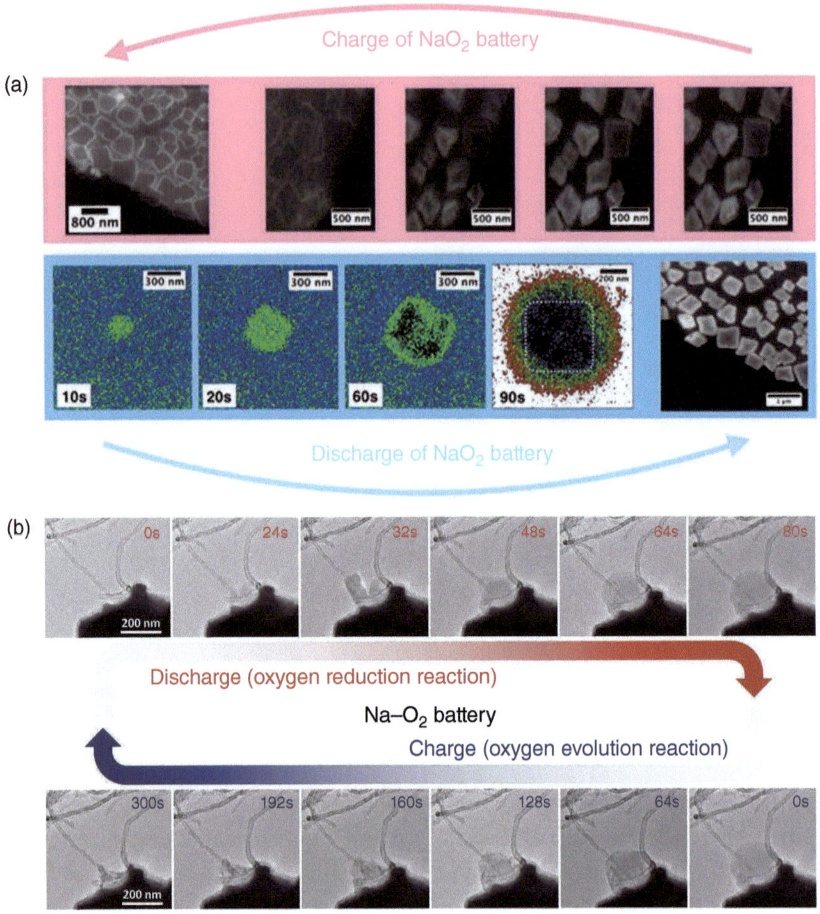

Figure 5.23 (a) Operando TEM images of the NaO_2 product during discharge and charge processes. Source: Lutz et al. [137] with permission of © American Chemical Society. (b) In-situ TEM images of the discharge/charge product using a Na–O_2 nanobattery based on the cathode. Source: Kwak et al. [95] with permission of © American Chemical Society.

and pulverization. As a result, the CuS-based cathode shows a maximum capacity over $3\,mAh\,cm^{-2}$ at $0.01\,mA\,cm^{-2}$ with a discharge cut-off voltage of 1.8 V. These results indicate that the size, distribution, and composition of the catalysts play an important role in controlling the morphology, chemical composition, and reversibility of discharge products. This study reveals that controlling the final reaction products by selecting the appropriate catalyst is an effective strategy to achieve high reversibility of the catalyst for ORR and OER of Na–O_2 battery.

5.7 Conclusion and Perspectives

The growing demand of energy accelerates the exploration of novel energy storage technologies with high energy density. The emergence of Na–O_2 batteries provides a new way for the realization of low cost, high energy density, and effective energy storage system. However, Na–O_2 batteries still face some serious challenges, such as slow ORR and OER kinetics, Na dendrite growth, formation of parasitic side products, electrolyte decomposition, short cycle life, poor rate performance, and low round-trip efficiency. These problems seriously hinder the further development of Na–O_2 batteries. In recent years, much efforts have been devoted to solving these problems and inspiring progress has been made. To further improve the performance of Na–O_2 batteries, current status and future challenges in terms of each component in Na–O_2 batteries can be summarized as follows:

(1) On the air cathode. To solve the slow ORR and OER kinetics issue of cathode, carbon materials, such as graphene, CNTs, and porous carbon, have been proposed as efficient cathode catalysts for Na–O_2 batteries due to their high specific surface area, high porosity, high electrical conductivity, and low cost. In addition, doping carbon materials with foreign atoms, such as N, B, S, could introduce more electrocatalytic active sites into the cathode and improve its intrinsic catalytic activity. However, the catalytic activity of carbon-based materials is still unsatisfied. Noble metals with high catalytic activity have been developed for improving the performance of Na–O_2 batteries, but the large-scale application of noble metal-based Na–O_2 batteries should be carefully evaluated due to the high cost. In recent years, the transition metals and their oxides as the substitute for noble metal have attracted great attention. Meanwhile, the development of bifunctional catalysts based on composite materials has also been shown to significantly improved ORR and OER kinetics of cathode. However, some problems remain: transition metal oxides usually have low electronic conductivity, which greatly limits the electrochemical performance improvement despite the catalytic activity improves. Currently, although the developed various catalysts have effectively improved ORR and OER kinetic of Na–O_2 batteries, the reported performance is still far from meeting the requirements of practical application. More importantly, the understanding of mechanism on electrochemical catalysis behind catalysts is still unclear. As a result, the rational design of efficient and low-cost cathode catalyst for Na–O_2 batteries is still highly desirable.
(2) On the anode. Various strategies have been proposed to solve the growth of Na dendrite, including: (i) modification of the separator to mechanically prevent dendrite penetration; (ii) surface modification of metallic Na by introducing artificial protective layer to protect Na metal anode; (iii) optimizing the electrolyte formulation or using additive to form a stable SEI; (iv) using solid electrolyte to avoid associate reactions between Na metal and organic solvent; (v) by electrostatic shielding effect to guide the Na uniform deposition; (vi) adopting carbon materials as host or current collector to facilitate Na uniform nucleation; (vii) decorating the carbon materials with sodiophilic functional groups to increase the affinity of substrate with sodium to guide the Na uniform nucleation; and (viii) developing alternatives to metallic Na. These strategies have shown to be feasible to solve the issue of Na dendrite growth. Although it is

effective at the laboratory stage, it is clear that most of the proposed strategies are difficult to implement or remain great challenges in low-cost and large-scale industrial applications. In this respect, solid electrolytes might be promising. However, research on all-solid-state Na–O_2 with SSEs battery is still in its infancy. The crucial issues, such as poor interface stability and high interface resistance between electrolyte and metallic sodium electrode, are still desired to be solved.

(3) On the electrolyte. The key parameters of electrolyte, such as solvent, electrolyte salt, concentration, and additive, have been demonstrated to significantly affect the morphology and composition of discharge products, deposition mechanism, solvation effect of ion pairs, parasitic reactions caused by electrolyte decomposition, and overall battery performance. However, the influence of electrolyte change during the charging and discharging processes on the battery performance is complex. Compared with the research on anode and cathode, there are relatively limited concerns on electrolyte and thus comprehensive studies are expected in the future. The key components of electrolyte, including solvent, electrolyte salts, and their concentration, have been shown to be closely related to the performance of Na–O_2 cells, and further exploration is needed to elucidate the correlation between the choice of these components and discharge capacity. Moreover, it is of great promise to realize high performance and high reversibility of Na–O_2 batteries by designing the optimal formulation of electrolyte.

(4) Mechanism studies. With the development of various in situ and real-time imaging techniques, the formation and decomposition evolution of discharge products and the catalytic mechanism of the catalyst for Na–O_2 batteries during charging and discharging process have been preliminarily studied. However, the relationship between the morphology, chemical composition, and structure of the discharge products and the discharge capacity and overpotential of the Na–O_2 battery needs to be further investigated. Although some studies have revealed that the irreversible formation of parasitic side products during charging and discharging is an important factor, which are responsible for the increased overpotential and the poor battery performance, it is still challenging to control and optimize these processes.

In conclusion, although Na–O_2 batteries are currently facing serious challenges, they are still one of the most attractive candidates for the next-generation energy storage system.

Acknowledgments

The authors are grateful for financial support from the Pearl River Talent Recruitment Program (2019QN01L096), the Fundamental Research Funds for the Central Universities (2018JQ06), the Guangdong Innovative and Entrepreneurial Research Team Program (2019ZT08L075), the Guangdong Science and Technology Program (2020B121201003), the "Young Talent Fellowship" Program through South China University of Technology, and the Natural Science Foundation of China (51802253).

References

1 Peng, Z., Freunberger, S.A., Chen, Y., and Bruce, P.G. (2012). *Science* 337: 563–566.
2 Amanchukwu, C.V., Harding, J.R., Shao-Horn, Y., and Hammond, P.T. (2015). *Chem. Mater.* 27: 550–561.
3 Liu, Y., Wang, L., Cao, L. et al. (2017). *Mater. Chem. Front.* 1: 2495–2510.
4 Yao, X., Dong, Q., Cheng, Q., and Wang, D. (2016). *Angew. Chem. Int. Ed.* 55: 11344–11353.
5 Ma, Z., Yuan, X., Li, L. et al. (2015). *Energy Environ. Sci.* 8: 2144–2198.
6 Zhao, G., Mo, R., Wang, B. et al. (2014). *Chem. Mater.* 26: 2551–2556.
7 Liu, C., Rehnlund, D., Brant, W.R. et al. (2017). *ACS Energy Lett.* 2: 2440–2444.
8 Ma, J.-L., Meng, F.-L., Xu, D., and Zhang, X.-B. (2017). *Energy Storage Mater.* 6: 1–8.
9 Matios, E., Wang, H., Wang, C., and Li, W. (2019). *Ind. Eng. Chem. Res.* 58: 9758–9780.
10 Lin, X., Sun, Q., Doyle Davis, K. et al. (2019). *Carbon Energy* 1: 141–164.
11 Peled, E., Golodnitsky, D., Mazor, H. et al. (2011). *J. Power Sources* 196: 6835–6840.
12 Sun, Q., Yang, Y., and Fu, Z.-W. (2012). *Electrochem. Commun.* 16: 22–25.
13 Wei, S., Choudhury, S., Tu, Z. et al. (2018). *Acc. Chem. Res.* 51: 80–88.
14 Xu, W., Wang, J., Ding, F. et al. (2014). *Energy Environ. Sci.* 7: 513–537.
15 Bi, X., Ren, X., Huang, Z. et al. (2015). *Chem. Commun.* 51: 7665–7668.
16 Xia, C., Black, R., Fernandes, R. et al. (2015). *Nat. Chem.* 7: 496–501.
17 Kim, J., Park, H., Lee, B. et al. (2016). *Nat. Commun.* 7: 10670.
18 Ottakam Thotiyl, M.M., Freunberger, S.A., Peng, Z. et al. (2013). *Nat. Mater.* 12: 1050–1056.
19 Lu, Y.-C., Gallant, B.M., Kwabi, D.G. et al. (2013). *Energy Environ. Sci.* 6: 750–768.
20 Gallant, B.M., Kwabi, D.G., Mitchell, R.R. et al. (2013). *Energy Environ. Sci.* 6: 2518–2528.
21 Hartmann, P., Bender, C.L., Vračar, M. et al. (2013). *Nat. Mater.* 12: 228–232.
22 Hartmann, P., Bender, C.L., Sann, J. et al. (2013). *Phys. Chem. Chem. Phys.* 15: 11661–11672.
23 Bender, C.L., Hartmann, P., Vračar, M. et al. (2014). *Adv. Energy Mater.* 4: 1301863.
24 Li, Y., Yadegari, H., Li, X. et al. (2013). *Chem. Commun.* 49: 11731–11733.
25 Liu, W., Sun, Q., Yang, Y. et al. (2013). *Chem. Commun.* 49: 1951–1953.
26 Kang, J.-H., Kwak, W.-J., Aurbach, D., and Sun, Y.-K. (2017). *J. Mater. Chem. A* 5: 20678–20686.
27 Jian, Z., Chen, Y., Li, F. et al. (2014). *J. Power Sources* 251: 466–469.
28 Kwak, W.-J., Chen, Z., Yoon, C.S. et al. (2015). *Nano Energy* 12: 123–130.
29 Kim, J., Lim, H.-D., Gwon, H., and Kang, K. (2013). *Phys. Chem. Chem. Phys.* 15: 3623–3629.
30 Song, K., Agyeman, D.A., Park, M. et al. (2017). *Adv. Mater.* 29: 1606572.

31 Khan, Z., Vagin, M., and Crispin, X. (2020). *Adv. Sci.* 7: 1902866.
32 Yadegari, H., Sun, Q., and Sun, X. (2016). *Adv. Mater.* 28: 7065–7093.
33 Sun, B., Kretschmer, K., Xie, X. et al. (2017). *Adv. Mater.* 29: 1606816.
34 Yadegari, H., Li, Y., Banis, M.N. et al. (2014). *Energy Environ. Sci.* 7: 3747–3757.
35 Kraytsberg, A. and Ein-Eli, Y. (2013). *Nano Energy* 2: 468–480.
36 Zhao, N., Li, C., and Guo, X. (2014). *Phys. Chem. Chem. Phys.* 16: 15646–15652.
37 Lin, X., Wang, J., Gao, X. et al. (2020). *Chem. Mater.* 32: 3018–3027.
38 Duan, J., Chen, S., Jaroniec, M., and Qiao, S.Z. (2015). *ACS Catal.* 5: 5207–5234.
39 Zhang, S., Wen, Z., Jin, J. et al. (2016). *J. Mater. Chem. A* 4: 7238–7244.
40 Sun, Q., Yadegari, H., Banis, M.N. et al. (2015). *Nano Energy* 12: 698–708.
41 Yadegari, H., Banis, M.N., Xiao, B. et al. (2015). *Chem. Mater.* 27: 3040–3047.
42 Ma, J.-l. and Zhang, X.-b. (2016). *J. Mater. Chem. A* 4: 10008–10013.
43 Shu, C., Lin, Y., Zhang, B. et al. (2016). *J. Mater. Chem. A* 4: 6610–6619.
44 Liang, J., Jiao, Y., Jaroniec, M., and Qiao, S.Z. (2012). *Angew. Chem. Int. Ed.* 51: 11496–11500.
45 Du, L., Shao, Y., Sun, J. et al. (2016). *Nano Energy* 29: 314–322.
46 Wang, Y.-J., Fang, B., Li, H. et al. (2016). *Prog. Mater Sci.* 82: 445–498.
47 Wang, Y., Ruiz Diaz, D.F., Chen, K.S. et al. (2020). *Mater. Today* 32: 178–203.
48 Black, R., Lee, J.-H., Adams, B. et al. (2013). *Angew. Chem. Int. Ed.* 52: 392–396.
49 Li, F., Zhang, T., and Zhou, H. (2013). *Energy Environ. Sci.* 6: 1125–1141.
50 Wang, Z.-L., Xu, D., Xu, J.-J., and Zhang, X.-B. (2014). *Chem. Soc. Rev.* 43: 7746–7786.
51 Zhang, S., Wen, Z., Rui, K. et al. (2015). *J. Mater. Chem. A* 3: 2568–2571.
52 Kumar, S., Kishore, B., and Munichandraiah, N. (2016). *RSC Adv.* 6: 63477–63479.
53 Lutz, L., Corte, D.A.D., Chen, Y. et al. (2018). *Adv. Energy Mater.* 8: 1701581.
54 Wu, F., Xing, Y., Lai, J. et al. (2017). *Adv. Funct. Mater.* 27: 1700632.
55 Liu, W.-M., Yin, W.-W., Ding, F. et al. (2014). *Electrochem. Commun.* 45: 87–90.
56 Khajehbashi, S.M.B., Xu, L., Zhang, G. et al. (2018). *Nano Lett.* 18: 3934–3942.
57 Sun, Q., Liu, J., Li, X. et al. (2017). *Adv. Funct. Mater.* 27: 1606662.
58 Liu, Y., Chi, X., Han, Q. et al. (2019). *J. Alloys Compd.* 772: 693–702.
59 Liu, Q., Yang, T., Du, C. et al. (2018). *Nano Lett.* 18: 3723–3730.
60 Liu, Q., Geng, L., Yang, T. et al. (2019). *Energy Storage Mater.* 19: 48–55.
61 Hu, Y., Han, X., Zhao, Q. et al. (2015). *J. Mater. Chem. A* 3: 3320–3324.
62 Li, N., Yin, Y., Meng, F. et al. (2017). *ACS Catal.* 7: 7688–7694.
63 Ma, J.-l., Li, N., Zhang, Q. et al. (2018). *Energy Environ. Sci.* 11: 2833–2838.
64 Tang, C., Min, Y., Chen, C. et al. (2019). *Nano Lett.* 19: 5577–5586.
65 Yadegari, H., Norouzi Banis, M., Lushington, A. et al. (2017). *Energy Environ. Sci.* 10: 286–295.
66 Ma, L., Zhang, D., Lei, Y. et al. (2018). *ACS Energy Lett.* 3: 276–277.
67 Wang, J., Gao, R., Zheng, L. et al. (2018). *ACS Catal.* 8: 8953–8960.
68 Tovini, M.F., Patil, B., Koz, C. et al. (2018). *Nanotechnology* 29: 475401.
69 Liu, Y., Chi, X., Han, Q. et al. (2019). *Nanoscale* 11: 5285–5294.
70 Zheng, X., Bommier, C., Luo, W. et al. (2019). *Energy Storage Mater.* 16: 6–23.
71 Zhao, Y., Adair, K.R., and Sun, X. (2018). *Energy Environ. Sci.* 11: 2673–2695.

72 Zhang, C., Wang, A., Zhang, J. et al. (2018). *Adv. Energy Mater.* 8: 1802833.
73 Luo, W. and Hu, L. (2015). *ACS Central Sci.* 1: 420–422.
74 Lin, D., Liu, Y., and Cui, Y. (2017). *Nat. Nanotechnol.* 12: 194–206.
75 Zhang, R., Li, N.-W., Cheng, X.-B. et al. (2017). *Adv. Sci.* 4: 1600445.
76 Zhao, Y., Liang, J., Sun, Q. et al. (2019). *J. Mater. Chem. A* 7: 4119–4125.
77 Ma, J.-L., Yin, Y.-B., Liu, T. et al. (2018). *Adv. Funct. Mater.* 28: 1703931.
78 Yang, H., Sun, J., Wang, H. et al. (2018). *Chem. Commun.* 54: 4057–4060.
79 Liu, K., Zhuo, D., Lee, H.-W. et al. (2017). *Adv. Mater.* 29: 1603987.
80 Ansari, Y., Virwani, K., Yahyazadeh, S. et al. (2018). *Adv. Energy Mater.* 8: 1802603.
81 Choudhury, S., Wei, S., Ozhabes, Y. et al. (2017). *Nat. Commun.* 8: 898.
82 Ma, M., Lu, Y., Yan, Z., and Chen, J. (2019). *Batter. Supercaps* 2: 663–667.
83 Luo, W., Lin, C.-F., Zhao, O. et al. (2017). *Adv. Energy Mater.* 7: 1601526.
84 Zhao, Y., Goncharova, L.V., Lushington, A. et al. (2017). *Adv. Mater.* 29: 1606663.
85 Zhao, Y., Goncharova, L.V., Zhang, Q. et al. (2017). *Nano Lett.* 17: 5653–5659.
86 Seh, Z.W., Sun, J., Sun, Y., and Cui, Y. (2015). *ACS Central Sci.* 1: 449–455.
87 Lutz, L., Alves Dalla Corte, D., Tang, M. et al. (2017). *Chem. Mater.* 29: 6066–6075.
88 Han, M., Zhu, C., Ma, T. et al. (2018). *Chem. Commun.* 54: 2381–2384.
89 Lee, Y., Lee, J., Lee, J. et al. (2018). *ACS Appl. Mater. Interfaces* 10: 15270–15280.
90 Wu, S., Qiao, Y., Jiang, K. et al. (2018). *Adv. Funct. Mater.* 28: 1706374.
91 Shi, Q., Zhong, Y., Wu, M. et al. (2018). *Angew. Chem. Int. Ed.* 57: 9069–9072.
92 Ruiz-Martínez, D., Kovacs, A., and Gómez, R. (2017). *Energy Environ. Sci.* 10: 1936–1941.
93 Zheng, J., Chen, S., Zhao, W. et al. (2018). *ACS Energy Lett.* 3: 315–321.
94 Gao, H., Xin, S., Xue, L., and Goodenough, J.B. (2018). *Chem* 4: 833–844.
95 Kwak, W.-J., Luo, L., Jung, H.-G. et al. (2018). *ACS Energy Lett.* 3: 393–399.
96 Medenbach, L., Bender, C.L., Haas, R. et al. (2017). *Energy Technol.* 5: 2265–2274.
97 Sun, B., Li, P., Zhang, J. et al. (2018). *Adv. Mater.* 30: 1801334.
98 Liu, S., Tang, S., Zhang, X. et al. (2017). *Nano Lett.* 17: 5862–5868.
99 Zhang, Q., Lu, Y., Zhou, M. et al. (2018). *Inorg. Chem. Front.* 5: 864–869.
100 Luo, W., Zhang, Y., Xu, S. et al. (2017). *Nano Lett.* 17: 3792–3797.
101 Wang, A., Hu, X., Tang, H. et al. (2017). *Angew. Chem. Int. Ed.* 56: 11921–11926.
102 Ye, L., Liao, M., Zhao, T. et al. (2019). *Angew. Chem. Int. Ed.* 58: 17054–17060.
103 He, X., Jin, S., Miao, L. et al. (2020). *Angew. Chem. Int. Ed.* 59 (38): 16705–16711.
104 Bender, C.L., Jache, B., Adelhelm, P., and Janek, J. (2015). *J. Mater. Chem. A* 3: 20633–20641.
105 Dilimon, V.S., Hwang, C., Cho, Y.-G. et al. (2017). *Sci. Rep.* 7: 17635.
106 Ma, J.L., Meng, F.L., Yu, Y. et al. (2019). *Nat. Chem.* 11: 64–70.
107 Lin, X., Sun, F., Sun, Q. et al. (2019). *Chem. Mater.* 31: 9024–9031.

108 Assary, R.S., Lu, J., Du, P. et al. (2013). *ChemSusChem* 6: 51–55.
109 Ren, X., Lau, K.C., Yu, M. et al. (2014). *ACS Appl. Mater. Interfaces* 6: 19299–19307.
110 Balaish, M., Kraytsberg, A., and Ein-Eli, Y. (2014). *Phys. Chem. Chem. Phys.* 16: 2801–2822.
111 Zhao, N. and Guo, X. (2015). *J. Phys. Chem. C* 119: 25319–25326.
112 Aldous, I.M. and Hardwick, L.J. (2016). *Angew. Chem. Int. Ed.* 55: 8254–8257.
113 Lutz, L., Yin, W., Grimaud, A. et al. (2016). *J. Phys. Chem. C* 120: 20068–20076.
114 Aetukuri, N.B., Jones, G.O., Thompson, L.E. et al. (2018). *ACS Energy Lett.* 3: 2342–2348.
115 Tatara, R., Leverick, G.M., Feng, S. et al. (2018). *J. Phys. Chem. C* 122: 18316–18328.
116 Yin, W.-W., Shadike, Z., Yang, Y. et al. (2015). *Chem. Commun.* 51: 2324–2327.
117 Yin, W.-W., Yue, J.-L., Cao, M.-H. et al. (2015). *J. Mater. Chem. A* 3: 19027–19032.
118 Frith, J.T., Landa-Medrano, I., Ruiz de Larramendi, I. et al. (2017). *Chem. Commun.* 53: 12008–12011.
119 Black, R., Shyamsunder, A., Adeli, P. et al. (2016). *ChemSusChem* 9: 1795–1803.
120 Yadegari, H., Norouzi Banis, M., Lin, X. et al. (2018). *Chem. Mater.* 30: 5156–5160.
121 He, M., Lau, K.C., Ren, X. et al. (2016). *Angew. Chem. Int. Ed.* 55: 15310–15314.
122 O'Connor, D.E. and Lyness, W.I. (1965). *The Journal of Organic Chemistry* 30: 1620–1623.
123 MacFarlane, D.R., Tachikawa, N., Forsyth, M. et al. (2014). *Energy Environ. Sci.* 7: 232–250.
124 Yan, Y., Gunzelmann, D., Pozo-Gonzalo, C. et al. (2017). *Electrochim. Acta* 235: 270–279.
125 Pozo-Gonzalo, C., Howlett, P.C., MacFarlane, D.R., and Forsyth, M. (2017). *Electrochem. Commun.* 74: 14–18.
126 Pozo-Gonzalo, C., Johnson, L.R., Jónsson, E. et al. (2017). *J. Phys. Chem. C* 121: 23307–23316.
127 Azaceta, E., Lutz, L., Grimaud, A. et al. (2017). *ChemSusChem* 10: 1616–1623.
128 Katayama, Y., Onodera, H., Yamagata, M., and Miura, T. (2004). *J. Electrochem. Soc.* 151: A59.
129 Mohd Noor, S.A., Howlett, P.C., MacFarlane, D.R., and Forsyth, M. (2013). *Electrochim. Acta* 114: 766–771.
130 Pearson, R.G. (1963). *J. Am. Chem. Soc.* 85: 3533–3539.
131 Allen, C.J., Hwang, J., Kautz, R. et al. (2012). *J. Phys. Chem. C* 116: 20755–20764.
132 Paul, R.C., Johar, S.P., Banait, J.S., and Narula, S.P. (1976). *J. Phys. Chem.* 80: 351–352.
133 Appetecchi, G.B., Montanino, M., Zane, D. et al. (2009). *Electrochim. Acta* 54: 1325–1332.
134 Zhang, Y., Ortiz-Vitoriano, N., Acebedo, B. et al. (2018). *J. Phys. Chem. C* 122: 15276–15286.

135 Ortiz-Vitoriano, N., Monterrubio, I., Garcia-Quintana, L. et al. (2020). *ACS Energy Lett.* 5: 903–909.

136 N. Banis, M., Yadegari, H., Sun, Q. et al. (2018). *Energy Environ. Sci.* 11: 2073–2077.

137 Lutz, L., Dachraoui, W., Demortière, A. et al. (2018). *Nano Lett.* 18: 1280–1289.

138 Han, S., Cai, C., Yang, F. et al. (2020). *ACS Nano* 14: 3669–3677.

6

Zn-Ion Battery

Gaoxue Jiang[1], Yurong Ren[2], Xiaobing Huang[3], and Jianmin Ma[1]

[1] *University of Electronic Science and Technology of China, School of Materials and Energy, Chengdu, 611731, P. R. China*
[2] *Changzhou University, School of Materials Science and Engineering, Jiangsu Collaborative Innovation Center of Photovolatic Science and Engineering, Changzhou 213164, Jiangsu, China*
[3] *Hunan University of Arts and Science, College of Chemistry and Materials Engineering, Changde 415000, China*

6.1 Introduction

In the field of energy storage, lithium-ion batteries (LIBs) have almost dominated the portable electronic equipment and electric vehicle markets due to their high energy density and long cycling life [1, 2]. However, excessive consumption of lithium resources, high costs, and safety issues limit their applications in the field of energy storage [3, 4]. Sodium-ion batteries (SIBs) and potassium-ion batteries (KIBs), as the alternatives to LIBs, have relatively abundant and cheap raw materials and exhibit similar chemical properties to lithium but have low energy density and use electrolytes that are highly toxic and flammable [5–9]. In addition, the high operating cost and safety issues also limit their practical applications. Aqueous rechargeable batteries have received extensive attention due to their low cost, high operational safety, and environmental friendliness properties [10]. In Table 6.1, we compare the characteristics of aqueous zinc-ion batteries (AZIBs) with non-aqueous Li/Na/K ion batteries, and the advantages of AZIBs in cost, volumetric capacity, capacity density, and safety are extremely obvious [11, 12].

In Zn ion batteries (ZIBs), aqueous electrolytes are safer and more environmentally friendly than flammable and toxic organic electrolytes [13–15]. In the preparation process of aqueous electrolytes, it is safer, cheaper, and easier to be handled and does not need to be carried out in an exact atmosphere [16, 17]. In addition, the multivalent AZIBs that allow the transfer of multiple electrons during the electrochemical reaction provide the opportunity to obtain high-energy and high-power densities simultaneously [18, 19]. As shown in Table 6.2, the aqueous electrolyte usually has a high ionic conductivity, which can provide the fast and rapid diffusion rate. This is beneficial for improving the high-rate performance of the ZIBs [20, 21, 23]. Therefore, AZIBs are expected to be used in grid storage systems. However, there are still many unsolved problems in AZIBs [10, 22, 24].

Battery Technologies: Materials and Components, First Edition. Edited by Jianmin Ma.

Table 6.1 Comparison of Zn^{2+} with other charge carrier ions [11, 12].

	LIBs	NIBs	KIBs	ZIBs
Ionic radius (Å)	0.76	1.02	1.38	0.75
Cost of metal anode (USD) (kg^{-1})	19.2	3.1	13.1	2.2
Volumetric capacity ($mA\,h\,m^{-3}$)	2042	1050	609	5857
Capacity density ($mA\,h\,cm^{-3}$)	2061	1129	610	5855
Safety		Low		High

Source: Refs. [11, 12].

Table 6.2 Comparison of organic and aqueous ZIBs [10, 20–22].

Electrolytes	Organic electrolytes	Aqueous electrolytes
Merits	Wide electrochemical window reversible Zn plating/stripping	Ability to solvate both the cation and anion High ionic conductivity Small solvation shell Low desolvation energy penalty Formation a pathway by H_2O for facile insertion of Zn^{2+}
Limitations	Low ionic conductivity Large solvation shell High toxicity High cost	Narrow electrochemical window Corrosion caused by water

Source: Refs. [10, 21, 22].

In this chapter, we will conduct a comprehensive review and generalization of AZIBs, including cathode materials, zinc anodes, and electrolytes as shown in Figure 6.1 [10], and summarize and analyze the problems in various parts of the development of ZIBs. Finally, we also present the challenges faced by aqueous Zn^{2+} and look forward to the future development of aqueous zinc batteries.

6.2 Fundamentals

In AZIBs, the cathode material and zinc metal anode are separated by a glass fiber or cellulose filter, which are immersed in the electrolyte that is composed of aqueous zinc salt solution. The schematic illustration of a ZIB is shown in Figure 6.2 [21]. During the charge–discharge process, Zn^{2+} ions strip from/plate on the zinc anode and insert into/extract from the cathode material. Due to the complex chemical reactions involved in AZIBs, the reaction mechanisms in lithium/sodium batteries are not completely different, and there are still controversies [10, 15]. Therefore, the detailed reaction mechanism is still controversial and needs to be further investigated in future.

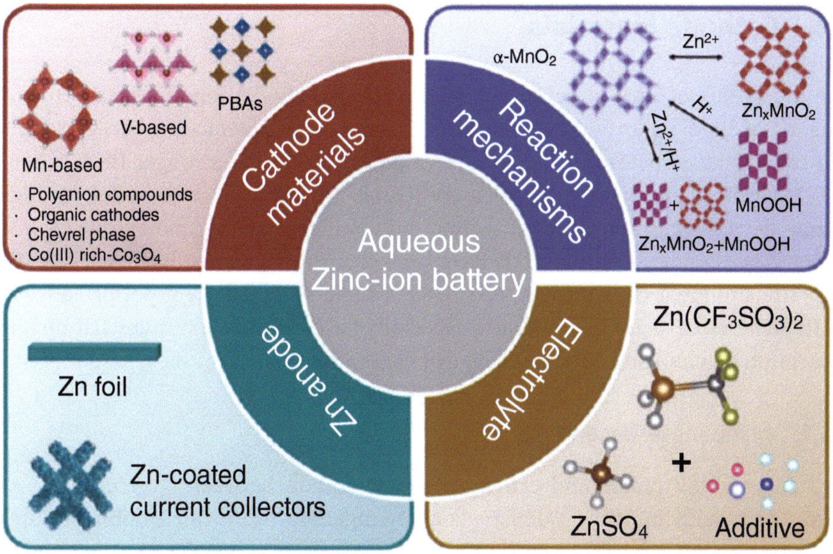

Figure 6.1 Challenges in various parts of AZIBs [15]. Source: Fang et al. [15]. Reproduced with permission of American Chemical Society.

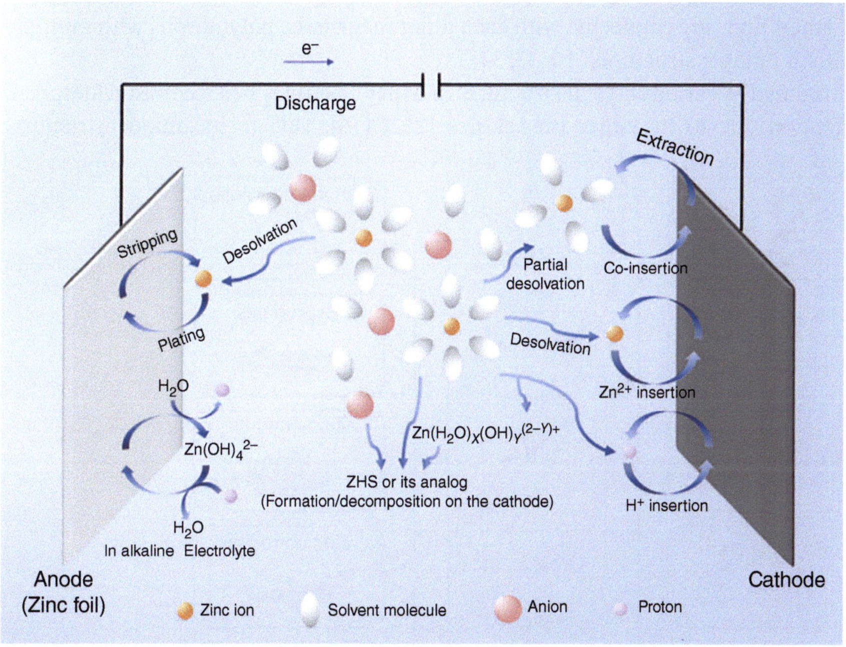

Figure 6.2 Schematic illustration of a ZIB with liquid electrolyte [21]. Source: Ye et al. [21]. Reproduced with permission of Elsevier.

6.3 Cathode Materials

The ion radius of Zn^{2+} is relatively small (0.75 Å), but Zn ions have the characteristics of high polarity. This makes the electrostatic interaction between Zn^{2+} and the crystal structure of the cathode material to exhibit stronger than Li ions [25, 26]. Although water molecules coexisting around Zn^{2+} can buffer its high charge density, it is still difficult to find suitable intercalating materials [11, 27]. Various types of cathode materials have been reported as shown in Figure 6.3, but there are still great challenges to be solved to ensure the excellent performance of AZIBs. Therefore, cathode materials, especially based on zinc ion intercalation and desorption mechanism, have received extensive attention.

6.3.1 Manganese-Based Materials

Zn^{2+} has a small radius and can travel along tunnels between large crystals. As a substance with multiple valence states, manganese oxide can exhibit multiple crystal structures [29]. Among them, manganese dioxide has the most abundant structure, which is divided into α-type, β-type, γ-type, δ-type, λ-type, R-type, Romanechite-type, and Todorokite-type [30, 31]. The octahedral unit of MnO_6 is composed of six oxygen atoms and one manganese atom is the basic crystal structure of manganese dioxide [32]. As shown in Figure 6.4, the different numbers of MnO_6 units are connected with each other to form the polymorphs with multiple tunnels or layer structures [10, 33, 34].

Among the various crystal structures of MnO_2, α-MnO_2 has received widespread attention due to its larger tunnel size [2*2] [16]. Due to its unique structural

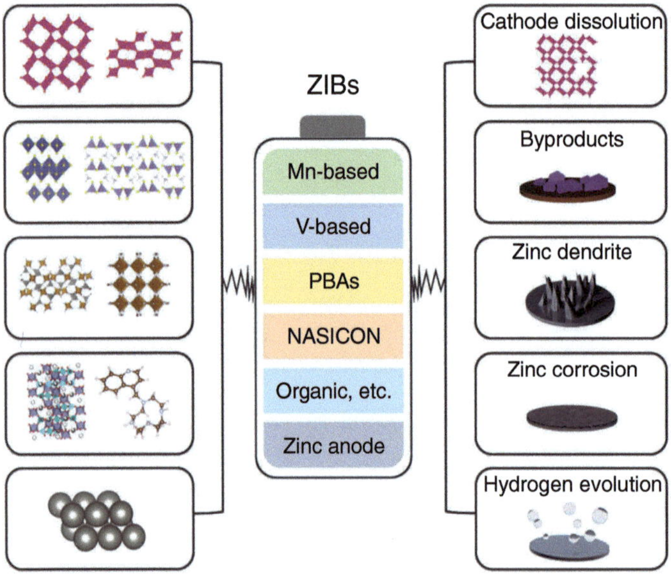

Figure 6.3 Main cathode materials of ZIBs and their issues [28]. Source: Liu et al. [28]. Reproduced with permission of American Chemical Society.

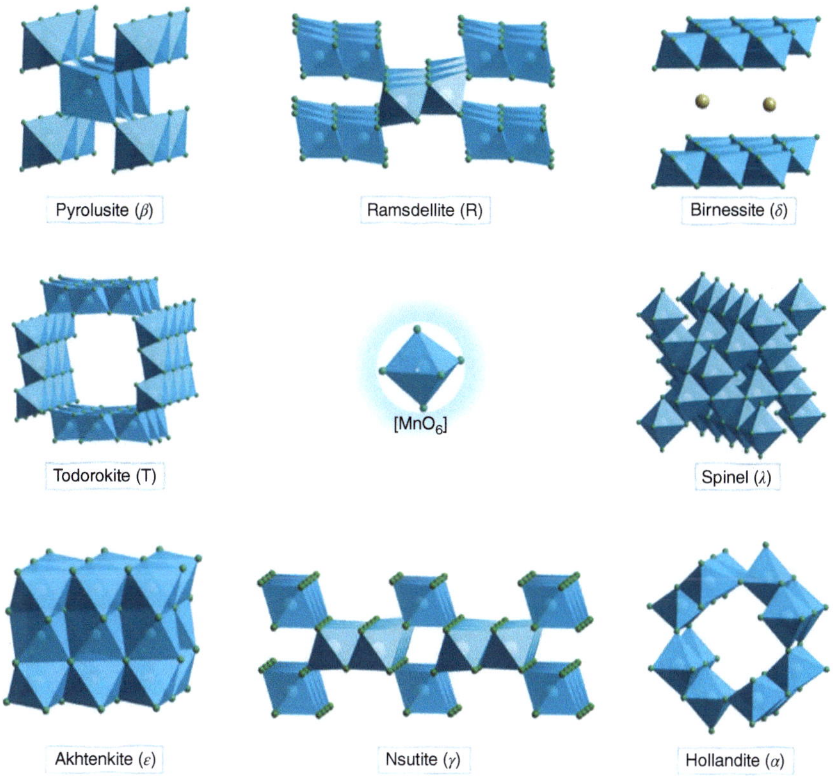

Figure 6.4 Schematic diagram of various crystal structures of manganese-based materials [10]. Source: Jia et al. [10]. Reproduced with permission of American Chemical Society.

advantages, α-MnO_2 can deliver a larger specific capacity under medium-voltage conditions [13]. However, the capacity will rapidly decay during the cycle, and the high-rate performance is poor. As shown in Figure 6.5, Xia and coworkers designed a polyaniline (PANI) intercalated MnO_2, which can greatly improve its cycling ability [35]. The presence of polymer enhances the layer distance of MnO_2 and eliminates the occurrence of phase changes. Compared with other crystalline manganese-based oxides, β-MnO_2 has the narrowest tunnel structure [1*1], which makes Zn^{2+} to insert difficultly, but morphology and structure-modified β-MnO_2 could exhibit excellent cycle performance [36, 37]. The structure of γ-MnO_2 is composed of randomly distributed [1*1] and [1*2] tunnels, which is equivalent to the coexistence of β- and R-MnO_2 [34]. Figure 6.6 demonstrates the reaction pathway of Zn^{2+} insertion in the γ-MnO_2 cathode. Due to its large tunnel size, it is suitable for Zn^{2+} insertion and has excellent electrochemical performance in AZIBs.

In addition to the tunnel-like crystal structure, the layered δ-MnO_2 has a wide interlayer distance and can accommodate water molecules or other cations in the middle layer, which leads to the expansion of the middle layer distance [38]. By

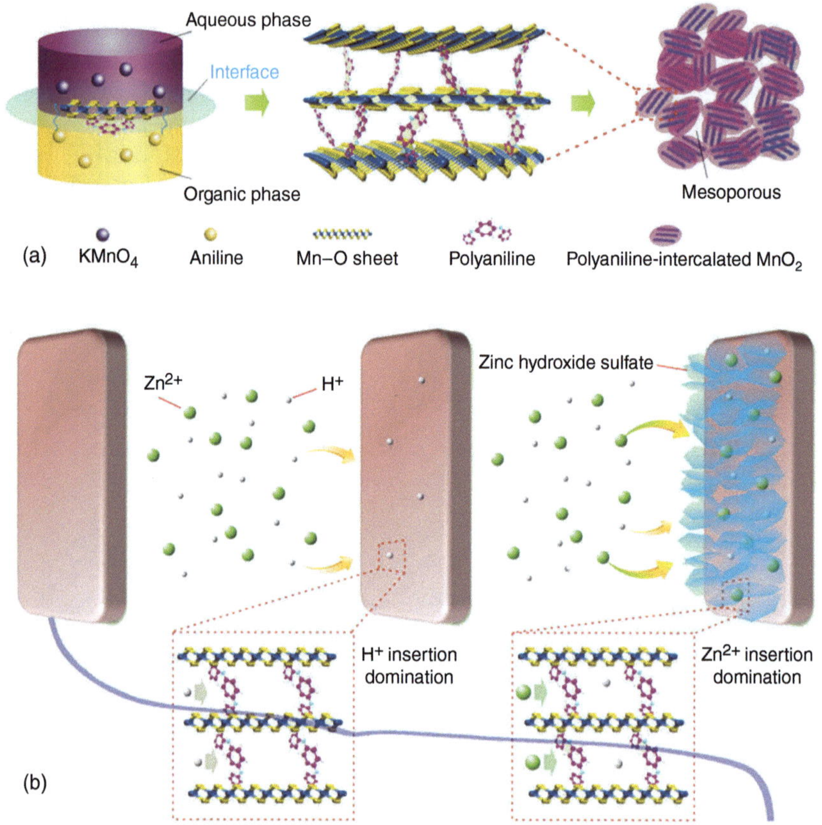

Figure 6.5 (a) Schematic illustration of expanded intercalated structure of polyaniline (PANI)-intercalated MnO_2 nanolayers; (b) diagram showing the sequential insertion of H^+ and Zn^{2+}. During the first discharge platform, H^+ insertion into polyaniline (PANI)-intercalated MnO_2 nanolayers dominates the electrode reaction, which gradually decreases H^+ concentration around the electrode. With a sustained decrease of H^+, Zn^{2+} insertion dominates the electrochemical reaction, raising the second discharge platform; meanwhile, the sustained decrease of H^+ concentration leads to the formation of zinc hydroxide sulfate on the electrode surface [35]. Source: Huang et al. [35]. Licenced under CC BY 4.0.

inserting various water molecules and other cations, different layered structures with different interlayer distances could be obtained [39]. Therefore, δ-MnO_2 is also an attractive candidate cathode. λ-MnO_2 is a metastable MnO_2 with a three-dimensional spinel structure [13]. Due to its limited three-dimensional tunnel and the smallest specific surface area, it exhibits poor electrochemical activity [14]. Romanecite-MnO_2 is composed of double and triple chains and exhibits the [2*3] tunnel structure [14], while Todorokite-MnO_2 has a larger [3*3] tunnel structure, it is occupied by a large amount of magnesium ions and therefore exhibits a lower specific capacity [40].

In addition to the commonly studied MnO_2, there are other manganese oxides, such as MnO [41], Mn_2O_3 [42], Mn_3O_4 [43, 44], and spinel type $ZnMn_2O_4$ [45].

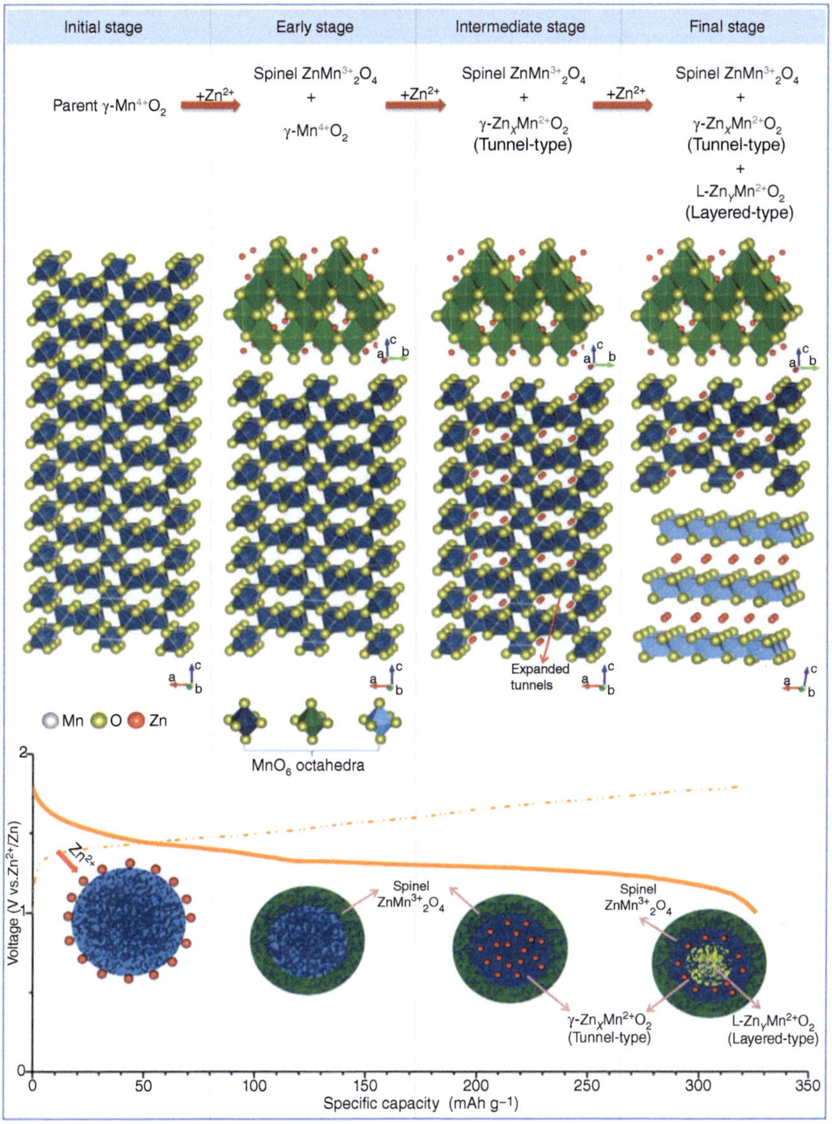

Figure 6.6 Schematic illustration of the reaction pathway of Zn insertion in the prepared γ-MnO_2 cathode [34]. Source: Alfaruqi et al. [34]. Reproduced with permission of American Chemical Society.

Although various manganese-based materials have high specific capacity and high energy density, as a low-conductivity material, the electrochemical performance of manganese-based materials is bound to be limited [30]. At the same time, severe phase changes will occur during the charge–discharge process, which leads to structural damage and volume changes. Moreover, the inevitable Mn dissolution causes rapid degradation of electrochemical performance, which affects the practical application of AZIBs [11, 46]. In order to solve these problems, many strategies have been developed, that is, nanostructure design, defect engineering, composites with conductive carbon, surface coating, etc.

6.3.2 Vanadium-Based Materials

Vanadium-based materials, especially vanadium oxides, have a large channel framework, which can provide space for Zn^{2+} storage [23]. Moreover, vanadium ions have a variety of oxidation valence states (e.g. V^{5+}, V^{4+}, V^{3+}, and V^{2+}), and each valence state can be converted mutually [47]. Due to the deformation of the V–O coordination polyhedron and the conversion of multiple oxidation states, when the vanadium-based compound is used as an energy storage electrode material, the specific capacity is higher. Vanadium oxides are mainly divided into tetrahedrons, triangular double cones, square cones, regular octahedrons, and twisted octahedrons, as shown in Figure 6.7 [48]. Different structures provide multiple ways for Zn^{2+} insertion and extraction. The vanadium ion in the tetrahedron is always V^{5+}, and it is difficult to realize the reversible intercalation/deintercalation of divalent ions without changing the structure. The vanadium ion can be V^{4+} or V^{5+} in the triangular bipyramidal and square pyramid structures, while the vanadium ion in the octahedron can be V^{5+} or V^{4+}, V^{3+}, or lower.

As shown in Figure 6.8, these polyhedrons are connected to each other to form a structure and exhibit different performance differences [49, 50]. Among them are mainly divided into the following seven categories: V_2O_5 and their derivatives [51–54]; monoclinic layered vanadates (AV_3O_8; A = Li, K, H_2) [55–57]; layered barnesite-type vanadates $B_2V_6O_{16}{\cdot}nH_2O$ (B = Na, K) [58–60]; monoclinic

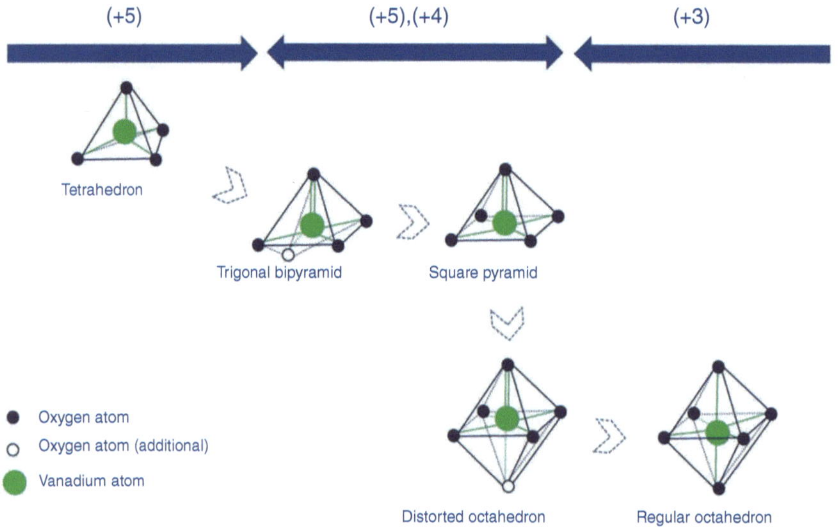

Figure 6.7 Metamorphosis of the vanadium coordination polyhedra [14]. Source: Song et al. [14]. Reproduced with permission of John Wiley & Sons.

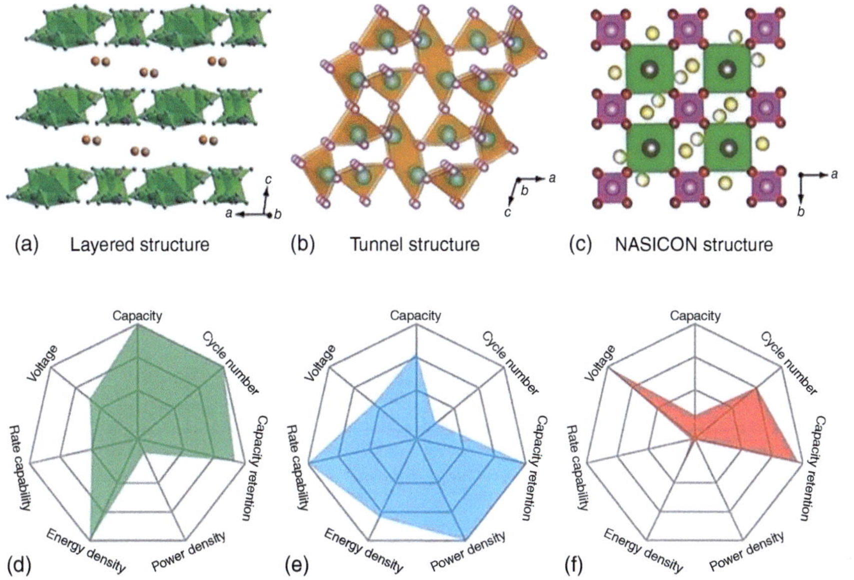

Figure 6.8 Three categories of typical vanadium-based cathodes in aqueous ZIBs and an overview of their electrochemical performance. Crystal structures of (a) layered $NaV_3O_8{\cdot}1.5\,H_2O$ (Na^+ exists in the form of a hydrated ion), (b) tunnel-type VO_2, and (c) NASICON-structured $Na_3V_2(PO_4)_2F_3$. An overview of the electrochemical performance of (d) layered, (e) tunnel-type, and (f) NASICON-structured vanadium-based cathodes [49]. Source: Wan et al. [49]. Reproduced with permission of John Wiley & Sons.

tunnel-type VO_2 [27, 61]; mixed-valent V_6O_{13} [62, 63]; transition metal vandates [64, 65]; and NASICON-type compounds [66–68]. As shown in Figure 6.9a, Wang et al. prepared $LiV_2(PO_4)_3$ (a NASICON-type compound) for AZIBs that can achieve ultrafast and reversible Zn^{2+} intercalation and deintercalation at a high working voltage [68].The high charge density of bivalent Zn^{2+} was effectively accommodated by the p–d hybridization between the V-d and O-p orbitals, while Zn^{2+} diffusion shares a facile pathway with Li^+, resulting in quick kinetics and high-power performance in Figure 6.9b,c. In this polyanion-based framework, the insertion/extraction of Zn^{2+} is highly reversible, undergoing only slight volume changes and yielding superior rechargeability.

However, V-based materials have low stability (especially in aqueous electrolytes), poor electrical conductivity, and low ion diffusion coefficient, which lead to poor long-term cycling performance. This greatly restricts their practical applications [69]. The introduction of guest species into V-based materials helps to alleviate structural deterioration caused by the "columnar effect." It can expand the crystal tunnel to absorb more ions, thereby improving the overall electrochemical performance [47, 70]. It is worth noting that water molecules play an important role in promoting the diffusion of Zn^{2+}. The water molecules embedded in the vanadium oxide intermediate layer can buffer the high charge density of Zn^{2+} and reduce electrostatic interaction. As shown in Figure 6.10, the inserted water molecules may

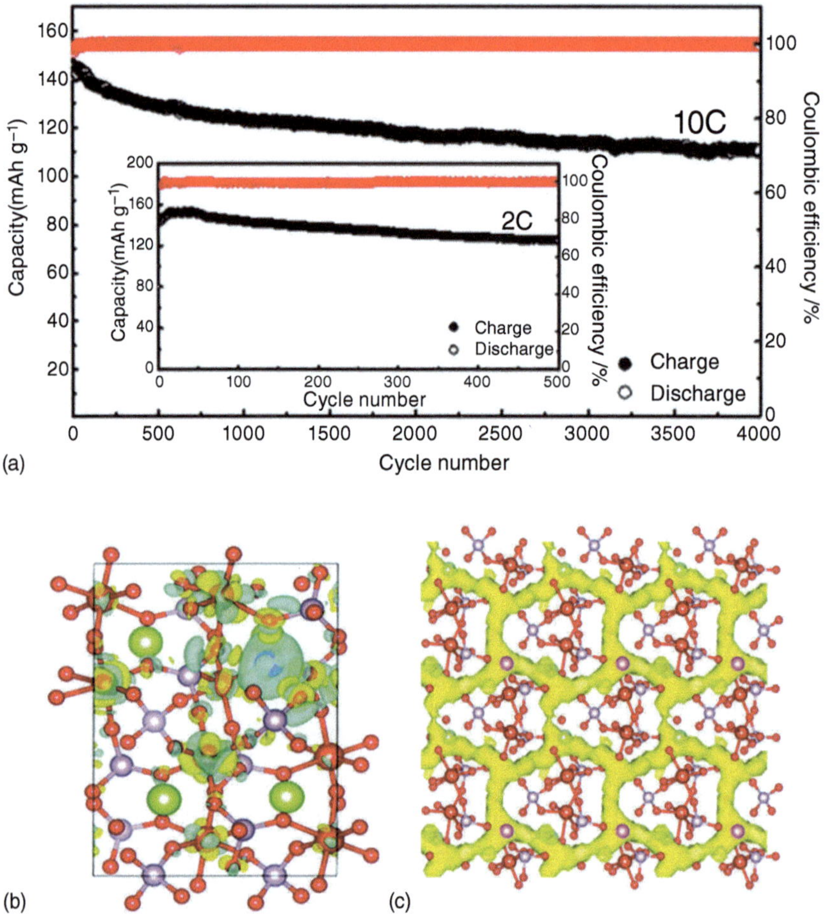

Figure 6.9 (a) The cycling performance and the corresponding coulombic efficiency at a high rate of 10C (inset: the low rate of 2C). (b) The schematic illustration of the change in charge distribution after Zn^{2+} insertion into the $LiV_2(PO_4)_3$ framework. (c) The structural insights and the Zn diffusional pathway within the $LiV_2(PO_4)_3$ framework [68]. Source: Wang et al. [68]. Reproduced with permission of Royal Society of Chemistry.

cause structural rearrangement, effectively expand the diffusion pathways for the rapid transport of Zn^{2+} [51, 54, 62]. In addition, the layered vanadium oxide and the interlayer Zn^{2+} in the structured water can support the layer spacing to stabilize the vanadium oxide framework [11, 51].

6.3.3 Prussian Blue Analogous

Prussian blue analogs (PBAs) are a series of compounds derived from Prussian blue (PB) after substitution and interstitial modification (as shown in Figure 6.11), which are roughly expressed as $A_xM_{1y}[M_2(CN)_6]_z{\cdot}nH_2O$ (A is an alkali metal; M_1 and M_2 are transition metals such as manganese, iron, cobalt, nickel, copper, and zinc).

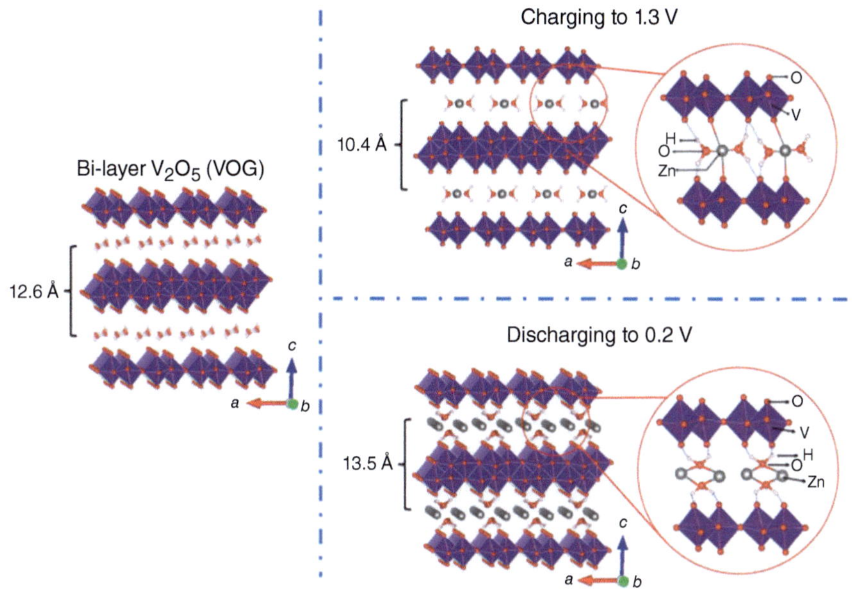

Figure 6.10 The pristine charge and discharge crystal structures of VOG [51]. Source: Yan et al. [51]. Reproduced with permission of John Wiley & Sons.

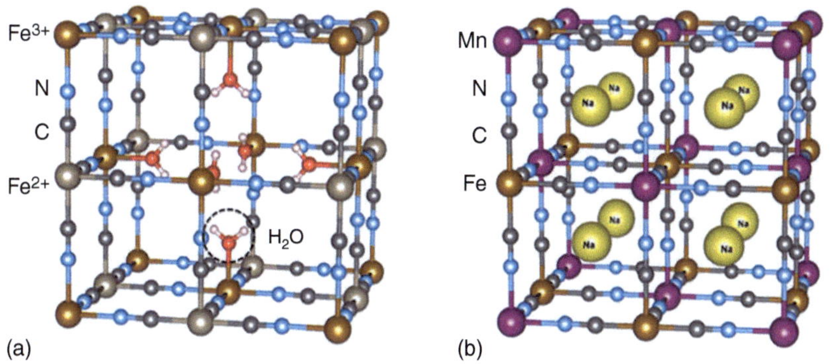

Figure 6.11 (a) Unit cell of the original Prussian blue (PB) $Fe_4[Fe(CN)_6]_3 \cdot nH_2O$ compound. (b) Prussian blue analogue (PBA) $Na_xMnFe(CN)_6$ structure [71]. Source: Paolella et al. [71]. Reproduced with permission of Royal Society of Chemistry.

The two metal ions (M_1 and M_2) are linked together by a cyanide (CN) ligand, and there is mobility alkali metal ions (A) [72, 73]. PBA usually has a face-centered cubic structure, a three-dimensional open framework, and larger ion intercalation sites, which contributes to rapid zinc ion transport, so it is more and more popular in energy storage devices [71, 74–76]. Although PBA as a ZIB cathode has excellent rate and cycle performance, its specific capacity is still low and needs to be further improved. This is mainly because the redox active sites of multivalent ions cannot be fully activated and utilized. In addition, uncontrolled phase changes during cycling can cause capacity degradation [73, 77–79]. In order to solve these problems,

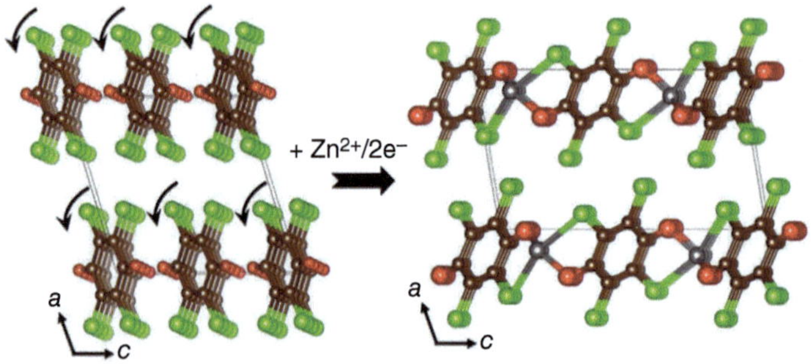

Figure 6.12 Structural transformation of p-chloranil after Zn^{2+} insertion as obtained from density functional theory (DFT) calculations [82]. Source: Kundu et al. [82]. Reproduced with permission of American Chemical Society.

high-pressure scanning activation [77], structural optimization [80], and composite material optimization [81] strategies can be used to solve these problems.

6.3.4 Other Types of Cathode Materials

Besides inorganic cathode materials, organic compounds also show excellent performance in AZIBs, since they can exhibit the light weight, abundant chemical substances, sustainability, cost-effectiveness, multiple electronic reactions, and adjustable electrochemical windows [82, 83]. As shown in Figure 6.12, through density functional theory (DFT) calculations, it is known that the molecular columns in *p*-chloranil undergo a twisted rotation to accommodate Zn^{2+}, thereby limiting the volume change (−2.7%) during the cycling process [82]. However, the development of organic compounds in AZIBs is still in its infancy, and further research is needed [84–86]. In addition to the above-mentioned materials, there are other types of materials that exhibit good performance in AZIBs, such as polyanion compounds [68, 87, 88], Co/Ni-based materials [89–91], Chevrel phase compounds [92–95], and transition metal dichalcogenides [96–98]. Ti-substituted tungsten molybdenum oxide (MTWO) is a kind of Chevrel phase compounds, and Figure 6.13 shows the rechargeable Zn/MTWO battery system and the possible Zn^{2+} intercalation/deintercalation tunnels. However, in the cathode material, the solvent of the cathode material, the slow diffusion rate, and the formation of undesirable byproducts often occur, which significantly affect the application of AZIBs [10, 11, 99]. Therefore, it is still necessary to continue to explore, solve the above problems, and promote the rapid development of cathode materials.

6.4 Zn Anode

Zn metal is considered to be the best anode for AZIBs because of its abundant sources, cheap price, chemical stability, low redox potential (−0.76 V vs. SHE (standard hydrogen electrode)), and easy processing [100]. However, the mass loading of Zn is too large, which leads to underutilization of Zn metal and relatively

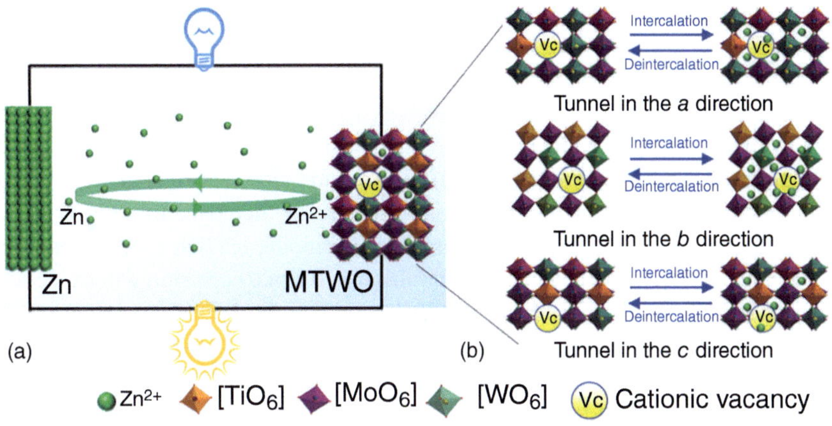

Figure 6.13 Zn/MTWO battery chemistry. (a) Schematic illustration of the rechargeable Zn/MTWO battery system. (b) Schematic illustration of possible Zn^{2+} intercalation/deintercalation tunnels in MTWO [95]. Source: Li et al. [95]. Reproduced with permission of John Wiley & Sons.

low energy density, which limits the practical application of AZIBs. In addition, as shown in Figure 6.14, Zn dendrite growth and hydrogen evolution also exist in charge–discharge process [101]. Uneven Zn deposition/exfoliation can lead to the production of Zn dendrites. Usually, brittle dendrites are needle-shaped, and the tip acts as a charge center in the subsequent reaction and triggers the tip effect, which

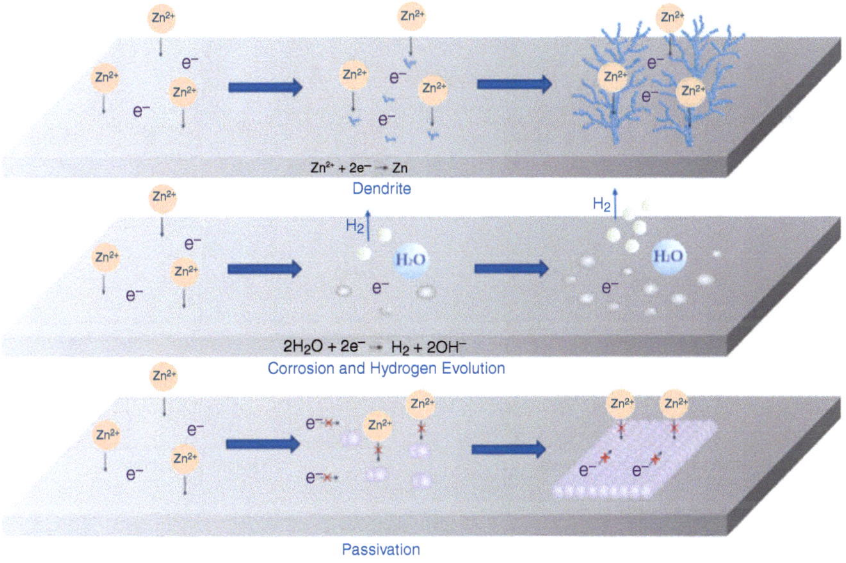

Figure 6.14 Schematic diagrams of the Zn dendrite, corrosion and hydrogen evolution, and passivation processes [101]. Source: Wang et al. [101]. Reproduced with permission of American Chemical Society.

leads to continuous charge accumulation, which further promotes the growth of Zn dendrites. Zn dendrites will reduce the utilization efficiency of Zn anode and cause a short circuit inside the battery and further lead to the battery to fail prematurely. Moreover, the growth of Zn dendrites is usually accompanied by electrode corrosion and the formation of byproducts. During the discharge process, the Zn anode peels off, the soluble Zn^{2+} ions enter the liquid phase, and side reactions occur in local high-energy areas. These will result in an increase in the concentration of OH^-. The Zn^{2+} and OH^- further react to passivate Zn anode, leading to a decrease in battery energy storage performance. In addition, electrode corrosion will cause the self-discharge of batteries, passivation will also increase the battery impedance, and the growth of Zn dendrites will increase the anode surface area. This will accelerate corrosion and side reactions at the Zn–electrolyte interface, resulting in a decrease in Coulomb efficiency and a decrease in cycle life. Therefore, in order to develop Zn anode with a long cycling life, the problems of dendrites, corrosion, and passivation must be solved.

6.4.1 Zinc Alloy Anode

The alloying of Zn with other metals is an effective strategy to improve the electrochemical performance of Zn anodes and inhibit the generation of H_2. The alloy components can effectively prevent the Zn anode from being corroded with high hydrogen evolution reaction (HER) overpotential to form a protective corrosion layer [102–104]. Jiang and coworkers designed the Zn/Al alloys in Figure 6.15

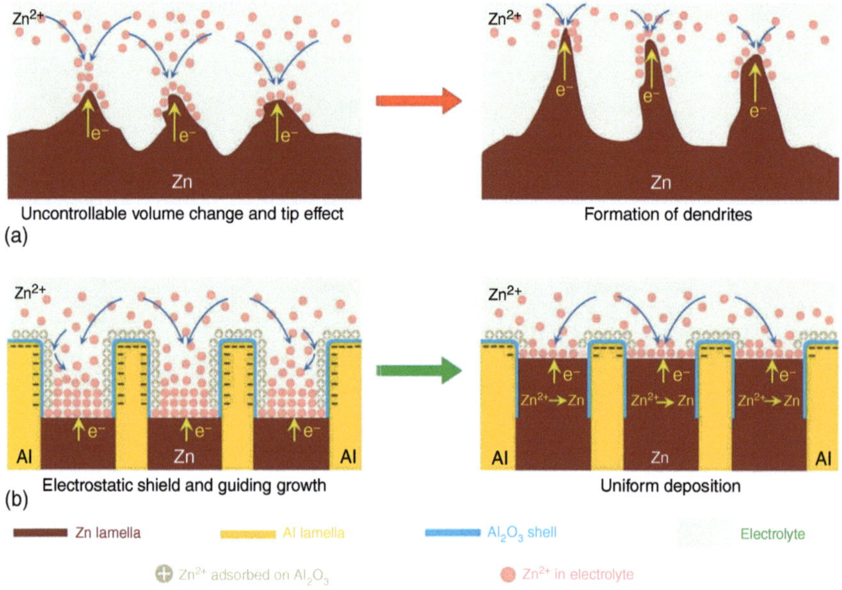

Figure 6.15 Schematic diagram of dendrite growth in (a) Zn anode and (b) Zn/Al alloys anode during the Zn stripping/plating processes [105]. Source: Wang et al. [105]. Licenced under CC BY 4.0.

[105]. Although the standard equilibrium potential of Al/Al^{3+} is lower than that of Zn/Zn^{2+}, the formation of Al_2O_3 can prevent Al from being dissolved, since the insulating Al_2O_3 layer prevents the electron transfer from Al to Zn^{2+}, and an electrostatic shielding protective layer is formed around the Al/ Al_2O_3 to guide the uniform deposition of Zn in Figure 6.15b, instead of forming Zn dendrites as shown in Figure 6.15a. Taking into account the environmental impact and cost of different metals; Al [105], Sn [106], Fe [107], Cu [108], and Ni [109] are the most promising alloying elements. However, the performance of Zn alloys in AZIBs in different electrolytes still lacks a comprehensive comparison. Therefore, it is still unable to provide comprehensive guidance for the design and use of zinc alloy anodes.

6.4.2 Surface Modification of Zn Anode

In AZIBs, surface modification can optimize the growth of Zn electrodeposition and inhibit the generation of Zn dendrites [110–112]. A continuous solid electrolyte interface with good structure between Zn anode and aqueous electrolyte is an effective method to inhibit the growth of Zn dendrites and prolong the cycling stability of AZIBs [11, 112]. Mai and his colleagues [113] coat Zn with a thin layer of TiO_2 by atomic layer deposition (ALD) technology as shown in Figure 6.16. TiO_2 acts as a protective layer, and the corrosion of the Zn anode is significantly suppressed, which result in less gas escape and the formation of $Zn(OH)_2$ byproducts in Figure 6.16b. The reduction of gas on the outer surface of the Zn plate allows the effective contact area between the electrolyte and the anode to be maintained and

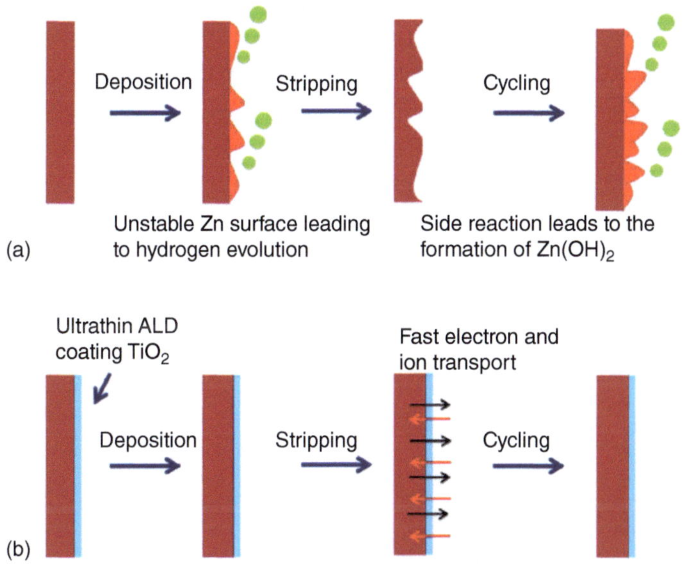

Figure 6.16 Schematic illustration of the stabilization of Zn anode without (a) and with TiO_2 coating (b) [113]. Source: Zhao et al. [113]. Reproduced with permission of John Wiley & Sons.

improves Coulomb efficiency. In addition, coating porous nanomaterials on the surface of Zn anode can also ensure uniform Zn deposition and peeling. Zhi and colleagues found that porous nano-$CaCO_3$ coating could serve as a buffer layer for Zn anodes to ensure uniform deposition of Zn [114]. The porous $CaCO_3$ coating on the Zn anode can guide the uniform deposition of Zn from bottom to top during charge–discharge process and inhibit the formation of Zn dendrites, which can improve the cycling stability of AZIBs.

6.4.3 Structural Optimization of the Zn Anode

Optimizing the Zn structure is also an effective strategy to improve the performance of aqueous Zn metal batteries (AZMBs) [103, 110, 112]. The 3D nanostructured Zn anode can expand the surface active area and generate uniform electric field flow lines. This can effectively minimize the size of the Zn core and ensure uniform Zn deposition [115, 116]. Zhou and colleagues deposited Zn on different substrates as the anode and demonstrated the Zn@Cu foam with the best performance [117]. This can be attributed to the uniform electric field distribution of the 3D foam structure, which can effectively enhance the uniform deposition of Zn. And effective electron transmission ensures the uniform deposition of Zn and good cycling stability due to the excellent electrical conductivity of the Zn@Cu foam anode. Lu and coworkers found that it effectively inhibited the growth of dendrites and prevented side reactions from occurring, when 3D carbon nanotubes were used as a base material for Zn plating due to its high specific surface area and good electrical conductivity [118]. Xia and his colleagues prepared Zn@ZIF-8-500 that can achieve uniform Zn deposition and peeling during charge–discharge process [119]. It can be seen that uniform Zn deposition can be achieved under different current densities in Figure 6.17a, which is due to the porous ZIF-8-500 that induces uniform zinc deposition in Figure 6.17b. In addition, the release of hydrogen is suppressed, and a coulombic efficiency close to 100% can be obtained.

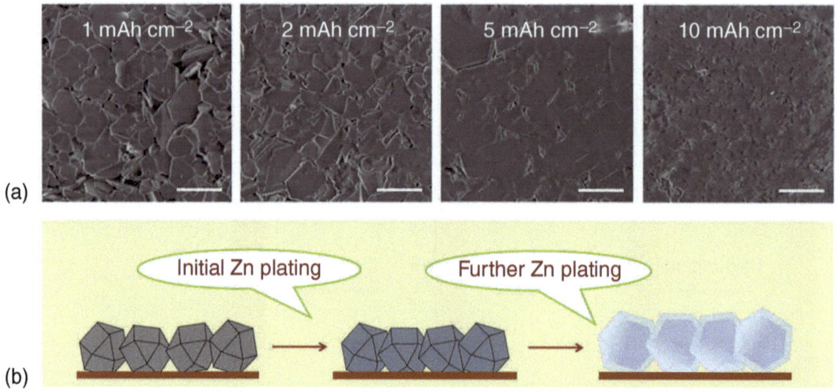

Figure 6.17 (a) Scanning electron microscopy (SEM) images of Zn deposits at a current density of 1.0 mA cm^{-2} for different capacities. Scale bars, 2 μm. (b) Schematic illustration of the Zn plating [119]. Source: Wang et al. [119]. Reproduced with permission of Elsevier.

6.5 Aqueous Electrolytes

As the main component of ZIBs, electrolyte is responsible for the transportation of Zn^{2+} between anode and cathode, ensuring a stable potential window (ESPW) and good electrochemical performance [21, 22]. Compared with organic electrolyte, aqueous electrolyte has the advantages of low cost, safety, environmental protection, and easy operation. In addition, aqueous electrolytes generally have higher ionic conductivity and exhibit higher rate performance [10, 120]. At present, $ZnSO_4$ and $Zn(CF_3SO_3)_2$ salt-based electrolytes are considered as promising electrolytes and are widely used in AZIBs. In addition to the exploration of electrode materials, electrolytes have also attracted widespread attention. The composition, concentration, and additives of the electrolyte also play a vital role in the electrochemical performance of ZIBs.

6.5.1 Types of Zinc Salts

In the electrolyte, anions can participate in the solvation process of Zn^{2+}. Therefore, the compatibility with the electrode and the coordination with Zn^{2+} are the main factors of the anionic nature and may be the key factors for affecting the performance of AZIBs. Many Zn salts have been reported, i.e. $ZnSO_4$, $Zn(NO_3)_2$, $ZnCl_2$, $Zn(ClO_4)_2$, ZnF_2, $Zn(CF_3SO_3)_2$, $Zn(TFSI)_2$, $Zn(CH_3COO)_2$, etc. As shown in Table 6.3, NO_3^- ion takes role of a strong oxidant, which will make Zn anodized, increase corrosion, and ultimately lead to a sharp increase in the pH of the electrolyte [79]. However, ClO_4^- ion can reduce the corrosion of Zn anode through forming a passivation layer of ZnO on its surface, resulting in slow reaction kinetics [121]. $ZnCl_2$ is also used as Zn salt due to its low oxidation, but the narrow electrochemical window limits its practical application [45]. SO_4^{2-} ion has become the most widely used electrolyte in AZIBs due to its good structural stability in Figure 6.18a and excellent compatibility with Zn

Table 6.3 The summary of comparison of different salts for optimization [17, 45, 79, 121, 122].

Salts	Advantages	Disadvantages
$ZnCl_2$	Low cost High solubility	Low anodic stability
$Zn(NO_3)_2$	Lost cost	Low anodic stability
$Zn(ClO_4)_2$	Lost cost	High overpotential Potential safety hazard
$ZnSO_4$	Lost cost High anodic stability	Accompany by zinc hydroxide sulfate compounds
$Zn(CF_3SO_3)_2$	High anodic stability Weakened solvation effect Wide electrochemical window	High cost

Source: Refs. [17, 45, 79, 121, 122].

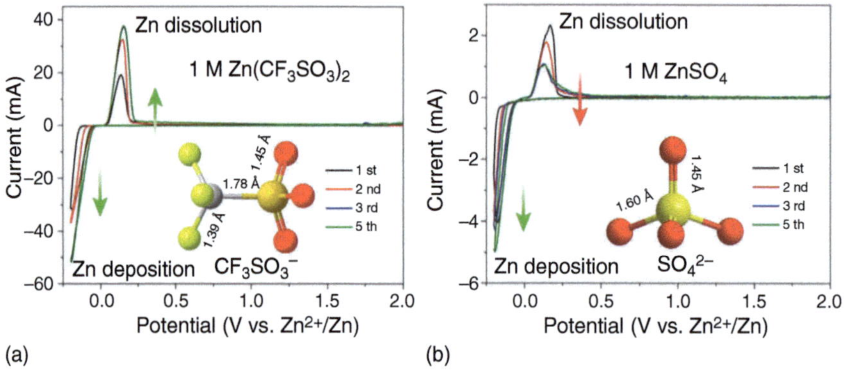

Figure 6.18 Cyclic voltammograms of Zn electrode in aqueous electrolyte of (a) 1 M $Zn(CF_3SO_3)_2$ and (b) 1 M $ZnSO_4$ at the scan rate of 0.5 mV s^{-1} between −0.2 and 2.0 V [45]. Source: Zhang et al. [45]. Reproduced with permission of American Chemical Society.

anodes [45, 81, 122]. However, side reactions and hydrogen release still occur during the charge–discharge process, which reduces the Coulombic efficiency of the batteries and hinders the further development of AZIBs. In recent years, Zn $(CF_3SO_3)_2$ has been popularized in AZIBs due to its unique advantages [45]. $CF_3SO_3^-$ ions can inhibit the formation of Zn dendrites, reduce the solvation of Zn^{2+} ions, and also promote the migration and charge transfer of Zn^{2+} ions. In addition, the electrochemical window is wider in the $Zn(CF_3SO_3)_2$-based electrolyte (Figure 6.18b), but its cost affects is relatively high [17, 45].

6.5.2 Concentration of Zinc Salt

In the electrolyte, Zn^{2+} and water molecules combine to form the salvation sheath of $(Zn(H_2O)_6)^{2+}$. Thus, the high-energy barrier needs to be overcome during the desolvation process, resulting in the reduction of the reversibility of the Zn anode [123, 124]. Increasing the concentration of the electrolyte can reduce the number of solvent molecules combined with Zn^{2+} and change the solvation structure and the transport behavior of anions and cations [28, 125, 126]. In various studies, the performance of AZIBs could be improved with the increase of Zn^{2+} concentration. Chen and his colleagues [45] found the Coulombic efficiency gradually increased to 100%, as the concentration of $Zn(CF_3SO_3)_2$ increased from 1 to 4 M, since side reactions related to water are reduced in high-concentration electrolytes, and the dissolution of active materials in the electrolyte can be reduced. In order to further reduce the number of water molecules near the Zn^{2+} in the electrolyte, the water-in-salt electrolyte is used as a high-concentration electrolyte in AZIBs. Wang et al. developed an electrolyte composed of 1 M $Zn(TFSI)_2$ and 20 M LiTFSI. From the molecular dynamic (MD) studies of the Zn^{2+} solvation structure in Figure 6.19, as the electrolyte concentration increases, the Zn^{2+} solvation sheath is occupied primarily by TFSI. Especially in the highly concentrated electrolyte, all six coordinating oxygens come from TFSI. Zn^{2+} ions are combined with the enriched $TFSI^-$ ions to reduce the occurrence of side reactions and effectively inhibit the generation

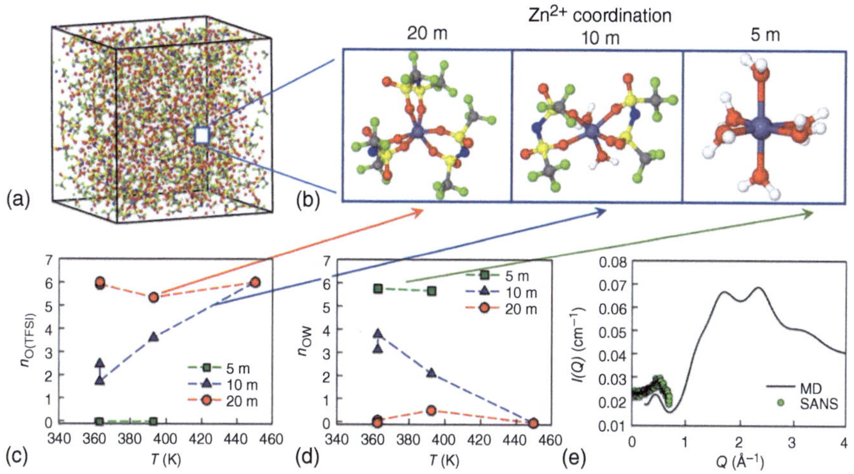

Figure 6.19 (a) Snapshot of the MD simulation cell for HCZE (1 M $Zn(TFSI)_2$ + 20 M LiTFSI) at 363 K. (b) Representative Zn^{2+} solvation structures in the electrolytes with 1 m $Zn(TFSI)_2$ and three concentrations of LiTFSI (5, 10, and 20 M). Zn^{2+}-O (TFSI) (c) and Zn^{2+}-O (water) (d) coordination numbers for aqueous electrolytes. (e) The experimental small-angle neutron scattering (SANS) curve (green circles) and the simulated form (black line) at 298 K for 1 M $Zn(TFSI)_2$ + 20 M LiTFSI electrolyte in D_2O [123]. Source: Wang et al. [123]. Reproduced with permission of Springer Nature.

of Zn dendrites and hydrogen in aqueous electrolytes [123]. However, there is still a problem of high cost. Then, Ji and coworkers explored the water-in-salt electrolyte with 30 M $ZnCl_2$ [127]. Zn^{2+} can combine with Cl^- ions near the negative electrode, thereby inhibiting the occurrence of side reactions for improving electrochemical performance. Subsequently, the electrolyte of 30 M $ZnCl_2$ and 5 M LiCl was prepared, in which Li^+ ions can further capture free water molecules, reduce the generation of hydrogen, and improve the electrochemical performance of the battery [128]. Nevertheless, the ionic conductivity of the electrolyte decreases and the viscosity of the solution increases as the increase in the concentrations. Therefore, it is of great significance for the electrolyte to find the best concentration.

6.5.3 Electrolyte Additives

The introduction of additives into electrolytes is considered to be a simple and efficient strategy and has played a significant role in optimizing the electrolytes [21, 28]. It is well known that the cathode materials will be slightly dissolved in the electrolyte, which will cause the capacity to decay rapidly and affect the practical application of the battery. Adding manganese ions to $ZnSO_4$ or $Zn(CF_3SO_3)_2$ electrolyte can reduce the dissolution of Mn^{2+} ions from the cathode and improve electrochemical performance [16, 129]. As shown in Figure 6.20, the MnO_2 electrodes with added Mn^{2+} electrolytes can maintain better stability and rate performance. This is because the pre-added manganese ions can balance the dissolution of Mn from the MnO_2 electrode [11, 130, 131].

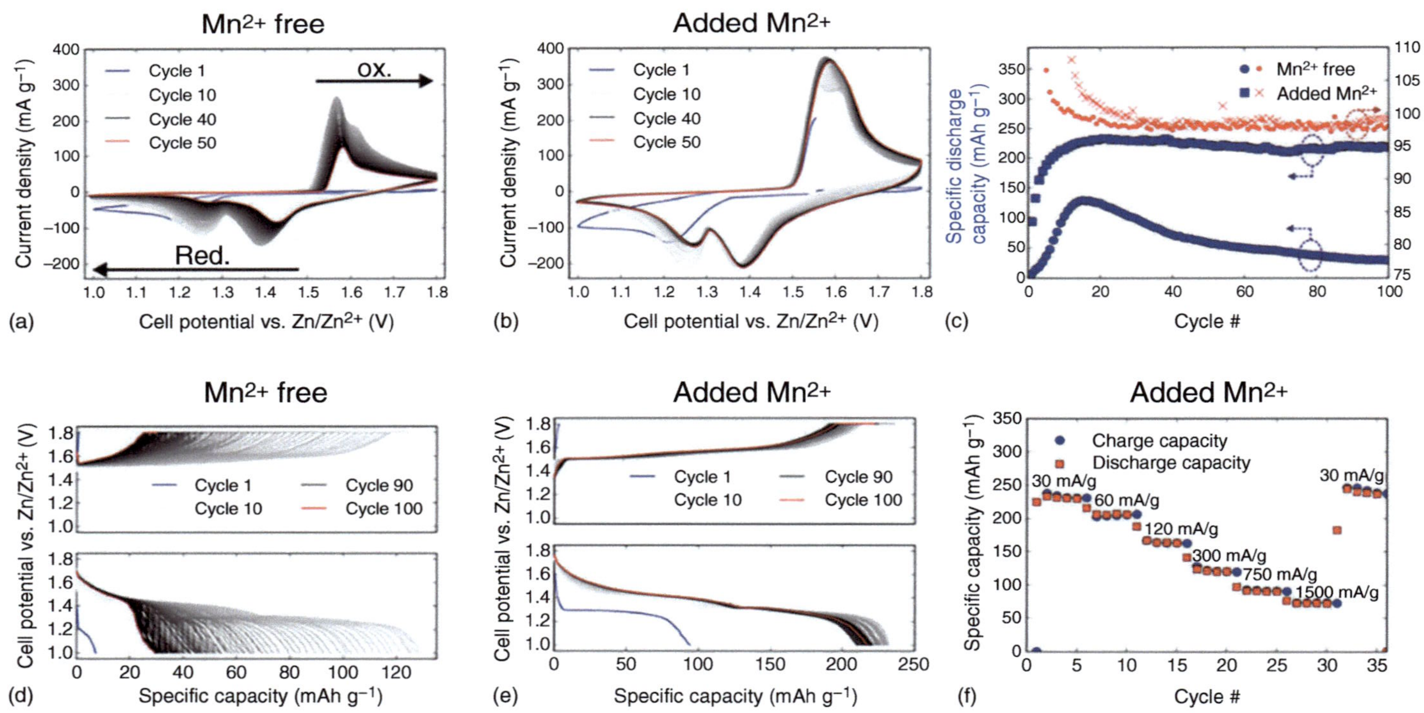

Figure 6.20 Cyclic voltammograms of the MnO_2 electrode in (a) Mn^{2+} free and (b) added Mn^{2+} electrolytes. (c) Cycling performance of the MnO_2 electrode in Mn^{2+} free and added Mn^{2+} electrolytes at 60 mA g^{-1}. Charge–discharge curves of MnO_2 electrodes in (d) Mn^{2+} free and (e) added Mn^{2+} electrolytes. (f) Rate capability investigation of the MnO_2 electrode with added Mn^{2+} [129]. Source: Chamoun et al. [129]. Reproduced with permission of Elsevier.

During the process of Zn deposition, Zn^{2+} ions tend to preferentially deposit at the tip part, which can lead to the growth of Zn dendrites. Some metal ion additives with lower reduction potential can be deposited preferentially to form an electrostatic shielding layer for inhibiting the formation of zinc dendrites [132]. Chen and coworkers found that Na_2SO_4-assisted $ZnSO_4$-based electrolyte could effectively avoid the growth of Zn dendrites in Figure 6.21a,b. Moreover, it can learn from Figure 6.21c,d, Na_2SO_4 can also prevent the dissolution of $NaV_3O_8{\cdot}1.5H_2O$, and sodium ions can form a positively charged electrostatic shielding layer around the initial protrusions of Zn, which avoid further the growth of Zn dendrites and significantly improve energy storage performance [56].

Some organic additives can also promote the uniform deposition of Zn. Wang and coworkers reported that an organic additive ether (Et_2O) could improve the electrochemical performance of AZIBs (Figure 6.22a) [133]. In the electrolyte, the highly polarized Et_2O molecules will preferably be adsorbed on the original Zn tips and prevent further deposition of Zn to the tip area (Figure 6.22b). Therefore, Et_2O can play the role of electrostatic shielding during the deposition process to facilitate the uniform deposition of Zn. In addition, the surfactants in aqueous electrolytes can also improve battery performance. Surfactants (e.g. sodium dodecyl sulfate, sodium carboxymethyl cellulose and sodium dodecylbenzene sulfonate) can increase the electrochemical stability window, inhibit the evolution of H_2 and O_2, and prevent Zn corrosion and the formation of Zn dendrites [75, 132].

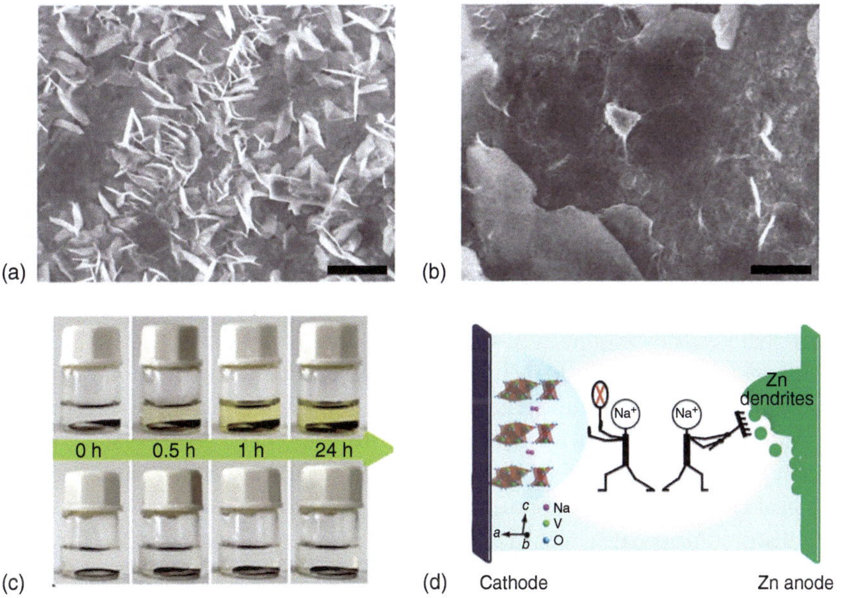

Figure 6.21 SEM images of the Zn electrode in (a) $ZnSO_4$ and (b) $ZnSO_4/Na_2SO_4$ electrolytes. Scale bars, 2 μm. (c) The optical images of $NaV_3O_8{\cdot}1.5H_2O$ electrodes in $ZnSO_4$ (above) and $ZnSO_4/Na_2SO_4$ (below) electrolytes for different periods. (d) Schematic diagram: Na_2SO_4 additive suppresses the dissolution of $NaV_3O_8{\cdot}1.5H_2O$ nanobelts and the formation of Zn dendrites [56]. Source: Wan et al. [56]. Licensed under CC BY 4.0.

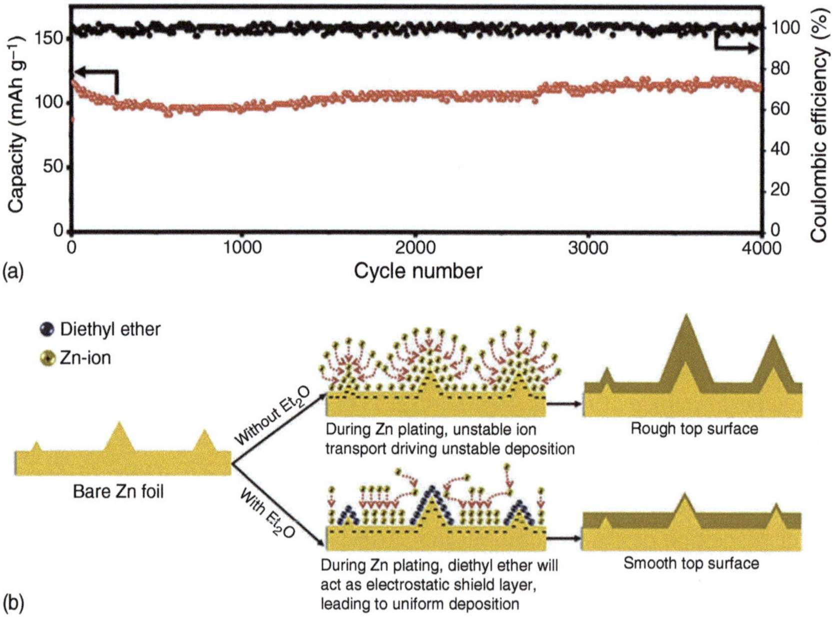

Figure 6.22 (a) Long-term cycling performance of Zn−MnO_2 battery with Et_2O additive at 5 A g^{-1}. (b) Schematics of morphology evolution for Zn anodes in mild aqueous electrolyte with and without Et_2O additive during Zn stripping/plating cycling [133]. Source: Xu et al. [133]. Reproduced with permission of Elsevier.

6.6 Challenges and Perspectives

As a potential substitute for LIBs, AZIBs have their unique advantages and attracted widespread attention. However, AZIBs also face a series of challenges, including narrow electrochemical windows in aqueous electrolytes, the formation and growth of Zn dendrites, the corrosion on the anode, the dissolution of cathode materials, and the slow kinetics. Although the research on the positive electrode, Zn negative electrode, and electrolyte has made some progress in recent years, there is still a lot of work that needs to be done.

AZIBs are popularized due to their low cost, high safety, and high ion conductivity. However, their narrow electrochemical window and the problem of side reactions with water limit the practical application of AZIBs. The generation of H_2 and O_2 will cause the battery to swell during the charge–discharge process, which is detrimental to the health of the battery. By optimizing the formulations of the electrolytes, electrolyte distribution can be screened out, thereby effectively inhibiting the occurrence of side reactions and reducing the generation of gas.

In the aspect of cathode materials, many feasible electrode materials have been proposed, but some problems are needed to be solved. Among Mn-based and V-based materials, the dissolution of Mn and V ions is the main problem for hindering their application. Although they can exhibit high capacity in the charge–discharge process, the irreversible loss caused by the dissolution of electrode materials leads to the rapid degradation of batteries, which seriously affects the service life. In addition, the strong electrostatic force of Zn^{2+} in the positive material will cause the structural collapse, resulting in a rapid decline in discharge capacity, slow diffusion, and unexpected byproducts. This will decrease the cycle life of AZIBs. The structural design of the cathode materials, surface coating, ion doping, and electrolyte optimization are effective strategies, but they still cannot solve the problem fundamentally. Therefore, there is still a long way to go for the research of cathode materials.

In AZIBs, Zn anode is considered to be the best choice due to its rich resources, low price, chemical stability, low redox potential, and easy processing. However, there is Zn dendrite growth for metallic Zn anode, along with the corrosion and passivation problems due to its thermodynamic activity. In addition, the growth, corrosion, and passivation of Zn dendrites will accelerate the side reactions at the interface between the Zn anode and the electrolyte. This will consume active Zn and reduce the Coulombic efficiency, which severely leads to premature battery failure. Although the surface modification and structural optimization of the Zn anode can improve the above-mentioned problems, the complex and expensive preparation process hinders its further promotion. Moreover, optimizing the structure will increase the contact area between the electrolyte and the electrode, which will lead to the increased corrosion. Therefore, it is still necessary to continue to explore, combine the design of the Zn anode with the optimization of the electrolyte, and maximize the strengths and avoid the weaknesses for solving the problem of the Zn anode.

As an important component of the AZIBs, the electrolyte is a bridge to connect the cathode and the anode. At the same time, most of the problems are related to the electrolyte. Therefore, the optimization of electrolytes is very important to improve battery performance. The variety of Zn salts in AZIBs provides us with more choices, but they all have their own shortcomings. The optimization of the electrolyte involves the type and concentration of the Zn salt and the selection of additives. Such a huge amount of workload requires a lot of manpower and material resources. Today, the emergence of machine learning provides a new path for our experimental exploration. If machine learning and experimentation can be effectively combined, the best electrolyte formula can be screened out faster, saving more time and energy.

During the charge–discharge process, complex chemical reactions occur. At present, our understanding of the reaction mechanism is still immature, and continuous efforts are needed to gradually know the reaction mechanisms. In this process, we need to rely on more in situ characterizations and a comprehensive theoretical model, but this part of the work has many shortcomings. Thus, this will become the future development direction of AZIBs.

References

1 Tarascon, J.M. and Armand, M. (2001). *Nature* 414: 359.
2 Armand, M. and Tarascon, J.M. (2008). *Nature* 451: 652.
3 Goodenough, J.B. and Kim, Y. (2010). *Chem. Mat.* 22: 587.

4 Liu, J., Bao, Z.N., Cui, Y. et al. (2019). *Nat. Energy* 4: 180.
5 Yabuuchi, N., Kubota, K., Dahbi, M., and Komaba, S. (2014). *Chem. Rev.* 114: 11636.
6 Rajagopalan, R., Tang, Y., Ji, X. et al. (2020). *Adv. Funct. Mater.*: 30, 1909486.
7 Hwang, J.Y., Myung, S.T., and Sun, Y.K. (2017). *Chem. Soc. Rev.* 46: 3529.
8 Hwang, J.-Y., Myung, S.-T., and Sun, Y.-K. (2018). *Adv. Funct. Mater.*: 28, 1802938.
9 Pramudita, J.C., Sehrawat, D., Goonetilleke, D., and Sharma, N. (2017). *Adv. Energy Mater.* 7: 1602911.
10 Jia, X., Liu, C., Neale, Z.G. et al. (2020). *Chem. Rev.* 120: 7795.
11 Tang, B., Shan, L., Liang, S., and Zhou, J. (2019). *Energy Environ. Sci.* 12: 3288.
12 Huang, S., Zhu, J.C., Tian, J.L., and Niu, Z.Q. (2019). *Chem. Eur. J.* 25: 14480.
13 Xu, C.J., Li, B.H., Du, H.D., and Kang, F.Y. (2012). *Angew. Chem. Int. Ed.* 51: 933.
14 Song, M., Tan, H., Chao, D., and Fan, H.J. (2018). *Adv. Funct. Mater.*: 28, 1802564.
15 Fang, G., Zhou, J., Pan, A., and Liang, S. (2018). *ACS Energy Lett.* 3: 2480.
16 Pan, H., Shao, Y., Yan, P. et al. (2016). *Nat. Energy* 1: 16039.
17 Zhang, N., Cheng, F., Liu, J. et al. (2017). *Nat. Commun.* 8: 405.
18 Huang, J., Guo, Z., Ma, Y. et al. (2019). *Small Methods*: 3, 1800272.
19 Zeng, X., Hao, J., Wang, Z. et al. (2019). *Energy Storage Mater.* 20: 410.
20 Xia, C., Guo, J., Lei, Y. et al. (2018). *Adv. Mater.*: 30, 1705580.
21 Ye, Z., Cao, Z., Lam Chee, M.O. et al. (2020). *Energy Storage Mater.* 32: 290.
22 Wang, L. and Zheng, J. (2020). *Mater. Today Adv.*: 7, 100078.
23 Kundu, D., Adams, B.D., Duffort, V. et al. (2016). *Nat. Energy* 1: 16119.
24 Liu, T., Cheng, X., Yu, H. et al. (2019). *Energy Storage Mater.* 18: 68.
25 Kundu, D., Hosseini, V.S., Wan, L. et al. (2018). *Energy Environ. Sci.* 11: 881.
26 Wang, F., Sun, W., Shadike, Z. et al. (2018). *Angew. Chem. Int. Ed.* 57: 11978.
27 Ding, J., Du, Z., Gu, L. et al. (2018). *Adv. Mater.*: 30, 1800762.
28 Liu, C., Xie, X., Lu, B. et al. (2021). *ACS Energy Lett.* 6: 1015.
29 Juran, T.R., Young, J., and Smeu, M. (2018). *J. Phys. Chem. C* 122: 8788.
30 Shi, W., Lee, W.S.V., and Xue, J. (2021). *ChemSusChem* 14: 1634. https://doi.org/10.1002/cssc.202002493.
31 He, P., Chen, Q., Yan, M. et al. (2019). *EnergyChem* 1: 100022.
32 Julien, C.M. and Mauger, A. (2017). *Nanomaterials* 7: 42.
33 Liu, B., Sun, Y., Liu, L. et al. (2018). *Adv. Funct. Mater.*: 28, 1704973.
34 Alfaruqi, M.H., Mathew, V., Gim, J. et al. (2015). *Chem. Mat.* 27: 3609.
35 Huang, J., Wang, Z., Hou, M. et al. (2018). *Nat. Commun.* 9: 2906.
36 Islam, S., Alfaruqi, M.H., Mathew, V. et al. (2017). *J. Mater. Chem. A* 5: 23299.
37 Han, M.M., Huang, J.W., Liang, S.Q. et al. (2020). *iScience* 23: 34.
38 Wei, C., Xu, C., Li, B. et al. (2012). *J. Phys. Chem. Solids* 73: 1487.
39 Han, S.D., Kim, S., Li, D.G. et al. (2017). *Chem. Mat.* 29: 4874.
40 Lee, J., Ju, J.B., Cho, W.I. et al. (2013). *Electrochim. Acta* 112: 138.
41 Zhu, C., Fang, G., Liang, S. et al. (2020). *Energy Storage Mater.* 24: 394.
42 Jiang, B., Xu, C., Wu, C. et al. (2017). *Electrochim. Acta* 229: 422.

43 Sun, M., Li, D.-S., Wang, Y.-F. et al. (2019). *ChemElectroChem* 6: 2510.
44 Hao, J., Mou, J., Zhang, J. et al. (2018). *Electrochim. Acta* 259: 170.
45 Zhang, N., Cheng, F., Liu, Y. et al. (2016). *J. Am. Chem. Soc.* 138: 12894.
46 Song, J., Xu, K., Liu, N. et al. (2021). *Mater. Today* https://doi.org/10.1016/j.mattod.2020.12.003.
47 Zhang, N., Dong, Y., Jia, M. et al. (2018). *ACS Energy Lett.* 3: 1366.
48 Zavalij, W. (1999). *Acta Crystallogr. B* 55: 627.
49 Wan, F. and Niu, Z. (2019). *Angew. Chem. Int. Ed.* 58: 16358.
50 Zhang, N., Chen, X., Yu, M. et al. (2020). *Chem. Soc. Rev.* 49: 4203.
51 Yan, M., He, P., Chen, Y. et al. (2018). *Adv. Mater.* 30: 1703725.
52 Guo, X., Fang, G., Zhang, W. et al. (2018). *Adv. Energy Mater.*: 8, 1801819.
53 Pang, Q., Sun, C., Yu, Y. et al. (2018). *Adv. Energy Mater.*: 8, 1800144.
54 Yang, Y., Tang, Y., Fang, G. et al. (2018). *Energy Environ. Sci.* 11: 3157.
55 Alfaruqi, M.H., Mathew, V., Song, J. et al. (2017). *Chem. Mat.* 29: 1684.
56 Wan, F., Zhang, L., Dai, X. et al. (2018). *Nat. Commun.* 9: 1656.
57 He, P., Quan, Y., Xu, X. et al. (2017). *Small*: 13, 1702551.
58 Sambandam, B., Soundharrajan, V., Kim, S. et al. (2018). *J. Mater. Chem. A* 6: 15530.
59 Liu, X., Zhang, H., Geiger, D. et al. (2019). *Chem. Commun.* 55: 2265.
60 Hu, F., Xie, D., Zhao, D. et al. (2019). *J. Energy Chem.* 38: 185.
61 Wei, T., Li, Q., Yang, G., and Wang, C. (2018). *J. Mater. Chem. A* 6: 8006.
62 Shin, J., Choi, D.S., Lee, H.J. et al. (2019). *Adv. Energy Mater.*: 9, 1900083.
63 Shan, L., Zhou, J., Zhang, W. et al. (2019). *Energy Technol.*: 7, 1900022.
64 Sambandam, B., Soundharrajan, V., Kim, S. et al. (2018). *J. Mater. Chem. A* 6: 3850.
65 Peng, Z., Wei, Q., Tan, S. et al. (2018). *Chem. Commun.* 54: 4041.
66 Li, G., Yang, Z., Jiang, Y. et al. (2016). *Nano Energy* 25: 211.
67 Zhang, X., Rui, X., Chen, D. et al. (2019). *Nanoscale* 11: 2556.
68 Wang, F., Hu, E., Sun, W. et al. (2018). *Energy Environ. Sci.* 11: 3168.
69 Mathew, V., Sambandam, B., Kim, S. et al. (2020). *ACS Energy Lett.* 5: 2376.
70 Tang, B., Zhou, J., Fang, G. et al. (2019). *J. Mater. Chem. A* 7: 940.
71 Paolella, A., Faure, C., Timoshevskii, V. et al. (2017). *J. Mater. Chem. A* 5: 18919.
72 Wang, B., Han, Y., Wang, X. et al. (2018). *iScience* 3: 110.
73 Zhang, L., Chen, L., Zhou, X., and Liu, Z. (2015). *Adv. Energy Mater.* 5: 1400930.
74 Pasta, M., Wessells, C.D., Huggins, R.A., and Cui, Y. (2012). *Nat. Commun.* 3: 1149.
75 Hou, Z., Zhang, X., Li, X. et al. (2017). *J. Mater. Chem. A* 5: 730.
76 Wang, R.Y., Wessells, C.D., Huggins, R.A., and Cui, Y. (2013). *Nano Lett.* 13: 5748.
77 Yang, Q., Mo, F., Liu, Z. et al. (2019). *Adv. Mater.* 31: 1901521.
78 You, Y., Wu, X.-L., Yin, Y.-X., and Guo, Y.-G. (2014). *Energy Environ. Sci.* 7: 1643.

79 Kasiri, G., Trócoli, R., Bani, H.A., and La Mantia, F. (2016). *Electrochim. Acta* 222: 74.

80 Kasiri, G., Glenneberg, J., Bani, H.A. et al. (2019). *Energy Storage Mater.* 19: 360.

81 Lu, K., Song, B., Zhang, Y. et al. (2017). *J. Mater. Chem. A* 5: 23628.

82 Kundu, D., Oberholzer, P., Glaros, C. et al. (2018). *Chem. Mat.* 30: 3874.

83 Zhao, Q., Huang, W., Luo, Z. et al. (2018). *Sci. Adv.*: 4, eaao1761.

84 Lee, B., Ko, Y., Kwon, G. et al. (2018). *Joule* 2: 61.

85 Shi, H.-Y., Ye, Y.-J., Liu, K. et al. (2018). *Angew. Chem. Int. Ed.* 57: 16359.

86 Wan, F., Zhang, L., Wang, X. et al. (2018). *Adv. Funct. Mater.*: 28, 1804975.

87 Hu, P., Zhu, T., Wang, X. et al. (2019). *Nano Energy* 58: 492.

88 Li, W., Wang, K., Cheng, S., and Jiang, K. (2018). *Energy Storage Mater.* 15: 14.

89 Ma, L., Chen, S., Li, H. et al. (2018). *Energy Environ. Sci.* 11: 2521.

90 Zeng, Y., Meng, Y., Lai, Z. et al. (2017). *Adv. Mater.*: 29, 1804975.

91 Gong, M., Li, Y., Zhang, H. et al. (2014). *Energy Environ. Sci.* 7: 2025.

92 Chae, M.S., Heo, J.W., Lim, S.-C., and Hong, S.-T. (2016). *Inorg. Chem.* 55: 3294.

93 Cheng, Y., Luo, L., Zhong, L. et al. (2016). *ACS Appl. Mater. Interfaces* 8: 13673.

94 Geng, L., Lv, G., Xing, X., and Guo, J. (2015). *Chem. Mat.* 27: 4926.

95 Li, H., McRae, L., Firby, C.J., and Elezzabi, A.Y. (2019). *Adv. Mater.*: 31, 1807065.

96 Liu, W., Hao, J., Xu, C. et al. (2017). *Chem. Commun.* 53: 6872.

97 Liang, H., Cao, Z., Ming, F. et al. (2019). *Nano Lett.* 19: 3199.

98 Li, H., Yang, Q., Mo, F. et al. (2019). *Energy Storage Mater.* 19: 94.

99 Li, H., Ma, L., Han, C. et al. (2019). *Nano Energy* 62: 550.

100 Li, W., Wang, K., Zhou, M. et al. (2018). *ACS Appl. Mater. Interfaces* 10: 22059.

101 Wang, T., Li, C., Xie, X. et al. (2020). *ACS Nano* 14: 16321.

102 Turney, D.E., Gallaway, J.W., Yadav, G.G. et al. (2017). *Chem. Mat.* 29: 4819.

103 Han, C., Li, W., Liu, H.K. et al. (2020). *Nano Energy* 74: 104880.

104 Tian, H., Li, Z., Feng, G. et al. (2021). *Nat. Commun.* 12: 237.

105 Wang, S.-B., Ran, Q., Yao, R.-Q. et al. (2020). *Nat. Commun.* 11: 1634.

106 Fashu, S., Gu, C.D., Zhang, J.L. et al. (2015). *Surf. Interface Anal.* 47: 403.

107 Kwak, B.S., Kim, D.Y., Park, S.S. et al. (2015). *Chem. Eng. J.* 281: 368.

108 Zhou, P., Erning, J.W., and Ogle, K. (2019). *Electrochim. Acta* 293: 290.

109 Kim, H.S., Jo, Y.N., Lee, W.J. et al. (2015). *Electroanalysis* 27: 517.

110 Cao, Z.Y., Zhuang, P.Y., Zhang, X. et al. (2020). *Adv. Energy Mater.* 10: 14, 2001599.

111 Zhou, M., Guo, S., Fang, G. et al. (2021). *J. Energy Chem.* 55: 549.

112 Zhang, Q., Luan, J.Y., Tang, Y.G. et al. (2020). *Angew. Chem. Int. Ed.* 59: 13180.

113 Zhao, K.N., Wang, C.X., Yu, Y.H. et al. (2018). *Adv. Mater. Interfaces* 5: 7, 1800848.

114 Kang, L., Cui, M., Jiang, F. et al. (2018). *Adv. Energy Mater.* 8: 1801090.

115 Qiu, W., Li, Y., You, A. et al. (2017). *J. Mater. Chem. A* 5: 14838.

116 Shi, W., Mao, J., Xu, X. et al. (2019). *J. Mater. Chem. A* 7: 15654.

117 Shi, X., Xu, G., Liang, S. et al. (2019). *ACS Sustainable Chem. Eng.* 7: 17737.
118 Zeng, Y., Zhang, X., Qin, R. et al. (2019). *Adv. Mater.* 31: 1903675.
119 Wang, Z., Huang, J.H., Guo, Z.W. et al. (2019). *Joule* 3: 1289.
120 Zhao, C., Wang, X., Shao, C. et al. (2021). *Sustain. Energy Fuels* 5: 332.
121 Venkata Narayanan, N.S., Ashokraj, B.V., and Sampath, S. (2010). *J. Colloid Interface Sci.* 342: 505.
122 Zhang, H., Liu, Q., Fang, Y. et al. (2019). *Adv. Mater.* 31: 1904948.
123 Wang, F., Borodin, O., Gao, T. et al. (2018). *Nat. Mater.* 17: 543.
124 Liu, Z., Cui, T., Pulletikurthi, G. et al. (2016). *Angew. Chem. Int. Ed.* 55: 2889.
125 Suo, L.M., Borodin, O., Gao, T. et al. (2015). *Science* 350: 938.
126 Yamada, Y., Furukawa, K., Sodeyama, K. et al. (2014). *J. Am. Chem. Soc.* 136: 5039.
127 Zhang, C., Holoubek, J., Wu, X. et al. (2018). *Chem. Commun.* 54: 14097.
128 Zhang, C., Shin, W., Zhu, L. et al. (2021). *Carbon Energy* 3: 339. https://doi.org/10.1002/cey2.70.
129 Chamoun, M., Brant, W.R., Tai, C.-W. et al. (2018). *Energy Storage Mater.* 15: 351.
130 Wan, F., Zhang, L.L., Dai, X. et al. (2018). *Nat. Commun.*: 9.
131 Soundharrajan, V., Sambandam, B., Kim, S. et al. (2020). *Energy Storage Mater.* 28: 407.
132 Ding, F., Xu, W., Graff, G.L. et al. (2013). *J. Am. Chem. Soc.* 135: 4450.
133 Xu, W., Zhao, K., Huo, W. et al. (2019). *Nano Energy* 62: 275.

7

Zn–Air Battery

J. Alberto Blázquez, Aroa R. Mainar, and Elena Iruin

CIDETEC, Basque Research and Technology Alliance (BRTA), Pº Miramón, 196, Donostia-San Sebastián, 20014, Spain

7.1 Introduction

European and global energy policies point toward a secure, competitive, and decarbonized energy system in the coming years. The European Green Deal [1, 2] aims for a climate-neutral Europe by 2050, and the Renewable Energy Directive issued in 2018 targets for at least 32% renewable energy production by 2030 [3]. The **use of renewable energy sources** becomes crucial as a sustainable tool for reducing the environmental impact [4]. However, renewable energy sources are dependent on time and location, and they require an efficient solution for energy storage to store and release high amount of energy when it is needed. In this context, the most suitable approaches for incorporating the stochastic renewable energy sources are the electrochemical energy storage systems, namely batteries [5].

Apart from stationary energy storage, the constant progress in **consumer electronics** also requires novel batteries, in a similar way that clean **electromobility** is asking for more compact, efficient, safe, and affordable electrochemical systems. Hence, it is required for the development of sustainable, inexpensive, and high energy density-based electrochemical energy storage devices. Unfortunately, there is not a unique electrochemical system which fulfills the requirements for all mentioned applications. For example, traditional rechargeable batteries (e.g. lead–acid, nickel–cadmium, or lithium-ion) present high cost, environmental concerns, and/or safety issues that limit their use in stationary energy storage applications.

In this context, metal–air batteries are considered promising technologies for high energy density applications (see Figure 7.1). These batteries use oxygen from the ambient as cathodic active material; hence, there is no need to incorporate heavy and high volume-based components in the positive electrode. Moreover, oxygen from the atmosphere is a free and abundant resource [7].

Battery Technologies: Materials and Components, First Edition. Edited by Jianmin Ma.

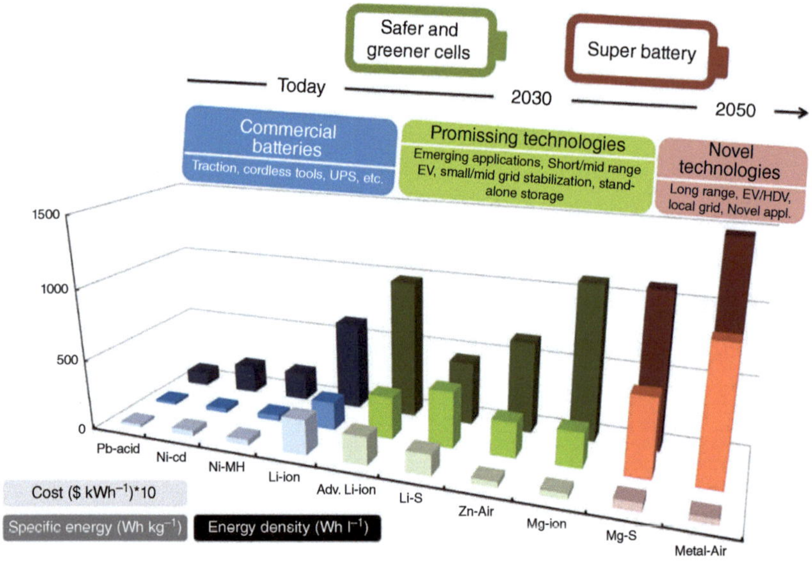

Figure 7.1 Commercial, promising, and novel battery technologies characteristics and development roadmap [6]. Source: European Magnesium Interactive Battery Community [6].

7.1.1 Metal–Air Batteries

Like other energy storage systems, metal–air batteries contain three main components: the cathode (positive electrode), the anode (negative electrode), and the electrolyte (see Figure 7.2a). The positive electrode is responsible for the attributed high specific energy and energy density of metal–air technologies, as they use oxygen from the surrounding ambient as cathodic active material. Hence, and on contrary to other technologies, these devices present lower volume and weight as cathodic active material is not stored in the positive electrode. The principal difference of resulting metal–air characteristics is related to the anode properties, as it is shown in Figure 7.2b.

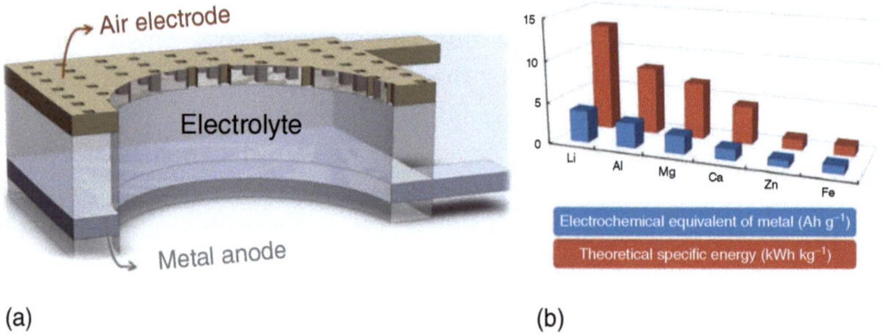

Figure 7.2 (a) Typical structure of metal–air batteries and (b) properties of metal–air cells [8]. Source: Mainar [8] / Universidad del País Vasco.

Among metal–air batteries, lithium–air, aluminum–air, and magnesium–air cells present the highest theoretical energy densities (13.0, 8.1, and 6.8 kWh kg^{-1}, respectively) as it is shown in Figure 7.2b. However, **magnesium–air** and **aluminum–air** batteries are still in an early stage of development facing many issues to demonstrate their practical feasibility [9]. On the other hand, **lithium–air** batteries have been widely studied due to the better electrochemical response compared to aluminum–air and magnesium–air batteries [10]. Compared to lithium, zinc presents much lower water reactivity (which enables the use of aqueous electrolytes [9]) and cost, as it is based on abundant, environmentally benign, and safe materials. Therefore, the **zinc–air** technology has attracted much of the scientific and industrial attention with higher theoretical energy density (1.3 kWh kg^{-1}) with respect to the currently used energy storage systems [11, 12].

Regarding the electrolyte system, it has to be pointed out that metal–air batteries can include different electrolyte natures and compositions in function of used metallic anode [9, 12, 13]. In this context, the electrolyte system can be aqueous, aprotic (organic), mixed aqueous/aprotic, or solid state, which present their own set of advantages and disadvantages. Figure 7.3 shows a classification of metal–air technologies taking into account different characteristics as used metallic anode, electrolyte nature, and rechargeability (i.e. primary [no rechargeable], secondary [electrically rechargeable], and mechanically recharged [based on the replacement of one component (metal anode) by fresh one, as it will be explained in Section 7.1.3.1] [14, 15]).

In general, the early-stage development of magnesium–air and aluminum–air technologies as well as the high cost and low safety of lithium–air battery [10] is

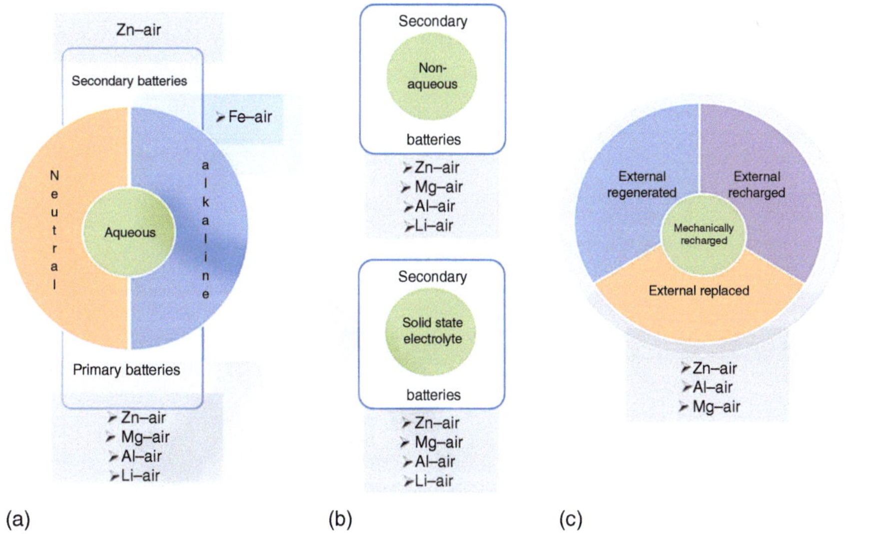

Figure 7.3 Classification of metal–air batteries: (a) aqueous electrolytes (neutral and alkaline), (b) non-aqueous electrolytes, and (c) mechanically recharged. Source: Mainar et al. [9]. Reproduced with permission of Elsevier.

resulting in an increased interest on the zinc–air technology. Moreover, among metal–air batteries, zinc–air is the only one which is commercially available as primary (non-rechargeable) battery since 1930. Therefore, the zinc–air technology is considered the most mature metal–air technology, which facilitates its study toward the feasibility of its rechargeability [16, 17]. However, many challenges need to be faced to obtain a durable and high-performance electrically rechargeable zinc–air battery, as it will be described in Section 7.1.3.3.

7.1.2 History of Zinc-Based Technologies

Since the first reported battery in 1800 (Volta cell) [18], zinc electrode has been widely used in different energy storage systems. Zinc-based technologies differ in the included electrolyte system and positive electrode coupled to it, as it is shown in Figure 7.4. In general, these technologies are characterized by the low cost, low toxicity, and high safety which has prompted the development and commercialization of diverse zinc-based battery technologies, composing nowadays one-third of the world battery market [20–25].

Concretely, the commercialization of **primary zinc–air batteries** (1930) supposed a low-cost and safe technology [26] implemented in many different applications, such as hearing aids or signaling [27, 28]. Commercially available primary zinc–air batteries present up to 1400 Wh l^{-1} [28], being able to be used in a wide range of temperatures, from −10 °C up to 55 °C. Moreover, if the air inlet is protected correctly, these batteries can be stored during a long period of time without losing its discharge capacity.

Taking into account the competitiveness of the primary zinc–air battery, its posterior study was based on the fabrication of **rechargeable or secondary zinc–air batteries**, which are considered as a proper energy storage system for various competing and emerging applications, as stationary energy storage.

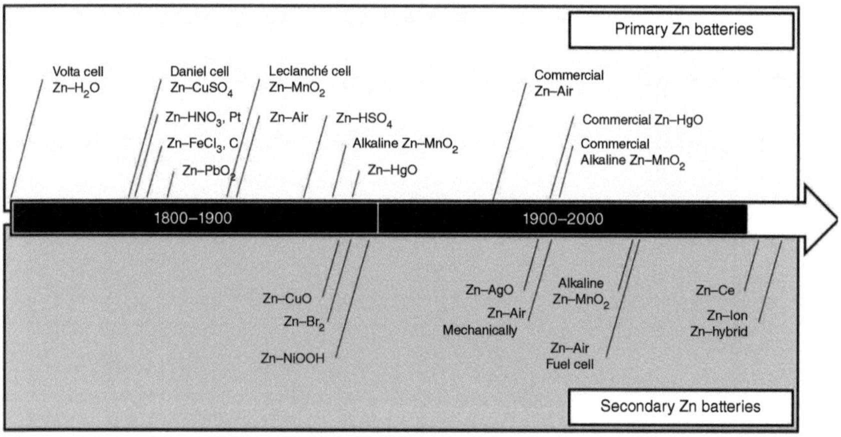

Figure 7.4 The historic evolution of zinc-based power source systems. Source: Mainar et al. [19]. Reproduced with permission of John Wiley & Sons.

7.1.3 Secondary Zinc–Air Batteries

7.1.3.1 Rechargeability

The rechargeability of secondary zinc–air batteries could be divided into three main groups: (i) mechanically, (ii) hydraulically, and (iii) electrically rechargeable, as it is shown in Figure 7.5.

Mechanically rechargeable zinc–air battery was proposed for first time at the beginning of 1960 based on the substitution of discharged zinc anode for a fresh one [29]. However, mechanically rechargeable zinc–air battery presents some limitations in terms of cost, time, and complexity.

Thereafter, **hydraulically rechargeable zinc–air batteries** were proposed in many different designs, which avoided the handling complications of a mechanical replacement of the zinc anode. This design is based on a continuous supply of the anodic system that is stored in an external deposit, as conventional internal combustion engines. This type of rechargeability would avoid many problems associated with the anodic system such as dendrite formation or shape change, and it became an attractive candidate in many applications where low cost and robustness are indispensable [30], as for example, stationary energy storage [31–33]. However, its use would be hardly competitive in electromobility as it needs a high volume of electrolyte and complex flow battery design. All limited their application due to performance, scale-up, and maintenance issues [34].

In general, in mechanically and hydraulically rechargeable zinc–air batteries, the substitution of the used materials by fresh ones is involved. In contrast, the **electrically rechargeable zinc–air battery** presents higher competitiveness as there is no need of renewing battery materials [35]. Different battery configurations have been proposed for electrically rechargeable zinc–air technology as planar or flexible. Among them, planar batteries have been most widely used due to their simplicity and low cost, as they are commonly fabricated by plastic plates,

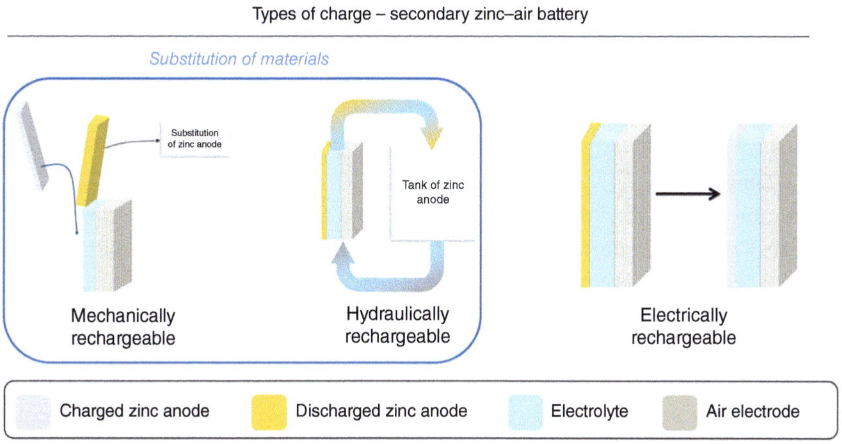

Figure 7.5 Secondary zinc–air batteries classification: mechanically, hydraulically, and electrically rechargeable.

Table 7.1 Zinc–air electrochemistry in different aqueous electrolytes at different pHs.

Aqueous electrolyte	Reaction on zinc electrode	E^0 (V)	Eq.	Reaction on air electrode	E^0 (V)	Eq.
Alkaline	$Zn + 2OH^- \leftrightarrow ZnO + H_2O + 2e^-$	1.250	(7.1)	$O_2 + 2H_2O + 4e^- \leftrightarrow 4OH^-$	0.401	(7.4)
Neutral	$Zn \leftrightarrow Zn^{2+} + 2e^-$	−0.762	(7.2)	$2H^+ + 1/2O_2 + 2e^- \leftrightarrow H_2O$	1.229	(7.5)
Acid	$Zn \leftrightarrow Zn^{2+} + 2e^-$	−0.762	(7.3)	$O_2 + 4H^+ + 2e^- \leftrightarrow H_2O$	1.229	(7.6)

Source: Mainar et al. [9]. Reproduced with permission of Elsevier.

chambers, and gaskets [28]. Recently, flexible batteries have also been investigated, as they have a great potential to be used in wearable devices [36].

Traditionally, electrically rechargeable zinc–air batteries have been developed with aqueous alkaline electrolytes, although the influence of different pH-based electrolytes was also reported. Table 7.1 summarizes the electrochemistry of zinc–air battery at different electrolyte pHs.

7.1.3.2 Industrial Approximations

Different industrial approximations have been performed for rechargeable zinc–air batteries since 1970, where a tubular zinc–air cell was developed by the French Compagnie Generale d'Electricite, with the possibility to be recharged either electrically or hydraulically [37]. In the year 2000, ZOXY and PSI developed a zinc–air system that could be electrically recharged for around 10 times but afterward the anode was mechanically replaced [37, 38]. One year later, electrically rechargeable zinc–air batteries developed by AER Energy Resources Inc. were used for portable computer notebooks for the first time, but without much success [39]. Recently, other companies and/or start-up research laboratories have reported advances in designing and manufacturing a secondary zinc–air battery, including Eos Energy Storage [40], Zinium [41], or ZincNyx Energy Solutions [42]. Nantenergy presented their prototype as a low-cost (<100 $ kWh^{-1}) zinc–air battery with thousands of charge–discharge cycles tested as an energy storage system for solar panels in remote areas of Africa and Asia [43].

7.1.3.3 Limitations

The commercialization of electrically rechargeable (hereinafter referred as secondary) zinc–air batteries was slowed down due to the low reversibility of the system, where many issues regarding zinc anode and bifunctional air electrode (BAE) have been identified to be responsible for the low durability of the battery.

Regarding the zinc anode, the main challenges when using aqueous alkaline electrolytes are related to (i) high zinc dissolution, (ii) corrosion, and (iii) passivation [44, 45], as it is shown in Figure 7.6. During the discharge of the battery, zinc dissolves in the electrolyte as zincate ($Zn(OH)_4^{2-}$), which is highly soluble in alkaline media. When the dissolution of zinc induces an excessive concentration of $Zn(OH)_4^{2-}$ surpassing the solubility limit, the formation of ZnO film occurs

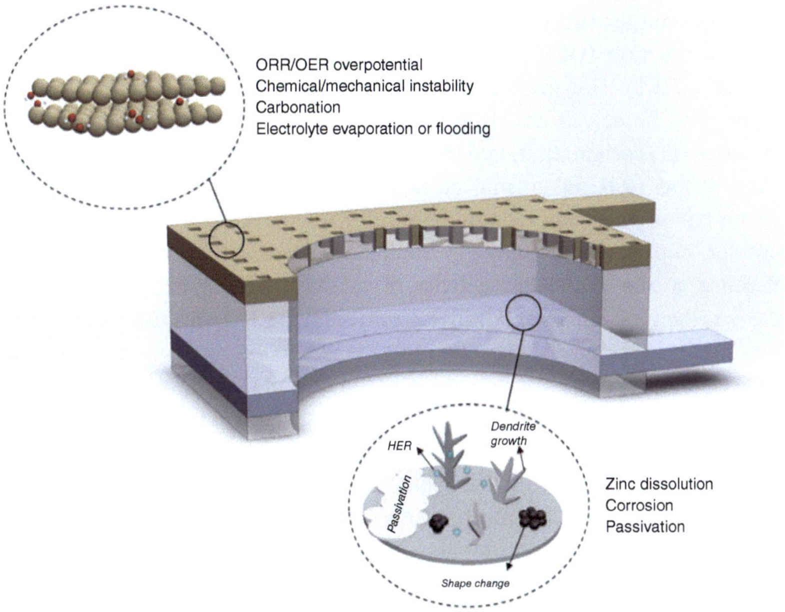

Figure 7.6 Main limitations of electrically rechargeable zinc–air batteries.

(related mechanisms are deeply defined in Section 7.2.1). On contrary, during the charge of the battery, zinc is redeposited as metallic zinc in the anode. However, the redeposition could not be homogeneous in the zinc electrode if the **high zinc dissolution** provokes different gradient concentrations in the electrolyte system, as it results in the presence of preferential areas for zinc deposition [28]. This nonuniform deposition promotes anode *shape changes* that reduce battery capacity. Different types of zinc deposits can be formed: mossy, heavy spongy, layered, boulder, or *dendrites* (particularly at high current densities), the latter being sharp and needle-like deposits that can perforate the separator and reach the BAE, causing a short circuit [46].

On the other hand, zinc metal is thermodynamically unstable in alkaline electrolytes leading to the parasitic evolution of hydrogen gas, known as **zinc corrosion**. The *hydrogen evolution reaction (HER)* (Eq. (7.7)) on the surface of the zinc anode affects the battery performance, increases the internal pressure of the cell, and can still occur over time leading to self-discharge. Moreover, HER consumes water from the electrolyte system (See Eq. (7.7)) [25].

$$Zn + 2H_2O \rightarrow Zn(OH)_2 + H_2 \uparrow \tag{7.7}$$

The last challenge related to zinc anode is the **passivation of zinc electrode**. The battery discharge could lead two types of ZnO films: (i) porous and reactive Type I and (ii) compact solid Type II. The latter is related to zinc passivation as it forms in the electrode/electrolyte interface, inactivating the zinc active material [9].

Regarding the BAE, the electrochemical reactions occurring there during discharge and charge of the battery are the oxygen reduction reaction (ORR) and oxygen evolution reaction (OER), respectively (related mechanisms are deeply defined in Section 7.3.1). However, BAE also presents some challenges (see Figure 7.6) that need to be solved as (i) ORR/OER overpotential, (ii) chemical/mechanical instability, (iii) carbonation, and (iv) electrolyte evaporation or flooding.

First, it has to be taken into account that the theoretical discharge voltage of zinc–air batteries with aqueous alkaline electrolytes is 1.651 V (Eqs. (7.1) and (7.4)). However, in practice, discharge of the battery occurs at around 1.2 V. This voltage difference is due to the low activity of catalysts. Hence, the secondary zinc–air battery presents **high overpotential between ORR and OER**, which results in a low battery efficiency [47]. Therefore, it is needed to develop bifunctional air catalysts with high catalytic activity in both ORR and OER processes.

Second, the BAE presents **chemical/mechanical instability** which limits the battery performance. The *chemical instability* is closely related to carbon corrosion (commonly used as a conductive agent) due to peroxide formation. ORR can occur through different reaction mechanisms (as it will be described in Section 7.3.1) and, depending on the reaction pathway, peroxide formation can occur [27], which involves carbon corrosion [28]. On the other hand, the *mechanical instability* of BAE is related to gas formation during OER, which can break the positive electrode.

Moreover, the aqueous alkaline electrolyte is highly sensitive to carbon dioxide (CO_2) existent in the surrounding air that reacts with hydroxyl groups (OH^-) of the electrolyte, leading to **bifunctional air electrode carbonation** (Eqs. (7.8) and (7.9)). Dissolved CO_2 (CO_3^{2-}) reacts with K^+ ions from KOH (most widely used as electrolyte salt) resulting in the formation of K_2CO_3. This species can precipitate in the porous structure of the air electrode, impeding the oxygen access and causing the electrode deterioration. In addition, the mobility of the carbonate anions is lower than OH^- groups; therefore, the carbonation process causes a reduction in the ionic conductivity and an increase in the aqueous alkaline electrolyte viscosity [9].

$$CO_2 + OH^- \rightarrow HCO_3^- \tag{7.8}$$

$$HCO_3^- + OH^- \leftrightarrow CO_3^{2-} + H_2O \tag{7.9}$$

Finally, as secondary zinc–air technology is an open system, the cell design plays an important role where water concentration should be controlled in the electrolyte system. The open cell design provokes a higher evaporation rate (water loss) or ambient moisture uptake (water gain) in aqueous electrolytes depending on the vapor pressure between the electrolyte and the ambient. In highly wet environments, the excess of water is accumulated generating **flooding** in the air electrode, which limits the oxygen entry. On the other hand, if ambient humidity is lower, the **loss of water** coming from the electrolyte provokes a contact loss between the two electrodes and the modification of the electrolyte concentration.

Taking into consideration the mentioned zinc anode and BAE limitations, scientists have carried out different approximations that reduced or suppressed the aforementioned problems. In this context, the development of each of the cell components will be described in the following sections.

7.2 Electrolyte System

7.2.1 Mechanisms for Zinc Dissolution

Traditionally, the electrolyte system for zinc–air batteries has been based on alkaline aqueous solutions such as KOH, NaOH, or LiOH, although the most widely used electrolyte is based on KOH due to the higher ionic conductivity of K^+ ion [9, 48], concretely, 30 wt% KOH [25]. In aqueous alkaline electrolytes, during the discharge process of the zinc–air battery, zinc is oxidized to $Zn(OH)_4^{2-}$ ions, which are dissolved in the electrolyte (Eq. (7.10)). Once the solution near the anodic surface is saturated with $Zn(OH)_4^{2-}$ ions, the formation of ZnO occurs (Eq. (7.11)).

$$\text{Zinc dissolution: } Zn + 4OH^- \leftrightarrow Zn(OH)_4^{2-} + 2e^- \tag{7.10}$$

$$\text{ZnO formation: } Zn(OH)_4^{2-} \leftrightarrow ZnO + H_2O + 2OH^- \tag{7.11}$$

$$\text{Overall: } Zn + 2OH^- \leftrightarrow ZnO + H_2O \tag{7.12}$$

However, there are controversial reaction mechanisms taking place for dissolution process, which are summarized in Figure 7.7. Although all of them agree on the formation of $Zn(OH)_4^{2-}$ (Eq. (7.10)), the differences are associated with the presence of different dissolved products, the number and type of intermediates, their state of solvation and absorption, and even their mobility [65].

Going into detail, Cachet et al. [49–51] proposed the formation of $Zn(OH)_4^{2-}$ ions through a previous reaction where Zn is oxidized to a monovalent intermediate (Zn*), probably hydroxylated such as $Zn(OH)_2^-$ and ZnOH.

Bockris et al. [52] described a multistep reaction for zinc dissolution in alkaline media, similar to that reported by Muralidharan and Rajagopalan [53], who stated that those reactions were only valid under transient conditions, and in alignment with the proposal described by Hendrikx et al. [54] for KOH concentrations up to 3 M. A simplified version of the Bockris model was reported by Chang and Prentice [55, 56] for the formation of $Zn(OH)_3^-$ [51].

Dirkse and Hampson [57–63] suggested different zinc dissolution mechanisms independent from the zincate concentration, which would justify why the zinc electrodes with different crystallographic orientations present disparate dissolution rates [54, 57–62, 66, 67].

Finally, Sharma and Reed suggested that three possible species can be formed from the reaction of $Zn(OH)_2$ with H_2O and/or OH^-: $Zn(OH)_2(H_2O)_2$, $Zn(OH)_3(H_2O)^-$, and $Zn(OH)_4^{2-}$ [64], which may later precipitate as ZnO. Alternatively, the presence of $Zn(OH)(H_2O)_3^+$ was also suggested by other authors [55, 68]. For concentrated solutions (>33 wt% KOH), Sharma and Reed claimed that the dissolution of $Zn(OH)_2$ is restricted to $Zn(OH)_4^{2-}$ formation due to lower hydration of K^+ and OH^- ions. However, for less concentrated solutions, the formation of $Zn(OH)_2(H_2O)_2$, $Zn(OH)_3(H_2O)^-$, and/or $Zn(OH)_4^{2-}$ can occur. Among them, only $Zn(OH)_4^{2-}$ appears to be the electroactive one [63], while both $Zn(OH)_3(H_2O)^-$ and $Zn(OH)_2(H_2O)_2$ slowly precipitate.

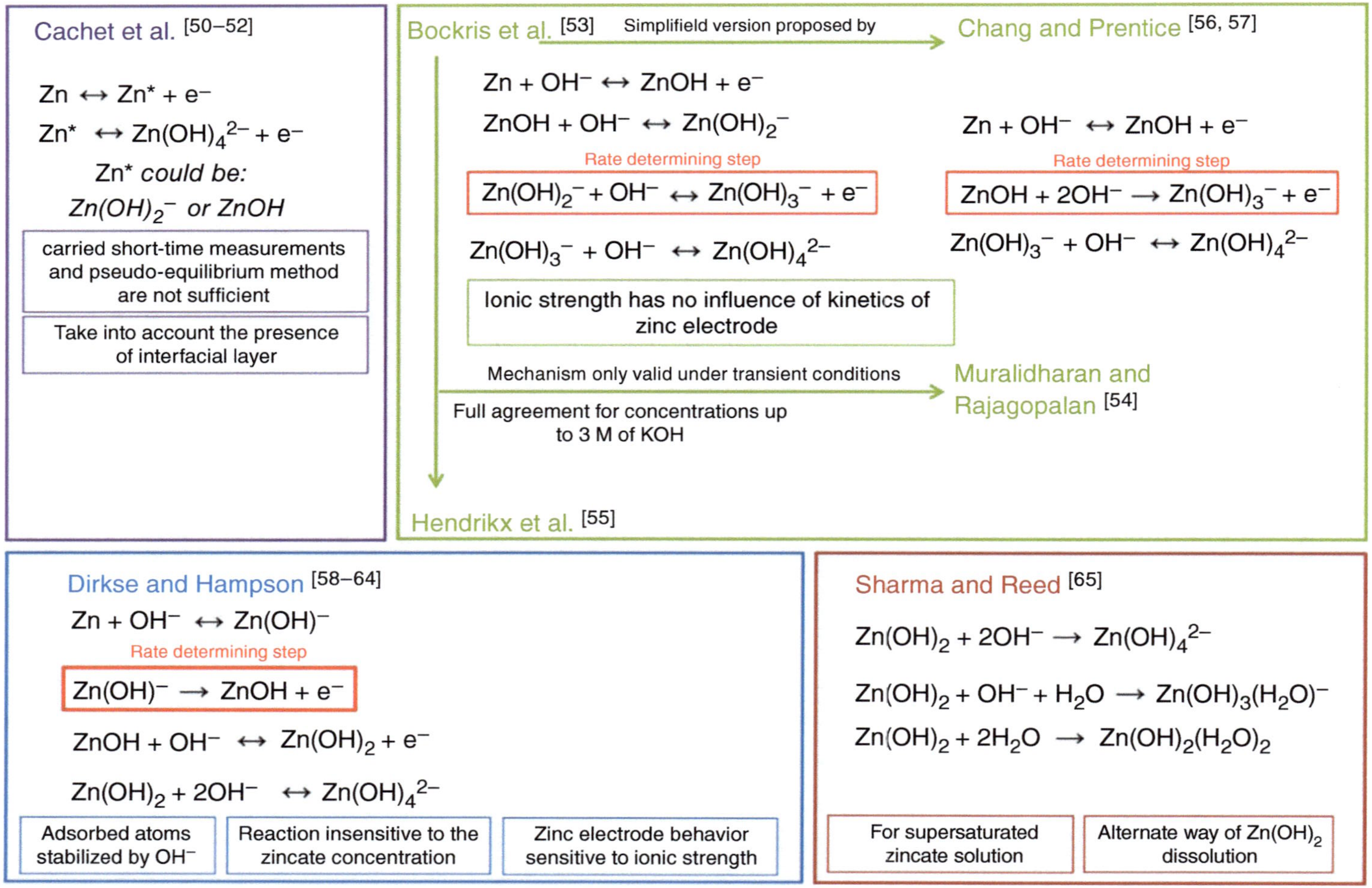

Figure 7.7 Proposed mechanisms for zinc dissolution. Source: Mainar et al. [44]. Reproduced with permission of John Wiley & Sons.

7.2.2 Strategies for Developing An Optimal Electrolyte System for Secondary Zinc–Air Batteries

7.2.2.1 Additives

Zinc as a negative electrode has been used in diverse technologies like manganese–zinc, silver–zinc, nickel–zinc, and zinc–air systems (as it was shown in Figure 7.4). These systems are based on alkaline electrolytes; hence, the electrochemistry of zinc is basically the same (Eqs. (7.10)–(7.12)). The incorporation of different additives to the electrolyte system has been one of the most known strategies in these zinc-based technologies, to reduce or avoid mainly the high dissolution of zinc through the electrolyte system and the corrosion. In that sense, used additives for aqueous alkaline zinc-based technologies, listed in Figure 7.8, could also be suitable for secondary zinc–air batteries. These additives have been classified in function of their principal role in the battery as corrosion inhibitors or their capability in reducing zinc shape changes and dendrite growth.

Among them, one of the most commonly used strategies is the incorporation of ZnO in aqueous alkaline electrolyte. This additive reduces the zinc dissolution

Carbonation inhibitor	
K_2CO_3	[70, 71]

Corrosion inhibitor	
Benzotriazole	[72]
Cd	
$In(CH_3COO)_2$	[73]
$In_2(SO_4)_3$	[73]
$In_3(NO_3)_2$	[73]
KF	[74]
K_2CO_3	[74]
$K_2Sn(OH)_6$	[73]
$Na_2Sn(OH)_6$	[73]
Sodium dodecyl benzene sulfonate	[72]
Tin acetate	[73]

Reduction of Zn shape change and dendrite growth			
Aluminate	[75]	Na_2CO_3	[25]
Benzotriazole	[72]	Na_2HBO_3	[25]
$BiCl_3$	[76]	Na_2HPO_4	[25]
$Ca(OH)_2$	[75]	Na_3BO_3	[25]
CdO	[77]	Na_3PO_4	[25]
Citric acid	[78]	$Pb(CH_3COO)_2$	[25]
Cs_2CO_3	[73]	PbO	[77]
$Cs_3C_6H_5O_7.2H_2O$	[73]	Perfluorosurfactants	[79, 80]
$CsCH_3CO_2$	[73]	Polyacrylic acid	[81]
CsF	[73]	Polyethylene glycol	[82]
H_3PO_4	[78]	Polyethyleneimine	[83]
In_2O_3	[77]	Polyoxyethylen alkyl phosphate ester acid	[82]
K_2BO_3	[84]	Sodium dodecyl benzene sulfonate	[72]
K_2CO_3	[70, 71]	Silicate ions	[75, 85, 86]
K_2HBO_3	[25]	SnO	[87]
K_2HPO_3	[25]	Sorbitol	[75]
K_3AsO_4	[25]	Stagnate	[75]
K_3BO_3	[71]	Succinic acid	[78]
K_3PO_4	[25]	Tartaric acid	[78]
KF	[70, 74]	Tetra-alkyl ammonium hydroxides	[88]
$KOH–CH_3OH$	[25]	Titanate	[75]
Li_3BO_3	[89]	Tungstate	[90]
LiF	[71]	$Zn(CH_3COO)_2$	[91–93]
LiOH	[94–97]	$ZnCl_2$	[98–102]
		ZnO	[82]

Figure 7.8 The different additives used for the aqueous alkaline electrolyte in zinc-based batteries.

because ZnO is dissolved as $Zn(OH)_4^{2-}$ in the aqueous alkaline electrolyte. Hence, the electrolyte gets close to the solubility limit where the migration of $Zn(OH)_4^{2-}$ coming from the dissolution of anodic active material is reduced [73]. Compounds based on arsenates, borates, carbonates, fluorides, and phosphates also reduce zinc dissolution in the electrolyte and hence, zinc shape change and dendrite growth [9].

On the other hand, organic compounds are well known for adsorbing in preferential areas where dendrites grow, impeding further growth at those locations. Thus, the presence of organic additives avoids zinc dendrite formation [102]. In this context, surfactants were used as electrolyte additives in Zn–MnO_2 alkaline batteries with the main aim of inhibiting zinc growth on preferential areas [79]. Lin-Feng Li reviewed and proposed in the patent published in 2010 [72] many other additives for the electrolyte system, such as quaternary ammonium compounds for influencing on zinc deposit morphology, bismuth soluble salts for a higher stability on the anode, SiO_3^{2-} for an enhanced zinc dissolution, or the incorporation of potassium acetate for increasing the ionic conductivity of the system.

However, even if all these compounds have demonstrated to avoid or reduce zinc anode-related challenges, it is important to evaluate their impact in a full system. Mainar et al. [73] selected KF, K_2CO_3, and ZnO as additives (previously reported in nickel–zinc technology), and they analyzed their impact in a secondary zinc–air battery. The work concluded that reported concentrations for nickel–zinc battery were not beneficial for the BAE performance. Hence, full cell system required a compromise between zinc anode and BAE to achieve an optimal electrochemical behavior of secondary zinc–air technology.

7.2.2.2 Alternatives to Alkaline Aqueous Electrolyte

The substitution of the alkaline aqueous electrolyte has been reported as promising strategy mainly for avoiding carbonate precipitation and reducing zinc dendrite formation [103]. For that, **aqueous near-neutral electrolytes** have been proposed to minimize the problems associated with the high alkalinity of the medium by the use of aqueous solutions in a pH range of 4–9. The most widely used near-neutral electrolytes have been based on chloride salts, such as $ZnCl_2$, NH_4Cl, NaCl, KCl, LiCl, or $MgCl_2$ [9]. Even if these electrolytes present the above-mentioned advantages [104], care should be taken to the different electrochemical reactions occurring in the zinc–air battery electrodes. As it was shown in the equations from Section 7.1.3 and Table 7.1, during the discharge of the battery, zinc dissolves in the electrolyte but, instead of forming $Zn(OH)_4^{2-}$ that would later produce ZnO (Eq. (7.1)), Zn dissolves in near-neutral electrolytes as Zn^{2+} (Eq. (7.2)), which could form irreversible species with the ions present in the electrolyte system [105]. On the other hand, it has to be taken into account that chloride evolution ($2Cl^- \rightarrow Cl_2(g) + 2e^-$) could be competing with the OER during the charge of the battery due to the proximity in the potential of both reactions [9]. Moreover, the incorporation of aqueous near-neutral electrolyte still faces many challenges associated with the aqueous nature of the medium, such as water evaporation, flooding, HER, or limited electrochemical window. Hence, secondary zinc–air batteries with near-neutral aqueous electrolytes need to be further investigated to give further insight into the mechanisms and the competitiveness of this alternative strategy. For this reason, non-aqueous systems were proposed, where requirements such as low toxicity, low cost, non-flammability, and low volatility are mandatory.

Solid polymer electrolytes (SPEs) are ionic conductive solids based on macromolecules containing heteroatoms that allow the dissolution of salts and enable their diffusion under an applied electric field. SPEs present some advantages over aqueous electrolytes as enhanced temperature range and reduced BAE flooding or electrolyte evaporation. Different SPEs have been studied, such as poly(ethylene oxide) (PEO) or poly(vinyl alcohol) (PVA) [106]. However, the principal challenges when using SPEs are related to the low ionic conductivity and/or the formation of passive layers between the SPE and the electrode [44] which inactivate zinc active material.

In this context, **gel polymer electrolytes (GPEs)** are presented as a promising alternative as their ionic conductivity is higher than SPEs. Generally, GPEs studied for zinc–air batteries have been based on polymeric matrixes such as PEO, poly(methyl methacrylate) (PMMA), poly(acrylonitrile) (PAN), or poly(vinylidene fluoride) (PVDF), and organic solvents such as propylene carbonate, ethylene carbonate (EC), or dimethyl sulfoxide (DMSO) [9, 48]. However, the mentioned organic solvents involve high volatility, and thus, more stable organic solvents have been investigated, such as oligomeric polyethers or the so-called *quasi-solid electrolytes*. Quasi-solid electrolytes are a type of GPEs generally based on the incorporation of a polymeric matrix (e.g. PVA, poly(acrylic acid) [PAA] or gelatin) to the aqueous alkaline electrolyte. These electrolytes have been widely studied in the development of flexible zinc–air batteries. However, even if performance and power density could be enhanced, the ionic conductivity and the electrode/electrolyte interface performance need to be further improved [107].

In the search for stable electrolytes, **room-temperature ionic liquids (RTILs) and deep eutectic solvents (DESs)** were reported as promising alternatives, as they are non-flammable and non-volatile compounds, and present higher ionic conductivity comparing to SPEs and GPEs. RTILs are ionic salts that have low melting points, mainly based on an organic cation and organic or inorganic anions [108]. The major advantages of RTILs are their low vapor pressure and high thermal and electrochemical stability. However, even if some approaches have been performed [109–111], these electrolytes still face some important issues, such as (i) their two-electron reaction pathway in the air electrode (which will be described in Section 7.3.1), (ii) their low power density due to high viscosity, (iii) their high cost due to the need of high purity compounds, and/or (iv) their sensitivity to air and water which limits their applications in zinc–air batteries under open air conditions [107].

Consequently, DESs have been proposed as a lower cost alternative to RTILs [9]. DESs are defined as solvents that contain a solid salt and a complexing agent that form a eutectic system with a melting point much lower than the individual components. The complexing agent is typically a hydrogen-bond donor, which interacts with the anion. As a result, the effective size increases and the interaction with the cation is isolated, decreasing the melting point. Many hydrogen-bond donors have been used in the literature for preparing DESs, such as amides [112, 113] (e.g. thiourea, acetamide), alcohols [114, 115] (e.g. ethylene glycol, glycerol), or acids [116, 117] (e.g. oxalic acid, malonic acid).

7.3 Bifunctional Air Electrode

7.3.1 Mechanism for Bifunctional Air Electrode

The high specific energy of zinc–air battery is related to cathodic active material, oxygen. It is presented in the surrounding air; hence, it is not stored inside the cell like traditional rechargeable batteries. The cathodic electrochemical reactions ORR and OER are very complex because they occur in the three-phase interphase between the liquid (electrolyte), solid (catalyst), and gas (oxygen) [118], and they involve multiple molecular adsorptions–desorptions, various elementary reaction steps, and also, the formation of different products [8].

During discharge of the zinc–air battery, there are two pathways for the ORR: (i) direct four-electron reduction pathway and (ii) two-electron reduction pathway. Four-electron reduction pathway directly forms OH^- (Eq. (7.13)) by the coordination of two oxygen atoms with the catalyst, while in the two-electron reduction pathway (Eqs. (7.14)–(7.16)), the catalyst is coordinated only to one oxygen atom and forms first peroxide and is further reduced to form OH^- [13, 118]. HO_2^- species are highly corrosive and accelerate the degradation of the BAE materials [119] as for example, the carbon commonly used as conductive agent (see Section 7.3.2.3). Thus, it is desirable to select a proper catalyst that promotes the four-electron pathway.

$$\text{Direct four-electron pathway: } O_2 + 2H_2O + 4e^- \rightarrow 4OH^- \tag{7.13}$$

$$\text{Two-electron pathway: } O_2 + H_2O + 2e^- \rightarrow HO_2^- + OH^- \tag{7.14}$$

$$\text{Peroxide reduction: } HO_2^- + H_2O + 2e^- \rightarrow 3OH^- \tag{7.15}$$

$$\text{Peroxide decomposition: } 2HO_2^- \rightarrow 2OH^- + O_2 \tag{7.16}$$

During the charge process of the battery, the electrochemical reactions associated with the OER involve O_2 formation from OH^- (Eqs. (7.17)–(7.20)), which require active, stable, and conductive catalysts at high potentials where OER takes place [119].

$$\text{Surface} + OH^- \rightarrow \text{Surface(OH)} + e^- \tag{7.17}$$

$$\text{Surface(OH)} + OH^- \rightarrow \text{Surface(O)} + H_2O + e^- \tag{7.18}$$

$$\text{Surface(O)} + OH^- \rightarrow \text{Surface(OOH)} + e^- \tag{7.19}$$

$$\text{Surface(OOH)} + OH^- \rightarrow \text{Surface} + O_2 + H_2O + e^- \tag{7.20}$$

The kinetics of these reactions related to oxygen depend on the binding strength of oxygen to the surface: if the binding strength is weak, Eq. (7.18) becomes the rate-determining step, while stronger binding of O_2 to the surface involves Eq. (7.19) be the limiting reaction [119].

7.3.2 Materials for Bifunctional Air Electrode

The BAE is based on two different layers: (i) an active layer (AL) and (ii) a gas diffusion layer (GDL). The **active layer** is in contact with the electrolyte and is composed of the catalyst, the conductive agent, and the binder (Figure 7.9). On the other hand, the **gas diffusion layer** is placed in the external part of the electrode. As it is in contact with the ambient air, the GDL material should be hydrophobic for avoiding electrolyte flooding as well as porous to allow oxygen accessibility [118, 120, 121]. Hence, the GDL is composed of a carbon-based material for a high electrical conductivity and a hydrophobic material such as polytetrafluoroethylene (PTFE) which adjusts the degree of hydrophobicity for allowing the permeation of air but not of the electrolyte system [122, 123]. On contrary to primary zinc–air batteries, the BAE should present high mechanical stability as the local pressure increases during OER as a result of the produced oxygen [124].

In this section, the relevant properties of each of the air electrode components as well as the different BAE designs carried out in the state of the art are summarized.

7.3.2.1 Catalysts

A BAE ranges from 0.7 V vs. reference hydrogen electrode (RHE) during discharge to more than 1.8 V vs. RHE during charge [118]. Therefore, catalysts are a key component of the BAE as their main aim is to favor the electrochemical reactions for obtaining a decreased overpotential between OER and ORR of the battery.

In this context, the main requirement for a bifunctional air catalyst is to present a high catalytic activity toward both the ORR and the OER. The catalytic activity can be enhanced by means of heteroatom doping and defect control, as these changes promote an electronic structure change, which reduces the adsorption energy of oxygen containing species. Porosity is also a relevant factor affecting the catalytic activity of the bifunctional air catalyst, where the presence of micro-/mesopores can increase the specific surface area and macropores can facilitate mass transfer. It has to be noted that many other requirements should be assessed, such as high conductivity, stability, selectivity toward the already mentioned four-electron reaction pathway, low cost, and environmental benignity [119]. Published catalysts can be

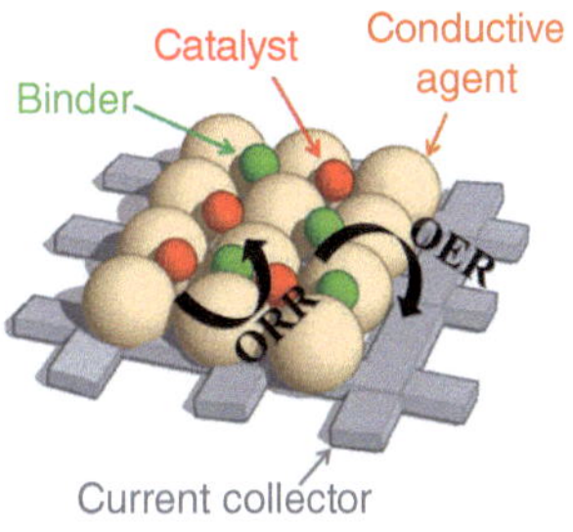

Figure 7.9 Bifunctional air electrode components.

classified into the following categories: (i) noble metals, (ii) transition metal oxides, and (iii) transition metals.

Noble metal catalysts, such as Pt or Ir, present high catalytic activity toward ORR or OER, respectively, but the bifunctionality of these catalysts is yet to be improved [119, 125, 126]. Moreover, the high cost and scarcity of these materials suppose the search of alternative bifunctional air catalysts [127].

On the other hand, **transition metal oxides** have been widely used as bifunctional air catalysts, which can be divided into three different categories: (i) single/binary transition metal oxides/hydroxides, (ii) spinels, and (iii) perovskites. *Single/binary transition metal oxides/hydroxides*, such as manganese oxides (Mn_xO_y), have been widely used as catalysts for zinc–air batteries due to their low cost, abundance, and environmental friendliness. Moreover, manganese oxide-based catalysts present high versatility, as they exist in many different crystallographic structures and polymorphic forms depending on the Mn valence state and preparation method [47]. In general, it has been stated that the catalytic activity for different manganese dioxides follows the order: $\alpha\text{-} \approx \delta\text{-} > \lambda\text{-} > \beta\text{-}MnO_2$ in alkaline aqueous electrolytes, mainly attributed to its crystal structure [126, 128].

Spinels [129] have also attracted great attention as catalysts for zinc–air batteries, which are based on the structure AB_2O_4, where A and B are metals. Spinels based on manganese and cobalt oxides have been widely studied, such as $MnCo_2O_4$ [130], $Co_3O_4/MnCo_2O_4$ [126, 131], Co_3O_4 [132–134], $NiCo_2O_4$ [135, 136], or Fe_3O_4 [137], due to their high catalytic activity. As an example, Ma et al. reported a voltage gap of ~0.75 V at 10 mA cm^{-2} for $NiCo_2O_4$ catalyst hybridized with carbon nanotubes (CNTs) [138].

Another group of transition metal oxides that has been evaluated is the *perovskites* (ABO_3), which are composed of a rare earth metal (A) and a transition metal (B). Perovskites have shown superior stability and corrosion resistance than spinels. Generally, lanthanum-based perovskites have been studied as bifunctional air catalysts, as it can be observed in Figure 7.10a.

Recently, **transition metals** have also been evaluated as bifunctional catalysts for zinc–air batteries (Figure 7.10b), combined with carbon-based materials, mainly Ni- and Co-based alloys such as FeCo, NiFe, and CoNi [145]. Additionally, transition metal sulfides (e.g. NiS_2/CoS_2, $NiCo_2S_4$) are attracting scientific interest due to their high electrical conductivity, tunable reactivities, and high stability. Transition metal nitrides (e.g. Co_3Mo_3N, CoN, Co_4N, Ni_3N, Ni_3FeN, and Fe_xN) have also been evaluated, mainly due to their anticorrosion resistance, high conductivity, and good OER performance [125].

7.3.2.2 Binder

The binder acts as a wet-proofing agent to maintain the hydrophobicity in the electrode. Hence, it is necessary for reducing the drying out of the electrolyte or flooding, which influences the battery performance [146]. In this context, the binder makes the electrode permeable to air but not to water [122, 123].

Both the GDL and the AL require a binder. However, in the AL, the surface should be partially wetted to allow the catalytic reactions to occur but need to partially

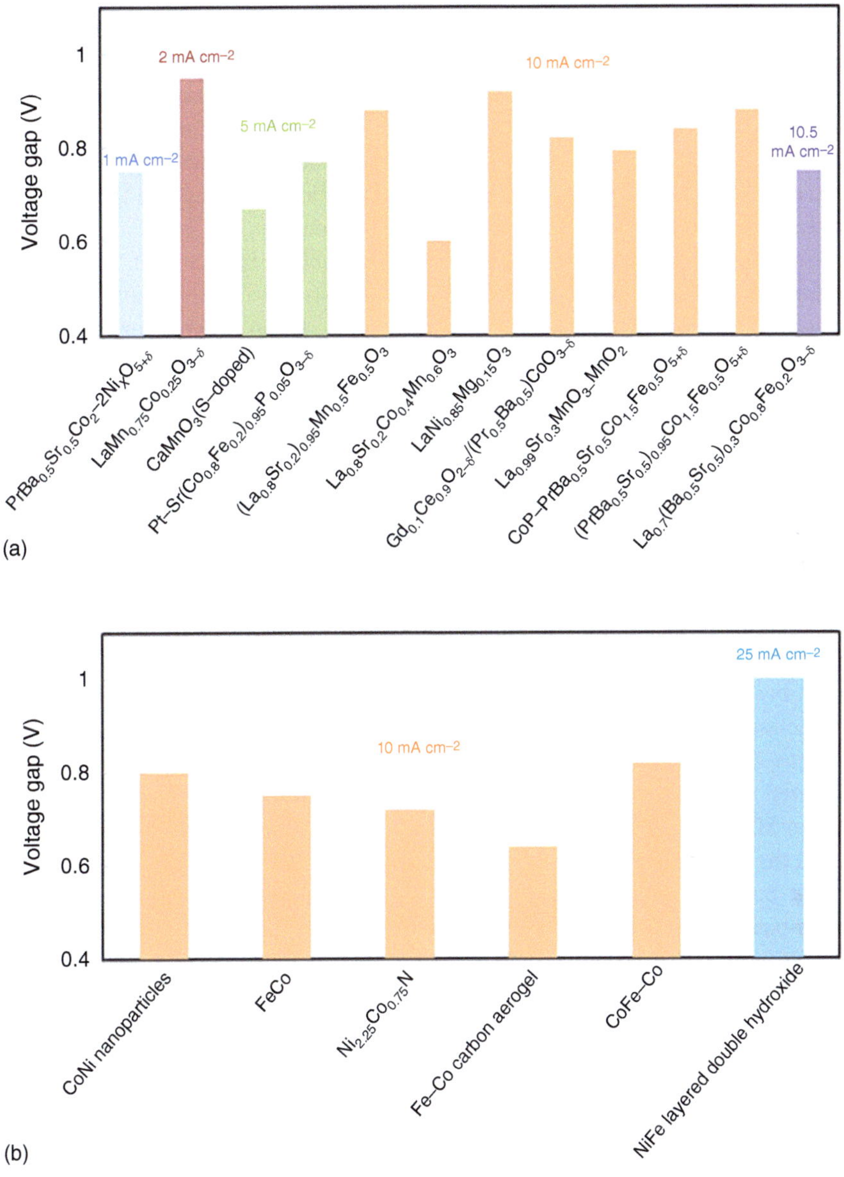

Figure 7.10 Obtained initial voltage gaps for (a) different perovskite-based catalysts and (b) different transition metal-based non-oxide catalysts. Source: Refs. [126, 139–144].

maintain hydrophobicity and porosity in the system for allowing gas accessibility [147, 148]. Therefore, the binder composition in GDL is generally higher than in AL. The most commonly used binder is PTFE [149, 150], and the ratios of binder for each layer have been determined in the state of the art, being the optimal composition 10–30% for AL and 30–70% for GDL [151].

7.3.2.3 Conductive Agents

Conductive agents are used for improving the electrical conductivity of the air electrode. In this context, many different carbon-based materials have been evaluated as conductive agents for the air electrode of zinc–air batteries. It has been demonstrated that the surface area, porosity, and catalytic activity of the carbonaceous material are relevant factors for an improved battery performance. For example, an increased surface area and porosity of the conductive agent allow an accessible three-phase (catalyst, oxygen, and electrolyte) site for ORR; however, too high surface area could accelerate carbon corrosion during the charge of the battery [118, 152].

Carbon materials are used as a substrate for catalyst, and also, their highly porous structure makes a diffusion path for oxygen. Hence, it can be used for both GDL and AL in a BAE. As it has been explained before, the hydrophobicity of the BAE plays an important role on battery performance. In this context, the used carbon material also affects this property and, as a result, hydrophobic carbon blacks such as acetylene carbon have been proposed for the GDL, while the combinations of different carbons with high and low surface areas have been proposed for the AL to control more effectively the electrolyte permeability and facilitate the three-phase interface [152–157].

On the other hand, it has to be taken into account that carbon-based materials such as carbon black, reduced graphene oxides (rGOs), or CNTs have demonstrated to present catalytic activity toward ORR [126, 158]. A comparison of different carbon-based materials was performed by Müller et al., where the activity for ORR followed the order: Ketjenblack > Vulcan as received > Vulcan graphitized and activated > Graphite HS-100 > Vulcan graphitized [159]. Later, Li et al. [160] compared the catalytic activity of activated carbon, graphite (SMGP), carbon black (Vulcan XC-72), and CNT, where CNTs resulted in a better electrocatalytic activity toward the ORR. However, these materials are still considerably distant from other catalyst materials described in Section 7.3.2.1. Generally, the most widely used carbon-based materials to act as conductive agents in the AL of the BAE of zinc–air batteries have been carbon blacks (particularly Vulcan XC-72) and CNTs [160, 161].

Even though carbon has been commonly used as conductive agent for the BAE [162–166], under charging voltages, carbon corrosion leads to the formation of CO_3^{2-}. Hence, it can hasten the BAE carbonation by the precipitation of K_2CO_3 in the air electrode pores blocking the catalyst active sites and limiting the oxygen diffusion [163, 167]. Moreover, the mechanical stability of BAE is affected by the pressures as a result of the gas formation (OER) during charging process. In this context, the development of carbon-free BAEs (metal-based air electrodes) turns a promising strategy to improve both, the mechanical and chemical stability of BAE [162, 163]. For example, nickel powder has also been used as conductive agent of BAE [168].

7.3.2.4 Current Collector

Different kinds of mesh or foams are commonly used as current collector materials for the BAE to allow the oxygen accessibility. The most widely used current collectors for BAE are Ni mesh, Ni foam, or stainless steel mesh [125]. Carbon-based materials have also been used as current collectors, in different morphologies such as fibers, cloths, mesh, aerogels, or films [169].

7.3.3 Electrode Structure

The BAE can be manufactured by means of three different designs: (i) single layered, (ii) double layered, and (iii) three electrode configurations. The simplest configuration is based on a **single layered electrode** (Figure 7.11a) commonly used in the state of the art, where the layer contains a bifunctional air catalyst, and hence, it is designed for both ORR and OER. However, the search of an advanced bifunctional air catalyst is more difficult than finding a unifunctional catalyst presenting high catalytic activity toward one of the two reactions (ORR and OER). For that reason, double layered and three electrode configurations have also been described in the state of the art [118] which present two different layers for each of the reactions associated with oxygen, ORR and OER (Figure 7.11b,c) [170].

The **double layered configuration** improves the stability and the electrochemical performance of the BAE as each layer includes the optimal materials for each reaction, ORR and OER. The layer responsible for OER is facing the electrolyte, while the ORR layer is oriented toward the air side (Figure 7.11b). This electrode requires the design of each layer with optimal concentrations to control porosity and wettability [118, 120]. However, it has a higher manufacturing complexity than the single layer-based BAE.

The **three electrode configuration** is composed of two different electrodes for each ORR and OER, each one located in each side of the zinc anode (Figure 7.11c). Compared to the double layered configuration, in this design, each electrode is not exposed to oxidative or reductive potentials because they are physically separated, where the zinc anode is connected to the ORR electrode during the discharge of the battery, while for the charging process, it switches to the OER electrode. However, apart from presenting a more complex manufacturing, the fabrication of two different electrodes increases the volume and weight of the system, thus decreasing its energy and power density [118, 120, 146, 171].

7.4 Zinc Anode

7.4.1 Zinc Electrode Configuration

The development of a reversible zinc anode plays an important role in secondary zinc–air batteries. Zinc electrodes can be classified into three different categories in function of their configuration: (i) zinc foil, (ii) zinc paste, and (iii) structured zinc electrode. Many approaches of secondary zinc–air studies have been performed with **zinc foil** as negative electrode [19] to analyze different strategies related to BAE, mainly the synthesis of catalysts.

However, two-dimension-based zinc anodes, as zinc foil, are not a practical electrode configuration for secondary zinc–air battery. As it was explained above

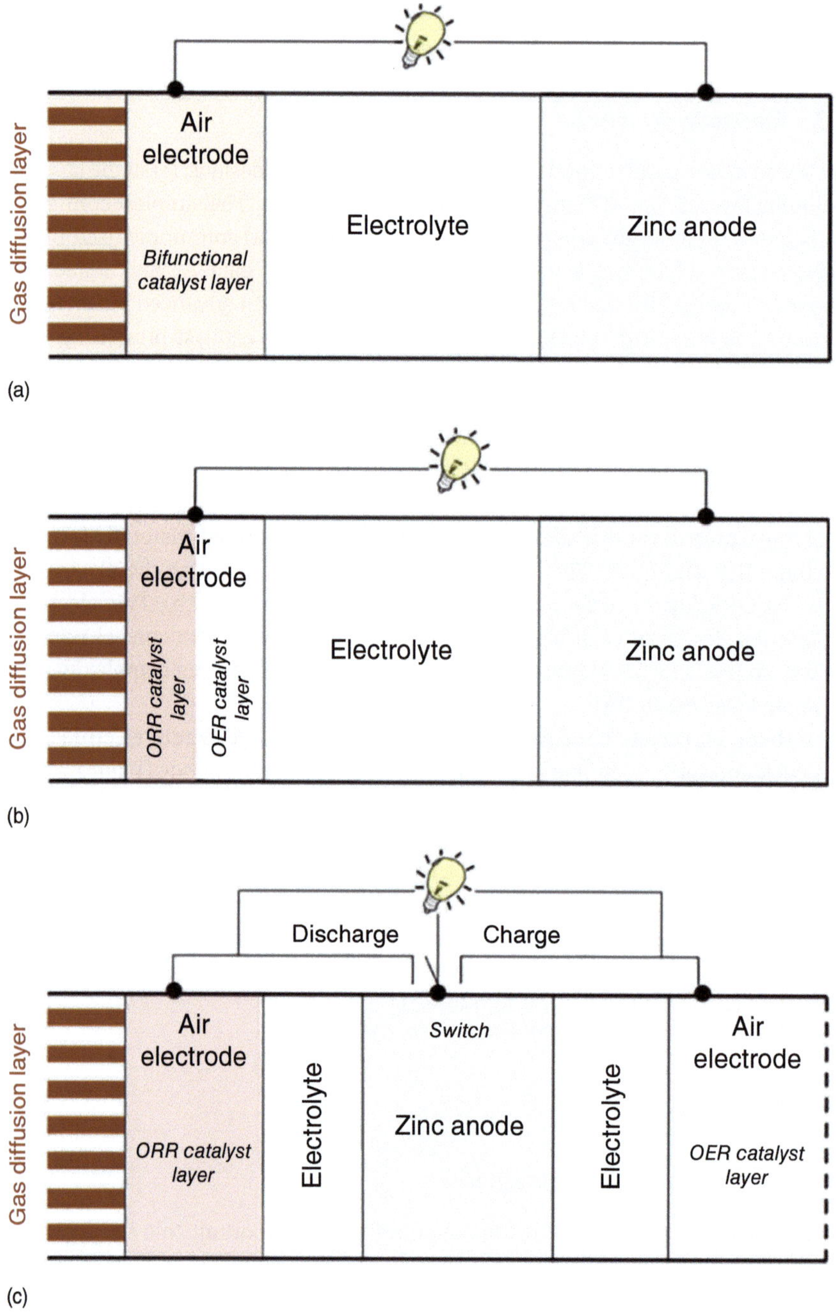

Figure 7.11 Different bifunctional air electrode configurations. (a) Single layered electrode. (b) Double layered electrode. (c) Three electrode configuration.

(Section 7.1.1), zinc–air battery characteristics are related to metallic zinc properties. In that sense, increasing the accessibility of zinc active material by means of the development of three-dimensional zinc anodes, as **zinc paste** or **structured zinc electrode**, turns mandatory to extract the maximum energy as it is done in commercial primary zinc–air batteries. On contrary to zinc foil, three-dimensional zinc anodes require different components which should be considered as gelling agent or binder, current collector or even, additives to improve zinc anode electrochemical performance.

7.4.2 Materials for Zinc Anode

7.4.2.1 Active Material

The main component of three-dimensional zinc anodes is the active material. The commercially available primary zinc–air batteries are based on a zinc paste as negative electrode. However, the rechargeability of zinc paste presents additional challenges that are no so critical in the primaries. For example, the zinc anode volume changes during the battery cycling as a consequence of the molar density difference between ZnO (discharge species, $14.5\,cm^3\,mol^{-1}$) and Zn (charged species, $9.15\,cm^3\,mol^{-1}$). It means that during discharging process, the metallic Zn is converted to ZnO (Eq. (7.1)); hence, the volume of the zinc anode increases. After that, during the charging process, the negative electrode is contracted again and as a consequence of previous cell deformation, the positive and negative electrodes could lose the contact [19]. In this context, the incorporation of ZnO in the zinc paste could bring some advantages to secondary zinc anode [73]. First, it accommodates the volume changes from Zn to ZnO; moreover, it could act as a discharge product reservoir that could be beneficial for battery performance.

Hence, in a secondary zinc–air battery, both metallic Zn and ZnO could be considered as active materials. Different active material ratios, shapes (e.g. atomized, flake, rod, etc.) [19] and sizes could be considered as they influence in the electrochemical response of secondary zinc–air technology. In general, an optimal electrode/electrolyte interface is desirable [19] and it should be stable during the cycling to improve the zinc anode reversibility.

7.4.2.2 Additives

The reversibility of secondary zinc anode is still hindered by technical challenges (see Section 7.1.3.3). In that sense, many strategies were followed up in the state of the art to improve the electrochemical performance of secondary zinc anode as the incorporation of additives through different procedures as physical mixture, particle coating, or alloying. The additives used in the literature have been divided into three main groups as it is shown in Figure 7.12; (i) carbon-based conducting additives, (ii) polymer additives, and (iii) inorganic additives. In general, each group has been reported for similar benefits in zinc-based technologies.

Carbon-based conducting additives are used to enhance the electronic conductivity of the zinc anode, as well as for reducing dendrite growth [88]. Among them, graphite, acetylene black, or graphene oxide have been widely used

Carbon based conducting additives	
Increase discharge performance, inhibit dendrite formation	
Acetylene black	[173]
Activated carbon	[174]
Carbon black	[175]
Graphite	[176, 177]
Graphene oxide (GO)	[178, 179]

Polymer additives	
Improve deposition and dissolution of Zn electrode, limit zinc corrosion	
α–hydroxy naphtoic acid	[180]
Agar binder	[181]
Anion exchange ionomer	[182–184]
Celullose	[185]
Cetyltrimethyl ammonium bromide	[186]
Ion containing polymer	[187]
Polyaniline	[188]
Poly(diallyldimethylammonium chloride)	[189]
Polyehtylene glycol	[186]
Polypirrole	[190]
Rubber polymers	[191]
SDS	[186]
Surfactant	[79, 192]
Thiourea	[186]

Inorganic additives			
Reduction of H_2 generation (Zn corrosion) and/or improve discharge performance			
Ag	[25]	$La(OH)_3$	[193]
Al_2O_3	[194, 195]	$La_2O_3/L(OH)_3$	[196]
BaO	[197]	$Li_2O–2B_2O_3$	[198]
$Ba(OH)_2$	[199]	$Nd(OH)_3$	[193]
Bi	[72, 200, 201]	Ni	[194]
Bi_2O_3	[147, 202]	Pb	[77, 203]
CaO	[202]	PbO	[77]
$Ca(OH)_2$	[204, 205]	Pb_3O_4	[77]
Cd	[200]	Sb_2O_4	[200]
Ga	[25, 200]	Sb_2O_3	[200]
Ga_2O_3	[201, 204]	SiO_2	[206]
Hg	[77]	Sn	[203]
HgO	[204]	SnO_2	[201, 207]
In	[197]	TiN	[208]
$In(OH)_3$	[197]	Tl_2O_3	[77]
In_2O_3	[147]		

Figure 7.12 List of additives and their effects on zinc anode. Source: Refs. [19, 88].

[172, 175, 177]. Concretely, graphite has been pointed out as a promising candidate due to its capacity of inhibiting zinc dissolution as well as increasing zinc anode capacity [175]. Moreover, the addition of Super-P carbon black demonstrated an increased capacity and power [174].

The zinc corrosion has been faced with **inorganic additives**. Metals such as Hg, Pb, Cd, and their oxides were first implemented many years ago, even if nowadays these materials have been substituted due to their high environmental impact [88]. In this context, environmentally benign metal oxides and hydroxides have also been considered for reducing HER as $Ca(OH)_2$ [207], Al_2O_3 [193], or Bi_2O_3 [146]. Some of the here reported inorganic additives also reduce the zinc solubility as for example, $Ca(OH)_2$ which accelerates $Zn(OH)_4^{2-}$ deposition [207] or even improve the electronic conductivity as Ag or TiN additives, which also promotes a higher homogeneity during zinc redeposition process [207].

Finally, the main characteristic of **polymeric additives** is the formation of a thin film on zinc surface. It alleviates the zinc corrosion and also reduces the migration of $Zn(OH)_4^{2-}$ through the electrolyte system favoring a homogeneous deposition of zinc and hence, reducing the dendrite growth [11].

7.4.2.3 Gelling Agents and Binders

Gelling agents and binders maintain zinc particle junction in the three-dimensional zinc electrodes. The main difference between them is the electrolyte immobilizing ability; while gelling agent stores the electrolyte system inside the zinc anode structure, the binder does not absorb the electrolyte [19].

Gelling agent is preferred for zinc anode as it absorbs the electrolyte systems inside zinc anode structure; hence, it reduces the total cell volume [19]. In this context, different polymeric materials as acrylate polymers [208], branched type (meth)acrylates [190], carboxymethylcellulose (CMC) [209], polyacrylic acid [193], carbopol [190], HPM [190], hydroxypropyl methylcellulose [208], methyl cellulose [190], polyacrylate [190], or polyethylene oxide [190] were used as gelling agents for zinc anode. However, deeper development should be performed to overcome current challenges as (i) the formation of a passive layer when high current densities are applied and (ii) electrode breaking due to the volume changes with repeated charge and discharge cycles (see Section 7.4.2.1) [19].

However, different **binders** have also been used in the state of the art as PTFE [208], PVA [200, 209], polycarbonate (PC) [210], polyethylene (PE) [208], rubber polymers [208], starch or derivates [208], polypropylene (PP) [208], rayon fibers [208], or sodium silicate [211]. The main limitation of binders is that they tend to coagulate during cycling [19].

7.4.2.4 Current Collector

The current collector, apart from being highly conductive and stable with electrolyte medium, should be a material nobler than zinc to avoid galvanic corrosion. However, it is complicated as materials with corrosion potentials close to zinc, stable in alkaline media, and highly electrically conductive have not yet been identified [212]. Copper or nickel current collectors have been widely used for zinc electrode, but advanced materials are required [212]. For that proposal, the electroplating of current collector with Ag, Ni, Pb, Cd, Sn, or even alloys like Pb–Sn, Sn–Zn, Cu–Sn–Zn is the most promising solution to reduce the galvanic corrosion [19]. On the other hand, the most appropriate configurations for current collector are mesh or foam to obtain a higher surface area of the zinc electrode [71, 212].

7.4.3 Zinc Anode Processing

The manufacturing process is critical to obtain a high-performance zinc anode where electrode thickness and porosity can be handled. In that sense, differences in zinc anode processing, as roll pressure, temperature, solution concentration, or time [19, 213], could result in different electrochemical behavior even if the same materials are used. Zinc electrode processing could be divided into two categories: (i) continuous or (ii) discontinuous.

In many of the used primary zinc–air batteries, the zinc anode process is **continuous** which is based on roll pressing. In this processing, the zinc paste is continuously deposited onto the current collector [19, 213]. On the other hand, **discontinuous processes** are divided into three main categories: (i) zinc anode compression, (ii) electrodeposition, and (iii) zinc anode sintering. The first one is similar to continuous processes described above; however, zinc paste or even dry zinc powder can be compressed discontinuously. Electrodeposition of zinc is based on a solution bath where zinc ions are electrodeposited in a matrix. The main advantage of this method is that binder or gelling agent is not necessary; hence, their challenges (see Section 7.4.2.3) are avoided. However, the electrodeposition must be optimized by means of optimizing the electrodeposition time, bath temperature, or even by adding additives. Finally, the zinc sintering processing is based on the transference of zinc powder and solution to a mold for a pressure and temperature treatment. The thermal sintering process typically converts metallic zinc powder into sponge electrodes by evaporating the solvent with high temperatures. Moreover, the porosity of the zinc electrode can be increased by means of the addition of porogens, which are materials that are part from the structure during the fabrication process but they are easily suppressed in the last part of the processing by, for example, temperature, leading free spaces (pores) in the anodic structure. As energy and time-consuming alternative to thermal sintering process, the cold sintering process has also been reported [214–217].

7.5 Membranes

In the present section, two types of membranes are described: (i) separators (placed between zinc anode and BAE) and (ii) CO_2 adsorption membranes (placed in the air side). The **separators** are used in batteries for avoiding electrical contact between the positive and negative electrodes, but they also have a relevant influence on the performance of the battery. For example, the selection of highly resistant separator could avoid dendrite perforation and, hence, the short circuit of the battery. However, as separators are embedded with the electrolyte system, they should present high stability with the strong alkaline nature of the traditional zinc–air battery electrolyte, high electrolyte adsorption capacity, low electrical conductivity, and high ionic conductivity.

Polymeric materials such as PP, PE, PVA, PAA, polyetherimide (PEI), or polyamide (PA) have been widely used for alkaline aqueous batteries and metal–air batteries [218–222]. The most employed separator in commercial primary zinc–air batteries is Celgard® 5550 [218, 223], which is composed of two different polymeric materials, PP and PE distributed in three polymeric layers. Two of them are based on PP which provides mechanical stability, while in the middle, a layer of PE is placed, to assure battery shut down if battery overheats. These membranes are often coated with surfactants to increase their hydrophilic nature and, as a result, increase the ionic conductivity [221, 224].

In the state of the art, some approaches have been carried out regarding the development of selective anion-exchange membranes. This strategy limits the zincate ion migration from the zinc anode, which affects to battery cycling efficiency [225, 226] as it was described in Section 7.1.3.3. For that proposal, only the free movement to OH^- ions is allowed since higher ions as $Zn(OH)_4{}^{2-}$ cannot cross. Dewi et al. [220] reported a polysulfonium-based membrane with an increased discharge capacity of zinc–air battery. Other strategy to prevent zincate ion migration is the coating of commercial Celgard 5550 separator with an anion exchange solution based on the copolymerization of ionic liquid monomers [227]. Recently, the methylation of polybenzimidazole (PBI) also demonstrated to reduce the permeability for zincate ions, which reduces shape change and dendrite growth [228]. Nevertheless, further work is still needed to find a suitable anion exchange membrane that presents a high stability in strong alkaline aqueous solutions [229].

Finally, zinc–air battery performance was enhanced by means of the incorporation of a **CO_2 adsorption membrane** close to air side of BAE to avoid the carbonation effect in the cathode [44]. Polystyrene (PS)/PEI fiber-based membranes have been reported as CO_2 adsorption membranes [230]. However, adsorbents such as LiOH or LiOH–$Ca(OH)_2$ have been used as filtering additives in the cathode, the latter demonstrating more than 20 times lower CO_2 concentration [231]. This effect is aligned with the chemisorption process, where LiOH produces Li_2CO_3 and water in the presence of CO_2. Another approximation is based on the high reaction rate of ammines with CO_2, which resulted in the use of piperazine (PZ), 2-(2-aminoethylamino)ethanol (AEEA), and monoethanolamine (MEA) as CO_2 filters, being PZ more effective for CO_2 adsorption [232].

7.6 Summary and Perspectives

The current changes in the society energy consumption habits have requested modifications of the worldwide policies to promote the use of renewals as sustainable energy sources. However, the intermittent power sources like wind, solar, and tide have sparked a need for better electrochemical energy storage systems (EESSs) to provide ancillary services and save excess energy for use at a later time. The development of battery storage systems for load-leveling applications requests safety, long-cycle life, cheap, abundant, and environmental-friendly materials.

Zinc–air battery technology is one of the suitable battery technologies to fulfill such requirements. Although the primary version is a worldwide well-established technology, its conversion to secondary system is still a challenge owing to the several issues related to the degradation of the electrodes during cycling, which results in a rapid reduction in (storage) capacity.

In the last years, many efforts have been focused on the development of BAEs as on contrary to primary zinc–air technology; it should be capable of catalyzing both ORR and OER. In this context, high-performance non-precious metal-based bifunctional electrocatalysts have been widely studied, mainly, those based on metal oxides (spinel, perovskite, etc.). These catalysts are commonly supported with carbon as conductive media due to their semiconducting nature. However, carbon support in the AL is severely attacked by the high reactive oxygen generated from the OER during charging which contributes to the low long-term stability of BAE. In that sense, the replacement of carbon by alternative conductive media like for example silver, nickel, or others joined to the incorporation of proper bifunctional catalyst is a key strategy to achieve long-term stability-based BAE.

On the other hand, zinc anode is also exposed to severe limitations (e.g. dendrite growth, shape changes, passivation, etc.) comparing to primary zinc–air technology. Most of them are related to the high dissolution of zinc through the aqueous alkaline electrolyte system. In this context, the electrolyte formulation and nature play a critical role on the electrochemical performance of secondary zinc anode. In that sense, several strategies have been proposed in the state of the art including the incorporation of non-aqueous alkaline-based electrolyte systems (e.g. solid electrolytes, aqueous neutral, etc.) or even the modification of electrolyte formulation following similar strategies than those used for commercially available secondary zinc-based technologies such as nickel–zinc. Joined to electrolyte development, the electrochemical performance of secondary zinc anode could be improved by choosing an adequate strategy for zinc anode as alloying, coating or even, the incorporation of suitable additives.

Finally, the membranes also play a critical role in the development of secondary zinc–air technology. Those placed between zinc anode and BAE should be high resistant separators to avoid dendrite perforation, and they should present low permeability to zincate ions to avoid zinc migration from the zinc anode. Moreover, the incorporation of selective membrane close to BAE could avoid the cell drying or even, the crossover of CO_2 to reduce the precipitation of insoluble carbonates in the porous structure of BAE which severely limits the reversibility of the technology.

Acronyms and Abbreviations

AEEA	2-(2-aminoethylamino)ethanol
AL	active layer
BAE	bifunctional air electrode
CNTs	carbon nanotubes
DESs	deep eutectic solvents
DMSO	dimethyl sulfoxide
EC	ethylene carbonate
EU	European union
GDL	gas diffusion layer
GPE	gel polymer electrolyte
HER	hydrogen evolution reaction
MEA	monoethanolamine
OER	oxygen evolution reaction
ORR	oxygen reduction reaction
PA	polyamide
PAA	poly(acrylic acid)
PAN	poly(acrylonitrile)
PBI	polybenzimidazole
PC	polycarbonate
PE	polyethylene
PEI	polyetherimide
PEO	poly(ethylene oxide)
PMMA	poly(methyl methacrylate)
PP	polypropylene
PS	polystyrene

PTFE	polytetrafluoroethylene
PVA	poly(vinyl alcohol)
PVDF	poly(vinylidene fluoride)
PZ	piperazine
RHE	reference hydrogen electrode
RTIL	room-temperature ionic liquid
SPE	solid polymer electrolyte

References

1 European Commission (2019). Communication on the European Green Deal. https://eur-lex.europa.eu/legal-content/EN/TXT/?qid=1588580774040&uri=CELEX%3A52019DC0640 (accessed 8 June 2020).

2 European Commission (2011). Communication from the Commission to the European Parliament, the Council, the European Economic and Social Committee and the Committee of the Regions. Energy roadmap 2050. https://eur-lex.europa.eu/LexUriServ/LexUriServ.do?uri=COM:2011:0885:FIN:EN:PDF (accessed 5 June 2020).

3 European Commission (2018). Europe leads the global clean energy transition: Commission welcomes ambitious agreement on further renewable energy development in the EU. https://ec.europa.eu/commission/presscorner/detail/en/STATEMENT_18_4155 (accessed 8 June 2020).

4 IRENA (2018). Renewable energy prospects for the European union. https://www.irena.org/-/media/Files/IRENA/Agency/Publication/2018/Feb/IRENA_REmap_EU_2018.pdf (accessed 8 June 2020).

5 M. Team (2014). Micro energy storage in buildings. https://www.brusselsretrofitxl.be/projects/mesb/ (accessed 5 June 2020).

6 European Magnesium Interactive Battery Community (2019). Europeanproject FET-Proactive 2018: E-MAGIC (Grant Agreement No. 824066). http://www.e-magic.eu/ (accessed 8 June 2020).

7 Caramia, V. and Bozzini, B. (2014). Material science aspects of zinc-air batteries: a review. *Mater. Renewable Sustainable Energy* 3: 28.

8 Mainar, A.R. (2017). Comprehensive study of secondary zinc-air batteries: from material development to full cell validation. Universidad del País Vasco.

9 Mainar, A.R., Iruin, E., Colmenares, L.C. et al. (2018). An overview of progress in electrolytes for secondary zinc-air batteries and other storage systems based on zinc. *J. Energy Storage* 15: 304–328.

10 Xu, M., Ivey, D.G., Xie, Z., and Qu, W. (2015). Rechargeable Zn-air batteries: progress in electrolyte development and cell configuration advancement. *J. Power Sources* 283: 358–371.

11 Kim, H., Jeong, G., Kim, Y.-U. et al. (2013). Metallic anodes for next-generation secondary batteries. *Chem. Soc. Rev.* 42: 9011–9034.

12 Koscher, G. and Kordesch, K. (2004). Can refillable alkaline methanol–air systems replace metal–air cells? *J. Power Sources* 136: 215–219.

13 Linden, D. and Reddy, T.B. (2002). *Handbook of Batteries*, 3e. McGraw Hill.
14 Wang, J., Li, Y., and Sun, X. (2013). Challenges and opportunities of nanostructured materials for aprotic rechargeable lithium–air batteries. *Nano Energy* 2: 443–467.
15 Rahman, M.A., Wang, X., and Wen, C. (2013). High energy density metal-air batteries: a Review. *J. Electrochem. Soc.* 160: A1759–A1771.
16 Abraham, K.M. (2008). A brief history of non-aqueous metal-air batteries. *ECS Trans.* 3: 67–71.
17 Wang, R.Y., Kirk, D.W., and Zhang, G.X. (2006). Effects of deposition conditions on the morphology of zinc deposits from alkaline zincate solutions. *J. Electrochem. Soc.* 153: C357–C364.
18 Dell, R. and Rand, D.A.J. (2001). *Understanding Batteries*. The Royal Society of Chemistry.
19 Mainar, A.R., Colmenares, L.C., Blázquez, J.A., and Urdampilleta, I. (2018). A brief overview of secondary zinc anode development: the key of improving zinc-based energy storage systems. *Int. J. Energy Res.* 42: 903–918.
20 Zhang, G.X. (2009). Zinc as an energy carrier for energy conversion and storage. *ECS Trans.* 16: 47–58.
21 Zhang, G.X. (2009). Zinc as an energy carrier for energy conversion and storage. *ECS Trans.* 16: 47–59.
22 Yan, J., Wang, J., Liu, H. et al. (2012). Rechargeable hybrid aqueous batteries. *J. Power Sources* 216: 222–226.
23 Xu, C., Li, B., Du, H., and Kang, F. (2012). Energetic zinc ion chemistry: the rechargeable zinc ion battery. *Angew. Chem. Int. Ed.* 51: 933–935.
24 Clarke, R.L., Dougherty, B., Harrison, S. *et al.* (2004). Cerium batteries. US 2004/0202925 A1.
25 McLarnon, F.R. (1991). The secondary alkaline zinc electrode. *J. Electrochem. Soc.* 138: 645–656.
26 Zakariah, A.A. and Folorunsho, A.D. (2012). Adsorption of zinc ion (Zn^{2+}) on manganese (IV) oxide (MnO_2) in a leclanche dry cell. *Der Chem. Sin.* 3: 182–191.
27 Hamlen R.P. and Atwater T.B. (1995). Chapter 38: Metal/air batteries. In: *Handbook of Batteries* (eds. D. Linden and T.B. Reddy). McGraw-Hill.
28 Chen, X., Zhou, Z., Karahan, H.E. et al. (2018). Recent advances in materials and design of electrochemically rechargeable zinc–air batteries. *Small* 14: 1–29.
29 Haas, O., Holzer, F., Müller, K., and Müller, S. (2010). Metal/air batteries: the zinc/air case. In: *Handbook of Fuel Cells Fundamentals Technology and Applications*, 1–27. https://onlinelibrary.wiley.com/doi/abs/10.1002/9780470974001.f104022.
30 Daniel, C. and Besenhard, J.O. (2009). *Handbook of Battery Materials*. Wiley-VCH.
31 Li, X., Ponce de León, C., Walsh, F.C. et al. (2015). Zinc-based flow batteries for medium- and large-scale energy storage. In: *Types and Applications*. Woodhead Publishing Series in Energy 1e, 293–315. Woodhead Publishing: Cambridge, GB. https://doi.org/10.1016/B978-1-78242-013-2.00008-X.

32 Garche, J., Dyer, C.K., Moseley, P.T. et al. (2009). *Encyclopedia of Electrochemical Power Sources*. Elsevier.

33 Chakrabarti, M.H., Hajimolana, S.A., Mjalli, F.S. et al. (2013). Redox flow battery for energy storage. *Arab. J. Sci. Eng.* 38: 723–739.

34 Arenas, L.F., Ponce de León, C., and Walsh, F.C. (2017). Engineering aspects of the design, construction and performance of modular redox flow batteries for energy storage. *J. Energy Storage* 11: 119–153.

35 Clark, N. and Kinoshita, K. (1994). *Zinc/air technology. Meeting report.*

36 Cai, X., Lai, L., Lin, J., and Shen, Z. (2017). Recent advances in air electrodes for Zn-air batteries: electrocatalysis and structural design. *Mater. Horiz.* 4: 945–976.

37 Zinc based energy storage systems for electric vehicles, 2009. https://www.evwind.es/2009/09/07/zinc-based-energy-storage-systems-for-electric-vehicles/1130 (accessed 5 June 2020).

38 Denton, T. (2004). *Automobile Electrical and Electronic Systems*, 3e. ElsevierButterworth-Heinemann.

39 Othman, R. and Saputra, H. (2012). Zinc-air battery – powering electric vehicles to smart active labels. *J. Eng. Sci. Technol. Dev.* 1: 1–5.

40 Eos Aurora® 150|600 DC battery system, 2020. https://eosenergystorage.com/wp-content/uploads/Eos-Energy-Storage-Product-Brochure.pdf.

41 Zinium - ZnR batteries (2020). https://znrblog.wordpress.com/ (accessed 8 June 2020).

42 Deing, J. (2018). ZincNyx Energy Solutions https://www.greentechmedia.com/articles/read/zincnyx-plansflow-battery-mass-production-within-a-year (accessed 8 June 2020).

43 Nantenergy (n.d.). Zinc-air Storage. https://nantenergy.com/zinc-air/ (accessed 8 June 2020).

44 Mainar, A.R., Leonet, O., Bengoechea, M. et al. (2016). Alkaline aqueous electrolytes for secondary zinc–air batteries: an overview. *Int. J. Energy Res.* 40: 1032–1049.

45 Zhao, Z., Fan, X., Ding, J. et al. (2019). Challenges in zinc electrodes for alkaline zinc-air batteries: obstacles to commercialization. *ACS Energy Lett.* 4: 2259–2270.

46 Shaigan, N., Qu, W., and Takeda, T. (2010). Morphology control of electrodeposited zinc from alkaline zincate solutions for rechargeable zinc air batteries. *ECS Trans.* 28: 35–44.

47 Mainar, A.R., Colmenares, L.C., Leonet, O. et al. (2016). Manganese oxide catalysts for secondary zinc air batteries: from electrocatalytic activity to bifunctional air electrode performance. *Electrochim. Acta* 217: 80–91.

48 Lee, D.U., Fu, J., and Chen, Z. (2016). Advanced materials for Zn-air rechargeable batteries. In: *Electrochemical: Advanced Materials and Technology*. Taylor & Francis Group, LLC.

49 Cachet, C., Saidani, B., and Wiart, R. (1992). The behavior of zinc electrode in alkaline electrolytes. II. A kinetic analysis of anodic dissolution. *J. Electrochem. Soc.* 139: 644–654.

50 Cachet, C., Stroder, U., and Wiart, R. (1982). The kinetics of a zinc electrode in alkaline zincate electrolytes. *Electrochim. Acta* 27: 903–908.

51 Cachet, C., Saïdani, B., and Wiart, R. (1991). The behavior of zinc electrode in alkaline electrolytes. I. A kinetic analysis of cathodic deposition. *J. Electrochem. Soc.* 138: 678–687.

52 Bockris, J., Nagy, Z., and Damjanovic, A. (1972). Deposition and dissolution of zinc in alkaline solutions. *J. Electrochem. Soc.* 119: 285–295.

53 Muralidharan, V. and Rajagopalan, K. (1978). Kinetics and mechanism of corrosion of zinc in sodium hydroxide solutions by steady-state and transient methods. *J. Electroanal. Chem.* 94: 21–36.

54 Hendrikx, J., Van Der Putten, A., Visscher, W., and Barendrecht, E. (1984). The electrodeposition and dissolution of zinc and amalgamated zinc in alkaline solutions. *Electrochim. Acta* 29: 81–89.

55 Chang, Y. and Prentice, G. (1984). A model for the anodic dissolution of zinc in alkaline electrolyte. Kinetics of initial dissolution. *J. Electrochem. Soc.* 131: 1465–1468.

56 Chang, Y. and Prentice, G. (1985). Anodic dissolution of zinc electrodes in alkaline electrolyte: mass transport effects. *J. Electrochem. Soc.* 132: 375–378.

57 Dirkse, T. and Hampson, N. (1972). Zinc(II)/zinc exchange reaction in potassium hydroxide solution. I. Exchange current density measurements using the galvanostatic method. *Electrochim. Acta* 17: 135–141.

58 Dirkse, T. and Hampson, N. (1972). Zinc(II)/zinc exchange reaction in potassium hydroxide solution. II. Exchange current density measurements using the double-impulse method. *Electrochim. Acta* 17: 383–386.

59 Dirkse, T. and Hampson, N. (1972). Zinc(II)/zinc exchange reaction in potassium hydroxide solution. III. Exchange current measurements using the potentiostatic method. *Electrochim. Acta.* 17: 1113–1119.

60 Dirkse, T. (1978). The behavior of the zinc electrode in alkaline solutions. I. The effect of ionic strength at the equilibrium potential. *J. Electrochem. Soc.* 125: 1591–1594.

61 Dirkse, T. (1979). The behavior of the zinc electrode in alkaline solutions. II. Reaction orders at the equilibrium potential. *J. Electrochem. Soc.* 126: 541–543.

62 Dirkse, T. (1980). The behavior of the zinc electrode in alkaline solutions. IV. The effect of ionic strength in the Tafel region. *J. Electrochem. Soc.* 127: 1452–1456.

63 Dirkse, T. (1981). The behavior of the zinc electrode in alkaline solutions. V. Supersaturated zincate solutions. *J. Electrochem. Soc.* 128: 1412–1415.

64 Sharma, S. and Reed, M. (1976). Raman study of zincate ions in concentrated alkaline solutions. *J. Inorg. Nucl. Chem.* 38: 1971–1972.

65 Zhang, X.G. (1996). *Corrosion and Electrochemistry of Zinc*. New York: Plenum Press.

66 Shivkumar, R., Kalaignan, G., and Vasudevan, T. (1997). Studies with porous zinc electrodes with additives for secondary alkaline batteries. *J. Power Sources* 75: 90–100.

67 Shivkumar, R., Kalaignan, G., and Vasudevan, T. (1995). Effect of additives on zinc electrodes in alkaline battery systems. *J. Power Sources* 55: 53–62.

68 El'kind, K., Mikhalenko, M., and Flerov, V. (1978). Nature of supersaturated potassium zincate solutions and mechanism of their aging. *Zh. Prikl. Khim.* 51: 1225–1228.

69 Schröder, D., Borker, N.N.S., König, M., and Krewer, U. (2015). Performance of zinc air batteries with added K_2CO_3 in the alkaline electrolyte. *J. Appl. Electrochem.* 45: 427–437.

70 Adler, T.C., McLarnon, F.R., and Cairns, E.J. (1998). Investigations of new famility of alkaline-fluoride-carbonate electrolytes for zinc/nickel oxide cells. *Ind. Eng. Chem. Res.* 37: 3237–3241.

71 Da, Y., Zhao, F., Shi, J., and Zhang, Z. (2020). Effects of ultrafine bismuth powder on the properties of zinc electrodes in zinc-air batteries. *J. Electron. Mater.* 49: 2479–2490.

72 Li, L. (2010). Non-toxic alkaline electrolyte with additives for rechargeable zinc cells. US 2010/0062327.

73 Mainar, A.R., Iruin, E., Colmenares, L.C. et al. (2018). Systematic cycle life assessment of a secondary zinc–air battery as a function of the alkaline electrolyte composition. *Energy Sci. Eng.* 6: 174–186.

74 Zhang, X.G. (2009). Secondary batteries – zinc systems | zinc electrodes: overview. In: *Encyclopedia of Electrochemical Power Sources* (ed. J. Garche), 454–468. Elsevier. ISBN: 9780444527455.

75 Wang, J.M., Zhang, L., Zhang, C., and Zhang, J.Q. (2001). Effects of bismuth ion and tetrabutylammonium bromide on the dendritic growth of zinc in alkaline zincate solutions. *J. Power Sources* 102: 139–143.

76 McBreen, J. and Gannon, E. (1981). The electrochemistry of metal oxide additives in pasted zinc electrodes. *Electrochim. Acta* 26: 1439–1446.

77 Lee, C.W., Sathiyanarayanan, K., Eom, S.W. et al. (2006). Novel electrochemical behavior of zinc anodes in zinc/air batteries in the presence of additives. *J. Power Sources* 159: 1474–1477.

78 Zhu, J., Zhou, Y., and Gao, C. (1998). Influence of surfactants on electrochemical behavior of zinc electrodes in alkaline solution. *J. Power Sources* 72: 231–235.

79 Ghavami, R., Rafiei, Z., and Tabatabaei, S. (2007). Effects of cationic CTAB and anionic SDBS surfactants on the performance of Zn–MnO_2 alkaline batteries. *J. Power Sources* 164: 934–946.

80 Lee, S.H., Park, D.J., Yang, W.G., and Ryu, K.S. (2017). Comparison of electrochemical performance for zinc anode via various electrolytes and conducting agents in Zn-air secondary batteries. *Ionics (Kiel)* 23: 1801–1809.

81 Ein-Eli, Y., Auinat, M., and Starosvetsky, D. (2003). Electrochemical and surface studies of zinc in alkaline solutions containing organic corrosion inhibitors. *J. Power Sources* 114: 330–337.

82 Hashemi, A.B., Kasiri, G., and La Mantia, F. (2017). The effect of polyethyleneimine as an electrolyte additive on zinc electrodeposition mechanism in aqueous zinc-ion batteries. *Electrochim. Acta* 258: 703–708.

83 Yuan, Y.F., Yu, L.Q., Wu, H.M. et al. (2011). Electrochemical performances of Bi based compound film-coated ZnO as anodic materials. *Electrochim. Acta* 56: 4378–4383.

84 Diomidis, N. and Celis, J.P. (2007). Anodic film formation on zinc in alkaline electrolytes containing silicate and tetraborate ions. *J. Electrochem. Soc.* 154: C711–C718.

85 Huot, J.Y. (1992). The effects of silicate ion on the corrosion of zinc powder in alkaline solutions. *J. Appl. Electrochem.* 22: 443–447.

86 Kim, H.-I. and Shin, H.-C. (2015). SnO additive for dendritic growth suppression. *J. Alloys Compd.* 645: 7–10.

87 Lan, C.J., Lee, C.Y., and Chin, T.S. (2007). Tetra-alkyl ammonium hydroxides as inhibitors of Zn dendrite in Zn-based secondary batteries. *Electrochim. Acta* 52: 5407–5416.

88 Yi, J., Liang, P., Liu, X. et al. (2018). Challenges, mitigation strategies and perspectives in development of zinc-electrode materials and fabrication for rechargeable zinc-air batteries. *Energy Environ. Sci.* 11: 3075–3095.

89 Wen, Y., Cheng, J., Zhang, L. et al. (2009). The inhibition of the spongy electrocrystallisation of zinc from doped flowing alkaline zincate solutions. *J. Power Sources* 193: 890–894.

90 Niu, W., Li, Z., Marcus, K. et al. (2018). Surface-modified porous carbon nitride composites as highly efficient electrocatalyst for Zn-air batteries. *Adv. Energy Mater.* 8: 1870002.

91 You, T.-H. and Hu, C.-C. (2018). Designing binary Ru–Sn oxides with optimized performances for the air electrode of rechargeable zinc–air batteries. *ACS Appl. Mater. Interfaces* 10: 10064–10075.

92 Zeng, S., Tong, X., Zhou, S. et al. (2018). All-in-one bifunctional oxygen electrode films for flexible Zn-air batteries. *Small* 14: 1803409.

93 Lee, S.H., Yi, C.W., and Kim, K. (2011). Characteristics and electrochemical performance of the TiO_2-coated ZnO anode for Ni-Zn secondary batteries. *J. Phys. Chem. C* 115: 2572–2577.

94 Zeng, D., Yang, Z., Wang, S. et al. (2011). Preparation and electrochemical performance of In-doped ZnO as anode material for Ni-Zn secondary cells. *Electrochim. Acta* 59: 4075–4080.

95 Parker, J.F., Chervin, C.N., Pala, I.R. et al. (2017). Rechargeable nickel–3D zinc batteries: an energy-dense, safer alternative to lithium-ion. *Science* 356: 415–418.

96 Lee, S.-H., Jeong, Y.-J., Lim, S.-H. et al. (2010). The stable rechargeability of secondary Zn-air batteries: is it possible to recharge a Zn-air battery? *J. Korean Electrochem. Soc.* 13: 45–49.

97 Wei, L., Karahan, H.E., Zhai, S. et al. (2017). Amorphous bimetallic oxide-graphene hybrids as bifunctional oxygen electrocatalysts for rechargeable Zn-air batteries. *Adv. Mater.* 29: 1701410.

98 Wu, X., Chen, F., Jin, Y. et al. (2015). Silver-copper nanoalloy catalyst layer for bifunctional air electrodes in alkaline media. *ACS Appl. Mater. Interfaces* 7: 17782–17791.

99 Li, T., Lu, Y., Zhao, S. et al. (2018). Co_3O_4-doped Co/CoFe nanoparticles encapsulated in carbon shells as bifunctional electrocatalysts for rechargeable Zn-air batteries. *J. Mater. Chem. A* 6: 3730–3737.

100 Fu, G., Wang, J., Chen, Y. et al. (2018). Exploring indium-based ternary thiospinel as conceivable high-potential air-cathode for rechargeable Zn-air batteries. *Adv. Energy Mater.* 8: 1802263.

101 Chen, B., He, X., Yin, F. et al. (2017). MO-Co@N-doped carbon (M = Zn or Co): vital roles of inactive Zn and highly efficient activity toward oxygen reduction/evolution reactions for rechargeable Zn–air battery. *Adv. Funct. Mater.* 27: 1700795.

102 Trudgeon, D.P., Qiu, K., Li, X. et al. (2019). Screening of effective electrolyte additives for zinc-based redox flow battery systems. *J. Power Sources* 412: 44–54.

103 Amendola, S., Johnson, L., Binder, M. et al. (2012). Electrically rechargeable, metal-air battery systems and methods. US2012021303.

104 Sumboja, A., Ge, X., Zheng, G. et al. (2016). Durable rechargeable zinc-air batteries with neutral electrolyte and manganese oxide catalyst. *J. Power Sources* 332: 330–336.

105 Clark, S., Latz, A., and Horstmann, B. (2017). Rational development of neutral aqueous electrolytes for zinc–air batteries. *ChemSusChem* 10: 4735–4747.

106 Liu, Z., Zein, S., Abedin, E., and Endres, F. (2014). Electrodeposition and stripping of zinc from an ionic liquid polymer gel electrolyte for rechargeable zinc-based batteries. *J. Solid State Electrochem.* 18: 2683–2691.

107 Chen, P., Zhang, K., Tang, D. et al. (2020). Recent progress in electrolytes for Zn–air batteries. *Front. Chem.* 8: 1–7.

108 Lahiri, A., Borisenko, N., and Endres, F. (2018). Electrochemical synthesis of battery electrode materials from ionic liquids. In: *Ionic Liquid II* (eds. B. Kirchner, E. Perlt), 55–83. Springer International Publishing.

109 Simons, T.J., Torriero, A.A.J., Howlett, P.C. et al. (2012). High current density, efficient cycling of Zn^{2+} in 1-ethyl-3-methylimidazolium dicyanamide ionic liquid: the effect of Zn^{2+} salt and water concentration. *Electrochem. Commun.* 18: 119–122.

110 Xu, M., Ivey, D.G., Xie, Z., and Qu, W. (2013). Electrochemical behavior of Zn/Zn(II) couples in aprotic ionic liquids based on pyrrolidinium and imidazolium cations and bis(trifluoromethanesulfonyl)imide and dicyanamide anions. *Electrochim. Acta* 89: 756–762.

111 Liu, S., Han, W., Cui, B. et al. (2017). A novel rechargeable zinc-air battery with molten salt electrolyte. *J. Power Sources* 28: 435–441.

112 Abbott, A.P., Capper, G., Davies, D.L. et al. (2003). Novel solvent properties of choline chloride/urea mixtures. *Chem. Commun.*: 70–71.

113 Narayanan, N.S.V., Ashokraj, B.V., and Sampath, S. (2010). Ambient temperature, zinc ion-conducting, binary molten electrolyte based on acetamide and zinc perchlorate: application in rechargeable zinc batteries. *J. Colloid Interface Sci.* 342: 505–512.

114 Abbott, A.P., Harris, R.C., Ryder, K.S. et al. (2011). Glycerol eutectics as sustainable solvent systems. *Green Chem.* 13: 82–90.

115 Kareem, M.A., Mjalli, F.S., Hashim, M.A., and Alnashef, I.M. (2010). Phosphonium-based ionic liquids analogues and their physical properties. *J. Chem. Eng. Data* 55: 4632–4637.

116 Nkuku, C.A. and LeSuer, R.J. (2007). Electrochemistry in deep eutectic solvents. *J. Phys. Chem. B* 111: 13271–13277.

117 Abbott, A.P., Boothby, D., Capper, G. et al. (2004). Deep eutectic solvents formed between choline chloride and carboxylic acids: versatile alternatives to ionic liquids. *J. Am. Chem. Soc.* 126: 9142–9147.

118 Jörissen, L. (2006). Bifunctional oxygen/air electrodes. *J. Power Sources* 155: 23–32.

119 Ren, S., Duan, X., Liang, S. et al. (2020). Bifunctional electrocatalysts for Zn-air batteries: recent developments and future perspectives. *J. Mater. Chem. A* 8: 6144–6182.

120 Pei, P., Wang, K., and Ma, Z. (2014). Technologies for extending zinc–air battery's cyclelife: a review. *Appl. Energy* 128: 315–324.

121 Müller, S., Striebel, K., and Haas, O. (1994). $La_{0.6}Ca_{0.4}CoO_3$: a stable and powerful catalyst for bifunctional air electrodes. *Electrochim. Acta* 39: 1661–1668.

122 Kim, J.H., Eom, S.W., Moon, S.I. et al. (2003). Effects of PTFE contents on characteristics of cathode for Zn-air batteries. *J. Korean Electrochem. Soc.* 6: 203–207.

123 Fang, Z.q., Hu, M., Liu, W.x. et al. (2006). Preparation and electrochemical property of three-phase gas-diffusion oxygen electrodes for metal air battery. *Electrochim. Acta* 51: 5654–5659.

124 Burchardt, T. and Becquet, A. (2012). Bifunctional air electrode. EP 2012/1977475.

125 Wu, M., Zhang, G., Wu, M. et al. (2019). Rational design of multifunctional air electrodes for rechargeable Zn–air batteries: recent progress and future perspectives. *Energy Storage Mater.* 21: 253–286.

126 Davari, E. and Ivey, D.G. (2018). Bifunctional electrocatalysts for Zn-air batteries. *Sustain. Energy Fuels* 2: 39–67.

127 Wang, Y.-J., Yuan, R., Ignaszak, A. et al. (2018). *Advanced Bifunctional Electrochemical Catalysts for Metal-Air Batteries.* CRC Press.

128 Shin, J., Seo, J.K., Yaylian, R. et al. (2020). A review on mechanistic understanding of MnO_2 in aqueous electrolyte for electrical energy storage systems. *Int. Mater. Rev.* 65 (6): 356–387.

129 Zhao, Q., Yan, Z., Chen, C., and Chen, J. (2017). Spinels: controlled preparation, oxygen reduction/evolution reaction application, and beyond. *Chem. Rev.* 117: 10121–10211.

130 Ge, X., Liu, Y., Goh, F.W.T. et al. (2014). Dual-phase spinel $MnCo_2O_4$ and spinel $MnCo_2O_4$/nanocarbon hybrids for electrocatalytic oxygen reduction and evolution. *ACS Appl. Mater. Interfaces* 6: 12684–12691.

131 Wang, D., Chen, X., Evans, D.G., and Yang, W. (2013). Well-dispersed Co_3O_4/Co_2MnO_4 nanocomposites as a synergistic bifunctional catalyst for oxygen reduction and oxygen evolution reactions. *Nanoscale* 5: 5312–5315.

132 Lee, D.U., Scott, J., Park, H.W. et al. (2014). Morphologically controlled Co_3O_4 nanodisks as practical bi-functional catalyst for rechargeable zinc-air battery applications. *Electrochem. Commun.* 43: 109–112.

133 Du, G., Liu, X., Zong, Y. et al. (2013). Co_3O_4 nanoparticle-modified MnO_2 nanotube bifunctional oxygen cathode catalysts for rechargeable zinc-air batteries. *Nanoscale* 5: 4657–4661.

134 An, T., Ge, X., Hor, T.S.A. et al. (2015). Co_3O_4 nanoparticles grown on N-doped vulcan carbon as a scalable bifunctional electrocatalyst for rechargeable zinc-air batteries. *RSC Adv.* 5: 75773–75780.

135 Prabu, M., Ketpang, K., and Shanmugam, S. (2014). Hierarchical nanostructured $NiCo_2O_4$ as an efficient bifunctional non-precious metal catalyst for rechargeable zinc-air batteries. *Nanoscale* 6: 3173–3181.

136 Pletcher, D., Li, X., Price, S.W.T. et al. (2016). Comparison of the spinels Co_3O_4 and $NiCo_2O_4$ as bifunctional oxygen catalysts in alkaline media. *Electrochim. Acta* 188: 286–293.

137 Zhang, Z., Zhou, D., Liao, J. et al. (2019). One-pot synthesis of Fe_3O_4/Fe/C by microwave sintering as an efficient bifunctional electrocatalyst for oxygen reduction and oxygen evolution reactions. *J. Alloys Compd.* 786: 134–138.

138 Ma, C., Xu, N., Qiao, J. et al. (2016). Facile synthesis of $NiCo_2O_4$ nanosphere-carbon nanotubes hybrid as an efficient bifunctional electrocatalyst for rechargeable Zn–air batteries. *Int. J. Hydrogen Energy* 41: 9211–9218.

139 Dai, Y., Yu, J., Cheng, C. et al. (2020). Mini-review of perovskite oxides as oxygen electrocatalysts for rechargeable zinc–air batteries. *Chem. Eng. J.* 397: 125516.

140 Wan, W., Liu, X., Li, H. et al. (2019). 3D carbon framework-supported CoNi nanoparticles as bifunctional oxygen electrocatalyst for rechargeable Zn-air batteries. *Appl. Catal., B* 240: 193–200.

141 He, Y., Liu, X., Yan, A. et al. (2019). Hybrid nanostructures of bimetallic NiCo nitride/N-doped reduced graphene oxide as efficient bifunctional electrocatalysts for rechargeable Zn-air batteries. *ACS Sustainable Chem. Eng.* 7: 19612–19620.

142 Chen, Y., Hu, S., Nichols, F. et al. (2020). Carbon aerogels with atomic dispersion of binary iron-cobalt sites as effective oxygen catalysts for flexible zinc-air batteries. *J. Mater. Chem. A.* 8: 11649–11655.

143 Zhu, P., Gao, J., and Liu, S. (2020). Facile in situ coupling CoFe/Co nanoparticles and N-doped carbon nanotubes/graphitic nanosheets as bifunctional oxygen electrocatalysts for rechargeable Zn-air batteries. *J. Power Sources* 449: 227512.

144 Wang, Q., Shang, L., Shi, R. et al. (2017). NiFe layered double hydroxide nanoparticles on Co,N-codoped carbon nanoframes as efficient bifunctional catalysts for rechargeable zinc–air batteries. *Adv. Energy Mater.* 7: 1–7.

145 Peng, Y. and Chen, S. (2018). Electrocatalysts based on metal@carbon core@shell nanocomposites: an overview. *Green Energy Environ.* 3: 335–351.

146 Li, Y. and Dai, H. (2014). Recent advances in zinc-air batteries. *Chem. Soc. Rev.* 43: 5257–5275.

147 Meng, H. and Shen, P.K. (2006). Novel Pt-free catalyst for oxygen electroreduction. *Electrochem. Commun.* 8: 588–594.

148 Chottiner, J. (1979). Size and weight graded multi-layered EAS diffusion electrodes. EP 1979/0110491 A2.

149 Buzzet, E.S. (1976). Metal/air cells and improved air electrodes for use therein. US 1976/3977901.

150 Henry, G. (1982). Air cathode for air depolarized cells. US 1982/4333993 A.

151 Lin, G. and Nguyenz, T.V. (2005). Effect of thickness and hydrophobic polymer content of the gas diffusion layer on electrode flooding level in a PEMFC. *J. Electrochem. Soc.* 152: A1942–A1948.

152 Shepard, V., Smalley, Y., and Bentz, R. (1994). Bifunctional metal-air electrode. US 1994/5306579.

153 Koshiba, N., HayakawaKeigo, H., and Ohta, M. (1984). Air cell electrode and process for preparing its catalyst. US 1984/4595643.

154 Golovin, N. (2002). Methods for making oxygen reduction catalyst using micelle encapsulation and metal-air electrode including said catalyst. US 2002/6428931.

155 Ndzebet, E., (2002). Manganese oxide based electrode for alkaline electrochemical system and method of its production. US 2002/6780347 B2.

156 Demczyk, B.G. and Liu, C.T. (1982). Potential distribution on a carbon-based bifunctional air electrode. *J. Electrochem. Soc.* 129: 1159–1164.

157 Liu, C.T. and Jackovitz, J.F. (1994). Bifunctional gas diffusion electrodes employing wettable, non-wettable layered structure using the mud-caking concept. US 1994/5318862.

158 Melchionna, M., Marchesan, S., Prato, M., and Fornasiero, P. (2015). Carbon nanotubes and catalysis: the many facets of a successful marriage. *Catal. Sci. Technol.* 5: 3859–3875.

159 Mueller, S., Holzer, F., Haas, O., and Arai, H. (1999). A study of carbon-catalyst interaction in bifunctional air electrodes for zinc-air batteries. *J. New Mater. Electrochem. Syst.* 2: 227–232.

160 Li, P.C., Hu, C.C., You, T.H., and Chen, P.Y. (2017). Development and characterization of bi-functional air electrodes for rechargeable zinc-air batteries: effects of carbons. *Carbon N. Y.* 111: 813–821.

161 Min, Y.J., Oh, S.J., Kim, M.S. et al. (2018). Effect of carbon properties on the electrochemical performance of carbon-based air electrodes for rechargeable zinc–air batteries. *J. Appl. Electrochem.* 48: 405–413.

162 Fu, J., Liang, R., Liu, G. et al. (2018). Recent progress in electrically rechargeable zinc-air batteries. *Adv. Mater.* 31: 1805230.

163 Cano, Z.P., Park, M.G., Lee, D.U. et al. (2018). New interpretation of the performance of nickel-based air electrodes for rechargeable zinc-air batteries. *J. Phys. Chem. C* 122: 20153–20166.

164 Toussaint, G., Stevens, P., Akrour, L. et al. (2010). Development of a rechargeable zinc-air battery. *ECS Trans.* 28: 25–34.

165 Seo, M.H., Park, M.G., Lee, D.U. et al. (2018). Bifunctionally active and durable hierarchically porous transition metal-based hybrid electrocatalyst for rechargeable metal-air batteries. *Appl. Catal., B* 239: 677–687.

166 Fu, G., Tang, Y., and Lee, J.-M. (2018). Recent advances in carbon-based bifunctional oxygen electrocatalysts for Zn-air Batteries. *ChemElectroChem* 5: 1424–1434.

167 Thippani, T., Mandal, S., Wang, G. et al. (2016). Probing oxygen reduction and oxygen evolution reactions on bifunctional non-precious metal catalysts for metal–air batteries. *RSC Adv.* 6: 71122–71133.

168 Lysgaard, S., Christensen, M.K., Hansen, H.A. et al. (2018). Combined DFT and differential electrochemical mass spectrometry investigation of the effect of dopants in secondary zinc–air batteries. *ChemSusChem* 11: 1933–1941.

169 Tomboc, G.M., Yu, P., Kwon, T. et al. (2020). Ideal design of air electrode—a step closer toward robust rechargeable Zn–air battery. *APL Mater.* 8: 050905.

170 Li, Y., Gong, M., Liang, Y. et al. (2013). Advanced zinc-air batteries based on high-performance hybrid electrocatalysts. *Nat. Commun.* 4: 1805.

171 Tsai, T. and Vartek, A. (2003). Rechargeable metal air electrochemical cell incorporating collapsible cathode assembly. WO 2003/061057 A1.

172 Huang, J., Yang, Z., Wang, R. et al. (2015). Zn-Al layered double oxides as high-performance anode materials for zinc-based secondary battery. *J. Mater. Chem. A* 3: 7429–7436.

173 Hongfei, L., Xu, C., Han, Y. et al. (2015). Enhancement on cycle performance of Zn anodes by activated carbon modification for neutral rechargeable zinc ion batteries. *J. Electrochem. Soc.* 162: A1439–A1444.

174 Masri, M.N. and Mohamad, A.A. (2013). Effect of adding carbon black to a porous zinc anode in a zinc-air battery. *J. Electrochem. Soc.* 160: A715–A721.

175 Othman, R., Yahaya, A.H., and Arof, A.K. (2002). A zinc-air cell employing a porous zinc electrode fabricated from zinc-graphite-natural biodegradable polymer paste. *J. Appl. Electrochem.* 32: 1347–1353.

176 Wang, L.-P., Li, N.-W., Wang, T.-S. et al. (2017). Conductive graphite fiber as a stable host for zinc metal anodes. *Electrochim. Acta* 244: 172–177.

177 Ozgit, D., Hiralal, P., and Amaratunga, G.A.J. (2014). Improving performance and cyclability of zinc-silver oxide batteries by using graphene as a two dimensional conductive additive. *ACS Appl. Mater. Interfaces* 6: 20752–20757.

178 Zhou, Z., Zhang, Y., Chen, P. et al. (2019). Graphene oxide-modified zinc anode for rechargeable aqueous batteries. *Chem. Eng. Sci.* 194: 142–147.

179 Devyatkina, T.I., Gun'ko, Y.L., and Mikhalenko, M.G. (2001). Development of ways to diminish corrosion of zinc electrode. *Russ. J. Appl. Chem.* 74: 1122–1125.

180 Masri, M.N. and Mohamad, A.A. (2009). Effect of adding potassium hydroxide to an agar binder for use as the anode in Zn-air batteries. *Corros. Sci.* 51: 3025–3029.

181 Mainar, A.R., Colmenares, L.C., Grande, H.J., and Blázquez, J.A. (2018). Enhancing the cycle life of a Zinc–air battery by means of electrolyte additives and zinc surface protection. *Batteries* 4: 46. https://doi.org/10.3390/batteries4030046

182 Stock, D., Dongmo, S., Walther, F. et al. (2018). Homogeneous coating with an anion-exchange ionomer improves the cycling stability of secondary batteries with zinc anodes. *ACS Appl. Mater. Interfaces* 10: 8640–8648.

183 Stock, D., Dongmo, S., Miyazaki, K. et al. (2018). Towards zinc-oxygen batteries with enhanced cycling stability: the benefit of anion-exchange ionomer for zinc sponge anodes. *J. Power Sources* 395: 195–204.

184 Lee, C.W., Sathiyanarayanan, K., Eom, S.W. et al. (2006). Effect of additives on the electrochemical behaviour of zinc anodes for zinc/air fuel cells. *J. Power Sources* 160: 161–164.

185 Sun, K.E.K., Hoang, T.K.A., Doan, T.N.L. et al. (2017). Suppression of dendrite formation and corrosion on zinc anode of secondary aqueous batteries. *ACS Appl. Mater. Interfaces* 9: 9681–9687.

186 Zhu, J. and Zhou, Y. (1998). Effects of ionomer films on secondary alkaline zinc electrodes. *J. Power Sources* 73: 266–270.

187 Jo, Y.N., Kang, S.H., Prasanna, K. et al. (2017). Shield effect of polyaniline between zinc active material and aqueous electrolyte in zinc-air batteries. *Appl. Surf. Sci.* 422: 406–412.

188 Gan, W., Zhou, D., Zhou, L. et al. (2015). Zinc electrode with anion conducting polyvinyl alcohol/poly(diallyldimethylammonium chloride) film coated ZnO for secondary zinc air batteries. *Electrochim. Acta* 182: 430–436.

189 Huang, J. and Yang, Z. (2015). A one-pot method to prepare a ZnO/Ag/polypyrrole composite for zinc alkaline secondary batteries. *RSC Adv.* 5: 33814–33817.

190 Trygve, B. (2006). Zinc electrode comprising an organic gelling agent and an organic binder. EP 1715536 A2.

191 Ghavami, R. and Rafiei, Z. (2006). Performance improvements of alkaline batteries by studying the effects of different kinds of surfactant and different derivatives of benzene on the electrochemical properties of electrolytic zinc. *J. Power Sources* 162: 893–899.

192 Zhu, J., Zhou, Y., and Yang, H. (1997). Effects of lanthanum and neodymium hydroxides on secondary alkaline zinc electrode. *J. Power Sources* 69: 169–173.

193 Lee, S.M., Kim, Y.J., Eom, S.W. et al. (2013). Improvement in self-discharge of Zn anode by applying surface modification for Zn-air batteries with high energy density. *J. Power Sources* 227: 177–184.

194 Hu, C. and Chang, C. (2004). Anodic stripping of zinc deposits for aqueous batteries: effects of anions, additives, current densities, and plating modes. *Mater. Chem. Phys.* 86: 195–203.

195 Yang, H., Meng, X., Yang, E. et al. (2004). Effect of La addition on the electrochemical properties of secondary zinc electrodes. *J. Electrochem. Soc.* 151: A389–A393.

196 Wang, R., Yang, Z., Yang, B. et al. (2014). Superior cycle stability and high rate capability of Zn–Al–In-hydrotalcite as negative electrode materials for Ni–Zn secondary batteries. *J. Power Sources* 251: 344–350.

197 Cho, Y.-D. and Fey, G.T.-K. (2008). Surface treatment of zinc anodes to improve discharge capacity and suppress hydrogen gas evolution. *J. Power Sources* 184: 610–616.

198 Zheng, Y., Wang, J., Chen, H. et al. (2004). Effects of barium on the performance of secondary alkaline zinc electrode. *Mater. Chem. Phys.* 84: 99–106.

199 Haas, O. and Wesemael, J.V. (2009). Secondary batteries – metal-air systems|zinc-air: electrical recharge. In: *Encyclopedia of Electrochemical Power Sources* (ed. J. Garche), 384–392. Elsevier. ISBN: 9780444527455.

200 Zhang, C., Wang, J., Zhang, L. et al. (2001). Study of the performance of secondary alkaline pasted zinc electrode. *J. Appl. Electrochem.* 31: 1049–1054.

201 Schmid, M. and Willert-Porada, M. (2018). Zinc particles coated with bismuth oxide based glasses as anode material for zinc air batteries with improved electrical rechargeability. *Electrochim. Acta* 260: 246–253.

202 Otani, T., Fukunaka, Y., and Homma, T. (2017). Effect of lead and tin additives on surface morphology evolution of electrodeposited zinc. *Electrochim. Acta* 242: 364–372.

203 Yuan, Y., Tu, J., Wu, H. et al. (2006). Effect of ZnO nanomaterials associated with $Ca(OH)_2$ as anode material for Ni–Zn batteries. *J. Power Sources* 159: 357–360.

204 Wang, R., Yang, Z., Yang, B. et al. (2014). A novel alcohol-thermal synthesis method of calcium zincates negative electrode materials for Ni-Zn secondary batteries. *J. Power Sources* 246: 313–321.

205 Schmid, M. and Willert-Porada, M. (2017). Electrochemical behavior of zinc particles with silica based coatings as anode material for zinc air batteries with improved discharge capacity. *J. Power Sources* 351: 115–122.

206 Yang, B., Yang, Z., and Wang, R. (2014). Facile synthesis of novel two-dimensional silver-coated layered double hydroxide nanosheets as advanced anode material for Ni-Zn secondary batteries. *J. Power Sources* 251: 14–19.

207 Caldeira, V., Rouget, R., Fourgeot, F. et al. (2017). Controlling the shape change and dendritic growth in Zn negative electrodes for application in Zn/Ni batteries. *J. Power Sources* 350: 109–116.

208 Yager, T. (2012). Rechargeable zinc air battery and anode. WO 2012/044319 A1.

209 Phillips, J., Mohanta, S., Maske, C. et al. (2010). Pasted zinc electrode for rechargeable nickel-zinc batteries. US 291, 439 A1.

210 Hilder, M., Winther-Jensen, B., and Clarka, N. (2009). Paper-based, printed zinc–air battery. *J. Power Sources* 194: 1135–1141.

211 Hilder, M., Winther-Jensen, B., and Clark, N. (2012). The effect of binder and electrolyte on the performance of thin zinc-air battery. *Electrochim. Acta* 69: 308–314.

212 Martin, A. (2012). High-performance zinc-base anodes for zinc-air battery application. Weinheim: Wiley.

213 Liu, Y., Yang, Z., Xie, X. et al. (2015). Layered double oxides nano-flakes derived from layered double hydroxides: preparation, properties and application in zinc/nickel secondary batteries. *Electrochim. Acta* 185: 190–197.

214 Ko, J.S., Geltmacher, A.B., Hopkins, B.J. et al. (2019). Robust 3D Zn sponges enable high-power, energy-dense alkaline batteries. *ACS Appl. Energy Mater.* 2: 212–216.

215 Parker, J.F., Chervin, C.N., Nelson, E.S. et al. (2014). Wiring zinc in three dimensions re-writes battery performance – dendrite-free cycling. *Energy Environ. Sci.* 7: 1117–1124.

216 Rolison, D.R., Parker, J.F., Long, J.W., and Ko, J.S. (2019). Zinc electrodes for batteries. WO 2019089505 A2 20190509.

217 Jayasayee, K., Clark, S., King, C. et al. (2020). Cold sintering as a cost-effective process to manufacture porous zinc electrodes for rechargeable zinc-air batteries. *Processes* 8: 1–12.

218 Kritzer, P. and Cook, J.A. (2007). Nonwovens as separators for alkaline batteries. *J. Electrochem. Soc.* 154: A481–A494.

219 Lewandowski, A., Skorupska, K., and Malinska, J. (2000). Novel poly(vinyl alcohol)-KOH-H_2O alkaline polymer electrolyte. *Solid State Ionics* 133: 265–271.

220 Dewi, E.L., Oyaizu, K., Nishide, H., and Tsuchida, E. (2003). Cationic polysulfonium membrane as separator in zinc-air cell. *J. Power Sources* 115: 149–152.

221 Wu, G.M., Lin, S.J., You, J.H., and Yang, C.C. (2008). Study of high-anionic conducting sulfonated microporous membranes for zinc-air electrochemical cells. *Mater. Chem. Phys.* 112: 798–804.

222 Saputra, H., Othman, R., Sutjipto, A.G.E., and Muhida, R. (2011). MCM-41 as a new separator material for electrochemical cell: application in zinc-air system. *J. Membr. Sci.* 367: 152–157.

223 Arora, P. and Zhang, Z. (2004). Battery separators. *Chem. Rev.* 104: 4419–4462.

224 Wu, G.M., Lin, S.J., and Yang, C.C. (2006). Preparation and characterization of high ionic conducting alkaline non-woven membranes by sulfonation. *J. Membr. Sci.* 284: 120–127.

225 Cheng, F. and Chen, J. (2012). Metal-air batteries: from oxygen reduction electrochemistry to cathode catalysts. *Chem. Soc. Rev.* 41: 2172–2192.

226 Lee, H.-J., Lim, J.-M., Kim, H.-W. et al. (2016). Electrospun polyetherimide nanofiber mat-reinforced, permselective polyvinyl alcohol composite separator membranes: a membrane-driven step closer toward rechargeable zinc-air batteries. *J. Membr. Sci.* 499: 526–537.

227 Hwang, H.J., Chi, W.S., Kwon, O. et al. (2016). Selective ion transporting polymerized ionic liquid membrane separator for enhancing cycle stability and durability in secondary zinc-air battery systems. *ACS Appl. Mater. Interfaces* 8: 26298–26308.

228 Konovalova, A., Stock, D., Schröder, S. et al. (2020). Partially methylated polybenzimidazoles as coating for alkaline zinc anodes. *J. Membr. Sci.* 610: 118254.

229 Yan-Jie, W., Jinli, Q., Ryan, B., and Jiujun, Z. (2013). Alkaline polymer electrolyte membranes for fuel cell applications. *Chem. Soc. Rev.* 42: 5768–5787.

230 Huang, C.-L., Wang, P.-Y., and Li, Y.-Y. (2020). Fabrication of electrospun CO_2 adsorption membrane for zinc-air battery application. *Chem. Eng. J.* 395: 125031.

231 Drillet, J., Holzer, F., Kallis, T. et al. (2002). Influence of CO_2 on the stability of bifunctional oxygen electrodes for rechargeable zinc–air batteries and study of different CO_2 filter materials. *Phys. Chem. Chem. Phys.* 3: 368–371.

232 Cheng, H.H. and Tan, C.S. (2006). Reduction of CO_2 concentration in a zinc/air battery by absorption in a rotating packed bed. *J. Power Sources* 162: 1431–1436.

8

Al-Ion Battery

David Muñoz-Torrero[1], Rebeca Marcilla[1], and Edgar Ventosa[1,2]

[1] *IMDEA Energy, Electrochemical Processes Unit, Av. Ramon de la Sagra 3, Móstoles, 28935, Spain*
[2] *University of Burgos, Department of Chemistry, Plaza Misael Bañuelos s/n, Burgos, E09001, Spain*

8.1 Introduction

Recently, aluminum (Al) has being attracting much scientific interest for rechargeable battery systems. This growing interest is triggered by its gravimetric charge storage capacity (2.98 Ah g^{-1}) that is comparable to that of Li (3.86 Ah g^{-1}), its volumetric capacity that is four times higher than that of Li (8.04 and 2.06 Ah cm^{-3}, respectively), and its abundance (the third most abundant element in the Earth's crust (8.1%) after oxygen [46.6%] and silicon [27.7%]) (Figure 8.1). Indeed, Al is used in a wide range of commercial applications due to its abundance, the wide extension of bauxite deposits ($Al(OH)_3$), and its low cost (30 times lower than the price of metallic Li) [1, 2]. Furthermore, Al is safe, environmentally friendly, and easy to handle. On the other hand, the standard redox potential of Al (−1.7 V vs. SHE) is more positive than that of other metals, e.g. Li (−3 V vs. SHE) and Na (−2.8 V vs. SHE) leading to lower cell voltage. Overall, the distinct features of Al-based batteries may be of high interest for a number of energy storage applications.

8.2 Historical Development of Aluminum Batteries

Due to the intriguing properties of Al metal, the scientific community has been interested in developing Al batteries for decades. The first attempts to design Al-based batteries dates from the 1940s. However, this technology picked up great interest in the 2010s decade becoming one of the hot topic in post-lithium battery research. Chronologically, primary batteries based on Al irrupted first. Over time, efforts have been focused on developing rechargeable systems based on Al. In this section, the evolution from the pioneering works on primary and non-rechargeable Al batteries to the development of state-of-the-art rechargeable Al-ion batteries is discussed.

Battery Technologies: Materials and Components, First Edition. Edited by Jianmin Ma.

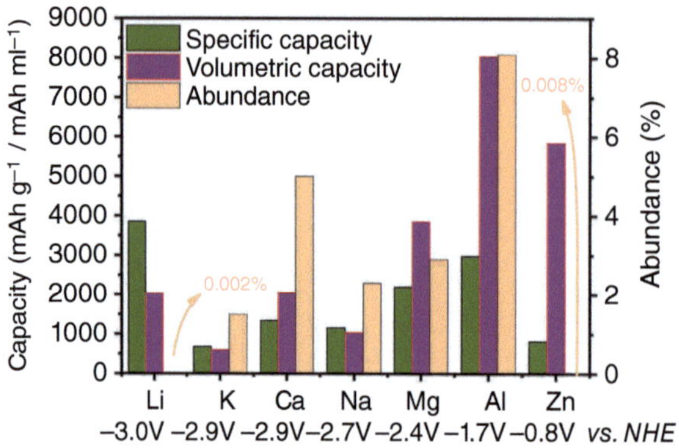

Figure 8.1 Theoretical properties and abundance of metal proposed as anode for metal-based batteries. Source: Muñoz-Torrero et al. [1]. Reproduced with permission of Royal Society of Chemistry.

8.2.1 Primary Aluminum Batteries: Aqueous Systems

As mentioned above, the first attempt to develop Al-based batteries dates from 1948 according to some works [2, 3]. This first Al-based battery was composed of a metallic Al anode and a chlorine cathode in an aqueous electrolyte having a medium-high voltage of 2.45 V. This proof of concept promoted the research of primary (no rechargeable) devices based on this technology. From 1951 to 1967, several patents were filed based on cell designs that use Al as anode, MnO_2 or carbon among others as cathode, and alkaline aqueous electrolytes [4–8]. However, the deployment of Al battery based on aqueous electrolytes has not been commercially successful due to several limitations of this technology. One of these limitations is the formation of a passivating oxide film onto the aluminum surface that leads the decay of the discharge voltage of the battery and, thus, the cell efficiency. Another limitation, and probably the most important one, was the impossibility of developing secondary batteries due to the hydrogen evolution reaction (HER) during the charging process, since the redox potential of Al electrodeposition occurs at much lower potential than that of the HER (−1.7 V and − 0.83 V vs. SHE, respectively, at pH 14). Therefore, the use of non-aqueous electrolyte for Al-based batteries attracted the attention of the scientific community in order to develop high-energy and low-cost rechargeable energy storage devices.

8.2.2 Rechargeable Aluminum Batteries: Non-aqueous Systems

One of the first approaches for developing rechargeable Al-based batteries dates from 1972. In this year, a research group from the Tycho Laboratories in Massachusetts designed a cell consisting of an Al anode and a vitreous carbon-positive electrode for the Cl_2 evolution [9, 10]. The electrolyte media was an eutectic molten salt of $AlCl_3$-KCl-NaCl as non-aqueous system at moderate temperature (90–150 °C). They used a mixture enriched in $AlCl_3$ since the molten salt is basic and dominated by Cl^- and $AlCl_4^-$ species when the concentration of $AlCl_3$ is below 50%, which prevents the electrodeposition of Al in this medium. However, the mixture become acidic when the concentration of $AlCl_3$ is higher, and the dominant species are $AlCl_4^-$ and $Al_2Cl_7^-$, which enable reversible Al electrodeposition following the next equation:

$$4Al_2Cl_7^- + 3e^- \Leftrightarrow Al + 7AlCl_4^- \tag{8.1}$$

In 1979, Fouletier and Armand reported for the first time the intercalation into graphite of the aluminum chloride compounds present in molten salts based on $AlCl_3$ as electrolyte [11]. They found that the intercalation behavior slightly varied as a function of the type of graphite used. However, in all cases, the intercalation and deintercalation reactions were found at 1.8–1.9 and 1.6–1.7 V, respectively, vs. $Al/AlCl_4^-$. The significance of this study relies on the fact that it was the first report on the intercalation of aluminum chloride compounds into graphite. Some years later, this concept attracted attention and was further studied. Mohandas et al. reported reversible intercalation at higher redox potentials of 2.1–2.2 and 1.8–1.7 V, respectively [12]. These redox potentials are very similar to those obtained in the state-of-the-art rechargeable Al batteries nowadays. The authors suggested that this reaction is related to the intercalation of Cl^- and $AlCl_4^-$ present in the electrolyte in the graphite [12].

In 1980, Koura et al. proposed another worth-mentioning concept in the development of Al-based electrochemical cells. In order to eliminate the production of toxic chlorine, an Al battery based on iron sulfide (FeS_2) as cathode and a molten salt as electrolyte was demonstrated [13]. The proposed battery had a discharge voltage of ~1.2 V. After that, some studies explored different reaction mechanisms using chloride-conjugated cathodes, such as Ni/NiCl, Mn/MnCl, or V/VCl [14].

However, the formation of Al dendrites during battery operation due to the instability of the metal chlorides and the concerns associated with the moderate temperature necessary for operating these types of batteries were acknowledged in some of these reports. In order to inhibit the dendrite formation and decrease the temperature for Al electrodeposition, Grjotheim [15] and Koura [16] added tetramethylammonium chloride (TMACl) and 1-butylpyridinium chloride (BPC) to the base molten salt, respectively. In the first case, the dendrite formation was inhibited successfully and the Al electroplating temperature was decreased. And, in the second case, charging process of the battery was performed at lower temperature (90 °C). These two reports, in which an organic halide mixed with $AlCl_3$-based molten salt was employed, led to the use of chloroaluminate ionic liquids (CILs) in Al-based batteries some years later [15, 16]. CILs are formed by adding $AlCl_3$ to an organic halide, such as 3-dialkylimidazolium chloride, alkylpyridinium chloride, or quaternary ammonium chloride.

One of the first and most representative examples using CILs as electrolyte was the Al-Cl_2 rechargeable battery developed by Gifford in 1988 [17], which used an acidic mixture between $AlCl_3$ and 1,2-dimethyl-3-propylimidazolium chloride (DMPrICl) (molar ratio, *r*, higher than 1). This cell showed a discharge voltage of 1.7 V and a specific capacity of 35–40 mAh g^{-1} (based in graphite mass) at 25 °C. Years later, a rechargeable Al-based battery system that used an Al anode and a $FeCl_3$ cathode infused in reticulated vitreous carbon was investigated using $AlCl_3$/EMImCl electrolyte at room temperature [18]. The battery showed a discharge voltage (1.4–1.7 V), but the dissolution of $FeCl_3$ in the electrolyte led to large battery self-discharge and limited rechargeability of the cell.

During early 1990s, polyaniline polymer was proposed as cathode for Al-based batteries based on a mixture between $AlCl_3$ and ethyl-methylimidazolium chloride (EMImCl), as electrolyte [19]. This rechargeable battery showed an average charge and discharge voltage of 1.5 and 1 V, respectively. However, the dissolution of polyanaline due to the acidic character of the electrolyte during cycling hindered the development of this system. This study constituted one of the first attempts at using polymer as cathode for Al-based batteries.

One of the most representative studies using graphitic cathode materials and CIL electrolytes in rechargeable Al batteries was reported in 2013 by J.V. Rani et al. [20]. This work was one of the first attempts to use fluorinated natural graphite (FNG) as cathode for rechargeable Al batteries using CILs. The interest in FNG for Al-based batteries originated from the high performance of fluoride carbons demonstrated in Mg and Li batteries [21, 22]. FNG has higher interlayer space between graphene sheets, higher electrical conductivity, and higher surface area than natural graphite. The discharge mechanism proposed by the authors can be summarized in the following equations:

$$\text{Anode reaction: } Al + AlCl_3BR^- - 3e^- \rightarrow Al_2Cl_6BR^- \tag{8.2}$$

$$\text{Cathode reaction: } Al_2Cl_6BR^- + C_n + e^- \rightarrow AlCl_3BR^- + C_n[Al] \tag{8.3}$$

The authors assumed that Al is reversibly electrodeposited in the negative electrode and Al^{3+} reversibly intercalates into the FNG structure. The assumption for the negative electrode was later shown to be correct, whereas for the positive electrode was not. The idea of the intercalation of Al^{3+} was refused in 2015 by Lin et al. [23]. In this article, the reaction involved in the positive electrode was demonstrated to be the reversible intercalation of $AlCl_4^-$ in the graphite, following the next equation [23]:

$$\text{Cathode reaction: } C_n[AlCl_4] \rightarrow AlCl_4^- + C_n + e^- \tag{8.4}$$

This work marked a turning point in the research of secondary Al batteries using graphitic materials as cathode. All of these advances, summarized in Figure 8.2, have driven the different Al-based battery technologies developed up to date.

8.3 Electrolytes for Al-Based Batteries

The development of electrolytes that are able to carry out the electrodeposition of Al reversibly at room temperature marked a turning point in the development of secondary Al-based batteries. One of the first groups of electrolytes for Al-based batteries operating at room temperature was the CILs. Nevertheless, the scientific community has continued to devote much effort in exploring alternative electrolytes that are able to reversibly carry out the Al electrodeposition at room temperature due

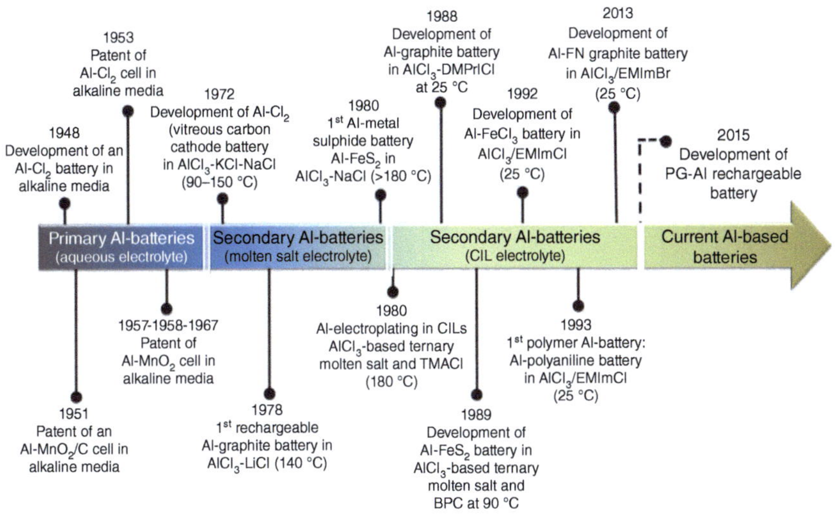

Figure 8.2 Historical highlights of the development of current Al-based batteries.

the high-cost and reactivity problems related to CILs. In this section, the CILs and the alternative electrolytes for Al electrodeposition are briefly discussed.

8.3.1 Al Electrodeposition in CILs and Their Use in Rechargeable Al-Based Batteries

CILs belong to the first generation of ionic liquids. They are formed by adding $AlCl_3$ to an organic halide such as 3-dialkylimidazolium chloride, alkylpyridinium chloride, or quaternary ammonium chloride. These CILs are liquid over a large range of temperatures including room temperature, and they have some promising properties in comparison with molten salts or solvents, such as very low vapor pressure or wide electrochemical potential window. The cost benefits related to the lower operation temperature of the resulting battery with respect to molten salts electrolytes has driven the interest in CILs as electrolyte for electrodeposition of Al at room or nearly room temperature [24–26]. Furthermore, these CILs show adjustable Lewis acidity over a wide range of the molar ratio $AlCl_3$ to organic halide (MCl) [27]. In this way, when the molar ratio, r, of $AlCl_3/MCl$ is lower than 1 (Lewis base), the predominant species in the electrolyte is the anion $AlCl_4^-$. In this media, the Al electrodeposition does not take place within the electrochemical stability window of the electrolyte [28]. On the contrary, when the molar ratio of $AlCl_3/MCl$ is higher than 1, the dominant species in the electrolyte is the $Al_2Cl_7^-$, which is electrochemically active for the Al electrodeposition following reaction (Eq. (8.1)).

In fact, these CILs electrolytes have been used in the majority of rechargeable Al-based batteries reported so far, being considered as the standard electrolyte for this technology. This is due to their good electrochemical performance, showing a high electrochemical stability window and low overpotential for the Al plating (electrodeposition) and stripping (oxidation). However, the use of this electrolyte in Al-based batteries entails some drawback: (i) CILs have a high hygroscopic character and high reactivity with ambient humidity, provoking the violent evolution of acidic vapors, (ii) CILs cost is currently high, constituting one of the costliest elements in Al-based batteries, [29] and (iii) the corrosive behavior of CILs as electrolyte makes necessary the use of scarce and expensive current collectors such as glassy carbon [23], tantalum (Ta) [30], molybdenum (Mo) [31], or tungsten (W) [32].

8.3.2 Al Electrodeposition Using Alternative Electrolytes and Their Use in Rechargeable Al-Based Batteries

The above-mentioned problems have driven the scientific interest in developing alternative electrolytes, which are presumably more stable. Such electrolytes are based in amides (such as urea or acetamide) [33], glymes [34, 35], bistrifluorosulfonyl-based ILs [36], and among others [37, 38]. Generally, these alternatives can be classified into three groups: (i) air- and water-stable ILs, (ii) organic-based electrolytes, and (iii) deep eutectic solvents (DESs) (Figure 8.3).

- Air- and water-stable ILs for Al electrodeposition. These non-haloaluminated ILs are formed solely by ions, and they are liquid at room temperature or nearly to room temperature (<100 °C). The main advantage of this family of ILs is the presence of anions such as $TFSI^-$, FSI^-, or $BF_4{}^-$, among others, that are less sensitive to moisture than the anions present in CILs. However, the necessary addition of $AlCl_3$ when used as electrolyte entails the formation of two phases or the solidification of the electrolyte, requiring separation processes or application of temperature. Moreover, their cost is higher than that of CILs [36].
- Organic-based electrolytes for Al electrodeposition. Some organic solvents, such as glymes, enable electrodeposition of Al when a certain molar fraction of $AlCl_3$ is added [34, 35]. Theoretically, this group of electrolyte possesses several advantages for their application in Al-based batteries such as their lower cost, lower reactivity, and safety properties. However, the use of this type of electrolytes in rechargeable Al-based batteries currently remains unexplored.
- DESs for Al electrodeposition. These electrolytes are obtained by mixing two or three generally cheap and safe components that are able to form a eutectic mixture through hydrogen bond interaction. In particular, DESs formed by the mixture between $AlCl_3$ and one amide (such as urea or acetamide) have shown promising results for Al electrodeposition [33]. In the case of $AlCl_3$:Urea electrolyte, the Al electrodeposition can be performed from the ($[AlCl_2{\cdot}urea_2]^+$) cation [33] as well as from the $Al_2Cl_7{}^-$ anion when the molar ratio, *r*, is higher than 1. Therefore, two pathways for the Al electrodeposition are possible:

$$4Al_2Cl_7^- + 3e^- \Leftrightarrow Al + 7AlCl_4^- \qquad (8.1)$$

$$2[AlCl_2 \cdot (Urea)_2]^+ + 3e^- \Leftrightarrow Al + AlCl_4^- + 4Urea \qquad (8.5)$$

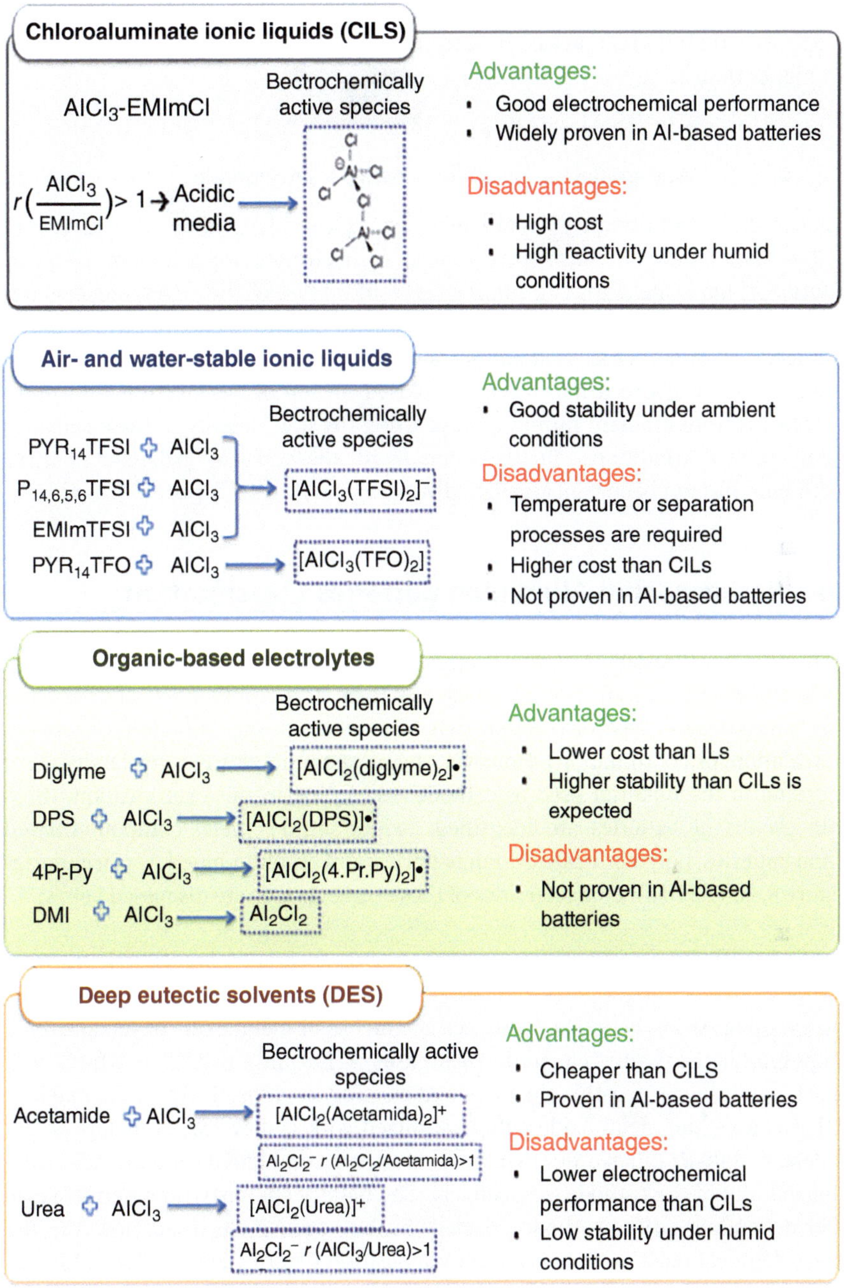

Figure 8.3 Electrolytes for Al electrodeposition.

Similarly, in $AlCl_3$:Acetamide, the Al electrodeposition can be performed from the ([$AlCl_2$·Acetamide]+) cation as well as from the $Al_2Cl_7^-$ anion when the molar ratio, *r*, is higher than 1:

$$4Al_2Cl_7^- + 3e^- \Leftrightarrow Al + 7AlCl_4^- \qquad (8.1)$$

$$2[AlCl_2 \cdot \text{Acetamide}]^+ + 3e^- \Leftrightarrow Al + AlCl_4^- + 2\text{Acetamide} \qquad (8.6)$$

Due to these promising results obtained for Al electrodeposition, $AlCl_3$:Urea and $AlCl_3$:acetamide mixture have been reported as electrolytes for secondary Al-based batteries at lab scale, showing promising results [39–41]. However, some issues such as high reactivity to humidity under ambient conditions are not avoided. For instance, A.P. Abbott et al. studied the Al electrodeposition using $AlCl_3$:Urea [42]. In such study, a hydrocarbon layer was placed on top of the electrolyte in order to protect it from ambient humidity, suggesting the low stability of the electrolyte under ambient conditions. On the other hand, the corrosive behavior of these electrolytes has not been well elucidated yet.

8.4 Rechargeable Aluminum Batteries Classification

There is not a well-established way to classify Al-based batteries. Some classification can be based on the electrolyte used such as CIL or DES. Another way of classifying is by analyzing the type of reaction that takes place in the cathode (conversion, intercalation, or coordination). However, the most common way is to classify them according to the material used as cathode. According to this classification, three main groups of batteries are identified: oxides/sulfides-metal compound-based Al-ion batteries, polymer-based Al-ion batteries, and graphite-based Al-ion batteries (Figure 8.4). The main characteristics of these three groups are discussed below.

8.4.1 Metal Oxide/Sulfide-Based Aluminum Batteries

This group of Al-ion batteries is featured by the use of metal oxide or metal sulfide compounds in the positive electrode. In all cases, Al foil and CIL $AlCl_3$:EMImCl with a $r > 1.1$ are used as negative electrode and electrolyte, respectively. In the majority of these cases, the metal oxide/sulfide is mixed with carbonaceous materials and binders, mainly PTFE, in order to improve electrical conductivity and provide structural stability of the cathode, respectively. The battery chemistry and performance differ depending on the cathode selected. In the negative electrode, however, the electrochemical reaction is the Al electrodeposition described in the Eq. (8.1), during the charge. There are two types of cathodes in this group: intercalation cathodes and conversion cathodes.

- Intercalation cathodes are defined as materials capable of intercalating Al^{3+} or Al-based complexes present in the electrolyte into their structure without losing their crystallographic order. This group includes materials such as vanadium dioxide (VO_2) nanorods [43], vanadium oxide/carbon composite

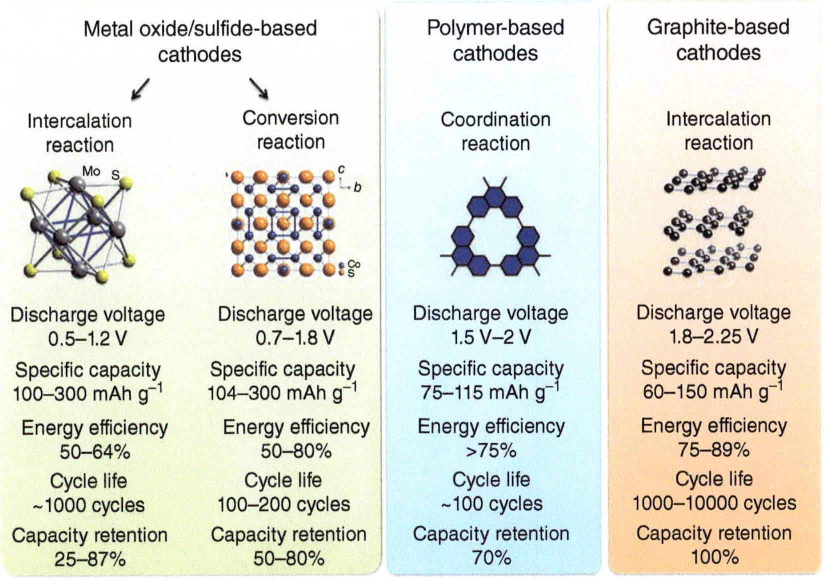

Figure 8.4 Cathode materials classification for Al-based batteries.

(V_2O_5/C) [44], chevrel phases molybdenum sulfide (Mo_6S_8) [45], reduced graphene oxide supported tin sulfide (rGO-SnS2) [46], molybden disulfide (MoS_2) [31], binder-free cobalt sulfide/carbon nanotube composite (Co_9S_8@CNT) [47] film, or flexible tin sulfide (SnS) [48] among others.

The resulting batteries show a charge and discharge voltages in the range of 0.7–1.8 V and 0.5–1.2 V, respectively. They can afford excellent values of capacity during the first cycle (from 100 to 300 mAh g^{-1}) with high coulombic efficiency (CE) (>90%). However, the voltage efficiency (VE) varies between 55% and 71%. This value is quite low and makes the energy efficiency (EE), estimated as the product of VE and CE, of the technology poor in comparison with other Al-based battery technologies. Furthermore, although the capacity obtained in the first cycle was high, these materials suffered from a strong capacity fading during cycling. The capacity fading varied from 0.025% to 1.6% per cycles in most of the cases. The poor capacity retention has limited the evaluation of long cycling performance to a couple of reports, in which 1000 cycles was achieved maintaining a commendable capacity [47, 48]. In addition, the C-rate capability of Al-based batteries based on intercalation metal oxide/sulfide compounds is rather poor, leading to significant differences between in capacities at different current densities. Table 8.1 shows a summary of the most relevant materials and their main electrochemical properties.

- Conversion cathodes are defined as materials that can react with Al-based complexes reversibly or can host Al^{3+} or Al-based complexes by altering its crystallographic structure. Materials such as Ni_2S_3/graphene nanoflakes

Table 8.1 Summary of intercalation cathodes in metal-based Al-ion batteries.

Material	Charge voltage (V)	Discharge voltage (V)	Specific capacity First cycle (mAh g^{-1})	Discharge positive reaction	Capacity retention (%)	Cycle life
VO_2 nanorod	0.7	0.5	100	$VO_2 + xAl^{3+} + 3e^- \rightarrow Al_xVO_2$	75	100
V_2O_5/C	1.25	0.75	150	$Al^{3+} + V_2O_5 + 3e^- \rightarrow AlV_2O_5$	53	30
Mo_6S_8	0.7	0.55/0.33	85	$8[Al_2Cl_7]^- + 6e^- + Mo_6S_8 \rightarrow Al_2Mo_6S_8 + 14AlCl_4^-$	82	50
rGO-SnS_2	1.8/1.25	1.6/0.7	275	$SnS_2[AlCl_4]_n + ne^- \rightarrow SnS_2 + nAlCl_4^-$	25	100
MoS_2	1.8	1	250	$MoS_2 + xAl^{3+} + 3e^- \rightarrow Al_xMoS_2$	28	100
Co_9S_8@CNT	1.4	0.8	120	No mechanism is proposed	87	6000
Flexible SnS	1.8	1.2	300	$SnS[AlCl_4]_n + ne^- \rightarrow SnS + nAlCl_4^-$	75	1000

composite [49], NiS nanobelts [30], Co_9S_8 nanosheets [50], CuS microsphere [51], $CoSe_2$/rGO [52], and $SbSe_2$ nanorods with N-doped rGO [53] belong in this group.

Generally, Al-based batteries based on conversion cathodes suffer from similar issues as those based on intercalation cathodes. However, higher charge and discharge voltages are generally obtained by using conversion cathode, in comparison with intercalation cathodes. In this case, the charge and discharge voltages range from 1 to 2.1 V and from 0.7 to 1.8 V, respectively. Despite this, VE of batteries using conversion cathodes (55–71%) is as low as the VE obtained with intercalation cathodes. However, it is important to highlight that some studies have shown voltage efficiencies above 80% improving the EE in comparison with the rest of materials proposed [30, 50]. Similarly to intercalation material-based Al-based batteries, the capacity of the first cycle using conversion cathodes is very high for every material studied, from 104 to 300 mAh g^{-1}, but the capacity fading during cycling (0.1–3.1% per cycle) is even higher than that for intercalation cathodes generally. As a result, Al-ion batteries based on conversion cathode suffer from poor cyclability (100–500 cycles) in comparison with other technologies such as Li-ion batteries (1000–3000 cycles). Table 8.2 shows a summary of the most relevant materials and their main electrochemical properties.

In conclusion, the main research objective in the field of metal oxide/sulfide-based cathode should be focused on searching new materials and/or strategies that improve the capacity retention as well as the VE. Although recent studies have reported promising results in this direction, important efforts should be made in order to deeply understand the degradation mechanisms that cause these problems.

8.4.2 Polymer-Based Aluminum Batteries

Although conductive polymers are one of the least explored cathode materials for Al-based batteries, recent results revealed very appealing properties. Polypyrrole and polythiophene [54], poly(nitropyrene-*co*-pyrene) [55], and redox-active triangular phenanthrenequinone-based macrocycles with graphite flakes (PQ-Δ-HY) [56] are the most representative examples of this group of materials. Despite the conductive character of these polymers, they are mixed with carbonaceous materials and binders in order to improve electrical conductivity and structural stability of the cathode. The negative electrode and electrolyte in all these reports are Al foil and CIL $AlCl_3$:EMImCl with a r between 1.3 and 1.5, respectively. The storage mechanism involved in the battery depends on the type of polymer used. But, generally, Al electrodeposition/stripping and coordination reaction between polymer and Al-based complexes from the electrolyte take place in the negative and positive electrode, respectively.

Depending on the material used, batteries using polymer-based cathodes present charge and discharge voltages in the range of 1.6–2.2 and 1.5–2 V, respectively. These values are higher than those for metal-based cathodes. Furthermore, the resulting batteries present good VE that ranges from 80% to 90% and high CE

Table 8.2 Summary of conversion cathodes in metal-based Al-ion batteries.

Material	Charge voltage (V)	Discharge voltage (V)	Specific capacity First cycle ($mAh\ g^{-1}$)	Discharge positive reaction	Capacity retention (%)	Cycle life
Ni_2S_3/graphene	1.9/1.5	1	350	$3NiS + 6NiCl_2 + Al_2S_3 + 6e^- \rightarrow Ni_2S_3 + 2Al^{3+} + 12Cl^-$	17	100
NiS nanobelts	1.7/1.15	0.9	104	$9NiS + 2Al^{3+} + 6e^- \rightarrow 3Ni_3S_2 + Al_2S_3$	90	100
Co_9S_8 nanosheets	1.4	0.8	250	No mechanism is proposed	100	250
CuS microsphere	1.5	0.7	240	$6CuS + 2Al^{3+} + 6e^- \rightarrow 3Cu_2S + Al_2S_3$	37.5	20
$CoSe_2$/rGO	2/1	1.8/0.9	300	$CoSe_2 + mAl^{3+} + 3me^- \rightarrow Al_mCo_nSe_2 + (1 - n)Co$	50	500
$SbSe_2$/N-rGO	2.1/1.6/1.2	1.7/1.5/0.7	300	$Sb_2Se_3 + Al^{3+} + 3e^- \rightarrow Al_2Se_3 + 2Sb$ at (0.7 V) $Se + 2e^- \rightarrow Se^{2-}$ at (1.7 and 1.5 V)	33	500

(>95%), which leads to good EE values of the batteries (>75% in most of the cases). In addition, relevant specific capacities in the range of 75–115 mAh g^{-1} during discharge process were reported depending on the polymer used. The capacity fading during cycling was found to be between 0.02% and 0.27% per cycle. Although some reports demonstrated more than 100 charge–discharge cycles [55], the cyclability of these materials is still quite limited in comparison with other technologies. Table 8.3 shows a summary of the most relevant materials and their main electrochemical properties. In spite of the promising properties of polymer-based Al-ion batteries, there are aspects that require significant improvements. The cycle stability is probably the most pressing issue, although improvements in specific capacity, C-rate capability, and up-scaling are necessary.

8.4.3 Graphite-Based Aluminum Batteries

The main characteristic of this group of Al-based batteries is the use of graphitic materials in the positive electrode. The negative electrode and the electrolyte of the battery are Al foil and CIL or DESs, respectively. The reaction mechanism of this class of batteries is the reversible Al electrodeposition onto the negative electrode or anode (Al foil) and the reversible intercalation of $AlCl_4^-$ anion in the interlayer space between graphene layers of graphite. Among all the types of Al-based batteries, graphite-based batteries have attracted most of the scientific interest, driven by the promising properties of graphite as cathode.

The graphite-based Al batteries using CILs as electrolytes have shown the highest discharge voltage, i.e. two plateaus located at 2.25 and 1.8 V with slight differences depending on the graphitic material used. Furthermore, these graphite-based Al-based batteries show the best VE (80–90%). The specific capacities of graphitic materials range from 60 to 150 mAh g^{-1} and the CE is high (~99%). Consequently, the EE usually surpasses that obtained by metal or polymer-based Al-based batteries. Besides, these batteries have an excellent C-rate capability even for higher current density (e.g. 5 C), which makes it capable of operating at high power rates. In addition, the cycle life of the battery surpasses 1000 cycles reaching impressive values (>10 000 cycle) in some reports [57, 58], which are the highest among all types of Al-based battery technologies.

There are two examples in literature for the use of DESs electrolytes, i.e. $AlCl_3$:Urea and $AlCl_3$:Acetamide, for graphite-based Al batteries. On the one hand, $AlCl_3$:Urea as electrolyte for Al-based batteries showed medium discharge voltages (1.9–1.6 V), moderate capacity (76 mAh g^{-1}), high values of CE (>99%), but low cycle life (200–500 cycles) [40, 41]. On the other hand, in 2018, N. Canvert et al. proposed the use of eutectic mixtures of $AlCl_3$:Acetamide as alternative electrolyte for Al-based batteries. In this case, the battery showed medium discharge voltages (2–1.6 V), moderate capacity (77 mAh g^{-1}), but low cycle life (50 cycles) and low values of CE (90%) [39]. It should be noted that the battery performance using DESs is lower than that using CILs (60–150 mAh g^{-1}, CE >99%, >1000 cycles and 2.2–1.8 V) when using the same cathode material [59].

Table 8.3 Summary of polymer-based Al-ion batteries.

Material	Charge voltage (V)	Discharge voltage (V)	Specific capacity First cycle (mAh g^{-1})	Discharge positive reaction	Capacity retention (%)	Cycle life
Polypyrrol	1.9	1.5	75	$3(\text{Polypyrrol}^+\text{AlCl}_4^-) + 3e^- \rightarrow 3\text{Polypyrrol} + 3\text{AlCl}_4^-$	73	100
Polythiophene	1.8/1.6	1.6/1	85	$3(\text{Polythiophene}^+\text{AlCl}_4^-) + 3e^- \rightarrow 3\text{Polythiophene} + 3\text{AlCl}_4^-$	82	100
Poly(pyrene)	1.9	1.7	100	$\text{Polypyrene}_x[\text{AlCl}_4] + e^- \rightarrow x\text{Polypyrene} + \text{AlCl}_4^-$	80	1000
PQ-Δ-HY	2.2/1.6	2/1.4	115	$\text{PQ} - \Delta + 3\text{Al}_2\text{Cl}_6 + 3e^- \rightarrow (\text{PQ} - \Delta^{3\cdot}) \cdot 3\text{AlCl}_2 + 3\text{AlCl}_4^-$ $\text{C}_n(\text{AlCl}_4^-) + e^-$	95	100

The historical findings and advances achieved for graphite-based batteries using CILs are revised in the following section, since it is the most widely studied system, and it delivers so far the best electrochemical performance.

8.5 Rechargeable Aluminum Batteries Based on Graphitic Cathodes

Graphite-based Al-ion batteries or graphite-Al batteries (GABs) are the battery technology based on Al that more scientific interest has attracted due to the promising properties of graphite as cathode. Research has focused on improving the battery capacity values by using graphitic materials with different graphitic degrees, surfaces areas, and structures (Figure 8.5).

In this section, the most important findings in this field are summarized, focusing on how the properties of the graphitic materials influence the battery performance (specific capacity, cycle life, C-rate capability, etc.). It should be noted that the electrolyte used in all of the research studies discussed in this section is a CIL, specifically $AlCl_3$:EMImCl mixtures. In fact, the vast majority of the reports on the secondary Al-based batteries reported use this CIL, which make it the standard electrolyte in Al-based batteries.

8.5.1 Carbon Paper

In 2015, Sun et al. used a graphite carbon paper (GCP) as cathode for Al-based batteries. Al foil and $AlCl_3$:EMImCl ($r = 1.3$) were used as negative electrode and electrolyte, respectively [60]. The galvanostatic charge–discharge of the resulting battery was performed between 0.2 and 2.35 V, showing clear plateaus of charge and discharge at 2.1 and 1.8 V, respectively. The battery achieved a promising and stable capacity value of 70 mAh g^{-1} during 100 cycles, maintaining a CE close to 100%.

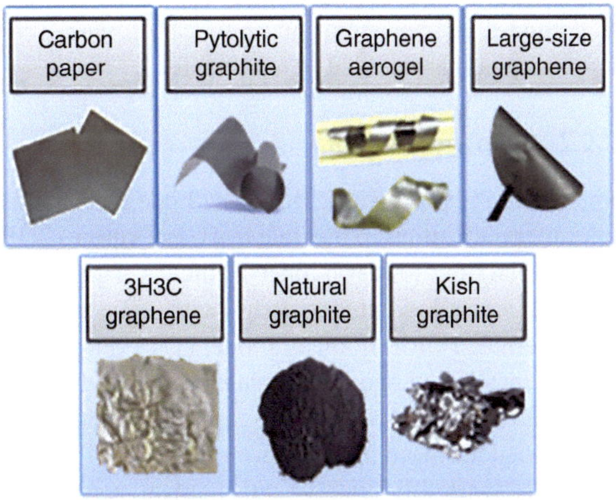

Figure 8.5 Most representative carbonaceous materials studied in Al-ion batteries based on graphitic cathodes.

The authors suggested that the storage mechanism of the battery was similar to the well-known "rocking-chair" for LIBs with the difference that, in this case, two kinds of ions participate in the reaction (Al^{3+} and $[Al_aCl_b]^-$). It was claimed that during the discharge the Al is oxidized into Al^{3+} in the anode and, together with the aluminum chloride coordination anions present in the electrolyte $[Al_aCl_b]^-$ ($AlCl_4^-$ and $Al_2Cl_7^-$), they are transported from anode to cathode. Then, these two types of ions intercalate simultaneously in the interlayer graphene sheets space, forming $Al_xCl_y^-$ anion. The global scheme of the battery during the discharge can be summarized in the next two equations:

$$\text{Anode reaction: Al} - 3e^- \rightarrow Al^{3+} \tag{8.7}$$

$$\text{Cathode reaction: } Al^{3+} + [Al_aCl_b]^- + C_n + e^- \rightarrow C_n[Al_xCl_y] \tag{8.8}$$

However, the proposed anode reaction differs from the plating mechanism of Al reported a decade before in similar CIL [61]. In such pioneering works, $Al_2Cl_7^-$ anion (not Al^{3+} cation) was identified as the only electroactive aluminum containing species in the electrolyte [61]. Although reaction mechanism suggested for the cathode was originally not well elucidated, the intercalation of big aluminum complexes was proven by the increment of the interlayer space of graphite (from 3.428 to 3.667 Å) during the cycling [60]. One year later, in 2016, the same research group constructed the first industrialized prototype of graphite-based Al-based battery using the same material as cathode [62]. The battery showed a capacity of around 1.3 Ah with CE close to 95% during 120 cycles. It is important to note that the authors proposed a different storage mechanism for the battery performance in that work. In good agreement with previous research studies on Al plating/stripping, the authors suggested that the Al dissolution in the anode includes the formation of $Al_2Cl_7^-$ complexes and proposed that the intercalation in the cathode involved $AlCl_4^-$ species. In fact, the same mechanism had been reported one year before by H. Dai's group [23]. Such article represents a turning point in the graphite-based Al-based battery research and provided some basic information about the technology, such as the reaction mechanism, the technology advantages, and the requirements of the cathode material. The materials employed as cathodes in such notorious publication were the pyrolytic graphite (PG) and graphitic foam as detailed in next section.

8.5.2 Pyrolytic Graphite

The use of PG was investigated as cathode for Al-based batteries is one of the most relevant publications in the field by H. Dai's group [23]. PG might be used as a "model" material since it was obtained by means of graphite pyrolysis and showed a highly graphitic crystallographic structure with almost negligible degree of defects. The battery assembled with Al foil as anode, PG as cathode, and $AlCl_3$:EMImCl ($r = 1.3$) as electrolyte showed two charge plateaus at 2 and 2.3 V and two discharge plateaus at 2.25 and 1.9 V (Figure 8.6a). The capacity was 60–66 mAh g^{-1} with 98% of efficiency maintained for more than 200 cycles.

In this work, the storage mechanism occurring at the cathode was deeply investigated for the first time by means of X-ray diffraction (XRD) (Figure 8.6b).

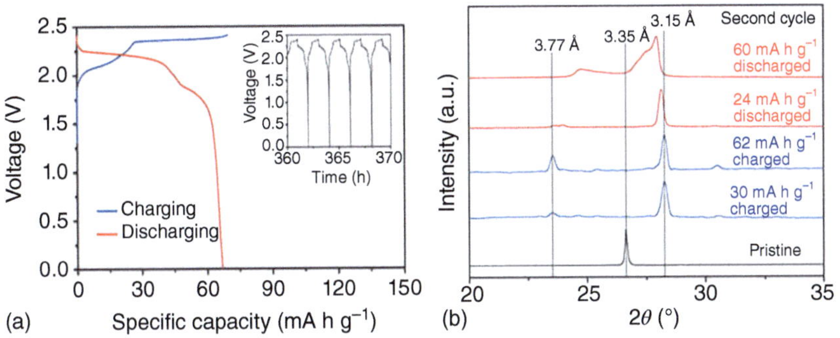

Figure 8.6 (a) Charge–discharge voltage profile for graphite-based Al-ion battery and (b) ex situ XRD graphite-based Al-ion battery during charge–discharge process. Source: Lin et al. [23]. Reproduced with permission of Springer Nature.

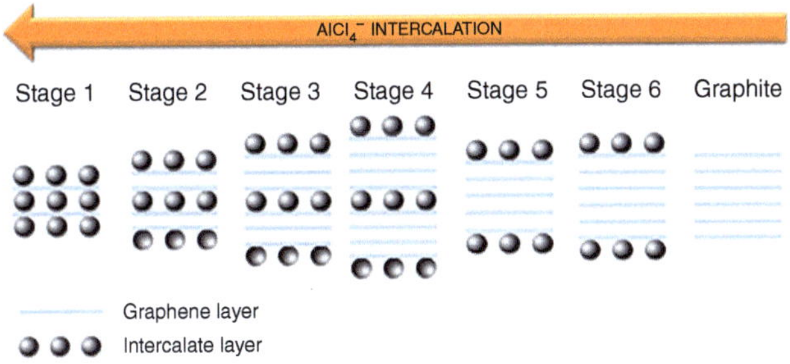

Figure 8.7 Visual representation of graphite intercalation compound (GIC).

During the charge step, the graphite peak (002) at θ = 26.55° (*d*-spacing = 3.35 Å) vanished while two new peaks appeared at θ = 28.2° (*d*-spacing = 3.15 Å) and at θ = 23.5° (*d*-spacing = 3.77 Å). This suggested that some graphene layers were strained and others are separated due to the intercalation of an anion between two graphene layers. Furthermore, the peak separation analysis suggested a stage 4 graphite intercalation compounds (GICs), which is the number with a gallery height (spacing between adjacent graphitic host layers) of ~5.7 Å, indicating that the $AlCl_4^-$ anions (size ~5.28 Å) were intercalated between graphene layers in a distorted state. The stage number is known as the number of graphene layers between two anions intercalated (see Figure 8.7). During discharge, the graphite recovered its characteristic peak but slightly shifted to a higher degree and with a broad shoulder indicating irreversible changes in the graphitic crystallographic structure.

Therefore, the reaction mechanism of the battery was established like the reversible Al electrodeposition in the negative electrode from $Al_2Cl_7^-$ anion species and the reversible intercalation of $AlCl_4^-$ anion in the graphite structure. The discharge reactions can be written as follows:

$$\text{Anode reaction: } Al + 4AlCl_4^- - 3e^- \rightarrow 7Al_2Cl_7^- \quad (8.9)$$

$$\text{Cathode reaction: } C_n[AlCl_4] \rightarrow AlCl_4^- + C_n + e^- \quad (8.4)$$

8.5.3 Graphitic Foam

In the same work discussed above, the authors also investigated graphitic foam as 3D intercalation cathode [23]. This material was synthesized by chemical vapor deposition (CVD) using nickel foam as template. This material had a high surface area with faster ion diffusion. This unique open and microporous structure improved the kinetics of the battery, achieving 7500 cycles at 75 C without any decay in the capacity ($60\,mAh\,g^{-1}$), maintaining a CE of ~97%. This specific capacity implied energy and power values as high as $40\,Wh\,kg^{-1}$ and $3000\,W\,kg^{-1}$, respectively [23].

Further confirmation of the intercalation mechanism was performed by X-ray photoelectron spectroscopy (XPS) using this graphitic foam. The XPS revealed the oxidation of the carbon graphite due to the intercalation of anions by the appearing of the sp^3 shoulder in the C 1s region in charged stage (Eq. (8.4)). Furthermore, the XPS performed for Al 2p and Cl 2p showed peaks in the charged graphitic foam corroborating the intercalation of $AlCl_4^-$ (Figure 8.8).

After this work, the mechanism proposed by H. Dai's group was settled on invariably and widely accepted by the research community for graphite-based Al-ion batteries. Following research studies have either deepened in the intercalation process focusing on the material used as intercalation cathode or developed new in situ or operando methods to get further insights about the intercalation process [32, 59, 63].

It is important to note that two different strategies have been implemented in order to improve the battery performance using graphitic foams. On the one hand, X. Yu et al. enhanced the specific capacity of the battery by developing a graphitic foam with nanosized pores created by Ar^+ plasma etching to increase the number of accessible edges (defects) [58]. On the other hand, Y. Wu et al. developed a graphitic foam in which the intercalation kinetics are improved due to the low degree of defects and functional groups as well as the vertical orientation of the graphene layers [64].

X. Yu et al. [58] developed a similar graphene foam called Graphene Nanoribbons on Highly Porous 3D Graphene (GNHPG) for its use in Al-ion battery. This graphitic material was also synthesized by CVD using nickel foam as template but, in this study, the GNHPG was additionally subjected to an Ar^+ plasma etching in order to create nanosized pores. The increasing number of accessible edges resulted in

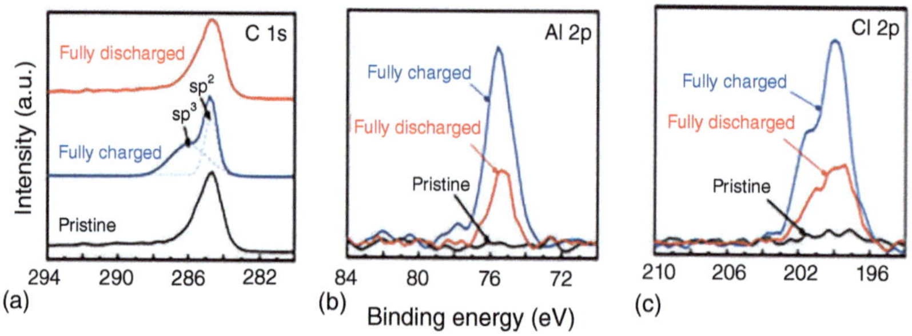

Figure 8.8 Ex situ XPS for pristine, charged and discharged graphite foam for (a) C 1s peak, (b) Al 2p peak, and (c) Cl 2p peak. Source: Lin et al. [23]. Reproduced with permission of Springer Nature.

enhanced capacity and C-rate capability of the material. Furthermore, the authors claimed that distribution of nanovoids across the entire surface of the cathode improves the easy percolation of $AlCl_4^-$ with a limited penetration depth. The battery showed an impressive capacity of 123 mAh g^{-1} during 10 000 cycles at 5 A g^{-1}, maintaining CEs of 98%.

In 2016, Y. Wu et al. proposed an alternative but complex method to CVD for synthesizing monolithic 3D graphitic foams (3DGFs). This material was made of vertically aligned few-layered graphene sheets oriented perpendicular to a current collector and having low density of defects or functional oxygen groups. In order to prepare it, PG electrode was fully charged by combining galvanostatic and long potentiostatic experiments. After that, a rapid thermal heating at 1000 °C provoked the expansion of chloroaluminate anion intercalated graphite (~120-fold expansion factor). And finally, an electrochemical expansion using the resulting material to carry out the HER was performed. Eventually, the obtained 3DGF presented an expansion of ~600-fold from the original PG thickness (from 17 μm to about 1 cm) [64]. Assembled batteries showed a capacity of 60 mAh g^{-1} at 12 A g^{-1} during 4000 cycles, maintaining a CE close to 100%. This demonstrated that the vertically oriented aligned graphene sheets with low-defect degree as well as low degree of functional oxygen groups facilitated the electrochemical diffusion of the anion involved in the positive reaction.

8.5.4 Graphene-Based Cathode

H. Chen et al. [65] proposed a novel defect-free graphene aerogels (GAs) electrode based on completely crystallized sp^2 carbons in the atomic structure cathode. Reduced GA was annealed to GA and compressed mechanically to get a flexible graphitic paper. Three annealing temperatures were studied: 2000, 2500, and 3000 °C, obtaining GA-2000, GA-2500, and GA-3000, respectively. The high-resolution transmission electron microscopy (HRTEM) revealed that the higher the annealing temperature, the lower the defects degree. In fact, the GA-3000 revealed the absence of defects. Authors reported several advantages of such GA cathode that is porous but yet presents continuous structure: (i) elimination of inactive defects which facilitates the fast intercalation of large anions, (ii) presence of more active sites for energy storage, and (iii) enhanced electrical conductivity. Al-based batteries assembled using GA as cathode showed that the lower the defects in the crystallographic structure, the better the ion diffusion. In this manner, the battery based on GA-3000 showed the highest capacity (~100 mAh g^{-1}), the best C-rate capability, and the lowest overpotential (Figure 8.9).

In this work, the effect of the cutoff charge voltage in the capacity and CE was for the first time analyzed. Cut-off voltages from 2.1 to 2.65 V were investigated using GA-3000 as cathode. It was observed that the capacity increased from 40 to ~110 mAh g^{-1}, following a logarithmic trend when the cutoff voltages increased from 2.1 to 2.65 V. The CE remained stable (~99%) up to 2.51 V which was selected as the ideal cut-off voltage due to high capacity and proper efficiency. At higher cut-offs than 2.51 V, the CE decreased drastically probably to current collector corrosion

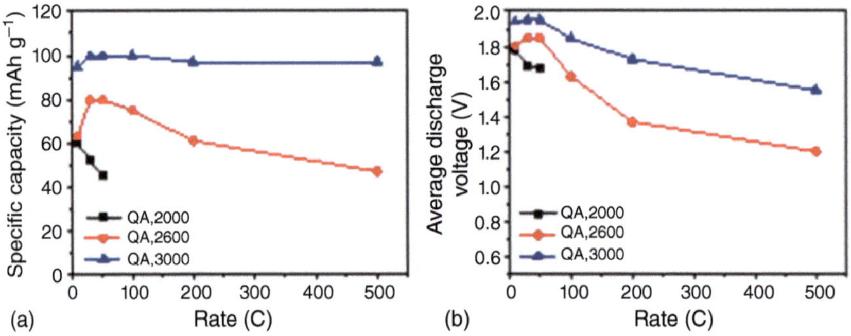

Figure 8.9 C-rate capability (a) and discharge average voltage (b) for GA cathodes materials. Source: Chen et al. [65]. Reproduced with permission of WILEY-VCH.

or electrolyte degradation (chlorine evolution reaction) that takes place at ~2.5 V vs. Al [29, 66]. Finally, GA-3000 showed a stable capacity as high as 100 mAh g^{-1} from 2 to 50 C with unprecedented high C-rate performance (97 mAh g^{-1}, 1.55 V at 500 C, and 74 mAh g^{-1} at 1000 C) and excellent cycle stability during 25 000 cycles. It is important to highlight that this research study demonstrated that GABs with very high capacity and good C-rate capability can be obtained by using GA-based cathodes with negligible defects.

In 2017, L. Zhang et al. evaluated how the size of highly graphitic materials in ab plane and c directions influences the battery performance, using large-size graphene (L-graphene), short-size graphene (S-graphene), large-size graphite (L-graphite), and short-size graphite (S-graphite) as cathode [57]. Results showed that large-size materials (L-graphene and L-graphite) reached higher capacities than short-size materials (S-graphene and S-graphite). In particular, L-graphene displayed the best performance, achieving a capacity of 87 mAh g^{-1} during 4500 cycles with a CE close to 100%. This better performance of L-graphene in comparison to graphite was related to two factors. Firstly, the L-graphene presented crystallographic structures with lower defects degree, and consequently, higher electrical conductivity. Secondly, few-layers graphene permitted to accommodate the anion intercalated within the interlayer space between graphene sheets easily, decreasing the activation energy barrier for the intercalation.

In this work, the intercalation process was investigated in detail through ex situ XRD analysis; taking samples at different stages of the charge–discharge process (see Figure 8.10). The main finding of this study was that during the first plateau, the separation peaks corresponds to the formation of the stage 5 and the second plateau is related to the conversion from stage 5 to stage 4. The intercalation gallery height of stage 4 GICs is ≈5.5 Å that is close to the size of $AlCl_4^-$ (5.28 Å), which is in good agreement with the study performed for PG although some differences in numbers can be found (Figure 8.6b).

Interestingly, the authors claimed that S-graphene-based cathode suffered an exfoliation process related to the (de)intercalation procedure during cycling. This process, which influences the long-term battery performance negatively, was found to be mitigated by using S-graphite or L-graphene materials as cathode.

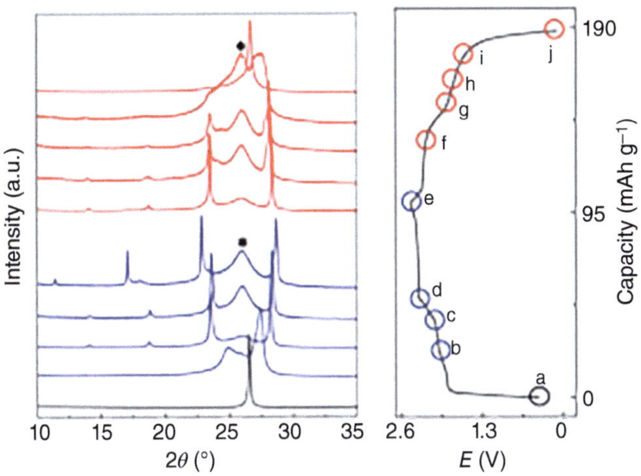

Figure 8.10 Ex situ XRD of L-graphene during charge–discharge cycle. Source: Zhang et al. [57]. Reproduced with permission of WILEY-VCH.

The authors concluded that decreasing the size of vertical dimension (*c* direction) strongly facilitates the kinetics of anions (de)intercalation while increasing the size of horizontal dimension (*ab* plane) improves the flexibility of graphitic materials enhancing cycling stability.

H. Chen et al. [67] developed a cathode called thihigh tricontinuous (3H3C) graphene film, to improve the C-rate capability, capacity, and lifetime of Al-based rechargeable batteries based on graphitic materials. The authors imposed four requirements that the cathode material must comply with: (i) highly crystallized defect-free graphene lattices, (ii) continuous electron-conducting matrix, (iii) a high mechanical strength to avoid the disintegration during cycling and, finally, (iv) interconnected channels for high electrolyte permeability and fast reaction. According to those requirements, the authors synthesized a flexible graphene film by chemical reduction of a graphene oxide (GO) film, followed by a high-temperature annealing (2850 °C) that led to the formation of channels due to the deoxygenating reaction during the annealing process. Al-ion batteries based on this cathode material, named 3H3C, showed an impressive C-rate capability with stable capacity values of about 120 mAh g^{-1} up to current densities as high as 100 A g^{-1}. Furthermore, the battery was able to retain such high capacity value for 250 000 cycles with CE as high as 98%. However, it is important to realize that, due to the high current densities employed during cycling (100 A g^{-1}), this quarter-million of cycles only translates in 300 hours of operation (12 days).

Other novel materials, such as graphene mesh [68], graphene microflowers [69], graphene nanoplatelets [70], graphene-coated activated carbon fiber cloth [71], or, more recently, defect-free soft carbon [72], have been also reported as cathode for Al-based batteries, which showed similar or lower performance than the materials formerly described.

8.5.5 Graphite Flakes-Based Cathodes

D.Y. Wang et al. explored for the first time the use of natural graphite flakes (NG) as cathode for Al-based batteries [59]. The resulting battery showed a discharge capacity of ~100 mAh g^{-1} (at 1.8 C) which remained stable during 1100 cycles with high CE (~99.5%). In comparison with other synthetic materials, this natural graphite showed high surface area as well as low-defect density which resulted in high-performance cathodes. Despite that the intercalation mechanism was previously proposed by the same group, in this report, such mechanism was deeply studied by both ex situ XRD and in situ Raman spectroscopy. Ex situ XRD (Figure 8.11a) revealed that, similar to PG [23], the graphitic peak (002) split up into two peaks during the charge, demonstrating the intercalation of $AlCl_4^-$. However, contrary to PG cathode, the position and the shape of the graphitic peak in NG cathode are completely recovered after the discharge. This indicates that the (de)intercalation reaction is highly reversible and structural disorder during charge–discharge cycling does not occur in NG. On the other hand, in situ Raman spectroscopy showed that the G band centered at 1586 cm^{-1} vanished during the first plateau of charging (0.5–2.37 V), while a new peak appeared at 1607 cm^{-1} indicating the graphite intercalation. In the course of the second plateau of charging, from 2.37 to 2.45 V, the peak located at 1607 cm^{-1} shifted to 1630 cm^{-1}. During the discharge stage, this process was found to be fully reversible (Figure 8.11b).

In 2017, M.V. Kovalenko and coworkers explored the use of the most basic and inexpensive form of graphite, NG flakes, as cathode for Al-based batteries [32]. In addition, ^{27}Al solid-state NMR was used as a new method to evidence the intercalation of $AlCl_4^-$. The charged flakes showed a single peak at 103 ppm, while the non-charge flakes showed no peak, demonstrating the incorporation of $AlCl_4^-$ species into the graphite structure during the charging process.

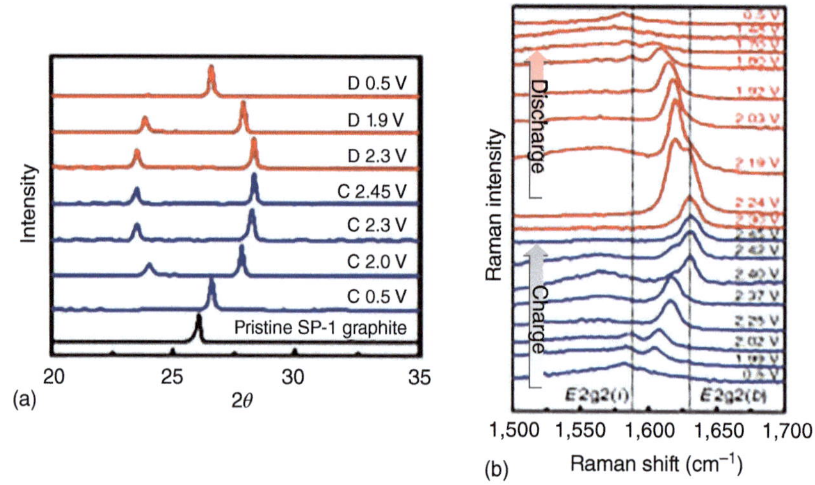

Figure 8.11 (a) Ex situ XRD and (b) in situ Raman spectroscopy of NG flakes during charge–discharge cycle. Source: Wang et al. [59]. Licensed under CC BY 4.0.

The GAB assembled with graphitic natural flakes-based cathode presented a capacity as high as 132 mAh g^{-1} with energy efficiencies higher than 92% when cycled at low current densities (100 mA g^{-1}). However, this material showed a moderate C-rate capability with specific capacity decreasing down to 40 mAh g^{-1} at moderate current densities of 2.5 A g^{-1}. To overcome this issue, the authors proposed a constant-current constant-voltage (CCCV) charging protocol in order to combine high capacities and high-power densities for the first time in this field. The authors claimed that with this protocol, which combines constant current with constant voltage charging steps, the secondary reactions that take place during the charge, presumably chlorine evolution reaction, are avoided. This charging method led to an improved discharge capacity of 150 mAh g^{-1} that remained stable from 50 mA g^{-1} up to 2.5 A g^{-1}. It is important to point out that this is the highest capacity value obtained for Al-based battery based on graphitic materials. In addition, due to the high mass loading of the electrode (10 mg cm^{-2}), this Al-based battery also showed an excellent areal capacity of 1.5 mAh cm^{-2}, being much higher than previous reported values (0.4–0.6 mAh cm^{-2}) [23, 59, 73].

S. Wang et al. [74] studied the usage of a highly crystalline synthetic graphite, which is a secondary product obtained from the steelmaking industry. This material called Kish graphite presents a flaky structure and can be easily produced at industrial scale. Large-sized Kish graphite flakes showed the highest crystalline structure among other graphitic materials studied the same work (natural long graphite flakes, natural short graphite flakes, highly oriented PG, potato-shaped graphite particles, and acetylene black) [74]. Batteries assembled using Kish graphite showed values of specific capacities varying from 125 to 142 mAh g^{-1} in a range of current densities of 0.05–10 A g^{-1}. Furthermore, the battery kept a stable capacity of 125 mAh g^{-1} during 200 cycles with EE close to 80% at 10 A g^{-1}. The values of areal and specific capacity are quite similar to those obtained in the previous work [32], but using an industrial secondary product.

8.6 Conclusions

The theoretical charge storage capacity and volumetric capacity of Al (2.98 Ah g^{-1} and 8.04 Ah cm^{-3}, respectively), its abundance, and its low cost have triggered the scientific interest in developing rechargeable batteries based on Al. Since the 1940s when the first systems of primary batteries based on Al emerged until now, the Al-based battery technology has become one of the most promising post-lithium battery technologies. One of the most important breakthroughs that marked a turning point in the development of secondary Al-based batteries was the use of CILs as electrolytes, since this type of electrolytes are able to carry out the electrodeposition of Al reversibly at room temperature. This fact permitted to avoid the use of molten salts at high temperature, commonly used as electrolyte from 1972 to 1988 in the first rechargeable Al-based battery prototypes. However, some problems related to the use of CILs as electrolytes such as their corrosive character or their high cost have driven research on alternatives electrolytes for carrying out the Al electrodeposition.

Among the alternatives, DESs have been used showing promising results. Mixtures based on urea or acetamide and $AlCl_3$ have been used in rechargeable Al-based batteries. Nevertheless, the electrochemical performance of batteries based on DESs has not surpassed that obtained by using CILs as electrolyte, and the main advances in Al-based technologies have been reached using CILs as electrolyte.

Different cathode materials have been explored. Oxides/sulfides-metal compound have been developed based on intercalation and conversion chemistries. This type of rechargeable Al-based batteries has shown an average discharge voltage of 0.5–1.2 V, a specific capacity that varies from 100 to 300 mAh g^{-1}, an energy efficiency of 50–64% and a cycle life of 1000 cycles in the case of intercalation cathodes, and an average discharge voltage that varies between 0.8 and 1.7 V, a specific capacity of 104 to 300 mAh g^{-1}, an energy efficiency of 50–80%, and a cycle life of 100–200 cycles in the case of conversion cathodes. Other research groups have explored polymer material as coordination cathodes obtaining an average voltage of 1.5–2 V, a specific capacity of 100–300 mAh g^{-1}, an energy efficiency of 50–64%, and a cycle life of 1000 cycles. And finally, the most important scientific efforts have been focused in the use of graphitic materials as intercalation cathodes. This type of batteries has shown the best electrochemical performance in comparison to the rest of groups. GABs show the highest average discharge voltage (1.8–2.25 V), a specific capacity that varies from 60 to 150 mAh g^{-1}, the highest energy efficiency (>75%), and a cycle life of more than 10 000 cycles.

In 2015, PG and graphitic foam were reported as cathodes for GABs, drawing the attention of the scientific community to this type of batteries [23]. This work, which is considered to be of outstanding interest in Al-based reported literature, showed the promising electrochemical characteristics of GABs. This battery that comprised of Al foil as anode, PG as cathode, and $AlCl_3$:EMImCl ($r = 1.3$) as electrolyte showed two discharge plateaus at 2.25 and 1.9 V, a capacity of 60–66 mAh g^{-1} with 98% of efficiency maintained for more than 200 cycles. One of the main findings of this report was the clarification of the battery mechanism by XPS and XRD tests. The reaction involved in the anode was found to be the electrodeposition of Al, whereas the reaction in the cathode was shown to be the intercalation of $AlCl_4^-$ into the graphitic structure forming the GICs.

After that publication, some reports based on graphene-based cathodes such as GAs, large-sized graphene, or 3H3C graphene as well as graphite power such as natural graphite and Kish graphite were studied as cathodes for graphite-based Al-ion batteries and influence the battery performance of the graphitic degree and textural properties such as geometric structure or particle size and shape. These studies have permitted to reach values of specific capacity as high as 150 mAh g^{-1} with high energy efficiency (>75%) with a negligible capacity fading along cycling.

Al-ion batteries and, in particular, graphite-based Al-ion batteries are a promising post-lithium batteries due to their values specific energy and power. However, due to the high corrosive character of the acidic electrolyte $AlCl_3$:EMImCl, only high cost and scarce metal such as molybdenum (Mo), tantalum (Ta), or tungsten (W) can be used as positive current collector. This makes the cost of the battery significantly higher than that of Li-ion batteries for a similar areal capacity in USD kWh^{-1}, hindering the commercial deployment of Al-ion batteries. Therefore, finding cheap and resistant positive current collector and increasing the areal capacity of Al-ion batteries is of paramount importance to develop this technology commercially [29, 75]. Another important aspect that requires significant scientific attention is the parasitic reactions that occur in the battery and how they can influence the battery performance, which can have a tremendous impact in sealed cells where the electrolyte volume is very low [1]. More specifically, the anodic decomposition of the electrolyte in the positive electrode during charging should be studied in detail. The thermodynamic potential or the mechanisms as well as the products formed, which can include highly corrosive chlorine, and the consequences of this reaction in the battery performance and its possible influence in the commercial development of this battery technology should be clarified. Finally, development of cost- and electrochemical- efficient electrolytes is of great interest since it could address simultaneously several challenges, e.g. energy density, corrosion of positive current collector, and side reactions.

References

1 Muñoz-Torrero, D., Palma, J., Marcilla, R., and Ventosa, E. (2019). A critical perspective on rechargeable Al-ion battery technology. *Dalton Trans.* 48: 9906–9911. https://doi.org/10.1039/C9DT02132A.

2 Li, Q. and Bjerrum, N.J. (2002). Aluminum as anode for energy storage and conversion: a review. *J. Power Sources* 110: 1–10. https://doi.org/10.1016/S0378-7753(01)01014-X.

3 Ambroz, F., Macdonald, T.J., and Nann, T. (2017). Trends in Aluminium-based intercalation batteries. *Adv. Energy Mater.* 7 https://doi.org/10.1002/aenm.201602093.

4 Stokes, J.J.J. (1967). Primary cell. Google Patents US3307976 A.

5 Stokes, J.J.J. (1958). Primary cell. Google Patents US2838591 A.

6 Stokes, J.J.J. (1957). Primary cell anode. Google Patents US2796456 A.

7 Samuel, R. (1953). Primary cell. Google Patents US2638489 A.

8 Sargent, D.E. (1951). Voltaic cell. Google Patents US2554447 A.

9 Holleck, G.L. and Giner, J. (1972). The aluminum electrode in $AlCl_3$-alkali-halide melts. *J. Electrochem. Soc.* 119: 1161. https://doi.org/10.1149/1.2404433.

10 Holleck, G.L. (1972). The reduction of chlorine on carbon in $AlCl_3$-KCl-NaCl melts. *J. Electrochem. Soc.* 119: 1158. https://doi.org/10.1149/1.2404432.

11 Fouletier, M. and Armand, M. (1979). Electrochemical method for characterization of graphite-aluminium chloride intercalation compounds. *Carbon N. Y.* 17: 427–429. https://doi.org/10.1016/0008-6223(79)90059-9.

12 Mohandas, K.S., Sanil, N., Noel, M., and Rodriguez, P. (2003). Electrochemical intercalation of aluminium chloride in graphite in the molten sodium chloroaluminate medium. *Carbon N. Y.* 41: 927–932. https://doi.org/10.1016/S0008-6223(02)00424-4.

13 Koura, N. (1980). A preliminary investigation for an Al/$AlCl_3$-NaCl/FeS_2 secondary cell. *J. Electrochem. Soc.* 127: 1529. https://doi.org/10.1149/1.2129943.

14 Weaving, J.S. and Walter Orchard, S. (1991). Experimental studies of transition metal chloride electrodes in undivided cells using molten $NaAlCl_4$ electrolyte. *J. Power Sources* 36: 537–546. https://doi.org/10.1016/0378-7753(91)80079-D.

15 Grjotheim, K., Matiašovský, K., Sørensen, P.E. et al. (1980). Some problems concerning aluminium electro-plating in molten salts. *Acta Chem. Scand.* 34a: 666–670. https://doi.org/10.3891/acta.chem.scand.34a-0666.

16 Takami, N. and Koura, N. (1989). Al/FeS_2 secondary cells using molten $AlCl_3$-MCl-l-butylpyridinium chloride electrolytes operated around 100 °C. *J. Electrochem. Soc.* 136: 730–731. https://doi.org/10.1149/1.2096719.

17 Gifford, P.R. and Palmisano, J.B. (1988). An aluminum/chlorine rechargeable cell employing a room temperature molten salt electrolyte. *J. Electrochem. Soc.* 135: 650. https://doi.org/10.1149/1.2095685.

18 Donahue, F.M., Mancini, S.E., and Simonsen, L. (1992). Secondary aluminium-iron (III) chloride batteries with a low temperature molten salt electrolyte. *J. Appl. Electrochem.* 22: 230–234. https://doi.org/10.1007/BF01030182.

19 Koura, N., Ejiri, H., and Takeishi, K. (1993). Polyaniline secondary cells with ambient temperature molten salt electrolytes. *J. Electrochem. Soc.* 140: 602. https://doi.org/10.1149/1.2056128.

20 Rani, J.V., Kanakaiah, V., Dadmal, T. et al. (2013). Fluorinated natural graphite cathode for rechargeable ionic liquid based aluminum-ion battery. *J. Electrochem. Soc.* 160: A1781–A1784. https://doi.org/10.1149/2.072310jes.

21 Nakajima, T., Koh, M., Gupta, V. et al. (2000). Electrochemical behavior of graphite highly fluorinated by high oxidation state complex fluorides and elemental fluorine. *Electrochim. Acta* 45: 1655–1661. https://doi.org/10.1016/S0013-4686(99)00389-8.

22 Kanakaiah, V., Latha, M., Sravan, B. et al. (2014). Rechargeable magnesium carbon-fluoride battery with electrolyte gel of ionic liquid and low molecular-weight gelator. *J. Electrochem. Soc.* 161: 1586–1592. https://doi.org/10.1149/2.0291410jes.

23 Lin, M.C., Gong, M., Lu, B. et al. (2015). An ultrafast rechargeable aluminium-ion battery. *Nature* 520: 325–328. https://doi.org/10.1038/nature14340.

24 Jiang, T., Chollier Brym, M.J., Dubé, G. et al. (2006). Electrodeposition of aluminium from ionic liquids: Part I - electrodeposition and surface morphology of aluminium from aluminium chloride ($AlCl_3$)-1-ethyl-3-methylimidazolium chloride ([EMIm]Cl) ionic liquids. *Surf. Coat. Technol.* 201: 1–9. https://doi.org/10.1016/j.surfcoat.2005.10.046.

25 Jiang, T., Chollier Brym, M.J., Dubé, G. et al. (2006). Electrodeposition of aluminium from ionic liquids: part II – studies on the electrodeposition of aluminum from aluminum chloride ($AlCl_3$) – trimethylphenylammonium chloride (TMPAC) ionic liquids. *Surf. Coat. Technol.* 201: 10–18. https://doi.org/10.1016/j.surfcoat.2005.12.024.

26 Zein El Abedin, S., Giridhar, P., Schwab, P., and Endres, F. (2010). Electrodeposition of nanocrystalline aluminium from a chloroaluminate ionic liquid.

Electrochem. Commun. 12: 1084–1086. https://doi.org/10.1016/j.elecom.2010.05.034.

27 Lai, P.K. and Skyllas-Kazacos, M. (1988). Electrodeposition of aluminium in aluminium chloride/1-methyl-3-ethylimidazolium chloride. *J. Electroanal. Chem. Interfacial Electrochem.* 248: 431–440. https://doi.org/10.1016/0022-0728(88)85103-9.

28 Carlin, R.T. and Osteryoung, R.A. (1989). Aluminum anodization in a basic ambient temperature molten salt. *J. Electrochem. Soc.* 136: 1409. https://doi.org/10.1149/1.2096931.

29 Muñoz-Torrero, D., Anderson, M., Palma, J. et al. (2019). Unexpected contribution of current collector to the cost of rechargeable Al-ion batteries. *ChemElectroChem* 6: 2766–2770. https://doi.org/10.1002/celc.201900679.

30 Yu, Z., Kang, Z., Hu, Z. et al. (2016). Hexagonal NiS nanobelts as advanced cathode materials for rechargeable Al-ion batteries. *Chem. Commun.* 52: 10427–10430. https://doi.org/10.1039/C6CC05974K.

31 Li, Z., Niu, B., Liu, J. et al. (2018). Rechargeable aluminum-ion battery based on MoS_2 microsphere cathode. *ACS Appl. Mater. Interfaces* 10: 9451–9459. https://doi.org/10.1021/acsami.8b00100.

32 Kravchyk, K.V., Wang, S., Piveteau, L., and Kovalenko, M.V. (2017). Efficient aluminum chloride-natural graphite battery. *Chem. Mater.* 29: 4484–4492. https://doi.org/10.1021/acs.chemmater.7b01060.

33 Abood, H.M., Abbott, A.P., Ballantyne, A.D., and Ryder, K.S. (2011). Do all ionic liquids need organic cations? Characterisation of $[AlCl_2 \cdot n\text{Amide}]+$ $AlCl_4(-)$ and comparison with imidazolium based systems. *Chem. Commun.* 47: 3523–3525. https://doi.org/10.1039/c0cc04989a.

34 Kitada, A., Nakamura, K., Fukami, K., and Murase, K. (2014). $AlCl_3$-dissolved diglyme as electrolyte for room-temperature aluminum electrodeposition. *Electrochemistry* 82: 946–948. https://doi.org/10.5796/electrochemistry.82.946.

35 Kitada, A., Nakamura, K., Fukami, K., and Murase, K. (2016). Electrochemically active species in aluminum electrodeposition baths of $AlCl_3$/glyme solutions. *Electrochim. Acta* 211: 561–567. https://doi.org/10.1016/j.electacta.2016.05.063.

36 El Abedin, S.Z., Moustafa, E.M., Hempelmann, R. et al. (2005). Additive free electrodeposition of nanocrystalline aluminium in a water and air stable ionic liquid. *Electrochem. Commun.* 7: 1111–1116. https://doi.org/10.1016/j.elecom.2005.08.010.

37 Pulletikurthi, G., Bödecker, B., Borodin, A. et al. (2015). Electrodeposition of Al from a 1-butylpyrrolidine-$AlCl_3$ ionic liquid. *Prog. Nat. Sci.: Mater. Int.* 25: 603–611. https://doi.org/10.1016/j.pnsc.2015.11.003.

38 Fang, Y., Yoshii, K., Jiang, X. et al. (2015). An $AlCl_3$ based ionic liquid with a neutral substituted pyridine ligand for electrochemical deposition of aluminum. *Electrochim. Acta* 160: 82–88. https://doi.org/10.1016/j.electacta.2015.02.020.

39 Canever, N., Bertrand, N., and Nann, T. (2018). Acetamide: a low-cost alternative to alkyl imidazolium chlorides for aluminium-ion batteries. *Chem. Commun.* 54: 11725–11728. https://doi.org/10.1039/c8cc04468f.

40 Jiao, H., Wang, C., Tu, J. et al. (2017). A rechargeable Al-ion battery: Al/molten $AlCl_3$ –urea/graphite. *Chem. Commun.* 53: 2331–2334. https://doi.org/10.1039/C6CC09825H.

41 Angell, M., Pan, C.-J., Rong, Y. et al. (2017). High Coulombic efficiency aluminum-ion battery using an $AlCl_3$-urea ionic liquid analog electrolyte. *Proc. Natl. Acad. Sci. U.S.A.* 114: 834–839. https://doi.org/10.1073/pnas.1619795114.

42 Abbott, A.P., Harris, R.C., Hsieh, Y.T. et al. (2014). Aluminium electrodeposition under ambient conditions. *Phys. Chem. Chem. Phys.*: 14675–14681. https://doi.org/10.1039/c4cp01508h.

43 Wang, W., Jiang, B., Xiong, W. et al. (2013). A new cathode material for super-valent battery based on aluminium ion intercalation and deintercalation. *Sci. Rep.* 3: 2–7. https://doi.org/10.1038/srep03383.

44 Chiku, M., Takeda, H., Matsumura, S. et al. (2015). Amorphous vanadium oxide/carbon composite positive electrode for rechargeable aluminum battery. *ACS Appl. Mater. Interfaces* 7: 24385–24389. https://doi.org/10.1021/acsami.5b06420.

45 Geng, L., Lv, G., Xing, X., and Guo, J. (2015). Reversible electrochemical intercalation of aluminum in Mo_6S_8. *Chem. Mater.* 27: 4926–4929. https://doi.org/10.1021/acs.chemmater.5b01918.

46 Hu, Y., Luo, B., Ye, D. et al. (2017). An innovative freeze-dried reduced graphene oxide supported SnS_2 cathode active material for aluminum-ion batteries. *Adv. Mater.* 29: 1606132. https://doi.org/10.1002/adma.201606132.

47 Hu, Y., Ye, D., Luo, B. et al. (2018). A binder-free and free-standing cobalt sulfide@carbon nanotube cathode material for aluminum-ion batteries. *Adv. Mater.* 30: 1–6. https://doi.org/10.1002/adma.201703824.

48 Liang, K., Ju, L., Koul, S. et al. (2019). Self-supported tin sulfide porous films for flexible aluminum-ion batteries. *Adv. Energy Mater.* 9: 1–7. https://doi.org/10.1002/aenm.201802543.

49 Wang, S., Yu, Z., Tu, J. et al. (2016). A novel aluminum-ion battery: Al/$AlCl_3$-[EMIm]Cl/Ni_3S_2@graphene. *Adv. Energy Mater.* 6 https://doi.org/10.1002/aenm.201600137.

50 Hu, Z., Li, Q., Zhao, Z. et al. (2019). Two-dimensionally porous cobalt sulfide nanosheets as a high-performance cathode for aluminum-ion batteries. *J. Power Sources* 440: 227147. https://doi.org/10.1016/j.jpowsour.2019.227147.

51 Wang, S., Jiao, S., Wang, J. et al. (2017). High-performance aluminum-ion battery with CuS@C microsphere composite cathode. *ACS Nano* 11: 469–477. https://doi.org/10.1021/acsnano.6b06446.

52 Cai, T., Zhao, L., Hu, H. et al. (2018). Stable $CoSe_2$/carbon nanodice@reduced graphene oxide composites for high-performance rechargeable aluminum-ion batteries. *Energy Environ. Sci.* 11: 2341–2347. https://doi.org/10.1039/c8ee00822a.

53 Guan, W., Wang, L., Lei, H. et al. (2019). Sb_2 Se_3 nanorods with N-doped reduced graphene oxide hybrids as high-capacity positive electrode materials for rechargeable aluminum batteries. *Nanoscale* 11: 16437–16444. https://doi.org/10.1039/c9nr06481h.

54 Hudak, N.S. (2014). Chloroaluminate-doped conducting polymers as positive electrodes in rechargeable aluminum batteries. *J. Phys. Chem. C* 118: 5203–5215. https://doi.org/10.1021/jp500593d.

55 Walter, M., Kravchyk, K.V., Böfer, C. et al. (2018). Polypyrenes as high-performance cathode materials for aluminum batteries. *Adv. Mater.* 30: 1–6. https://doi.org/10.1002/adma.201705644.

56 Kim, D.J., Yoo, D.J., Otley, M.T. et al. (2019). Rechargeable aluminium organic batteries. *Nat. Energy* 4: 51–59. https://doi.org/10.1038/s41560-018-0291-0.

57 Zhang, L., Chen, L., Luo, H. et al. (2017). Large-sized few-layer Graphene enables an ultrafast and long-life aluminum-ion battery. *Adv. Energy Mater.* 7 https://doi.org/10.1002/aenm.201700034.

58 Yu, X., Wang, B., Gong, D. et al. (2017). Graphene nanoribbons on highly porous 3D graphene for high-capacity and ultrastable Al-ion batteries. *Adv. Mater.* 29: 1–8. https://doi.org/10.1002/adma.201604118.

59 Wang, D.Y., Wei, C.Y., Lin, M.C. et al. (2017). Advanced rechargeable aluminium ion battery with a high-quality natural graphite cathode. *Nat. Commun.* 8: 1–7. https://doi.org/10.1038/ncomms14283.

60 Sun, H., Wang, W., Yu, Z. et al. (2015). A new aluminium-ion battery with high voltage, high safety and low cost. *Chem. Commun.* 51: 11892–11895. https://doi.org/10.1039/C5CC00542F.

61 Liu, Q.X., El Abedin, S.Z., and Endres, F. (2006). Electroplating of mild steel by aluminium in a first generation ionic liquid: a green alternative to commercial Al-plating in organic solvents. *Surf. Coat. Technol.* 201: 1352–1356. https://doi.org/10.1016/j.surfcoat.2006.01.065.

62 Jiao, S., Lei, H., Tu, J. et al. (2016). An industrialized prototype of the rechargeable Al/$AlCl_3$-[EMIm]Cl/graphite battery and recycling of the graphitic cathode into graphene. *Carbon, N. Y.* 109: 276–281. https://doi.org/10.1016/j.carbon.2016.08.027.

63 Pan, C.-J., Yuan, C., Zhu, G. et al. (2018). An operando X-ray diffraction study of chloroaluminate anion-graphite intercalation in aluminum batteries. *Proc. Natl. Acad. Sci. U.S.A.* 115: 5670–5675. https://doi.org/10.1073/pnas.1803576115.

64 Wu, Y., Gong, M., Lin, M.C. et al. (2016). 3D graphitic foams derived from chloroaluminate anion intercalation for ultrafast aluminum-ion battery. *Adv. Mater.* 28: 9218–9222. https://doi.org/10.1002/adma.201602958.

65 Chen, H., Guo, F., Liu, Y. et al. (2017). A defect-free principle for advanced graphene cathode of aluminum-ion battery. *Adv. Mater.*: 1605958. https://doi.org/10.1002/adma.201605958.

66 Wang, S., Kravchyk, K.V., Filippin, A.N. et al. (2018). Aluminum chloride-graphite batteries with flexible current collectors prepared from earth-abundant elements. *Adv. Sci.* 5: 1–6. https://doi.org/10.1002/advs.201700712.

67 Chen, H., Xu, H., Wang, S. et al. (2017). Ultrafast all-climate aluminum-graphene battery with quarter-million cycle life. *Sci. Adv.* 3 https://doi.org/10.1126/sciadv.aao7233.

68 Yang, G.Y., Chen, L., Jiang, P. et al. (2016). Fabrication of tunable 3D graphene mesh network with enhanced electrical and thermal properties for high-rate aluminum-ion battery application. *RSC Adv.* 6: 47655–47660. https://doi.org/10.1039/C6RA06467A.

69 Chen, H., Chen, C., Liu, Y. et al. (2017). High-quality graphene microflower design for high-performance Li–S and Al-ion batteries. *Adv. Energy Mater.* 7: 1–9. https://doi.org/10.1002/aenm.201700051.

70 Uemura, Y., Chen, C.-Y., Hashimoto, Y. et al. (2018). Graphene nanoplatelet composite cathode for a chloroaluminate ionic liquid-based aluminum secondary battery. *ACS Appl. Energy Mater.* 1: 2269–2274. https://doi.org/10.1021/acsaem.8b00341.

71 Tsuda, T., Uemura, Y., Chen, C.-Y. et al. (2017). Graphene-coated activated carbon fiber cloth positive electrodes for aluminum rechargeable batteries with a chloroaluminate room-temperature ionic liquid. *J. Electrochem. Soc.* 164: A2468–A2473. https://doi.org/10.1149/2.0981712jes.

72 Qiao, J., Zhou, H., Liu, Z. et al. (2019). Defect-free soft carbon as cathode material for Al-ion batteries. *Ionics (Kiel)* 25: 1235–1242. https://doi.org/10.1007/s11581-019-02896-8.

73 Elia, G.A., Ducros, J.-B., Sotta, D. et al. (2017). Polyacrylonitrile separator for high-performance aluminum batteries with improved Interface stability. *ACS Appl. Mater. Interfaces* 9: 38381–38389. https://doi.org/10.1021/acsami.7b09378.

74 Wang, S., Kravchyk, K.V., Krumeich, F., and Kovalenko, M.V. (2017). Kish graphite flakes as a cathode material for an aluminum chloride-graphite battery. *ACS Appl. Mater. Interfaces* 9: 28478–28485. https://doi.org/10.1021/acsami.7b07499.

75 Muñoz-Torrero, D., Palma, J., Marcilla, R., and Ventosa, E. (2020). Al-ion battery based on semisolid electrodes for higher specific energy and lower cost. *ACS Appl. Energy Mater.* https://doi.org/10.1021/acsaem.9b02253.

9

Al-Air Batteries

Pengyu Meng, Jianmin Ren, Min Jiang, and Chaopeng Fu

Shanghai Jiao Tong University, School of Materials Science and Engineering, Shanghai 200240, P.R. China

9.1 Introduction

Aluminum with an atomic number of 13 is located in IIIA group, and its melting point is 660 °C, boiling point is 2327 °C, and the relative molecular weight is 26.98. Al is a silvery-white, soft, nonmagnetic, and ductile metal in the boron group. Al is the most abundant metal and the third most abundant element in the Earth's crust by mass.

Aluminum–air (Al–air) batteries have received tremendous attentions due to the high energy density (8100 Wh kg^{-1}), light weight, low cost, and abundant reserve of Al anodes. The operation of Al–air battery is safe and does not produce any toxic or harmful substances during use. In addition, the electrolytes of Al–air batteries can be recycled to further reduce the cost.

The cathode material of Al–air battery is oxygen in air, while the anode material is Al or its alloys. The electrolyte is commonly either sodium hydroxide or sodium chloride aqueous solution. Oxygen in air can continuously diffuse to the electrochemical reaction interface through air cathode to react with Al anode to release electrons. According to the electrolyte characteristics, the discharge reactions are slightly different.

Under neutral conditions, the electrochemical reaction is shown as follows:

$$\text{Anode: } Al + 3OH^- \rightarrow Al(OH)_3 + 3e^-$$

$$\text{Cathode: } O_2 + 2H_2O + 4e^- \rightarrow 4OH^-$$

$$\text{Over All: } 4Al + 3O_2 + 6H_2O \rightarrow 4Al(OH)_3$$

Under alkaline conditions, the electrochemical reaction is shown as follows:

$$\text{Anode: } Al + 4OH^- \rightarrow Al(OH)_4{}^- + 3e^-$$

$$\text{Cathode: } O_2 + 2H_2O + 4e^- \rightarrow 4OH^-$$

$$\text{Over All: } 4Al + 3O_2 + 6H_2O + 4OH^- \rightarrow 4Al(OH)_4{}^-$$

Battery Technologies: Materials and Components, First Edition. Edited by Jianmin Ma.

Meanwhile, there is a side reaction of hydrogen evolution occurring under both conditions:

$$2Al + 6H_2O \rightarrow 2Al(OH)_3 + 3H_2 \uparrow$$

Al–air batteries with the characteristics of high energy density and low cost can be used in various fields. Al–air batteries can be considered as power sources for electric vehicles, and the high energy density of Al–air battery enables long driving distance of electric vehicles. Non-flammable solvents can ensure good safety. Phinergy, an Al–air battery producer from Israel, reported a 1600 km's driving distance of their electric vehicle powered by the Al–air battery. Another feather is that Al–air batteries can be mechanically charged, meaning that it is more welcome than Li-ion batteries at off-grid areas. For example, Al–air batteries can be used to support marine industrial activities, including marine exploration vessels, offshore oil development, maritime rescue. Al–air batteries can also be used as backup power supplies for telecommunication equipment, hospitals, and Al–air batteries can be stored for more than 10 years as standby power supply.

9.2 Aluminum Anodes

Aluminum–air batteries have been highlighted as one of next-generation power sources because of the high theoretical energy density (8.1 Wh kg^{-1}), abundant reserve, high negative standard potential, low production cost, and light weight of Al. However, two main issues lead to the mismatch between theoretical and practical energy capacity, hindering the commercial application of Al–air batteries.

(1) Electrode polarization: Aluminum is a very active metal, and it is easy to form a compact oxide film on surface during discharge process. The presence of oxide film inhibits electrochemical activity of Al and leads to a decrease in operating voltage.
(2) Self-corrosion: Al anode undergoes parasitically self-corrosion leading to hydrogen evolution as follows:

$$2Al + 6H_2O = 2Al(OH)_3 + 3H_2$$

The self-corrosion sharply reduces the utilization of anode resulting in a lower energy density than the theoretical value. The Al anode performance is highly related with its composition and microstructure.

9.2.1 Al Alloying Elements

The alloy elements in Al anode play key roles in determining electrochemical performance. Extensive studies have shown that adding specific alloy elements to Al matrix is an effective method to address the above problems. The main functions of alloy elements in Al alloys are as follows: (i) some alloy elements can destroy the continuity of the oxide film on Al surface to activate Al anode; (ii) some alloy elements with high hydrogen evolution overpotentials can reduce hydrogen evolution corrosion; (iii) some alloy elements can eliminate negative effects of harmful impurities (such as Fe, Si) in Al anodes.

Ga, In, Sn elements can activate Al anode mainly by damaging the dense oxide film. These alloy elements damage the oxide film in two ways. One is that these alloy element atoms with larger atomic radius form solid solution with Al matrix, which results in lattice distortion of Al or precipitation at the grain boundaries. These defects prevent the formation of continuous or uniform passivation film on Al surface. The other is that these alloy elements can react with Al atoms in the matrix and deposit on surface to damage the passive film [1].

Alloy elements with a low melting point (Pb and Bi) can combine with other alloy elements in Al alloy (such as Ga) to form low eutectics with good fluidity. These low eutectics can promote the shedding of oxidation film on Al surface, thereby eliminating the polarization of Al anode. However, if the amount of Pb or Bi exceeds the solid solubility limit, Pb or Bi tends to accumulate at the grain boundaries to form a second phase, which in turn increases the self-corrosion rate of Al anode. Additionally, Bi solid-soluble in Al matrix distorts and expands the Al lattice, resulting in an increase in the solubility of other elements, such as Sn in Al matrix, which further activates Al anodes.

Mn can combine with harmful impurity elements (such as Si and Fe) in Al matrix to prevent these points from becoming active sites of corrosion reaction [2]. Furthermore, Mn can improve the mechanical properties of Al anode. However, an excess Mn can accelerate the self-corrosion of Al anode.

Mg with a more negative electrode potential can form galvanic cell with Al to alleviate self-corrosion. Moreover, an appropriate amount of Mg can work together with Si to form uniform alloy structure to alleviate corrosion of Al anode.

9.2.2 Research Progress of Al Anodes

9.2.2.1 Aluminum Microalloying

Microalloying is an effective method to improve electrochemical performance of Al anodes. Typical Al alloy systems currently used in Al–air batteries are Al–Zn, Al–In, Al–Ga, and Al–Sn. The anode behavior of Al–Sn binary alloy is involved in dissolution and deposition mechanism. The upper limit of tin fraction in the binary Al alloy anode is 0.12 wt%. In the ternary and quaternary aluminum alloys, the highest solubility of tin varies. The upper limit of indium fraction in Al–In alloy anode is 0.16%, which is close to the solid solubility limit of indium in aluminum at a heat treatment temperature of 640 °C. In addition to the above binary alloys, ternary and quaternary Al alloy anodes also show good battery performance. Al-0.5Mg-0.1Sn-0.02Ga-0.1Si alloy displays good electrochemical performance in a 2 M NaCl solution, while Al-0.5 Mg-0.1Sn-0.02 In-0.1 Si alloy displays good electrochemical performance in a 4 M NaOH solution, and the addition of Si to the Al-0.5Mg-0.1Sn-based alloy can effectively reduce the self-corrosion rate of the alloy and increase the utilization rate of the anode [3]. The addition of In to Al–Zn anode can significantly reduce the formation of passivation film and improve discharge performance of the Al–air battery. When Ga fraction is the same, Al–In–Ga displays a larger corrosion resistance than Al–Zn–Ga [4].

9.2.2.2 Heat Treatment of Al Anodes

Heat treatment is an effective way to adjust the distribution of alloying elements in Al alloy, which can eliminate segregation. Srinivas et al. studied the microstructure and corrosion behavior of Al-0.5Mg-0.08Sn-0.08Ga anode processed at different solid solution temperatures [5]. Thermal analysis results showed that Sn is soluble in Al alloy at 400–550 °C, while microstructure analysis showed that Sn displayed a maximum solubility at above 450 °C. As a result, the Al alloy anode after heat treatment at 450 °C showed a lower corrosion and hydrogen evolution rate; however, the corrosion rate increased at a higher heat treatment temperature due to the segregation of Ga.

9.2.2.3 Processing of Al Anodes

The forming technology of Al alloys can adjust microstructure and defects of Al anodes and distribution of precipitated phases, which can further affect the electrochemical performance of Al anodes.

The electrochemical performance of casted and rolled Al anodes is significantly different. Compared with as-cast Al alloy anodes, the Al–air battery with rolled Al alloy anodes can display larger battery voltage, higher anode efficiency, and larger energy density in both alkaline and neutral electrolytes, as the rolling process produced more dispersed segregation phases and more grain boundaries.

3D printing of Al represents the future direction of Al production with designed properties. Al anodes produced by 3D printing are also great of interest for Al–air batteries. Studies show that laser sintering can effectively remove the organic solvent in the slurry and significantly improved the conductivity of Al anode, and the Al–air battery assembled with 3D laser-sintered Al anode and gel electrolyte displayed a discharge capacity of 239 mAh g^{-1} and a working voltage of 0.95 V [6].

9.2.2.4 Surface coating on Al anodes

Surface coating is also employed to boost Al anode performance. Pino et al. revealed that the Al anode discharge performance was improved by coating a carbon layer on Al anode, as the carbon layer on the surface can avoid the gelation of aluminate on the surface of anode (demonstrated in Figure 9.1) [7]. Copper plating on Al anode was also studied, and the copper layer can not only improve the utilization efficiency, but also protect Al from corrosion reaction by forming a separator without limiting the battery reaction [8].

9.3 Air Cathodes

Air cathodes need to provide oxygen from air while preventing seeping out of liquid electrolyte during the discharge process. Therefore, the structure and configuration of air cathodes play vital roles in determining the electrode performance. The

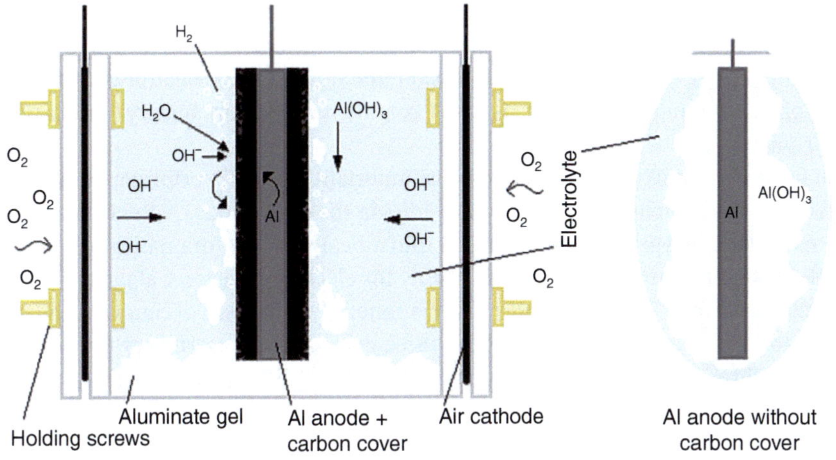

Figure 9.1 Schematic representation of aluminate accumulation of Al–air in a neutral electrolyte with and without carbon coating [7]. Source: Pino et al. [7]. Reproduced with permission of Elsevier.

cathode reaction of Al–air battery is oxygen reduction reaction (ORR), and electrocatalysts are necessary to promote the sluggish kinetics of ORR. In this section, ORR electrocatalysts and air cathode structure of Al–air batteries are introduced.

9.3.1 Structure of Air Cathodes

The power capability of Al–air batteries is strongly determined by air electrodes. Air cathode is generally composed of three layers including current collector, gas diffusion layer, and catalyst layer. The air cathode architecture plays a decisive role in improving battery performance, and the ORR mainly takes place at the triple-phase boundary illustrated in Figure 9.2, where the solid catalyst is interfaced with the electrolyte and oxygen. A large triple-phase interface where oxygen

Figure 9.2 The triple-phase boundary of the air cathode [9]. Source: Li and Lu [9]. Reproduced with permission of American Chemical Society.

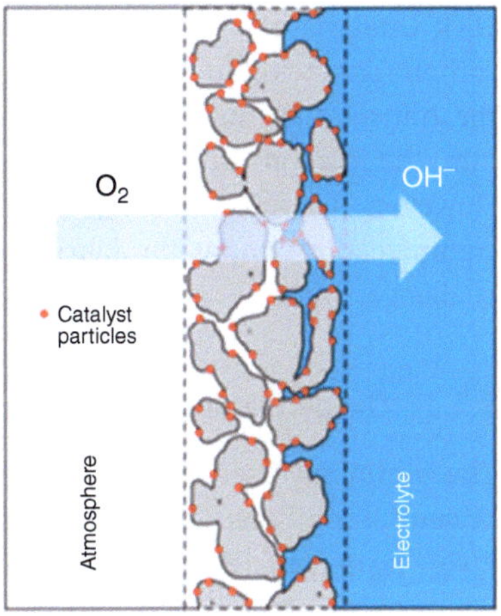

can diffuse throughout and electrolyte can be adsorbed is essential to facilitate ORR. Therefore, developing highly efficient electrocatalysts and designing air electrode structures to promote gas diffusion and enhance gas/electrolyte/catalyst triple-phase boundaries would greatly benefit the Al–air battery discharge performance.

An optimized thickness of air cathode is important for ORR performance, as the thickness strongly relates to oxygen diffusion via the air cathode. A thick air cathode may restrict oxygen diffusion, while a thin air cathode may dramatically reduce the electrode life. Current collector is mainly for electron transfer. Catalyst layer is responsible for ORR, and the catalyst layer is generally composed of catalyst, binder, and conductive carbon. The ratio of the three components determines the porosity, wettability, and electrical conductivity of the catalyst layer, which are vital for ORR. Rational design of air cathode structure can enhance catalytic activity and shelf life.

9.3.2 Integrated Cathode

The typical sandwich-type air cathode is a complex part, involving gas diffusion, electrocatalytic process, and outer waterproof. Air cathodes are traditionally fabricated by slurry casting of catalyst, conductive carbon, and binder onto current collector (metal mesh, foam, or porous carbon paper), and conductive carbon and polymeric binder are necessary to form the catalytic layer and gas diffusion layer. The fabrication procedure of air cathodes is relatively complex, and the usage of conductive carbon and binder increases the cost. Therefore, it is promising to develop integrated and binder-free cathodes, which can also meet the requirements for flexible devices. Integrated and binder-free cathodes can be prepared by directly growing electrocatalysts on conductive substrates, which possesses many intrinsic advantages, such as high conductivity, high porosity, large surface area, and desirable mechanical strength [10].

To simplify the design and fabrication of air cathodes for Al–air batteries, ORR catalysts can be integrated with porous carbon-based supports. For example, ORR catalysts can be directly grown on a 3D nanostructured carbon aerogel or carbon foam without binders, which reduces the contact resistance and facilitates the transport of mass and electrons [11]. Moreover, the catalysts encapsulated into the carbon supports can prevent them from aggregation or corrosion. However, the integrated air cathodes for Al–air batteries are still away from commercialization, as the fabrication of a large area integrated cathode is still a challenge.

9.3.3 Oxygen Reduction Reaction

The commonly used electrolytes for Al–air batteries are mainly alkaline solutions or neutral solutions. ORR consists of multiple adsorption/desorption reaction steps involving oxygen diffusion, oxygen absorption on the catalyst, weakening and breaking of oxygen bonds, and removal of the hydroxyl ion product from the catalyst surface to electrolyte [12]. ORR can occur via either a four-electron pathway or a two-electron pathway.

Direct four-electron pathway:

$$O_2 + 2H_2O + 4e^- \rightarrow 4OH^-$$

Two-electron pathway:

$$O_2 + 2H_2O + 2e^- \rightarrow HO_2^- + OH^-$$

$$HO_2^- + H_2O + 2e^- \rightarrow 3OH^-$$

During the discharge process, oxygen is firstly adsorbed on active sites of the electrocatalyst, and then the O—O bond is weakened and broken by accepting electrons from the anode. Finally, the hydroxyl ions are generated from the catalyst surface. The activities of electrocatalysts are essential to accelerate the kinetics of ORR.

9.3.4 Electrocatalysts

Extensive studies focus on developing high-performance electrocatalysts for ORR. A variety of materials, such as precious metals (Pt), transition metal oxides, modified carbon, and single atoms, have been researched as ORR electrocatalysts for Al–air batteries.

9.3.4.1 Precious Metals and Alloys

Precious metals (Pt, Pd) have been intensively studied as ORR catalysts over decades due to superior electrocatalytic activity and good stability. Generally, Pt is often employed as a benchmark material in current studies of alternative non-Pt catalysts. However, the scarcity and high cost of Pt make it necessity to further increase the specific activity and reduce mass loading of Pt-based catalysts. Manipulation of morphology and size of Pt-group catalysts can further enhance ORR performance [13]. Alloying of Pt with transition metals is an effective way not only to boost activity but also to reduce Pt loading. Alternatively, preparation of Pt monolayer on suitable transition metal substrates can also enhance catalytic activity and reduce Pt loading. It is demonstrated that Pt_3Co displayed the highest specific activity among the various PtM alloys [14], and Pt_3Ni (111) showed a 10 times higher specific activity than the Pt(111) and 90 times higher than commercial Pt/C catalysts for ORR in 0.1 M $HClO_4$ solution [15]. Additionally, Pt-skin catalyst with open-framework structure displayed much better ORR catalytic activity than commercial Pt/C catalyst. Electrocatalytic trends in ORR have been established in terms of a relationship between the surface composition, specific activity, and the electronic structure of Pt_3M (M = Fe, Co, Ni, Ti, V) surfaces [14].

The relationship between adsorption energy of reactive intermediate and specific catalytic activity of PtM alloys shows volcano-type behavior. The theoretical and experimental studies have further revealed the volcano relationship (Figure 9.3). Therefore, the oxygen adsorption energy is a good descriptor for ORR activity.

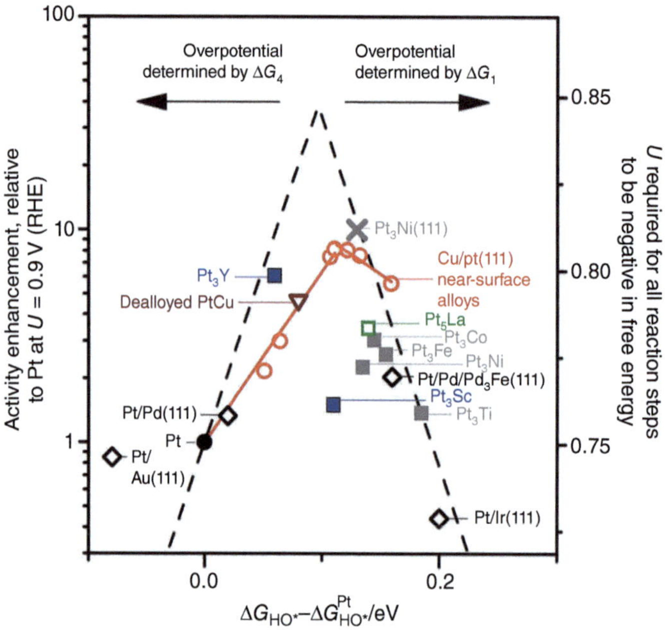

Figure 9.3 Volcano plots of the oxygen reduction kinetic current density (j_k) as a function of the calculated oxygen adsorption energy (ΔE_o) [16]. Source: Stephens et al. [16]. Reproduced with permission of Royal Society of Chemistry.

9.3.4.2 Transition Metal Oxides

Transition metal oxides with low cost and large-scale availability have received tremendous attention for Al–air batteries. Manganese oxides are the most studied ORR electrocatalysts. The catalytic activity of manganese oxides strongly depends on morphology, size, specific surface area, and crystallographic structure. Various MnO_x catalysts were prepared as air cathodes for Al–air batteries. Furthermore, defect engineering of manganese oxides with oxygen vacancy without introducing foreign additives is a newly proposed strategy for enhancing ORR performance, and Figure 9.4 illustrates the structure of manganese oxides with vacancies. Heat treatment and Ar-plasma leading to oxygen nonstoichiometry are usually employed for defect engineering [18]. An appropriate amount of oxygen vacancy in metal oxides enables a more positive reduction potential, larger current density, and lower peroxide yield. The theory computational simulation further reveals that the introduction of oxygen vacancies enhances the interaction between oxygen-containing species and MnO_2 surfaces and reduces the kinetic barrier.

Doping is another strategy to boost ORR performance of MnO_x. The doped MnO_x with a low-valent element (e.g. Mg, Ca, and Ni) exhibited a higher activity than non-doped MnO_x [19]. The main reason is that the doping metal cations may lead to generation of MnOOH and stabilize the intermediate Mn^{III}/Mn^{IV} species. Cobalt oxide is another type of promising alternative of non-precious ORR catalysts in alkaline media due to its high electrocatalytic activity [20]. The previous work pointed

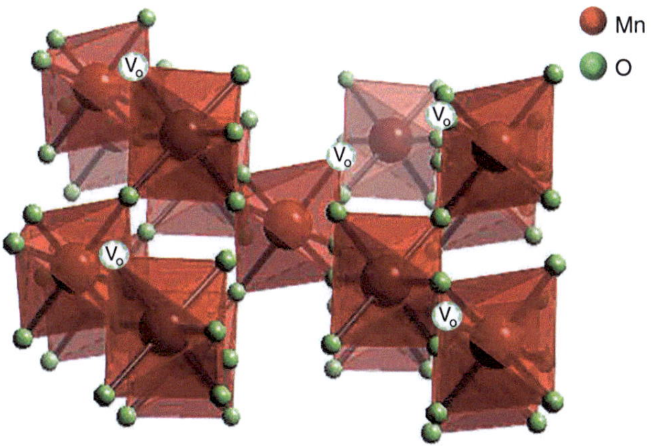

Figure 9.4 Structure illustration of manganese oxides with vacancy. [17]. Source: Cheng et al. [17]. Reproduced with permission John Wiley & Sons.

out that Co^{3+} ion in Co_3O_4 crystal structure played a key role in ORR catalysis, and it is therefore deduced that increasing the content of Co^{3+} ion is an effective approach to enhance the catalytic activity for Al–air batteries.

The combination of two or three metal oxides will lead to a series of ternary or quaternary metal oxide composites. Mixed-metal oxides with spinel or perovskite structure are widely used for ORR. Spinel is a group of oxides with the formula AB_2O_4, where A is a divalent metal ion (such as Mg, Fe, Co, Ni, Mn, or Zn) and B is a trivalent metal ion (such as Al, Fe, Co, Cr, or Mn). Spinel-type metal oxides are one of the most popular ORR catalyst for Al–air batteries owning to the simple synthetic route and good stability. In the spinel structure, A^{2+} and B^{3+} cations occupy part or all of the tetrahedral and octahedral sites, respectively, and the content of A^{2+} or B^{3+} cations in the formula can be varied to adjust the catalysis performance. Similar to spinel oxides, perovskite-type oxides have the general formula of ABO_3. The cubic perovskite crystal lattice can be tuned by partially replacing A or B cations with other metals, giving rise to interesting properties. Substituted perovskites can generally be described with the formula $A_{1-x}A'_xB_{1-y}B'_yO_3$, where A or A′ is a rare-earth or alkaline-earth metal and B or B′ is a transition metal. $La_{1-x}Ca_xMO_3$ (M = Ni, Mn, Co) perovskite-type oxides receive much interests owning to the efficient electrocatalytic activities and stability. The electrocatalytic performance can be tuned by altering the oxidation state of a surface site, and the increase in phase purity and decrease in crystallite size can also enhance the activity.

9.3.4.3 Carbon-Based Catalysts

Carbon nanomaterials, including 1D carbon nanotubes, 2D graphene, and three-dimensional (3D) carbon can offer new opportunities for development of metal-free catalysts. Metal-free carbon materials have been studied extensively for metal–air batteries. However, the electrocatalytic performance is far below the standard benchmark (Pt/C). Heteroatom doping of carbon (N, B, P, S, Cl, Br, and I) can tune electronic structure and electrochemical properties of the carbon species, and doping can increase the intrinsic activities of the carbon materials [21]. Dai's group demonstrated that vertically aligned nitrogen-doped carbon nanotubes (VA-NCNTs) can act as a metal-free electrode with a much better electrocatalytic activity and long-term operation stability than platinum for oxygen reduction [22], and the enhanced electrocatalytic activity is attributed to the change of the electronic structure during doping of the carbon nanotubes.

Additionally, the doping of carbon with B, S, or P atoms can also enhance the graphitization and shift the Fermi level, thereby leading to an improved catalytic activity for ORR [23]. The electrocatalytic performance can be further improved by co-doping with two or more heteroatoms due to the synergistic effects. The main reason is that the introduction of heteroatoms into the carbon can effectively modulate the electronic structure of the surrounding carbon atoms and tune the local charge density distribution, thereby resulting in an improvement in chemical reactivity and catalytic activity.

9.3.4.4 Single-Atom Catalysts

Downsizing metal nanoparticles into nanoclusters and even single atoms can obviously increase the catalytic activity and selectivity toward ORR as illustrated in Figure 9.5. Single-atom catalysts (SACs) with the maximum atomic utilization efficiency and unsaturated coordination environment have recently emerged as a new frontier in catalysis science and have attracted extensive research attention [25]. However, high surface free energy of single atoms makes their aggregation easy, and hence the isolated atoms should be anchored on various substrates to form a stable configuration [26]. Carbon materials with high surface areas as supports can provide abundant anchoring sites for single metal atoms and manipulate the charge density and electronic structure of the metal atoms owning to the

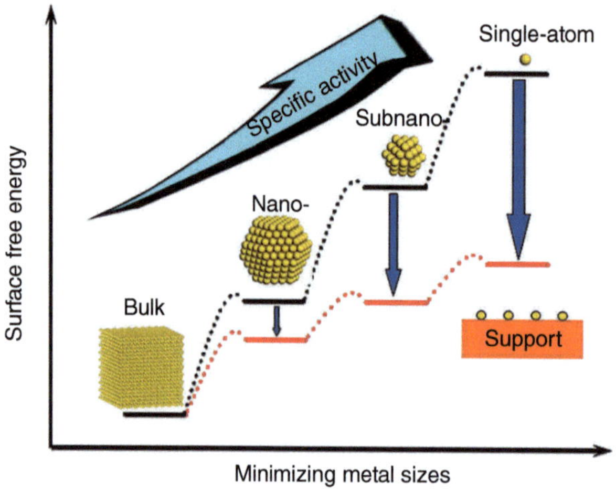

Figure 9.5 Schematic illustrate the changes of surface free energy and specific activity per metal atom with metal particle size and the support effects on stabilizing single atoms [24]. Source: Yang et al. [24]. Reproduced with permission of American Chemical Society.

strong metal–support interfacial interactions. The strong metal–support interfacial interaction also plays a key role in lowering the energy barrier of ORR and further determining catalytic performance.

Several approaches have been proposed for the synthesis of SACs over the past few years. Nevertheless, great challenges still exist, such as low mass loading of single atoms and difficulty in large-scale production.

9.4 Electrolytes

9.4.1 Aqueous Electrolytes

Aqueous electrolytes are widely used because of low cost, good safety, widespread availability, and high ionic conductivity. Alkaline solutions (NaOH, KOH) are the most widely used electrolytes in Al–air batteries. In alkaline electrolytes, the passive layer on anode can be dissolved to maintain a large discharge current, also ORR reaction is more favorable with faster reaction kinetics and lower overpotentials compared with those in neutral and acidic electrolytes. However, a drawback of alkaline aqueous electrolytes is carbonate precipitation due to the reaction between carbon dioxide and hydroxide, which may cause the clogging of air cathode pores. Acid electrolytes such as sulfuric acid and hydrochloric acid can effectively avoid carbonate formation on cathode. However, Al anode is not stable in most acidic electrolytes, leading to invalid consumption of aluminum anode. In neutral electrolytes, the main problem is the formation of passive layer on Al anode, leading to a low discharge current. Also the generally used neutral electrolyte is NaCl solution, and the Cl^- can block the active sites for oxygen reduction. Therefore, the power density of Al–air battery with neutral electrolyte is relatively low.

9.4.2 Corrosion Inhibitors

The self-corrosion of Al anode is the main challenge for commercialization of Al–air batteries. Adding inhibitors to electrolytes is another effective strategy to relieve the self-corrosion problem.

The inhibitors of Al–air batteries can be divided into inorganic corrosion inhibitors and organic corrosion inhibitors. Zinc oxide (ZnO) and sodium stannate (Na_2SnO_3) are the widely used inorganic inhibitors. ZnO is advantageous to prevent Al self-corrosion through the deposition of a Zn layer on Al anode, and Na_2SnO_3 can negatively shift the open-circuit potential. Various aromatic carboxylic acids, plant extracts, amino acids and surfactants also have been tried as organic corrosion inhibitors. The inhibitory effects are attributed to the surface adsorption via clusters/functional groups of carbon, oxygen, nitrogen, and sulfur. Generally, the denser the surface film formed by organic inhibitors, the better the inhibition effect. Inorganic and organic inhibitors can be combined together (complex inhibitors) to achieve higher suppression efficiency and larger output voltage.

9.4.3 Polymer Electrolytes

Apart from liquid electrolytes, the rapid development of wearable and flexible electronics stimulates the exploration of polymer electrolytes. Traditional liquid electrolytes tend to easily suffer from leakage, which will lead to both battery failure and environment pollution, even hazard to human health. Concerning the serious stability and safety problems, it is highly essential to develop solid and flexible electrolytes, which can work both as separator and as ion conductor. More importantly, gel polymer electrolytes (GPEs) can suppress self-corrosion of Al anodes to a certain extent.

GPEs including poly (vinyl alcohol) (PVA), poly (ethylene glycol) (PEO), poly (acrylic acid) (PAA) considered as quasi-solid state have been widely studied in various power sources. PVA is a water-soluble synthetic polymer, which is easy to form gel. The gel has good mechanical properties, and the tensile strength of the gel polymer increases with the increase in degree of polymerization. In addition, it has good biocompatibility and is nontoxic.

PEO is a crystalline, thermoplastic, water-soluble polymer. Due to the existence of C—O—C bond, it is usually flexible and can react with electron acceptors or certain inorganic electrolytes. However, owing to a poor mobility of OH^- ion, it has a high interfacial resistance and a low ionic conductivity. To enhance the electrochemical performance of GPE based on PEO, many efforts have been dedicated to preparing copolymers with PEO and polymeric host. For instance, a modified PVA/PEO GPE was proposed for flexible Al–air battery. The GPE was designed to include PVA, PEO, and KOH solution with a high ionic conductivity of $0.18\,S\,cm^{-1}$, and the obtained GPE could be easily bent or stretched. The resulting Al–air battery delivered a specific capacity of $935\,mA\,h\,g^{-1}$ and a high energy density of $1168\,Wh\,kg^{-1}$ [27]. PAA is an anionic polymer, and its side chains can lose protons and acquire negative charges. PAA-based GPE can absorb and retain water, which leads to a relatively high ionic conductivity.

9.5 Al–Air Battery Structure Design

Al–air batteries are mainly composed of Al anode (negative electrode), air electrode (positive electrode), electrolyte, and battery casing. The structure of Al–air battery, namely the arrangement and shape of anode, cathode, and electrolyte, relates to the different application fields. For liquid electrolytes, the common Al–air battery configuration is shown in Figure 9.6, and both the anode and cathode immersing in a liquid electrolyte are arranged face to face. To meet the requirements of wearable electronics, flexible batteries are needed. There are usually two types of configurations. One is the planer type and generally a polymer electrolyte is sandwiched between flexible anode and cathode to allow flexibility of the whole battery, as illustrated in Figure 9.7. Another type is fiber shape. Compared with planer flexible battery, the textiles composed of fibers can undergo various deformations, including twisting, stretching, and bending (illustrated in Figure 9.8), and so building chemical fibers is an efficient way to realize the high performance of wearable devices.

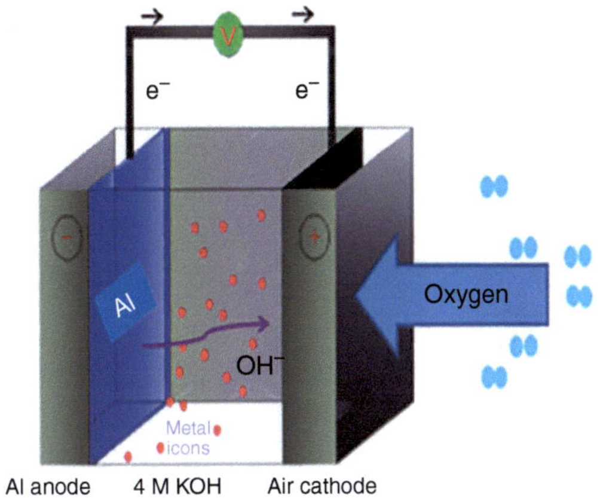

Figure 9.6 Schematic illustration of Al–air battery. Source: Jiang et al. [18]. Reproduced with permission of Elsevier.

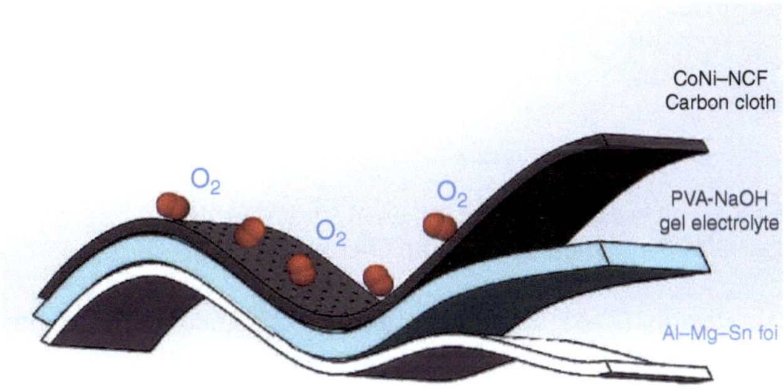

Figure 9.7 Flexible configuration of Al–air battery. Source: Jiang et al. [28]. Reproduced with permission Elsevier.

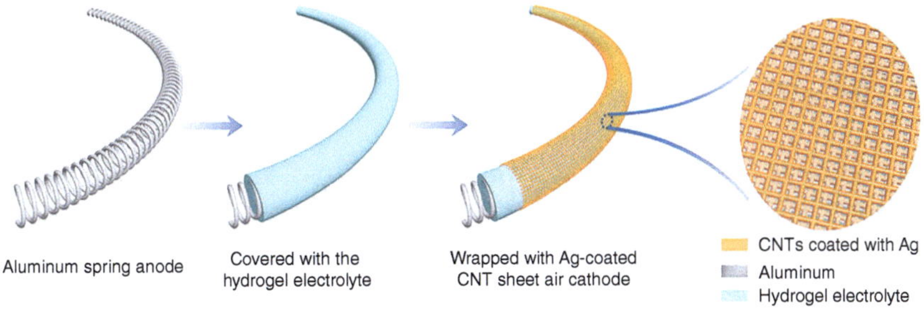

Figure 9.8 Schematic of fiber-shaped Al–air battery [27]. Source: Xu et al. [27]. Reproduced with permission John Wiley & Sons.

Apart from these conventional battery structures, some unique structures are also designed. An Al–air scavenger as shown in Figure 9.9 is proposed by Wang et al. [29]. Hydrogel electrolytes were combined with air cathodes, while the energy was extracted from the Al anode, which was put under the hydrogel without having to carry the anode material on-board.

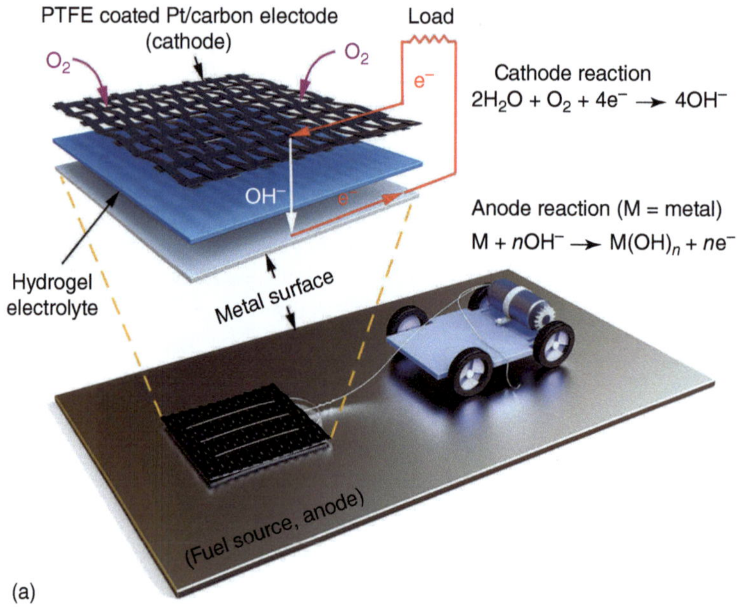

Figure 9.9 Illustration of an Al–air scavenger powering an electric vehicle [29]. Source: Wang et al. [29]. Reproduced with permission of American Chemical Society.

Another unique configuration of Al–air battery is also designed to solve the self-corrosion problem. In this novel configuration, the separator allows for the reversible displacement of electrolyte from the negative electrode surface with an oil when the battery is on standby [30]. The nonconducting displacing oil dramatically reduces the diffusion rate of electrolyte to the negative electrode surface, enabling ultralow open-circuit corrosion. By displacing the oil with electrolyte for battery discharge, high power and energy density were achieved.

9.6 Recycle of Al–Air Batteries

Al is considered as the energy carrier in Al–air batteries, and aluminum hydroxide is generated after discharge. The obtained aluminum hydroxide after washing can be converted to alumina after thermal treatment. Alumina can then be recycled in the aluminum factory, enabling a closed and sustainable life cycle. Alternatively, alumina is also a widely used commercial feedstock to produce firebrick, flame retardant, and so on. Additionally, alumina with an ultrahigh purity has great economic added value.

9.7 Rechargeable Al–Air Batteries

Al–air batteries have been intensively investigated with the features of high energy density, low cost, and abundant in the earth crust of Al. However, the most studied

Al–air batteries are primary ones, as the reduction of aluminum hydroxide in aqueous electrolytes is impossible. To further extend the application of Al–air batteries, rechargeable cells are developed when nonaqueous electrolytes are employed. This section focuses on the introduction of rechargeable Al–air batteries.

Compared with primary Al–air batteries, rechargeable Al–air battery has one more charge process. Namely, the discharge product should be reduced to Al metal. Ionic liquid is a good choice that can allow the reversible deposition and stripping of Al. Meanwhile, the ORR catalytic mechanism on cathode is also different from that in aqueous electrolytes. Moreover, water-based primary batteries do not involve oxygen evolution reaction (OER) reaction on air cathodes. Therefore, it is key to develop air cathodes with both high ORR and OER activities in ionic liquids in order to improve electrochemical performance of Al–air secondary batteries. Currently, there are just very a few studies on the compatibility between catalysts and corresponding ionic liquids, and the catalysts used in Al–air secondary batteries are mainly graphite and MnO_2.

The charge and discharge reactions of Al|$AlCl_3$/EMIm Cl|MnO_2 aluminum–air rechargeable batteries are proposed as follows:

$$\text{Electrolyte: } AlCl_3 + Cl^- \rightleftharpoons AlCl_4$$

$$AlCl_3 + AlCl_4^- \rightleftharpoons Al_2Cl_7^-$$

$$\text{Anode: } Al + 7AlCl_4^- \rightleftharpoons 4Al_2Cl_7^- + 3e^-$$

$$\text{Cathode: } O_2 + e^- \rightleftharpoons O_2^-$$

During charge and discharge processes, the reaction on anode is dissolution and deposition of Al in $AlCl_3$/EMIMCl ionic liquid. The pH of $AlCl_3$/EMImCl with different molar ratios is different. When the molar ratio of $AlCl_3$: EMImCl>1, the main ion in the electrolyte is $Al_2Cl_7^-$, and aluminum can deposit in the ionic liquid. On the other hand, the ORR reaction occurs on the air cathode, and oxygen is reduced to superoxide species.

The earliest research on rechargeable Al–air batteries is traced back to 2013, Hibino et al. investigated Sb-doped SnP_2O_7 as a solid electrolyte for rechargeable Al–air batteries [31], and the battery delivered a relatively high discharge capacity of 800 mA h g^{-1} during the initial discharge to 0 V over a wide range of current densities. The discharge product aluminate species could be reduced to aluminum by charging process. More recently, Mori reported a rechargeable Al–air battery with stable electrochemical reactions achieved by mixing $AlCl_3$, urea, carboxymethyl cellulose, and glycerin as an electrolyte [32]. The charge–discharge curves and cyclic voltammograms demonstrated the reversible charge–discharge reactions. When TiN was used as an air cathode catalyst, even if a deep co-solvent-based electrolyte was used as the electrolyte, the typical products of aluminum–air batteries, such as Al_2O_3 or $Al(OH)_3$ were not experimentally evidenced.

The key issue of rechargeable Al–air batteries is the selection of electrolytes, in which the discharge species must be reduced to Al. Ionic liquids are good choices as electrolytes, whose wide electrochemical window allows efficient and

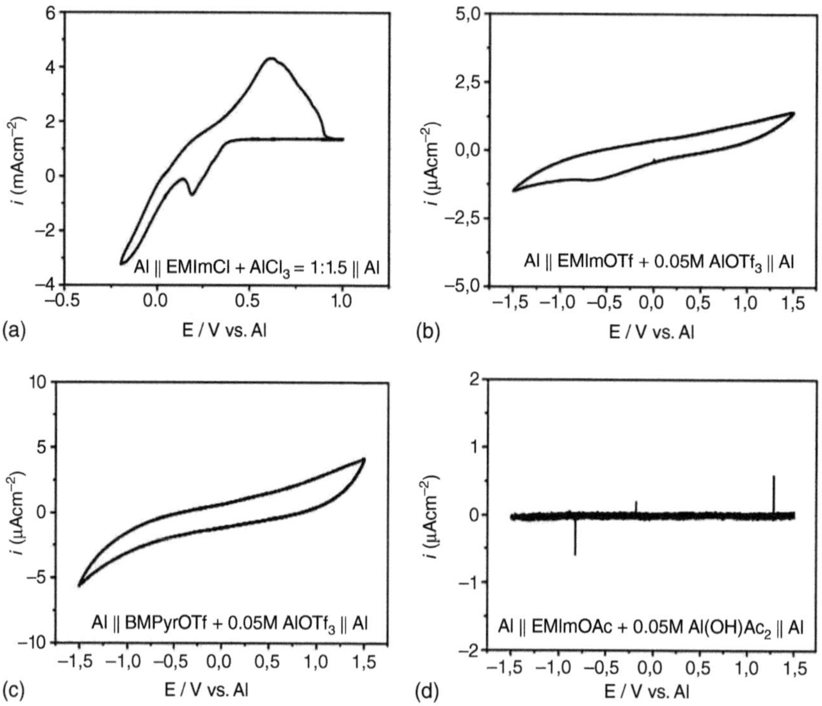

Figure 9.10 CVs for Al deposition/stripping in different ionic liquids [35]. Source: Bogolowski and Drillet [35]. Reproduced with permission of IOP Publishing.

reversible charge–discharge cycles for rechargeable Al–air batteries. Gelman et al. first reported an Al–air battery utilizing $EMIm(HF)_{2.3}F$ as an electrolyte, and Al_2O_3 was found to be the most possible discharge product. The proposed battery delivered reversible energy densities of $2300\,W\,h\,kg^{-1}$ and $6200\,W\,h\,l^{-1}$ [33]. Then, another new Al–air couple in $AlCl_3$/EMImCl ionic liquid electrolyte was fabricated by Revel et al. [34], and this battery delivered a reversible capacity of $71\,mAh\,cm^{-2}$ with a low-self discharge rate. Charging of such a system is possible, it is noteworthy that the limitation is the air electrode that generates a higher overpotential than the potential stability of the electrolyte. Bogolowski et al. [35] compared the properties of several ionic liquids ($Al(OTF)_3$/EMImOTF, Al(OH)Ac/$EMImOAc_2$, $Al(OTF)_3$/BMPyrOTF, and $AlCl_3$/EMImCl), and the results showed that the activities of triflate- and acetate-based IL electrolytes for aluminum deposition/dissolution were clearly lower than those obtained in chloride-based ones (Figure 9.10). Similarly, the activities of different $AlCl_3$-based electrolytes for rechargeable Al–air batteries were compared by Bogolowski et al., and the results revealed satisfactory activities of all studied electrolytes for Al stripping/deposition with a certain disparity in overpotential and stability (Figure 9.11) [36].

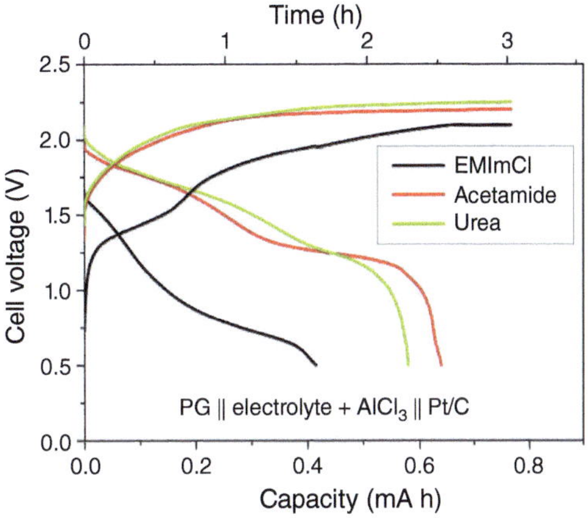

Figure 9.11 Cell voltage behaviors during the fifth charge/discharge cycle of different Al–air cells at 100 μA cm^{-2}. Source: Bogolowski and Drillet [36]. Reproduced with permission of Elsevier.

9.8 Summary and Outlook

Al–air batteries with the merits of high energy density and low cost are promising as power sources for electric vehicles and other energy storage systems. However, there are still some issues that need to be addressed for commercialization. The self-discharge caused by self-corrosion of Al–air batteries requires more attention for practical application. Meanwhile, extensive efforts are still needed to boost anode utilization rate, which is expected to be 80% for commercialization. Furthermore, the cell design and optimization, which are currently more or less neglected, need more attention for the sake of practical application. However, the defects cannot obscure the virtues, and it is still believed that Al–air battery systems have a promising potential for next-generation power supply.

References

1 (a) Sun, Z. and Lu, H. (2015). *J. Electrochem. Soc.* 162: A1617–A1623. (b) Gudić, S., Smoljko, I., and Kliškić, M. (2010). *J. Alloys Comp.*, vol. 505, 54–63.
2 Moghanni-Bavil-Olyaei, H. and Arjomandi, J. (2016). *RSC Adv.* 6: 28055–28062.
3 Ma, J., Wen, J., Ren, F. et al. (2016). *J. Electrochem. Soc.* 163: A1759–A1764.
4 El Shayeb, H.A., Abd El Wahab, F.M., and Zein El Abedin, S. (2001). *Corros. Sci.* 43: 655–669.
5 Srinivas, M., Adapaka, S.K., and Neelakantan, L. (2016). *J. Alloys Compd.* 683: 647–653.
6 Yu, Y., Chen, M., Wang, S. et al. (2018). *J. Electrochem. Soc.* 165: A584–A592.
7 Pino, M., Herranz, D., Chacon, J. et al. (2016). *J. Power Sources* 326: 296–302.
8 Mutlu, R.N. and Yazıcı, B. (2019). *J. Solid State Electrochem.* 23: 529–541.
9 Li, Y. and Lu, J. (2017). *ACS Energy Lett.* 2: 1370–1377.
10 Ma, T.Y., Dai, S., and Qiao, S.Z. (2016). *Mater. Today* 19: 265–273.
11 (a) Wu, K., Zhang, L., Yuan, Y. et al. (2020). *Adv. Mater.* 32: 2002292. (b) Jiang, M., Fu, C., Cheng, R. et al. (2020). *Adv. Sci.* 7: 2000747.

12 Cao, R., Lee, J.-S., Liu, M., and Cho, J. (2012). *Adv. Energy Mater.* 2: 816–829.
13 Bing, Y., Liu, H., Zhang, L. et al. (2010). *Chem. Soc. Rev.* 39: 2184–2202.
14 Stamenkovic, V.R., Mun, B.S., Arenz, M. et al. (2007). *Nat. Mater.* 6: 241–247.
15 Stamenkovic, V.R., Fowler, B., Mun, B.S. et al. (2007). *Science* 315: 493.
16 Stephens, I.E.L., Bondarenko, A.S., Grønbjerg, U. et al. (2012). *Energy Environ. Sci.* 5: 6744.
17 Cheng, F., Zhang, T., Zhang, Y. et al. (2013). *Angew. Chem. Int. Ed.* 52: 2474–2477.
18 Jiang, M., Fu, C., Yang, J. et al. (2019). *Energy Storage Mater.* 18: 34–42.
19 (a) Bezdička, P., Grygar, T., Klápště, B., and Vondrák, J. (1999). *Electrochim. Acta* 45: 913–920. (b) Klápště, B., Vondrák, J., and Velická, J. (2002). *Electrochim. Acta* 47: 2365–2369. (c) Roche, I., Chaînet, E., Chatenet, M., and Vondrák, J. (2007). *J. Phys. Chem. C* 111: 1434–1443. (d) Roche, I. and Scott, K. (2008). *J. Appl. Electrochem.* 39: 197–204.
20 (a) Hamdani, M., Singh, R.N., and Chartier, P. (2010). *Int. J. Electrochem. Sci.* 556: 556. (b) Esswein, A.J., McMurdo, M.J., Ross, P.N. et al. (2009). *J. Phys. Chemi. C* 113: 15068–15072.
21 (a) Dai, L., Xue, Y., Qu, L. et al. (2015). *Chem. Rev.* 115: 4823–4892. (b) Zhang, J. and Dai, L. (2015). *ACS Catal.* 5: 7244–7253.
22 Gong, K., Du, F., Xia, Z. et al. (2009). *Science* 323: 760.
23 (a) Han, W., Bando, Y., Kurashima, K., and Sato, T. (1999). *Chem. Phys. Lett.* 299: 368–373. (b) Mondal, K.C., Strydom, A.M., Erasmus, R.M. et al. (2008). *Mater. Chem. Phys.* 111: 386–390.
24 Yang, X.-F., Wang, A., Qiao, B. et al. (2013). *Acc. Chem. Res.* 46: 1740–1748.
25 (a) Chen, Y., Ji, S., Chen, C. et al. (2018). *Joule* 2: 1242–1264. (b) Zhang, B.-W., Wang, Y.-X., Chou, S.-L. et al. (2019). *Small Methods* 3: 1800497.
26 (a) Liu, L. and Corma, A. (2018). *Chem. Rev.* 118: 4981–5079. (b) Qin, R., Liu, P., Fu, G., and Zheng, N. (2018). *Small Methods* 2: 1700286. (c) Hu, M., Zhang, J., Zhu, W. et al. (2018). *Nano Res.* 11: 905–912.
27 Xu, Y., Zhao, Y., Ren, J. et al. (2016). *Angew. Chem. Int. Ed.* 55: 7979–7982.
28 Jiang, M., Yang, J., Ju, J. et al. (2020). *Energy Storage Mater.* 27: 96–108.
29 Wang, M., Joshi, U., and Pikul, J.H. (2020). *ACS Energy Lett.* 5: 758–765.
30 Hopkins, B.J., Shao-Horn, Y., and Hart, D.P. (2018). *Science* 362: 658.
31 Hibino, T., Kobayashi, K., and Nagao, M. (2013). *J. Mater. Chem. A*: 1.
32 Mori, R. (2019). *RSC Adv.* 9: 22220–22226.
33 Gelman, D., Shvartsev, B., Ein-Eli, Y., and Mater, J. (2014). *Chem. A* 2: 20237–20242.
34 Revel, R., Audichon, T., and Gonzalez, S. (2014). *J. Power Sources* 272: 415–421.
35 Bogolowski, N. and Drillet, J.-F. (2017). *ECS Trans.* 75: 85–92.
36 Bogolowski, N. and Drillet, J.-F. (2018). *Electrochim. Acta* 274: 353–358.

10

Dual-Ion Battery

Haitao Wang, Luojiang Zhang, and Yongbing Tang

Advanced Energy Storage Technology Research Center, Institute of Technology of Carbon Neutrality, Shenzhen Institutes of Advanced Technology, Chinese Academy of Sciences, 1068 Xueyuan Avenue, Shenzhen 518055, China

10.1 Cation–Anion Dual-Ion Battery

10.1.1 Introduction

Lithium-ion batteries (LIBs) based on the "rocking-chair" mechanism have dominated the electric vehicles and consumer electronics markets attributed to their safety, high energy density, and stable cycling performance over the past few decades [1, 2]. However, the quickly increasing consumption of the cathode materials (especially $LiCoO_2$ and $LiNi_xCo_yMn_zO_2$) results in soaring price and rapid depletion of the reserves of cobalt and nickel [3]. Therefore, it is an urgent demand to design a novel potential battery system with low cost, high performance, and environmental friendliness. Dual-ion battery (DIB), in which both the cations and anions involved in the battery reaction, has attracted considerable attention in recent years due to its low cost, high working voltage, and raw materials sustainability [4, 5]. In conventional LIB, the cost of cathode material (e.g. $LiCoO_2$ and $LiNi_xCo_yMn_zO_2$) accounts for c. 1/3 of the total cost. However, in DIB, graphite can be adopted as both the cathode and anode materials, leading a sharp reduction of cost by about 70% [6]. Consequently, the DIB has an obvious advantage of low cost compared to LIB.

Figure 10.1 displays the schematic of the charge/discharge process of a typical DIB. During the charging process, the cations (e.g. Li^+) are intercalated into the anode and the anions (e.g. PF_6^-) are inserted into the cathode, while for discharge process, the cations/anions diffuse back from anode/cathode to electrolyte, respectively. Thus, the specific electrode reactions in the dual-graphite cell could be summarized as follows:

$$\text{Cathode: } C + xA^- \leftrightarrow AxC + xe^- \tag{10.1}$$

$$\text{Anode: } C + xLi^+ + xe^- \leftrightarrow Li_xC \tag{10.2}$$

$$\text{Overall cell reaction: } xLi^+ + xA^- + C + C \leftrightarrow Li_xC + A_xC \tag{10.3}$$

Battery Technologies: Materials and Components, First Edition. Edited by Jianmin Ma.

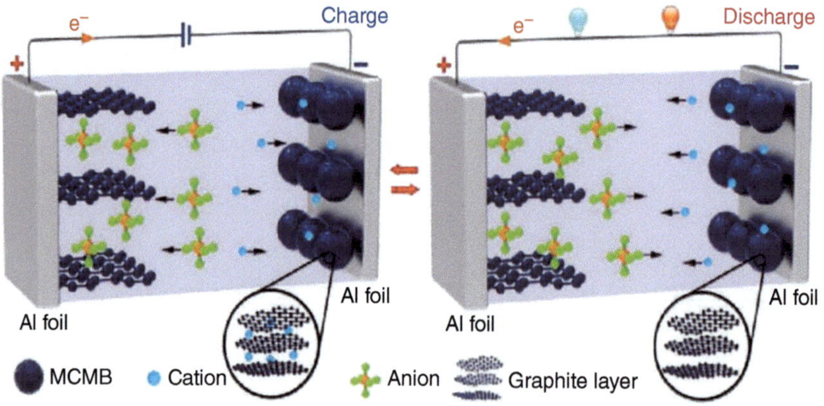

Figure 10.1 Schematic of the charge/discharge process of a typical dual-ion battery.

where A^- stands for anions in the electrolyte. The intercalation of anions leads to high voltage, which is favorable for improved energy density. Another obvious difference from LIB is that the active ions come from electrolytes. Hence, electrolytes are also regarded as active materials in DIBs. It is essential to develop suitable electrolytes with a wide electrochemical window. Moreover, concentrated electrolytes are crucial to gain higher energy density for the graphite DIB.

Rüdorff and Hofmann discovered the phenomenon of anion intercalate into graphite in concentrated sulfuric acid for the first time as early as 1938 [7]. However, it was hampered because of the potential safety issue brought from a high concentration acid solution as the electrolyte. Until 1989, McCullough et al. reported a dual-graphite battery based on cation and anion intercalation in a non-aqueous electrolyte [8]. In 1994, Carlin et al. reported a dual-graphite battery based on room-temperature ionic liquids (ILs) and anion intercalated graphite cathode [9]. In 2000, Dahn and coworker investigated the PF_6^- intercalation mechanism by the in situ XRD technique [10]. They discovered that a variety of staged phases of graphite exist in the intercalation process. The concept of the DIB was first introduced by Winter and coworkers in 2012 [11]. Then, in 2014, Read et al. proposed a reversible dual-graphite battery with simultaneous intercalation/deintercalation of Li^+ and PF_6^- into/from graphite based on a fluorinated solvent and additive [12]. The battery displayed a capacity of 90 mAh g^{-1} and an operating voltage of 5.2 V. In the same year, Winter and coworkers reported a dual-graphite cell based on IL Pyr_{14}TFSI-LiTFSI with additive ethylene sulfite (ES), which exhibited a high working voltage of 5.0 V vs. Li/Li^+ and a discharge capacity of 97 mAh g^{-1} [13]. In 2016, Tang and coworkers proposed Al-graphite DIB using Al as both the anode active material and current collector [5]. This DIB exhibits a reversible capacity of 104 mAh g^{-1}, which suggest its potential to be a novel low-cost energy storage device. In 2018, a room-temperature rechargeable Ca-based DIB was designed by Tang and coworkers [14]. Figure 10.2 summarizes the brief development history of DIB. In recent years, a number of novel rechargeable batteries based on abundant

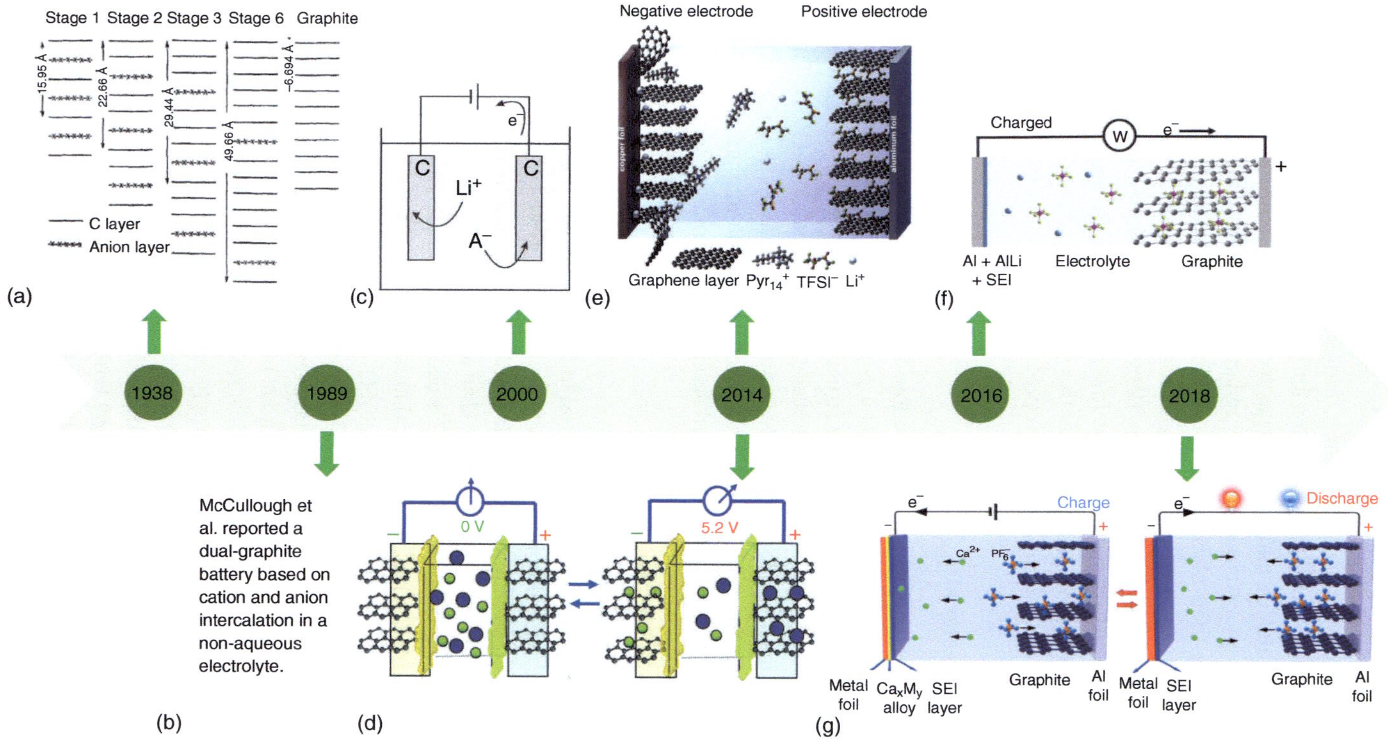

Figure 10.2 A brief development history of DIB: (a) Rüdorff and coworker demonstrated the phenomenon of anion intercalated into graphite in concentrated sulfuric acid and developed the first dual-graphite cell. Source: Rüdorff and Hofmann [7]. Reproduced with permission from John Wiley & Sons. (b) McCullough et al. introduced and applied for the first dual-carbon battery patent [8]. (c) Dahn and coworker investigated the PF_6^- intercalation mechanism. Source: Seel et al. [10]. Reproduced with permission of the Electrochemical Society. (d) The first study on the high-voltage electrolyte for DIBs by Read et al. Source: Read et al. [12]. Reproduced with permission of Royal Society of Chemistry. (e) Winter and coworkers reported a dual-graphite cell based on ionic liquid. Source: Rothermel et al. [13]. Licensed under CC BY 3.0. (f) Tang and coworkers proposed Al-graphite dual-ion battery. Source: Zhang et al. [5]. Reproduced with permission of WILEY-VCH. (g) The first Ca-ion-based DIBs used the Sn-graphite system. Source: Wang et al. [14]. Reproduced with permission of Springer Nature.

elements (e.g. Na^+, K^+, Mg^{2+}, Ca^{2+}, and Al^{3+}) with similar DIB mechanism were also developed [4, 15–18].

To date, it should be pointed out that the DIBs are still in their preliminary stage compared to the state-of-the-art LIBs. More efforts still remain to be made to develop novel host materials and electrolyte systems for DIBs with high performance. Furthermore, more strategies and characterize techniques should be employed to achieve an in-depth understanding of the fundamental reaction mechanisms based on DIBs to improve their electrochemical performance. Hence, in the following sections, we will discuss the principle and progress of electrode material design and electrolytes optimization involved in DIBs in detail. Moreover, the current challenges and effective strategies toward low-cost DIBs with high performance will also be summarized. We also hope that this chapter can attract more scientists to the research field of DIBs and boost their practical applications in the future.

10.1.2 Cathode Materials

In a typical cation–anion DIB, the intercalation/deintercalation of anions into/from cathode materials during the charging/discharging process contributes to enhanced capacity and relatively high voltage. The structure and characteristic of cathode materials play a crucial role in anion intercalation/deintercalation (10.6). Therefore, developing highly stable and high-performance cathode materials is crucial for designing an advanced DIB. In the following section, fundamental issues in some typical cathode materials and corresponding solutions to improve their performances in DIBs will be discussed.

10.1.2.1 Graphitic Materials

Graphite is widely used as an anode material in LIBs and is also the most common material for the cathode electrode in DIBs due to its unique layered structure stacked together by π–π interaction of the electronic network [19]. Graphite, which contains polyaromatic rings with sp^2 carbon atoms, is an amphoteric material that can function as a host in both oxidative and reductive chemical intercalation reactions.

In the past decades, the reversible intercalation/deintercalation of various anions including PF_6^-, BF_4^-, ClO_4^-, $AlCl_4^-$, $CF_3SO_3^-$, and $C_6H_5CO_2^-$ etc. into/from graphite was discovered in DIBs [20]. The graphite has gained more attention, which is also attributed to its environmental friendliness, low cost, and abundance compared to conventional metal oxide cathode in metal-ion batteries.

The anion intercalation is the most typical reaction, which determines the electrochemical performance of DIBs. Accordingly, an in-depth understanding reaction mechanism of anionic intercalation in graphite cathodes is greatly conducive to further develop the DIBs. The interaction situations of cations/anions into graphite can be characterized by the "stage," namely the number of graphite layers that lie between alternate intercalated layers. This process was also proposed by Rüdorff and Hofmann in 1940 [21]. For instance, stage 4 means that adjacent intercalant layers are separated by four graphene layers. Different stage numbers of graphite cathode are schematically displayed in Figure 10.3. It should be pointed out that the

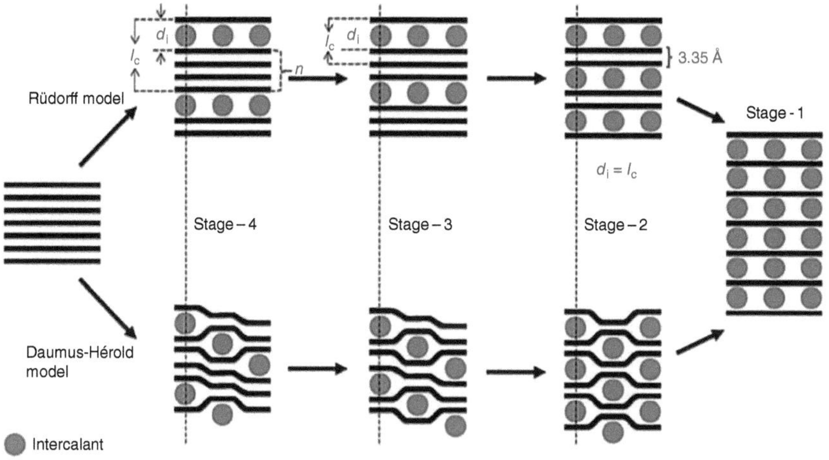

Figure 10.3 Schematic staging mechanism of Rüdorff and Daumas-Herold models for ion intercalation into a graphite cathode along the stacking direction; with periodically repeated distance, I_c, intercalate gallery height, d_i, gallery expansion, Δd. Source: Zhou et al. [20]. Reproduced with permission of Wiley-VCH.

schematic is based on the assumption of defect-free graphite lattice. The periodically repeating distance (I_c), the intercalant gallery height (d_i), and the gallery expansion (Δd) are determined by the following equation:

$$I_c = d_i + (n-1) \times 3.35\,\text{Å} = \Delta d + n \times 3.35\,\text{Å} = l \times d_{obs} \tag{10.4}$$

where l is the index of (00l) planes oriented in the stacking direction, and d_{obs} is the observed value of the spacing between two adjacent planes. Initially, the adjacent intercalant layers maybe four graphene layers or even more. As the battery is further charged, the voltage increases and it is easier to form stage 1 or stage 2 orders. However, in fact, the process is much more complicated. Daumas and Herold proposed a more realistic model, in which ions were inserted into graphene layers simultaneously, then deforming the layers around them [22]. This model can explain the coexistence of domains with different stages in the same crystallite.

The types and structures of graphite have a significant impact on the electrochemistry of anion intercalation. Great effort has been taken to consider the correlations between the graphitic structure and DIBs' performance. In fact, many types of graphite (e.g. KS4, KS6L, KS10, KS15, KS25, KS44, SFG6L, T44, SLP30, SLP50, SFG44, SFG75, and MCMB) can act as host for anion (e.g. $TFSI^-$) intercalation [23]. The surface properties of these graphites are displayed in Table 10.1, including the D90 values for the particle size distribution, the Brunauer–Emmett–Teller (BET) and density functional theory (DFT)-specific surface areas, as well as basal plane and "non-basal plane" surface areas and the associated ratios of these areas. Table 10.2 presents the electrochemical performance of $TFSI^-$ anion intercalation into various types of graphite at 20 °C. It can be seen that the discharge capacity achievable at 20 °C varies from 18.3 mAh g^{-1} for the spherical MCMB graphite to 52.1 mAh g^{-1} for the KS4 graphite. The MCMB sample contains a surface coating

Table 10.1 D90 values of the particle size distribution, BET surface areas, DFT surface areas, basal plane surface areas, "non-basal plane" surface areas, and ratios of the basal plane to the "non-basal plane" surface area of the investigated graphites.

Graphite type	D90 (μm)	BET surface area ($m^2 g^{-1}$)	DFT surface area ($m^2 g^{-1}$)	Basel plane surface area ($m^2 g^{-1}$)	"Non-basal plane" surface area ($m^2 g^{-1}$)	Ratio of basal to "Non-basal plane" surface area ($m^2 g^{-1}$)
KS4	4.9	24.6	28.1	10.5	17.6	0.60
KS6L	8.5	19.0	20.5	11.2	9.3	1.20
KS10	11.4	13.3	15.3	6.3	9.0	0.70
KS15	13.9	13.0	15.1	6.7	8.4	0.80
KS25	18.2	10.9	12.4	5.6	6.8	0.82
KS44	38.5	8.3	9.3	4.4	4.9	0.90
SFG6L	6.8	14.9	16.5	8.5	8.0	1.06
T44	44.7	8.4	9.8	4.8	5.0	0.96
SLP30	32.0	6.9	7.1	3.5	3.6	0.97
SLP50	45.0	5.3	5.7	2.3	3.4	0.68
SFG44	56.9	3.3	3.9	2.9	1.0	2.90
SFG75	74.5	2.9	3.5	1.2	2.3	0.52
El	44.6	1.9	2.2	1.2	1.0	1.20
MCMB	26.9	2.0	2.2	1.5	0.7	2.14

Source: Placke et al. [23]. Reproduced with permission of the Electrochemical Society.

that is more likely to hinder the intercalation from the large $TFSI^-$ anion. In general, the discharge capacity decreases with a reducing surface area (BET and "non-basal plane") and with increasing particle size. This provides an insight to design graphite with a more suitable structure for anion intercalation.

Although intercalation reactions have been well studied for the metal cations (such as Li^+, K^+, Mg^{2+}, Al^{3+}, etc.), it faces the remarkable challenge to accomplish the same task with anions, due to the much larger size and relatively sluggish kinetics [24]. A primary challenge of graphite cathode is the relatively high working voltages, resulting from an anion's intercalation/deintercalation process. Such potential may lead to exfoliation of the graphite and electrolyte decomposition, resulting in inferior performance. Surface modification is regarded as an effective tool to solve the above issues. Wu and coworkers constructed artificial solid electrolyte interphase (SEI) on the graphite surface through the constant current pre-discharging/charging processes of five cycles in Li-graphite DIB [25]. The artificial SEI could widen the stable window of electrolyte from E_g to E_g', so that electrolyte decomposition would be alleviated (Figure 10.4). Moreover, the SEI protector can mitigate the solvation effect of anions, leading to much improved cycling performance at high operating voltages. This work provides a new route to enhance the performance of graphite cathode in DIB.

Table 10.2 Coulombic efficiency (CE) at different cycles and discharge capacity (50th cycle) of various graphite cathodes acting as host for $TFSI^-$ anion intercalation.

Graphite type	CE in 1st cycle (%)	CE in 2nd cycle (%)	CE in 3rd cycle (%)	CE in 50th cycle (%)	Discharge capacity in 50th cycle ($mAh\,g^{-1}$)
KS4	71.7	92.8	94.7	99.3	52.1
KS6L	71.8	93.2	95.3	99.4	50.1
KS10	72.2	94.2	96.6	99.4	49.5
KS15	73.0	94.4	96.3	99.4	47.9
KS25	70.1	93.6	96.2	99.5	44.3
KS44	68.6	90.6	94.3	99.7	39.6
SFG6L	70.3	94.2	95.9	99.3	47.0
T44	86.2	95.9	97.5	99.6	38.4
SLP30	59.8	89.9	93.7	99.5	38.8
SLP50	73.2	94.0	95.9	99.4	42.1
SFG44	74.9	94.7	96.6	99.6	35.5
SFG75	77.5	95.3	96.8	99.6	33.4
El	68.5	93.6	96.9	99.7	24.5
MCMB	66.1	91.1	95.1	99.7	18.3

CE: Coulombic efficiency.
Source: Placke et al. [23]. Reproduced with permission of the Electrochemical Society.

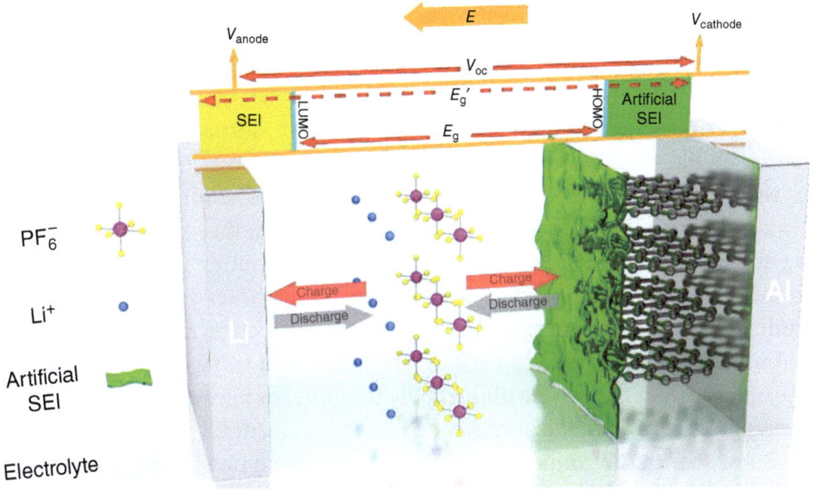

Figure 10.4 Schematic of the working mechanism of DIB using SEI-modified graphite cathode and Li anode. Source: Li et al. [25]. Reproduced with permission of WILEY-VCH.

However, the DIB is still in the preliminary stage because the energy density of such graphite-based DIBs lags behind the rocking-chair LIBs in the cell level. Moreover, the high intercalation voltage of anions into graphite may lead to the exfoliation of graphite layers and decomposition of organic electrolytes. Thus, searching for other anion host materials with high structural stability and high capacity is urgently desirable for developing advanced DIB.

10.1.2.2 Organic Materials

Organic materials have attracted considerable attention as promising cathode candidates for anion intercalation attributed to their structure/property diversity, mild operating voltage, and good renewability without any transition-metal elements. Some organic materials that contain carbonyl or anhydride functional groups exhibit the potential to be ideal cathode materials because these groups can reversibly incorporate and dissociate ions in a faradaic process. For example, Poizot et al. synthesized layered aromatic amine dilithium 2,5-(dianilino)terephthalate (Li_2DAnT) by conversion of 2,5-(dianilino)terephthalic acid to its lithium salt [26]. The Li_2DAnT initially accommodates the solvent molecules into its host structure, which facilitates the subsequent reversible electrochemical anion ($TFSI^-$) intercalation. The organic structure can react reversibly at high operating potential (3.22 V vs. Li^+/Li) with acceptable cycling performance in DIBs using Li as the anode.

Coronene, a polycyclic aromatic hydrocarbon (PAH), can exhibit highly reversible PF_6^- anion-storage properties. The photograph and chemical structure are depicted in Figure 10.5a. Coronene crystals exhibit a monoclinic structure (space group: $P2_1/a$) with lattice constants of 16.1 Å (a), 4.7 Å (b), and 10.1 Å (c). Moreover, the much lower density of coronene (1.37 g cm^{-3}) than graphite (2.26 g cm^{-3}) facilitates the incorporation of bulky anions [27]. A DIB constructed with coronene cathode and Li anode in an electrolyte consisting of 1 M $LiPF_6$ exhibits excellent long-term cyclability of 92% retention after over 960 cycles (Figure 10.5b). During the charging process, PF_6^- intercalates into the coronene structure, forming $(coronene)_xPF_6$ compound. While, the deintercalation of PF_6^- anion extract from the cathode during the discharging process. Though the capacity of coronene cathode for PF_6^- anion storage is low compared to graphitic cathodes, this work opens a new window to seek alternative cathodes with a moderate operating voltage in organic electrolyte.

Organic material composed of sulfur in the chain can also be applied as cathode material in the DIB. Esser and coworkers reported a high-voltage organic cathode material of three thianthrene-substituted polynorbornenes for the DIB [28]. During the charging process, the radical nitroxide groups are oxidized to oxoammonium cations. Subsequently, the anions such as PF_6^- are inserted into the polymer to balance the positive charge. As shown in Figure 10.5c, Li^+ ions are intercalated into graphite or deposited onto lithium foil on the anode side. Two pairs of redox peaks are detected in cyclic voltammetry (CV) curves, which suggest a two-step oxidation process during charging. The first oxidation potential of thianthrene is located at 0.84 V vs. Fc/Fc^+, corresponding to 4.09 V vs. Li/Li^+, which makes it a superior candidate for high-voltage organic cathode material. The organic polymer electrode

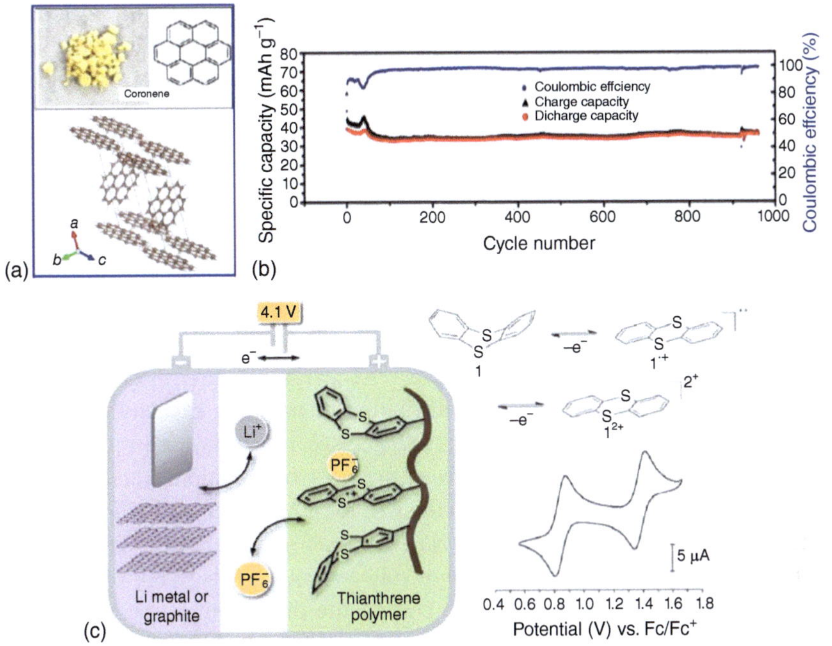

Figure 10.5 (a) Digital image of coronene solid with its molecular structure (top) and a monoclinic unit cell. (b) Long-term cycling of the half-cell for over 960 cycles at 20 mA g^{-1}, along with Coulombic efficiency. Source: Rodríguez-Pérez et al. [27]. Reproduced with permission of ©American Chemical Society. (c) Schematic illustration of a DIB by using thianthrene polymer as the cathode (left); twofold oxidation of thianthrene during cell reaction and the cyclic voltammogram of thianthrene (right). Source: Speer et al. [28]. Reproduced with permission of Royal Society of Chemistry.

demonstrated no detectable solubility in the electrolyte solutions. Furthermore, constant current investigations exhibited a maximum capacity of 66 mAh g^{-1}, which accounts for 90% of the theoretical capacity.

Polytriphenylamine (PTPAn) can be employed as a cathode material in potassium DIB. Lu and coworkers developed a potassium-based DIB with PTPAn as the cathode, graphite as anode, and KPF_6 as electrolyte [29]. In this DIB, the K^+ cations in the electrolyte migrate to the graphite anode and further intercalate into the graphite to form C_xK compounds during the charging process. Meanwhile, the active N atoms in the PTPAn cathode will lose electrons by forming N^+, and then the PF_6^- anions could be captured by the N^+ in the PTPAn cathode and form a metastable compound $N^+PF_6^-$. During the following discharge process, the C_xK compounds in the anode will lose electrons by releasing K^+ cations to the electrolyte and become electrically neutral graphite. Simultaneously, the N^+ ions in the $N^+PF_6^-$ compound would accept the electrons though releasing PF_6^- anions from the cathode to turn back into electrically natural N atoms. This full cell delivers a high reversible capacity of 60 mAh g^{-1} at a median discharge voltage of 3.23 V. Moreover, the DIB exhibits superior rate performance and long-term cycling stability over 500 cycles with a capacity retention of 75.5%.

Many organic cathode materials have been investigated for anion intercalation in recent years [30–33]. However, the organic materials are hampered by their low capacity, poor electronic conductivity, and limited cycle life. Thus, some strategies, such as coating and doping techniques, are proposed to improve organic electrodes in DIBs to enhance their conductivity and reversibility. On the other hand, searching for new host materials for anion intercalation with high capacity, low cost, and reversibility is also an urgent requirement for designing high-performance DIBs.

10.1.2.3 Other Materials

Besides the graphite and organic cathode materials, other materials have also emerged as promising cathode candidates for anion storage. Layered metal-oxide materials (e.g. $LiNi_xCo_yMn_zO_2$, $LiNi_{0.8}Co_{0.15}Al_{0.05}O_2$) are the most common cathode materials in commercial LIBs attributed to the high capacity, excellent cycling performance, and high working voltage. Recently, a layered metal-oxide material P3-$Na_{0.5}Ni_{0.25}Mn_{0.75}O_2$ was investigated as both cationic and anionic co-(de)intercalation cathode materials in the DIB [34]. As presented in Figure 10.6, during the initial charging (Stage I), the Na^+ cations de-intercalated from P3-$Na_{0.5}Ni_{0.25}Mn_{0.75}O_2$. When charged to the high-voltage plateau (Stage II), the anions (ClO_4^-) intercalated into the interlayer. During the discharging process (Stages II′ and I′), the ClO_4^- extracted from P3-$Na_{0.5}Ni_{0.25}Mn_{0.75}O_2$, and Na^+ intercalated into P3-$Na_{0.5}Ni_{0.25}Mn_{0.75}O_2$. The energy storage is provided by both the cationic (Na^+) (de)intercalation and the intercalation/de-intercalation of anions (ClO_4^-). Correspondingly, the cationic (Ni^{2+} Ni^{4+}) and anionic (O^{2-} O^-) reversible redox compensates the charge balance.

However, the intercalation of anions into two-dimensional (2D) layered compounds may result in the separation of layers and poor cycling stability. Recently, metal–organic frameworks (MOFs) have been proposed as a promising cathode

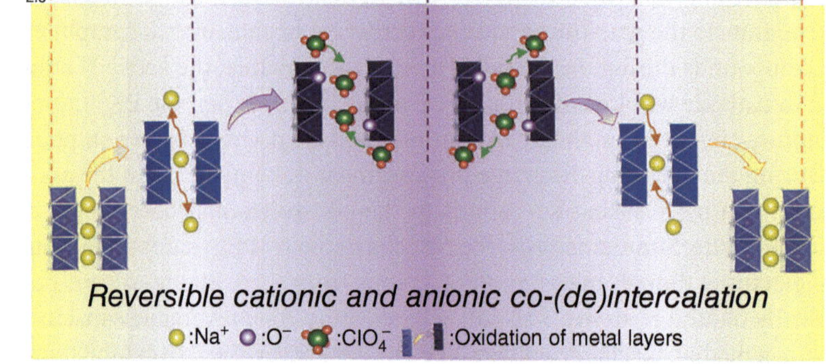

Figure 10.6 Schematic of cationic and anionic co-(de)intercalation mechanism in the layered metal-oxide P3-$Na_{0.5}Ni_{0.25}Mn_{0.75}O_2$. Source: Li et al. [34]. Reproduced with permission of Elsevier.

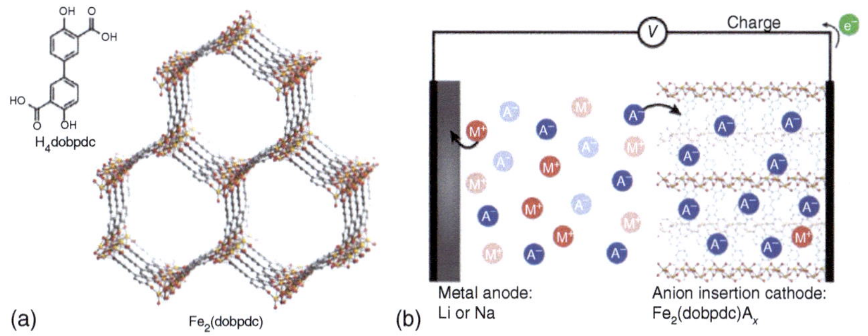

Figure 10.7 (a) Structure of the Fe_2(dobpdc) framework. (b) Schematic of a DIB by using Fe_2(dobpdc) as the cathode. A^-, can be tetrafluoroborate, hexafluorophosphate, bis(trifluoromethylsulfonyl)imide, or tetrakis(perfluorophenyl)borate, $0 < x < 2$, M^+, can be lithium, sodium, or potassium. Source: Aubrey et al. [36]. Reproduced with permission of American Chemical Society.

for reversible intercalation/deintercalation of anions attributed to their unique 3D porous structure and variable mixed-valence states [35]. Long et al. designed a redox-active MOF Fe_2(dobpdc) ($dobpdc^{4-}$ = 4,4′-dioxidobiphenyl-3,3′-dicarboxylate) with larger pore size, as shown in Figure 10.7a [36]. The Fe_2(dobpdc) can be applied as cathode materials in DIB, which is shown to undergo reversible intercalation of BF_4^- and PF_6^- (Figure 10.7b). During charging/discharging processes, a minor lattice change and a broad intervalence charge-transfer band can be detected, which suggests charge mobility [36]. The anion insertion cathode exhibits high initial reduction potential and further stability of over 50 cycles in DIB-based sodium anode. Placke et al. reported Cu(TCNQ) MOF as the cathode material for reversible storage of PF_6^- anions [37]. The Cu(TCNQ) MOF composed of Cu(I) ions and radical anionic ligands of 7,7,8,8-tetracyanoquinodimethane (TCNQ) exhibits a potential above 3.75 V vs. Li^+/Li during the transition into a new crystal phase, rendering a high energy density. Further investigation suggests that the Cu(TCNQ) MOF belongs to both “conversion” and “insertion” materials.

Apart from layered metal-oxide and MOF material, sodium manganese oxides (NMOs), which are widely used cathode in sodium-ion battery, have also been reported as cathode material in DIB. Hou and coworkers investigated an aqueous DIB using NMO as the cathode, BiF_3 as the anode, and NaF solution as electrolyte [38]. The reaction mechanism is different from the common cation–anion DIB. During the charging/discharging process, sodium ions de-intercalate/intercalate from/into NMO cathode, and BiF_3 anode release/capture the fluoride ions in aqueous electrolyte (Figure 10.8). NMO is either acceptor or a donor of sodium ions in the charging/discharging process. The aqueous battery can be allowed to assemble in the open-air environment, which provides a promising alternative to the traditional battery system.

10.1.3 Anode Materials

Aside from the cathode materials, it is also quite necessary to explore high-performance anode materials as they can greatly codetermine the electrochemical

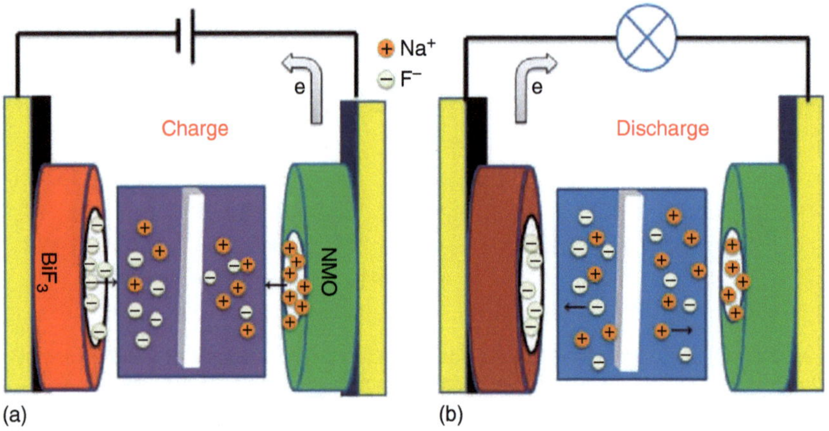

Figure 10.8 Schematics of the reaction mechanism in aqueous DIB during the (a) charging and (b) discharging process. Source: Zhang et al. [38]. Reproduced with permission of Royal Society of Chemistry.

performance of DIBs (i.e. specific capacity, rate capability, cycling stability, etc.) by storing/releasing cations during charge/discharge process based on the same working mechanism in rocking-chair batteries. A broad range of active cations reported in published papers for DIBs, including monovalent ones (e.g. Li^+, Na^+, K^+, etc.) or multivalent ones (e.g. Zn^{2+}, Mg^{2+}, Ca^{2+}, Al^{3+}, etc.) [39]. In principle, any type of host materials, which is able to reversibly store a specific type of the cation, can be considered to serve as anode materials. Considering that the main reaction mechanism between cations and anode, including metal plating/stripping, alloying/de-alloying, intercalation/de-intercalation, and conversion, the novel anode materials will be introduced here based on the categories of metallic materials, alloying-type materials, intercalation-type materials, and conversion-type materials.

10.1.3.1 Metallic Materials

Metallic materials (i.e. Li, Na, K, Zn, Ca, Al) can be directly used as anode materials and/or current collectors due to their high theoretical capacities, suitable redox potential, and excellent conductivity [40]. Li-graphite DIB based on Li metal anode was investigated by Winter and coworkers with the general term of "dual-ion battery" proposed for the first time, in which the $TFSI^-$ intercalated into the graphite cathode and the lithium deposited onto the metallic lithium anode in the IL electrolyte [41]. The battery performance was studied in terms of cut-off voltage, temperature, cycling stability, self-discharge, and rate performance. However, the formation of Li dendrite and rupturing of SEI layers caused by the repeated Li plating/stripping process is an unavoidable issue for Li metal anode [42], leading to the excessive consumption of electrolyte, enhanced electrode polarization, rapid capacity fade, internal short circuit, and other security breaches, which impede the practical application of the Li metal anode in DIBs [43]. Strategies have been made on surface modification of Li anode to express the Li dendritic growth, such as artificial fabrication of robust SEI layer, the introduction of other conductive materials, and so on [44, 45]. A self-supporting film of carbon nanofibers (CNFs) with large specific surface area, high conductivity, and surface stability was covered and pressed on Li anode by Wu and coworkers [45]. The assembled corresponding Li-graphite DIBs exhibited dendrite-free deposition of Li anode during cycling and delivered much enhanced cycle life (86.4% capacity retention over 2000 cycles) as well as more stable and smoother values of discharge capacity and coulombic efficiency (CE), compared with the bare Li anode during cycling (Figure 10.9). The formation of dendrites could also be observed in some other metallic materials anode-based batteries or DIBs (e.g. Na [46], K [47], Zn [48], Al [49], etc.), and much efforts have been devoted to the development of methods to limit the dendrites formation, such as modification of common electrolyte, protective coating over the metallic anode, incorporating a ceramic separator, and so on [40, 50–52].

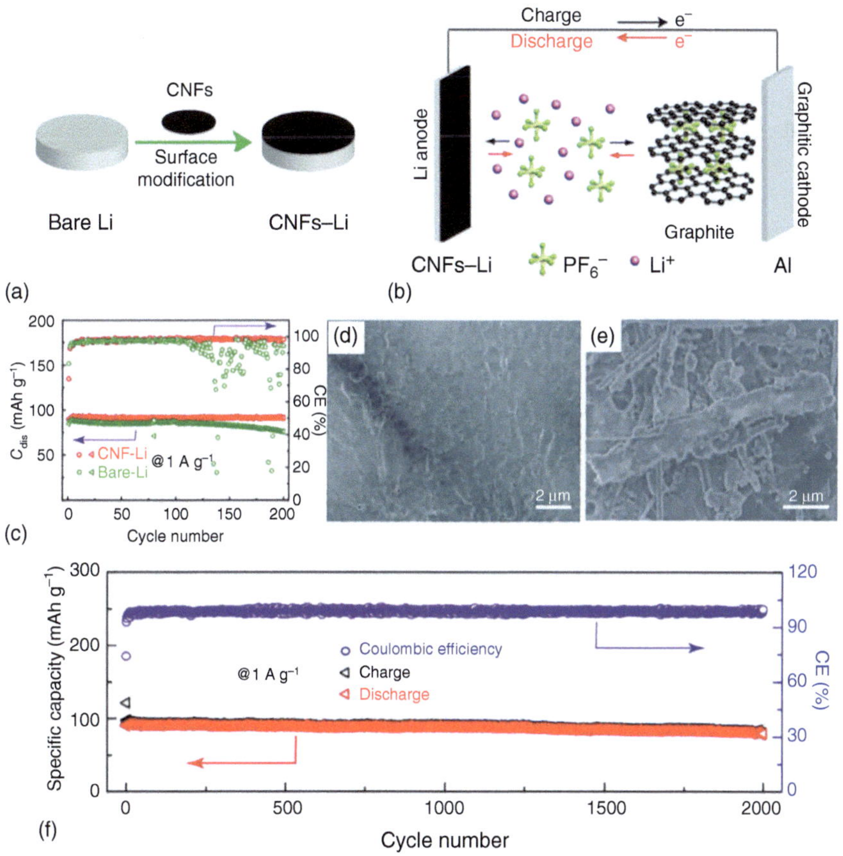

Figure 10.9 (a) Schematic of Li surface modification by CNFs film. (b) Working illustration of the CNFs modified Li-graphite DIB. (c) The cycle performance of Li-graphite DIBs at a current density of $1\,A\,g^{-1}$. The surface SEM images of CNFs-Li (d) and bare Li (e) after 200 cycles. (f) The long-term cycle performance of improved Li-graphite DIB over 2000 cycles at $1\,A\,g^{-1}$. Source: Xi et al. [45]. Reproduced with permission of Royal Society of Chemistry.

For metallic Ca anode-based DIBs, the development is hindered by the incompatibility between electrodes and electrolytes and the formation of nonconductive passive layer for Ca^{2+} diffusion, and thus the poor reversibility of Ca plating/stripping. The pioneering work of Aurbach et al. revealed that the deposition of calcium on a metallic Ca anode was impossible in organic electrolytes at room temperature [53]. Palacin and coworkers observed calcium electrodeposition at 75–100 °C in $Ca(BF_4)_2$-based organic electrolyte and confirmed its reversibility for more than 30 cycles accompanied by significant side reactions including the formation of CaF_2 [54]. When using organic electrolyte $Ca(BH_4)_2$ in tetrahydrofuran (THF), the CaH_2 by-product layer formed to protect calcium from further reaction with the electrolyte at open circuit [55]. The calcium can be plated and stripped at room temperature with capacities of 1 mAh cm^{-2} at a rate of 1 mA cm^{-2} over 50 cycles.

Although great efforts have been made to optimize metallic materials for improving the performance of metal anode-based DIBs, the security issues, as well as metal anode–electrolyte solution compatibility, hamper the implementation of the metal anode in commercially viable DIBs, which prompts the researcher to seek for other kinds of anode material candidates, such as alloying materials and intercalation materials.

10.1.3.2 Alloying-Type Materials

There exist many elements which are able to form alloys with active cations in electrolytes. Typically, alloying materials exhibit a high outstanding ability for storing metal cations, low cost, environmental friendliness, and safe operation potentials [6, 20]. The electrochemical alloying/dealloying process of alloying-type anode material is shown as follows:

$$xA + yM^{n+} + nye^- \leftrightarrow A_xM_{ny} \tag{10.5}$$

Where A is the alloying element (e.g. Al, Sn, Si, etc.) [31], M^{n+} represents the active cations (e.g. Li^+, Na^+, K^+, Zn^{2+}, Mg^{2+}, Ca^{2+}, Al^{3+}, etc.) in electrolyte [56], and x and y are the stoichiometric coefficients for the chemical formula of the alloy.

Al, an inexpensive and environmentally benign element with a superior electrical conductivity (3.5×10^7 m^{-1} Ω^{-1}) [57], shows the great promise in alloying materials: a high theoretical capacity of 993 mAh g^{-1} is achieved with the phase of LiAl [58]. Besides, a flat and wide plateau can be obtained from the charge/discharge curves, indicating a steady power output [59]. Furthermore, two other stable intermetallic compounds, Li_3Al_2 and Li_9Al_4 with the maximum theoretical capacity of 2235 mAh g^{-1} [20], can be formed at room temperature according to the Li–Al binary phase diagram [60]. An Al foil acting as both the anode and current collector was first reported in DIB with graphite as the cathode material in an ethyl-methyl carbonate (EMC) electrolyte containing $LiPF_6$ [5]. A flat plateau at c. 0.22 V (vs. Li/Li^+) and c. 0.52 V (vs. Li/Li^+) during Li–Al alloying/dealloying process was exhibited in the charge/discharge curve of a Li–Al half-cell. A rough layer could be observed on the surface of Al anode after charging with the composition of F, C, and O elements, confirming the formation of the SEI layer and the Li–Al alloying process, which would protect the Al anode from destruction during the cyclic alloying/dealloying process. The packaged Al-graphite DIB was estimated to deliver an energy density of ≈150 Wh kg^{-1} at a power density of ≈1200 W kg^{-1}, which was much higher than most commercial LIBs. However, the lithiation of aluminum faced a ~100% volume expansion, which would tear up the anode SEI layer and cause the pulverization of the active material, leading to the rapid capacity fading, loss of CE, or even the death of cells in a worse situation [61].

Several methods, like surface modification and structure design, are proposed to optimize the electrochemical performance of Al anode-based DIBs [62–64]. Tang et al. developed a 3D porous Al foil coated with a carbon layer for use as the anode and current collector in DIB. The 3D porous structure of Al not only alleviated the mechanical stress caused by Al volume change during the continuous electrochemical cyclic process but also shortened the ion diffusion length. Simultaneously, the carbon-coated layer helped buffer the Al volume change and alleviate undesirable surface reactions. An assembled DIB with graphite as the cathode material exhibited excellent cycling stability over 1000 cycles with a reversible capacity of 93 mAh g^{-1} at 2 C (corresponding to 200 mA g^{-1}) with 89.4% capacity retention. They also presented a core/shell Al@carbon nanosphere (nAl@C) anode material for DIB [62], in which the nAl@C can accommodate mechanical strain and stress better than flat electrodes, while the carbon layer was beneficial for ions and electrons transfer, leading to superior long-term cycling stability with a capacity retention of 94.6% after 1000 cycles at a high current rate of 15 C (1500 mA g^{-1}) (Figure 10.10).

Sn is another well-studied alternative alloying anode material for DIBs [65, 66]. High theoretical capacities of 991 and 847 mAh g^{-1} can be achieved with the phase of $Li_{4.4}Sn$ and $Na_{15}Sn_4$ alloys, respectively [67, 68]. In addition, a great number of other intermediate phases can be formed according to the corresponding equilibrium phase diagrams (e.g. $NaSn_6$ and $NaSn_4$, etc., in the Na–Sn system [69]; K_4Sn_{23} and KSn, etc., in the K–Sn system [70]; Ca_7Sn_6 and CaSn, etc., in the Ca–Sn system [71]). The working mechanism of Sn foil anode in Na-ion-based DIB using graphite as cathode material and $NaPF_6$ salt-based organic electrolyte was first reported by Tang's group [72]. The diffraction peaks of NaSn were detected from the X-ray diffraction (XRD) pattern of Sn foil in the charged state, confirming the formation of an alloy on Sn anode, which also appeared in Na-ion battery investigations [73, 74]. Although the shallow sodiation of NaSn alloy possessed much lower theoretical capacity (226 mAh g^{-1}) than the full conversion alloy $Na_{15}Sn_4$ (847 mAh g^{-1}), it generated much smaller volume variation during the continuous charge/discharge process, resulting in an improved cycling performance with 94% capacity retention for 400 cycles. Subsequently, they studied the Sn foil as anode material and current collector for K-ion-based DIB [75]. A pair of flat plateau observed in the charge/discharge curves represented the K–Sn alloying/dealloying process, and the final product was identified to be K_2Sn with the space group of R-3m. Furthermore, a Ca-ion-based DIB utilizing Sn foil as the anode and graphite as the cathode was also investigated [14]. According to the XRD results of the Sn anode in different charged states during the first cycle, the reaction mechanism on Sn anode can be attributed to the reversible alloying/dealloying process of Sn to Ca_7Sn_6 which provided a high capacity of 526 mAh g^{-1} with a volume expansion of 136.8%. Four

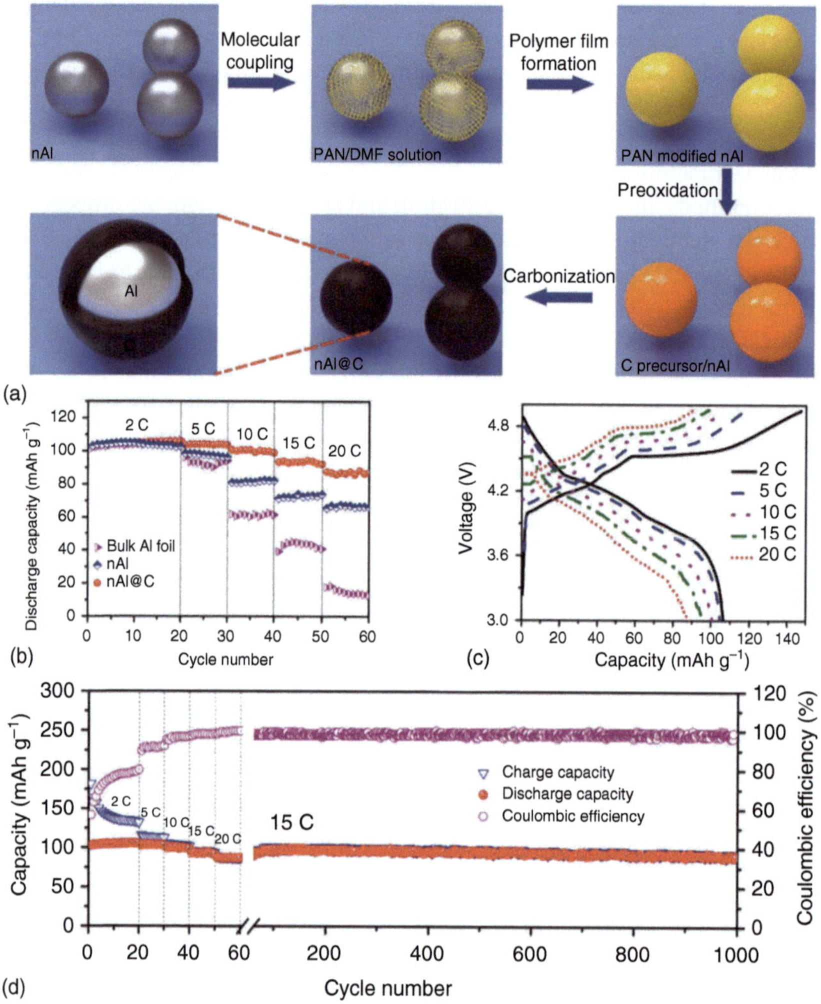

Figure 10.10 (a) Schematic illustration of the nAl@C fabrication process. (b) Discharge capacities of the nAl@C-graphite DIB at a current rate of 2, 5, 10, 15, and 20 C, with DIBs based on the nAl and bulk Al foil anodes shown for comparison. (c) The corresponding charge–discharge curves of the nAl@C-graphite DIB at different current rates. (d) The long-term cycling stability of the nAl@C- graphite DIB at a high current rate of 15 C for 1000 charge–discharge cycles. Source: Tong et al. [62]. Reproduced with permission of Wiley-VCH.

different bonding situations between Ca and Sn in the Ca_7Sn_6 alloy were explored by DFT calculation, indicating the feasibility of the formation of Ca_7Sn_6 phase thermodynamically. In addition, the stress evolution of the Sn anode during battery reactions was monitored in the first cycle. As Ca–Sn alloying proceeded during the charging process, compressive stress increased slowly up to around −10 MPa at a voltage of 4.34 V and then more rapidly until reaching −48.13 MPa, while during the

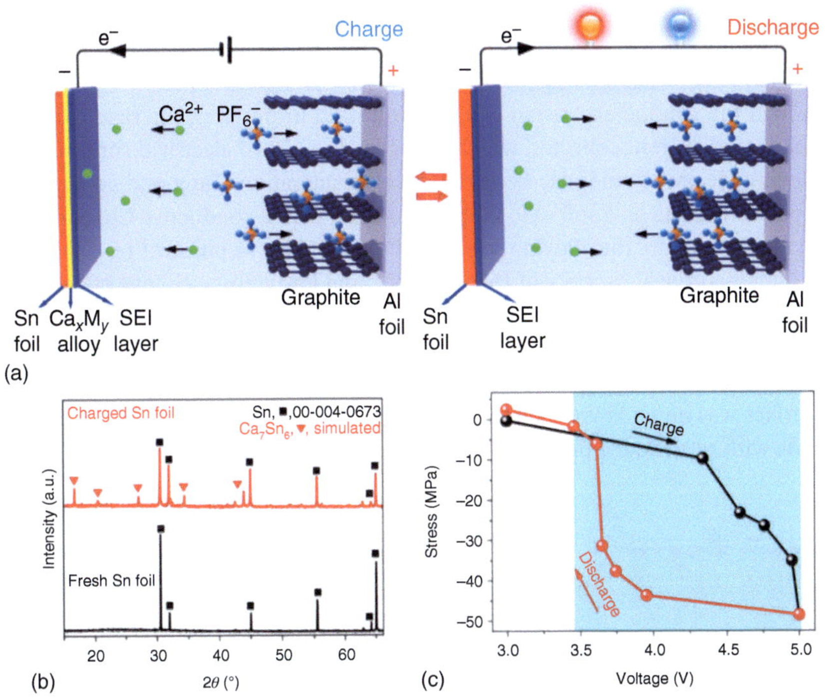

Figure 10.11 (a) Work schematics of the proposed Ca-ion based DIB using Sn foil that could form an alloy with Ca as both current collector and anode. Green, blue, and orange balls represent Ca, F, and P atoms, respectively. (b) XRD patterns of a fresh and charged Sn anode (after 300 cycles). (c) In situ stress measurement of an Sn anode during charge–discharge during the first cycle. The shaded area indicates the stress-dominant area. Source: Wang et al. [14]. Reproduced with permission of Springer Nature.

discharge step, the compressive stress gradually decreased as the Ca–Sn de-alloyed (Figure 10.11).

Si, an abundant material in nature, is an attractive anode material for LIBs due to its high theoretical capacity with the phase of $Li_{22}Si_5$ alloy (4200 mAh g^{-1}) [76]. However, the pulverization and capacity fading caused by the large volume expansion problem (~400%) during the alloying/dealloying process, as well as the low electronic conductivity, greatly limit its application. Only when the problems have been overcome to some extent, can Si be served as anode material for DIBs. The useful idea to optimize Si anode in LIBs such as nanostructure design [77] combination with high conductivity agent [78] can be borrowed. Shao and coworkers reported a Si/C core–shell-structured anode prepared by interfacial adhesion between nano-silicon and graphite with the help of pitch for DIB [79]. The reaction mechanism of Si/C was investigated by the electrochemical tests of half-cell against Li metal, in which a cathodic peak at 0.12 V originated from the insertion of lithium and the formation of the amorphous Li_xSi phase, and an anodic peak at 0.35 V arose due to the extraction of lithium from Li–Si alloy. Benefiting from the carbon layer, the binder pitch and the volume buffer on the integrity of the core–shell-like structure of Si nanospheres in the Si/C anode, the assemble DIB using expanded graphite as the cathode material delivered highly attractive energy densities of 252–222.6 Wh kg^{-1} at power densities of 215–5420 W kg^{-1}, as well as good cycling stability over 1000 cycles.

Although the volume expansion can be minimized in this way, the rigid contact between Si and current collector, which could lead to loss of electrical contact and fast capacity decay, should also be taken into consideration. Tang and coworkers developed a Si anode on a soft nylon fabric modified with a conductive Cu–Ni transition layer to regulate the alloying stress of Si for DIB [80]. A phase of $Li_{12}Si_7$ alloy with a high theoretical capacity of 1636 mAh g^{-1} and low volume change of c. 117% was detected by XRD patterns during the charging process of Si anode. The significant stress relaxation was confirmed by the finite element method (FEM) and in situ stress test. The maximum compressive stress (−182 MPa) for Si anode with a flexible interface was much lower than the rigid one (−724 MPa), which endowed the Si electrode with remarkable flexibility and stability over 50 000 bends (Figure 10.12).

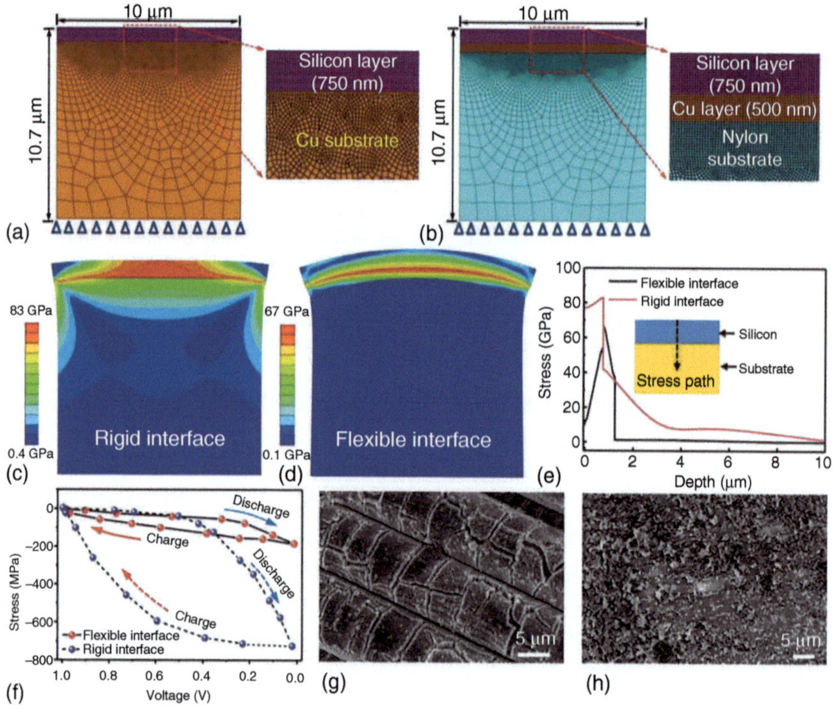

Figure 10.12 The FEM models, mesh, and boundary conditions of silicon anode with (a) rigid and (b) flexible interfaces. (c,d) Contours of the von Mises stress (GPa) generated in silicon anodes with rigid (c) and flexible (d) interfaces. (e) von Mises stress path distribution of silicon anodes with flexible and rigid interfaces along the symmetric axis from the top surface to the substrate. (f) Stress evolution with different charging/discharging voltages of silicon anodes with flexible and rigid interfaces (the rigid interface is constructed on the stainless steel substrate). (g, h) Surface SEM images of silicon anode with a flexible interface (after 2000 cycles at 10 C) (g) and a rigid interface (after 850 cycles at 10 C) (h). Source: Jiang et al. [80]. Reproduced with permission of Wiley-VCH.

Besides these alloying materials, some elements like germanium (Ge) [81], lead (Pb) [75], and stibium (Sb) [72] also exhibit the ability of alloying/dealloying with active cations in electrolyte for DIBs. However, their reaction mechanism and electrochemical performance should be further clarified and optimized. And, many other materials (e.g. bismuth [Bi] [82], phosphorus-based materials [Sn_4P_3] [83], cobalt- and nickel-based materials [$CoSn_2$, Ni_3Sn_2] [84], etc.) which have been demonstrated in metal-ion batteries based on alloying mechanism will have potential applications as alloying anode materials in DIBs.

10.1.3.3 Intercalation-Type Materials

MoS_2, a typical layered metal sulfide, has been regarded as a promising anode for Na^+ insertion/extraction with a high theoretical Na storage capacity of 670 mAh g^{-1} [87]. Strategies such as nanostructure design and combination with a conductive agent have also been taken to adjust its poor conductivity and severe aggregation during the charge/discharge process [88–90]. Tang and coworkers prepared penne-like MoS_2/C nanotube composite via a two-step microwave-solvothermal and carbonization method [18]. The hierarchical MoS_2/C nanotube provided expanded (002) interlayer spacing of 2H-MoS_2, which facilitated fast Na^+ insertion/extraction reaction kinetics, thus contributing to improved Na-ion-based DIB performance with a reversible capacity of 65 mAh g^{-1} at 2 C in the voltage window of 1.0–4.0 V, as well as good cycling performance for 200 cycles and 85% capacity retention.

The active cations intercalate into this kind of material upon charging and then migrate back to the electrolyte during the discharge process. Since LIBs were first commercialized by Sony in 1991, graphite has become the dominant anode material because of its low cost, high safety, and excellent structural stability. The first DIB prototype "dual-graphite cell" based on cation and anion intercalation of non-aqueous electrolyte in graphite was reported in 1989 by McCullough et al. [8]. The lithium intercalation into graphite can be described as follows:

$$6C + Li^+ + e^- \leftrightarrow LiC_6 \tag{10.6}$$

The reaction was proved experimentally and theoretically as a staging phenomenon including the formation of Li_xC_6 ($0 < x < 1$) at different potential ranges: [20] stage 4 was the most stable phase for $x \leq 1/6$, then it transformed into a mixture of stages 3, 2, and 1 for $1/6 < x < 1/3$, followed by a mixture of stage 2 and stage 1 for $1/3 < x < 1/2$, and finally into a fully stage 1 for $x > 1/2$. However, the formation of electrolyte decomposition SEI layer would lead to loss of CE in the first cycle and reduced capacity in the following cycles [85]. A feasible solution is to adjust the lower cut-off potential to a value higher than the SEI formation potential of electrolytes [86]. And, it is necessary to find some compatible intercalation-type anode materials which could fully release their performance with enhanced lower cut-off potential.

Especially, the orthorhombic Nb_2O_5 received great attention due to its layered structure, open intrinsic framework, and good chemical stability [16, 94]. A Nb_2O_5 nanomaterial with a hierarchical urchin-like structure assembled by nanowires designed by Tang and coworkers was fully explored as an anode based on the hybrid potassiation/depotassiation mechanism of intercalation-pseudocapacitive

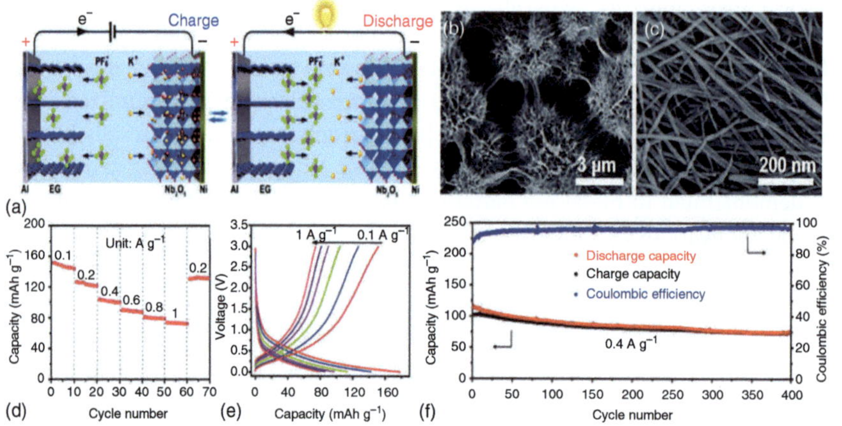

Figure 10.13 (a) Schematic illustration of the working mechanism of a DIB based on the Nb_2O_5 anode and expanded graphite (EG) cathode. (b, c) SEM images of the Nb_2O_5 with different magnifications. (d) Charge/discharge curves of the Nb_2O_5 electrode at different current densities from 0.1 to 1 A g^{-1}. (e) Corresponding charge capacities at different current densities. (f) Cycling performance of the Nb_2O_5/K half-cell for 400 cycles at 0.4 A g^{-1}. Source: Heidrich et al. [85]. Reproduced with permission of Elsevier.

behavior in K-ion based DIB [16]. The Nb_2O_5 nanowires delivered a reversible capacity of 104 mAh g^{-1} at a current density of 0.4 A g^{-1} and low decay of 0.068% per cycle for 400 cycles (Figure 10.13). In addition, a few bimetallic oxides are also available as anode materials for DIBs, which include spinal-structure $Li_4Ti_5O_{12}$ (operating at ~1.55 V vs. Li/Li$^+$), NASICON material $LiTi_2(PO_4)_3$ (showing redox activities at ~2.5 V vs. Li/Li$^+$ for the Ti^{4+}/Ti^{3+} redox couple) [95], and perovskite nanocrystal ($NaNbO_3$ [96], $NaTaO_3$ [97], etc.). However, the unavoidable mismatch of cation radii and the interlayer spacing of intercalation-type anode materials greatly affect the stability of the corresponding DIBs. And, the continuous intercalation/deintercalation could trigger their exfoliation, which leads to the degradation of cycling and overall performance.

10.1.3.4 Conversion-Type Materials

Yoshio and coworkers reported a number of metal oxides such as Nb_2O_5 [91], MoO_3 [92], and TiO_2 [93] serving as anode for DIBs with the cathode material of graphite, all of which showed improved CE and charge capacity.

Due to the high theoretical capacities, some transitional metal oxide and sulfides (i.e. Co_3O_4, Ni_3S_2) showed their potential as conversion-type anode materials in DIBs [98, 99]. Generally, these conversion-type materials will convert to Li_2O/Li_xS and elemental metal by reacting with active cation (i.e. Li$^+$) in the electrolyte during cycling [100] (Figure 10.14). Obviously, the anode materials based on this mechanism suffer the problem of serious volume variation during the charging/discharging process, which would result in poor stability. Decreasing particle size and combination with the conductive agent are effective ways to solve the problem for achieving improved performance [98].

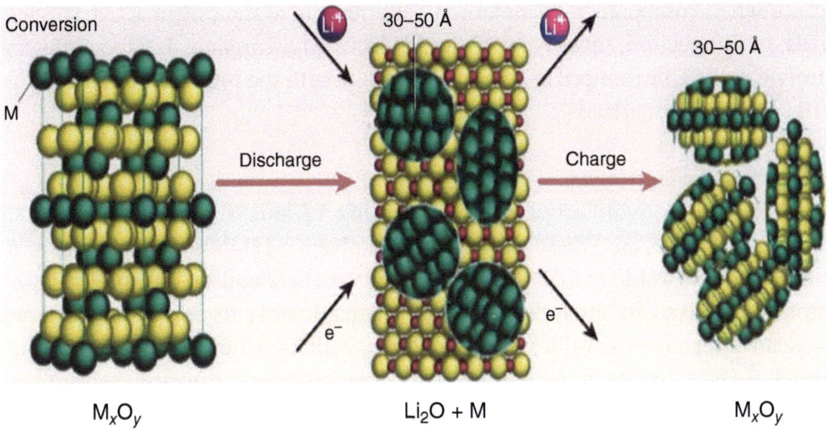

Figure 10.14 Schematic of the reaction mechanism based on the conversion reaction (M_xO_y and M represent transitional metal oxides and elemental metal, respectively). Source: Zheng et al. [100]. Licensed under CC BY 4.0.

For example, an integrated electrode with Co_3O_4 nanosheets deposited directly on carbon fiber paper (Co_3O_4/CFP) was acted as the anode in DIB [98]. A half-cell using Li foil as both counter and reference electrodes showed an initial CE of ~87.4% and an increased CE of 98.6% after the third cycle. It also exhibited good stability with a capacity retention of 90% after 40 cycles at a current density of 200 mA g^{-1}.

Intense research has been done on finding suitable alternative anode materials in DIBs to improve the overall performances. However, a desirable anode material with superior capacity, good conductivity, excellent structural, and SEI stability, as well as high security, is still a major challenge for practical application of DIBs.

10.1.4 Electrolyte

Based on the storage mechanism, the working potential (ΔV) of DIBs can be expressed as follows when taking Li^+ and A^- as the cation and anion in the electrolyte, respectively [31]:

$$\Delta V = \frac{\mu_{\text{Li}} + \mu_{\text{A}} - \mu^0_{\text{Li}^+} - \mu^0_{\text{A}^-} - kT\ln[\text{Li}^+] - kT\ln[\text{A}^-]}{\text{e}} \tag{10.7}$$

in which μ_{Li} and μ_{A} represent the chemical potential of Li and A inserted into the anode and cathode, respectively, $\mu^0_{\text{Li}^+}$ and $\mu^0_{\text{A}^-}$ stand for the chemical potential of Li^+ and A^- in 1 M solution, respectively, $[Li^+]$ and $[A^-]$ are the concentration of Li^+ and A^- in the electrolyte, respectively, k is the Boltzmann constant, and T is the temperature in kelvins. Obviously, it can be concluded that capacity and working potential of DIBs depend to a large extent on the electrolyte formulation (i.e. type of intercalated anions and solvent, salt concentration, additives). Suitable electrolytes which afford stable operation at high potentials are quite important for DIBs as the anion intercalation into host materials usually occurs at a high oxidizing potential. A much larger amount of electrolyte is needed to be stored in the pores of anode, cathode, and separator when compared with metal-ion batteries due to the role of active species in DIBs. In this section, three types of liquid electrolytes (organic, IL, and aqueous electrolyte) will be introduced and strategies to deal with the high working potential of DIBs will be summarized.

10.1.4.1 Organic Electrolyte

Organic electrolyte, with active salts (containing various anions such as PF_6^-, BF_4^-, ClO_4^-, $AlCl_4^-$, etc.) dissolved in organic solvents, exhibits its feasibility and potential practical value in DIBs owing to their varieties, ability to form SEI layer for protection and so on. First, the selections of organic solvents are quite important as they can determine the values of $\mu^0_{Li^+}$ and $\mu^0_{A^-}$, which further affects the working potential of DIBs. Taking a page from LIBs, the most popular organic solvents in DIBs are carbonates involving ethylene carbonate (EC), propylene carbonate (PC), EMC, dimethyl carbonate (DMC), etc. [101, 102], and each solvent plays a specific role in determining the electrochemical performance of DIBs. It was reported that electrolytes containing EC could form a protective film on anode materials to prevent continuous decomposition of electrolyte and thus achieve a relatively high CE; PC would endow the discharge capacities with less fluctuation [103]; EMC showed high capacities when alloying-type anode materials were used [72], while DMC can lower the charge voltage plateau, resulting in improved reaction kinetics [104]. Consequently, the combination of different carbonates should be more applicable as solvents for organic electrolytes in DIBs [14, 18, 65, 105], and it is still difficult to identify the most suitable mixtures of carbonates for DIBs due to the diversities of electrodes and electrolytes. Tang and coworkers investigated the electrochemical performance of Ca-ion-based DIB in different electrolyte formulations with mixed solvents containing 0.8 M $Ca(PF_6)_2$ [14]. Various binary solvents such as EC–DMC (v/v = 1 : 1), EC–EMC (v/v = 1 : 1), PC–DMC (v/v = 1 : 1), PC–EMC (v/v = 1 : 1), ternary solvents such as EC–DMC–EMC (v/v/v = 1 : 1 : 1), PC–DMC–EMC (v/v/v = 1 : 1 : 1), EC–PC–DMC (v/v/v = 1 : 1 : 1), EC–PC–EMC (v/v/v = 1 : 1 : 1), and quaternary solvents (i.e. EC–PC–DMC–EMC (v/v/v/v = 1 : 1 : 1 : 1) and EC–PC–DMC–EMC (v/v/v/v = 2 : 2 : 3 : 3)) were taken as examples. As a result, the composition of mixed solvents EC–PC–DMC–EMC (v/v/v/v = 2 : 2 : 3 : 3) containing 0.8 M $Ca(PF_6)_2$ was chosen as the most reasonable electrolyte in the proposed Sn-graphite DIB due to the best charge/discharge behavior when compared to other electrolytes.

Second, the concentration of active salts in organic solvents is another emphasis in DIBs as it was proposed that increasing the electrolyte concentration could benefit for the intercalation of anion into the cathode and thus improve the electrochemical properties [106]. The implementation of highly concentrated electrolytes (i.e. 4 M $LiPF_6$:DMC) in DIBs with better capacity and CE than diluted electrolytes (i.e. 1 M $LiPF_6$:DMC) was presented by Placke and coworkers [107]. Besides, the highly concentrated electrolytes provided highly reversible Li^+ uptake/release at the graphite anode and remarkably increased the reductive stability due to the synergistic effects of the solvent molecules and the salt in forming an effective SEI. And, it also provides increased stability against oxidative electrolyte decomposition at the corresponding cathode, accompanied by suppressed dissolution of Al (current collector) with the salt solvent combinations. However, it should be noted that the introduction of more active salts would lead to higher costs and viscosity of the electrolyte as well as reduce the ionic conductivity.

Third, various electrolyte additives should be taken into consideration as they can help improve the ionic conductivity of the electrolyte, form a stable SEI layer to protect the electrode materials from dissolution, and so on. Vinylene carbonate (VC) is the most common additive used in DIBs [79, 80, 92, 93, 108, 109]. It has been disclosed that the addition of 2 wt% VC in the organic electrolyte (3.5 M $LiPF_6$:EMC) to dual-graphite-based DIB led to an initially higher capacity of >90 mAh g^{-1} and remarkably improved the 90th cycle capacity retention of ~61% compared to VC-free electrolyte (~17%) (Figure 10.15) [85]. However, it should be kept in mind that a proper concentration of VC in the organic electrolyte is crucial to good electrochemical performance [62]. Besides, fluorinated material such as tris(hexafluoro-iso-propyl)phosphate (HFIP) was also served as a useful additive to improve the cycling behavior of dual-graphite-based DIB operating at a high potential between 4.0 and 5.2 V (vs. Li/Li^+) and enhance the specific capacity [12].

In addition, except for carbonate solvents, sulfone-based [110], ester-based [111], or furan-based [112] solvents also showed potential ability in DIBs due to their high oxidation voltage. A cut-off of 5.5 V and a charge capacity of c. 140 mAh g^{-1} can be achieved when using 2 M $LiPF_6$:ethyl methyl sulfone (EMS) electrolyte in "dual-graphite" cell [10]. To further improve the viscosity of sulfone-based solvent at ambient temperature, sulfolane was mixed with EMC, which can efficiently avail active PF_6^- anions insertion into graphite cathode in Li/graphite DIBs [110].

10.1.4.2 Ionic Liquid Electrolyte

The relatively poor oxidative stability of organic solvents at high working potentials (4.4–4.5 V vs. Li/Li^+) prompts the researcher to develop other novel electrolytes. ILs, with the virtues of low flammability and volatility, high stability against oxidative decomposition, broad electrochemical stability window and high thermal stability, have been served as promising electrolyte candidates for high-voltage DIBs since the first successful application of ILs as electrolytes in dual-graphite batteries was reported by Carlin et al. in 1994 [9]. A series of ILs using 1-ethyl-3-methylimidazolium (EMI^+) or 1,2-dimethyl-3-propylimidazolium ($DMPI^+$) as the cation and $AlCl_4^-$, BF_4^-, PF_6^-, $CF_3SO_3^-$, or $C_6H_5COO^-$ as the anion were investigated. Pure 1-butyl-1-methylpiperidinium bis(trifluoromethylsulfonyl) imide ($PP_{14}TFSI$) IL was also developed by Yuan and coworkers to serve as both electrolyte and charge carrier in a DIB using composite graphite as anode and cathode [113]. A well-defined specific capacity of 82 mAh g^{-1} and excellent reversibility of ~100% capacity retention for 600 cycles can be obtained for the DIB (Figure 10.16). In addition, the appropriate viscosity and conductivity of ($PP_{14}TFSI$) IL were a benefit for retarding the self-discharge process, resulting in high CE.

The high oxidative stability of IL electrolyte in DIBs was further verified by much subsequent research work [11, 41]. However, the compatibility of many ILs and graphite anodes was found to be too poor to form an effective SEI on the electrode

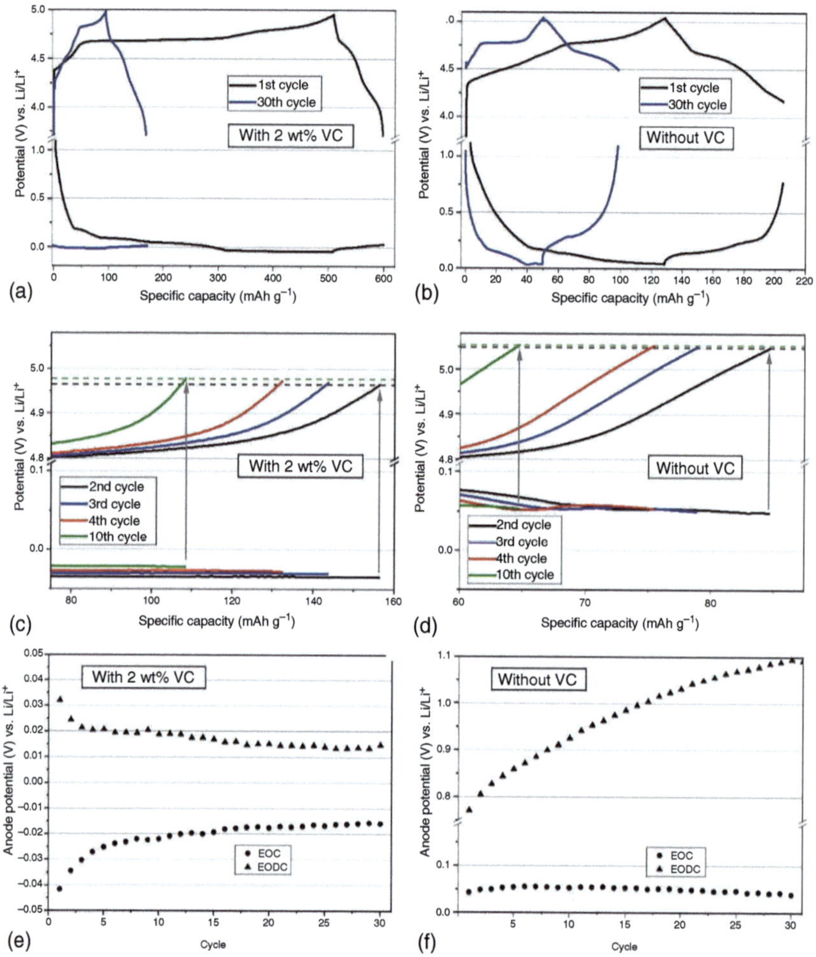

Figure 10.15 (a, b) Cathode and anode potentials vs. specific capacity. (c, d) Shift of the anode and cathode end of charge (EOC) potentials within the first 10 cycles. (e, f) Anode potentials at the end of charge (EOC) and end of discharge (EODC) vs. cycle number. All figures correspond to dual-graphite-based DIBs using 3.5 M $LiPF_6$:EMC as the electrolyte, with (a, c, e) and without (b, d, f) 2 wt% VC as electrolyte additive. Batteries are cycled in a voltage range of 3.4–5.0 V. Source: Heidrich et al. [85]. Reproduced with permission of Elsevier.

surface, leading to the co-intercalation of the IL cations into graphite anodes, which might result in graphite exfoliation and undesirable cycling stability [114]. The combination of ILs with other electrolyte additives could be an effective method to solve the aforementioned problems. Tang and coworkers investigated the role of mixtures of *N*-butyl-*N*-methylpyrrolidinium bis(trifluoromethanesulfonyl) imide (Pyr_{14}TFSI) IL and LiTFSI salts with different concentrations under the presence of fluoroethylene carbonate (FEC) additive on the electrochemical performance

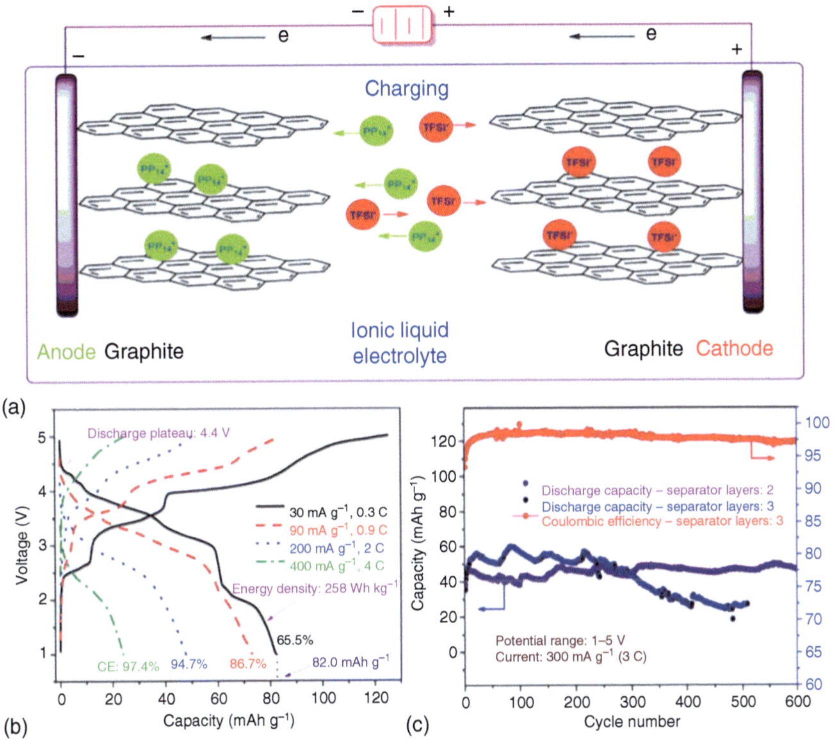

Figure 10.16 (a) Schematic illustration of the PP_{14}TFSI DIB. (b) Charge/discharge profiles of PP_{14}TFSI DIB at a series of current densities. (c) Long-term cycling performance of PP_{14}TFSI DIB with two or three layers of separators. Source: Fan et al. [113]. Reproduced with permission of Royal Society of Chemistry.

of Al/graphite DIBs [115]. When the concentration of LiTFSI was 1.5 M, the corresponding DIB showed a reversible capacity of 98 $mAh\,g^{-1}$ at 0.5 C and good cycling stability at a high cut-off voltage of 4.8 V with negligible capacity decay after 300 charge/discharge cycles. ES was another useful additive to improve the ions of Pyr_{14}TFSI-LiTFSI electrolyte intercalation into graphite electrodes. The charge capacity of Li/graphite half-cell with 2 wt% ES-containing electrolyte (97 $mAh\,g^{-1}$) was much higher than that of pure IL electrolyte ($\approx$50 $mAh\,g^{-1}$) [13]. And meanwhile, the system exhibited a highly stable cycle life, in which a capacity of 74 $mAh\,g^{-1}$ retained after 500 cycles charge/discharge process at a current density of 50 $mA\,g^{-1}$.

10.1.4.3 Aqueous Electrolyte

In view of the potential safety hazards caused by organic electrolyte and the relatively high cost of IL electrolyte, aqueous electrolyte with the advantages of low cost, high safety, and conductivity enters researchers' vision accompanied by the sacrifices of high working potential for high energy density. Tremendous efforts have been made to store anions in materials in aqueous electrolyte DIBs [38, 116]. A simple NH_4NO_3 aqueous electrolyte was employed in Mn_3O_4 cathode-based DIB under a three-electrode system [117]. It was revealed that the NO_3^- anion can be reversibly fast inserted into Mn_3O_4 with a high capacity of 183 mAh g^{-1} at a current density of 0.1 A g^{-1}. Ex-situ high resolution transmission electron microscopy (HRTEM) and corresponding energy-dispersive X-ray (EDX) mapping results showed that NO_3^- insertion de-crystalized the structure of Mn_3O_4 by creating amorphous nanoregions with portions of the nanocrystalline structure intact. However, the cut-off potential in this DIB three-electrode system was only 1.0 V (vs. Ag/AgCl), which was much lower than those in DIBs based on organic or IL electrolytes.

"Water-in-salt" electrolytes (WiSEs), a new class of aqueous electrolytes formulated with superconcentrated lithium salts in water (>21 m, mol kg^{-1}), have been developed to extend the stable potential window of aqueous electrolytes from 1.9 to 4.9 V (vs. Li/Li^+) [118, 119]. In WiSE, the existed water molecules can serve as part of ion solvation shells for the active salts and the high concentrated salts can help form dense SEI layers. Inspired by this, Miyazaki and coworkers reported the intercalation of bis(fluorosulfonyl)amide (FSA) anion aqueous electrolyte with a high concentration of 19 m into graphite host material [120]. The electrode potential can reach to 1.7 V (vs. Ag/AgCl) under the application of a constant current. Further, hybrid aqueous/nonaqueous WiSEs were explored by researchers for DIBs [94, 95]. By mixing a WiSE (21 m LiTFSI and 7 m lithium trifluoromethane sulfonate [LiOTF] in water) and organic component of 9.25 m LiTFSI in DMC, a hybrid electrolyte was developed for DIB using graphite and Nb_2O_5 as the cathode and anode, respectively [94]. No obvious signal from free water was observed in 3150–3500 cm^{-1} from Raman characterization, indicating that the water activity of the hybrid electrolyte was sufficiently suppressed. Besides, the hybrid electrolyte also showed improved safety when compared with the highly flammable property of DMC and good wettability to the cathode and anode of the DIB (Figure 10.17). The assembled DIB showed a stable working potential window of 0–3.2 V, a high initial capacity of 47.6 mAh g^{-1}, and acceptable capacity retention of 29.6 mAh g^{-1} after 300 cycles.

In a word, finding suitable electrolytes which can keep stable at high working potential during the anion insertion process in DIBs is an important decisive factor in achieving satisfactory cycling performance. The key point is to balance the important parameters such as activity, stability, viscosity, ion conductivity, wettability, and so on. The type of intercalated anions and solvent, salt concentration, and additives should be taken into consideration.

10.2 Multi-Ion Battery

A multi-ion strategy based on more than two types of active ions is proposed to further utilize the merits of dual-ion strategies. Additional cations or anions may

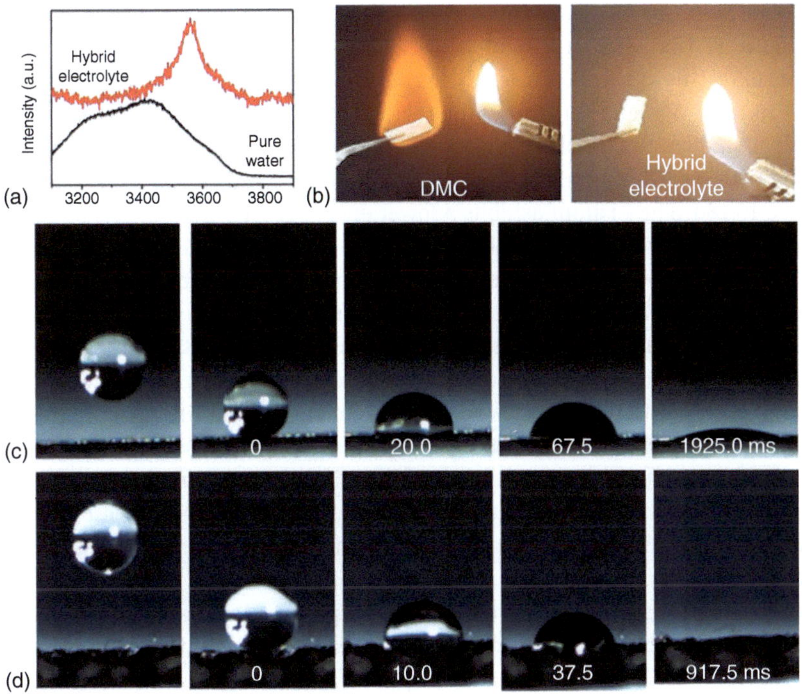

Figure 10.17 (a) Raman spectra of the hybrid electrolyte and pure water. (b) Photographs of the combustion test for DMC and the hybrid electrolyte. Snapshots showing contacting and infiltrating of 5 μl hybrid electrolyte on the (c) positive electrode and (d) negative electrode. Source: Zhu et al. [94] Reproduced with permission of Wiley VCH.

participate in the reaction in DIB to provide energy. Multi-ion strategy can further enhance the kinetics performance and improve the working voltage and energy density at the cell level, which will be discussed in detail in the following section.

10.2.1 Triple-Ion Battery

Triple-ion battery, involving "dual-cation-anion battery" and "dual-anion-cation battery," exhibits great feasibility to enhance the electrochemical properties and can be used to tackle the critical issues in the DIB.

10.2.1.1 Dual Cation–Anion Battery

Dual cation–anion battery, involving two types of cations and one type of anion, can exhibit further enhanced kinetics performance. Tang and coworkers proposed a multi-ion configuration-based sodium-ion battery, which exhibits enhanced rate performance and high capacity retention [121]. In this multi-ion battery, expanded graphite was applied as the cathode and Sn foil was used as both anode and current collector. The electrolyte was composed of Na^+, Li^+, and PF_6^- hybrid ions (Figure 10.18a). The density of states (DOS) of NaSn displays an energy gap of about 0.5 eV near the Fermi level, suggesting a semiconductor characteristic (Figure 10.18b). While Li_2Sn_5 shows the relatively high DOS near the Fermi level, indicating a typical metal characteristic (Figure 10.18c). Moreover, two possible diffusion paths (Path A and Path B) of Li and Na atom in Sn crystal are presented in Figure 10.18d. The Li atom exhibits a lower energy barrier than the Na atom in the Sn crystal, leading to faster mobility of the Li atom (Figure 10.18e). Thus, the addition of Li ions could enhance the rate performance resulting from improved ion transfer efficiency. The mechanical evolution of Sn anode in the multi-ion battery was also investigated by an in situ electrochemical stress testing system. As shown in Figure 10.18f, the stress of Sn anode almost recovers to its original value during the second cycle at 2 C, indicating excellent mechanical reversibility. As a result, the sodium-ion-based multi-ion full battery exhibits superior rate performance from 5 to 30 C (Figure 10.18g). Furthermore, the multi-ion full battery displays much more stable cycling retention of 95% after 500 cycles (Figure 10.18h). These results indicate dual cation–anion strategy endows sodium-ion battery with superior reaction kinetics.

Calcium-ion batteries (CIBs) have shown great promise to be applied in energy storage devices due to the merits of natural abundance, bivalent-ion characteristic of calcium, and similar standard reduction potential (−2.87 V) to Li/Li^+ (−3.04 V). However, the development of CIBs is hampered by lacking of suitable electrode materials and sluggish kinetics performance at room temperature. The multi-ion strategy is an effective tool to broaden the option of electrode materials and enhance electrochemical performance. A calcium-ion-based triple-ion battery with appropriate ratios of Ca^{2+}, Li^+, and PF_6^- was rationally designed, as shown in Figure 10.19a [122]. In this triple-ion battery, expanded graphite coated on the Al foil acts as the cathode and Sn foil was chosen as both the anode material and current collector. During the charging process, Ca^{2+} and Li^+ transport through the electrolyte to the Sn metal anode and form alloys. Meanwhile, PF_6^- ions intercalated into the graphite cathode. The discharging process proceeds reversibly. Bader charge analysis suggests that Ca and Li atoms donate electrons to Sn atoms to form $Ca_{31}Sn_{20}$ and Li_2Sn_5 phases, respectively (Figure 10.19b). The calcium-ion-based triple-ion battery exhibits superior rate performance at the current rate increased from 2 to 15 C (Figure 10.19c). The high capacity retention of 97% can be achieved at 15 C. Moreover, when the rate was set back, the discharge capacity at 2 C can return to the original value, indicating excellent rate reversibility. As a result, the multi-ion strategy can provide an effective solution to develop high-performance calcium-ion batteries. For the dual cation–anion strategy, the co-alloying reaction of two cations with Sn metal anode significantly alleviates the volume expansion and improves the diffusion kinetics of ions and the electron transfer in anode material; meanwhile, a relatively high working voltage can be achieved owing to the anion intercalation into the cathode.

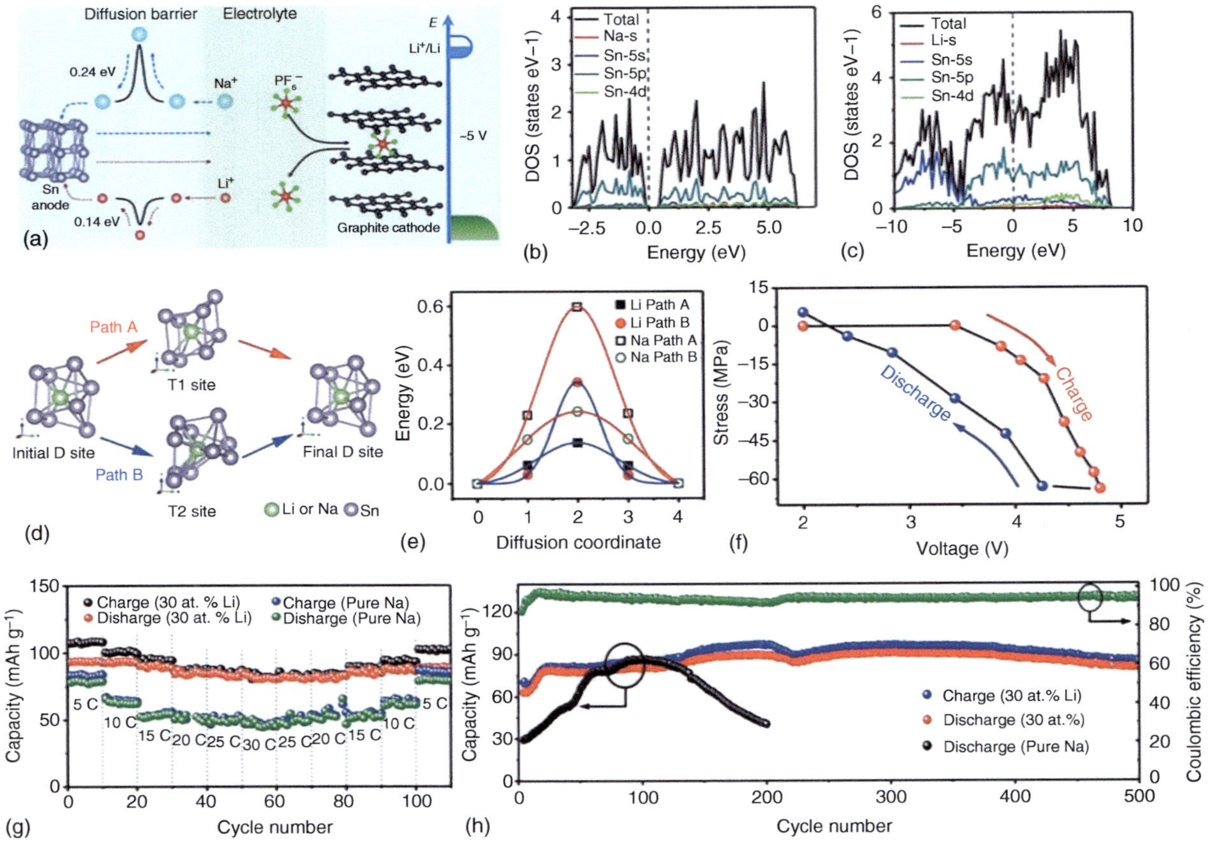

Figure 10.18 (a) Schematic illustration of the sodium-ion- based triple-ion full battery, which uses Sn metals as both the anode and current collector, expanded graphite coated on the Al foil as the cathode, and a hybrid electrolyte with Na^+, Li^+, and PF_6^- ions. (b, c) The calculated total and partial density of states of NaSn (b) and Li_2Sn_5 (c), and the vertical dot-line indicates the Fermi level. (d) Diffusion paths (Path A and Path B) for Li and Na atom in Sn crystal. (e) Diffusion energy barriers for Li and Na atom along path A and path B, respectively. (f) Stress evolution with voltage during the second cycle at 2 C. (g) Rate capacities of the sodium-ion based multi-ion full battery with 30 at.% Li and pure Na in the electrolyte at current rates ranging from 5 to 30 C. (h) Cycling performance comparison of the sodium-ion-based multi-ion full battery with 30 at.% Li and pure Na in the electrolyte at a current rate of 5 C. Source: Jiang et al. [121]. Reproduced with permission from John Wiley & Sons.

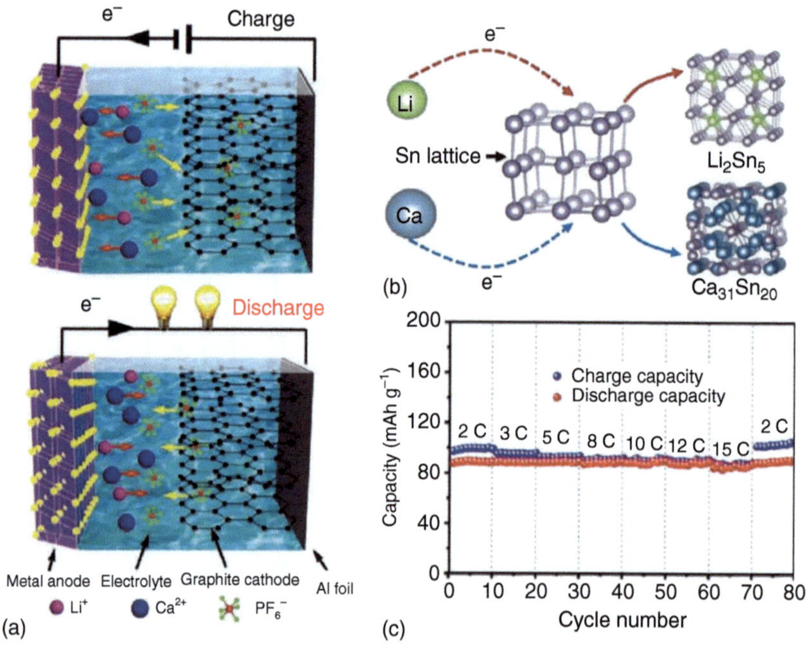

Figure 10.19 (a) Schematic illustration of the calcium-ion-based triple-ion full battery, which uses Sn metals as both the anode and current collector, expanded graphite coated on the Al foil as the cathode, and a hybrid electrolyte with Ca^{2+}, Li^{+}, and PF_6^- ions. (b) Schematic illustration of the electron transfer during the formation of Li_2Sn_5 and $Ca_{31}Sn_{20}$ phases. (c) Rate capacities of the optimized calcium-ion based triple-ion full battery at varied current rates (2–15 C). Source: Lang et al. [122]. Reproduced with permission from John Wiley & Sons.

10.2.1.2 Dual Anion–Cation Battery

Apart from the dual cation–anion battery, another triple-ion strategy called dual anion–cation battery, in which one type of cation and two types of anions take part in the electrochemical reactions, was also investigated. In the cation–anion DIB, the intercalation/deintercalation of large anions usually leads to slow ion diffusion kinetics and large volume expansion of graphite cathodes, which result in unsatisfactory rate performance and poor cycling stability. The anion-hybridization strategy by combining PF_6^- with BF_4^- in the electrolyte was demonstrated to greatly enhance the rate performance and cycling stability [109].

The schematic illustration of the dual anion–cation battery is shown in Figure 10.20a. Natural graphite is applied as the cathode material, and Al foil acts as both the anode material and current collector, Li^+, PF_6^-, and BF_4^- hybrid ions are used as the electrolyte. During the charging process, PF_6^- and BF_4^- anions co-intercalate into the graphite cathode; meanwhile, Li^+ cations transport to the Al foil anode to form an AlLi alloy. The discharging process occurs reversibly. The proportion of the PF_6^-/BF_4^- content in the hybrid electrolyte was demonstrated to be crucial to improve electrochemical performance. Hybrid electrolytes with various $LiBF_4$ contents of 0%, 5%, 10%, and 20% (at.%) were investigated. As shown

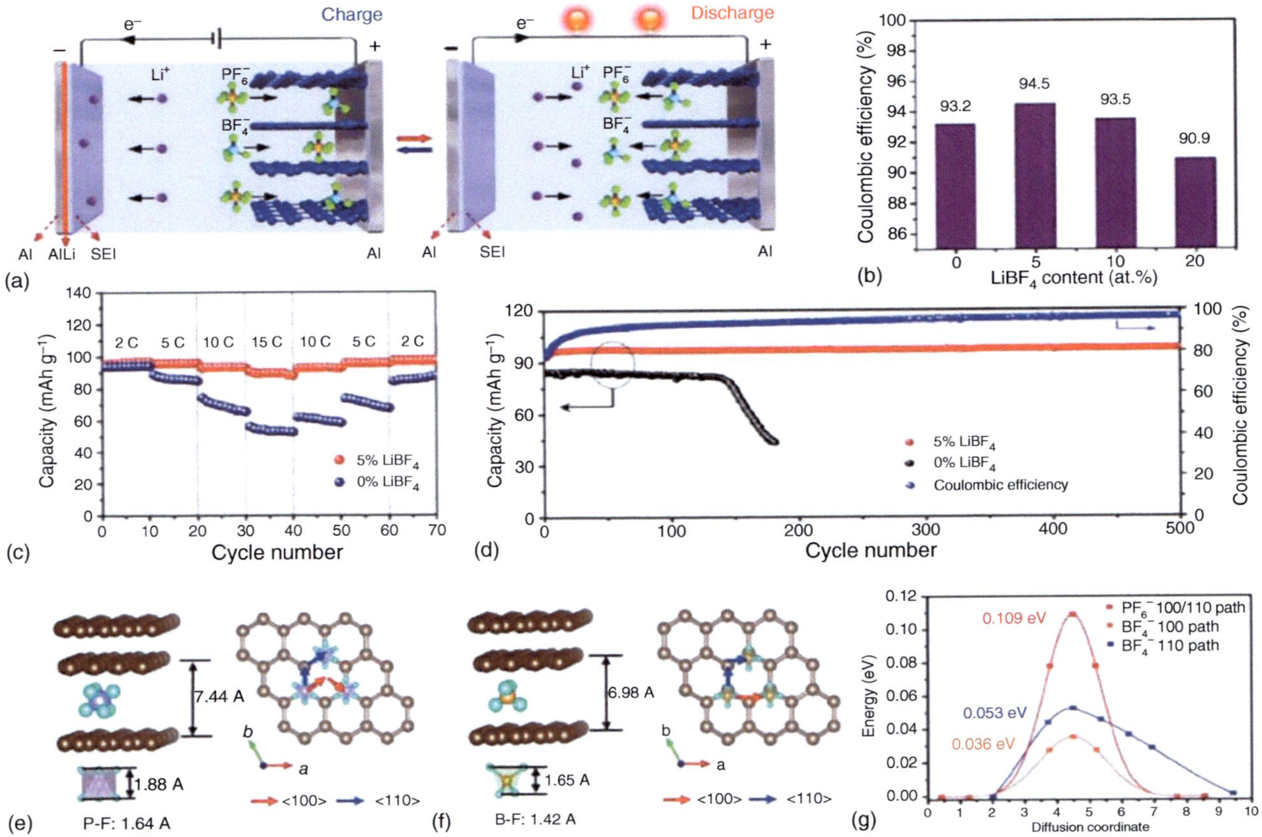

Figure 10.20 (a) Schematic illustration of the configuration. (b) Coulombic efficiency of the aluminum–graphite dual anion–cation battery with various $LiBF_4$ contents in electrolytes at 2 C. (c,d) Rate and cycling performance of the aluminum–graphite dual anion–cation battery based on 5% $LiBF_4$ electrolyte at various current rates and the battery with 0% $LiBF_4$ for comparison. Diffusion paths of PF_6^- (e) and BF_4^- (f) in the graphite layer. (g) Calculated diffusion energy of PF_6^- and BF_4^- in different paths. Source: Shin et al. [109]. Reproduced with permission of Royal Society of Chemistry.

in Figure 10.20b, the dual anion–cation battery with 5% $LiBF_4$ contents displays the highest CE of 94.5% at 2 C, which is probably due to the small dissociation constant of BF_4^- [109]. According to electrochemical performance, 5 at.% $LiBF_4$ content-based hybrid electrolyte is chosen as the optimal electrolyte.

The aluminum–graphite dual anion–cation battery with 5% $LiBF_4$ electrolyte exhibits superior rate performance at current rate increased from 2 to 15 C (92.5% capacity retention) (Figure 10.20c). Furthermore, when the current rate was set back, the discharge capacity at 2 C can recover to the original value, suggesting excellent rate reversibility. Moreover, the cycling performance at 5 C remains superior stability for over 500 cycles and high CE to 97%, as presented in Figure 10.20d. DFT was employed to investigate the diffusion kinetics of dual anions in graphite. As shown in Figure 10.20e,f, the interlayer distance of graphite for intercalation of PF_6^- and BF_4^- was demonstrated to be 0.744 and 0.698 nm, respectively. It is discovered that the anions are not solvated in the electrolyte due to their large sizes. To further investigate the reaction mechanism of the dual anions, the climbing-image nudged elastic band (CI-NEB) method was applied to calculate the energy barriers for PF_6^- and BF_4^- diffusion in graphite cathode. According to calculation results, the diffusion energy barriers of PF_6^- (0.109 eV) are much higher than those of BF_4^- via the path (100) (0.036 eV) and the path (110) (0.053 eV), indicating that BF_4^- exhibits faster diffusion kinetics than PF_6^- in graphite cathode (Figure 10.20g). Consequently, appropriate BF_4^- addition can alleviate the volume expansion of the graphite cathode and facilitate ion diffusion kinetics in graphite. These results suggest that the dual anion–cation hybridization strategy provides a significant feasibility to enhance the electrochemical properties of rechargeable batteries.

10.2.2 Quadruple-Ion Battery

In a similar manner, quadruple-ion battery, composed of two types of cations and two types of anions, could be expected to simultaneously enhance the kinetics performance of cations and anions in anode and cathode, respectively. Jiao and coworkers designed a quadruple-ion battery using graphite as the cathode, Al foil as both the anode and current collector, and $Li^+/Al^{3+}/AlF_4^-/PF_6^-$ hybrid electrolyte [123]. Figure 10.21a presents the reaction mechanism of the quadruple-ion battery during the charging/discharging process. During the charging process, the intercalation of AlF_4^- and PF_6^- into graphite cathode, meanwhile, Li^+ and Al^{3+} cations migrate to the Al foil anode to form an $AlLi_x$ alloy. During the discharge process, all cations and anions released back to the electrolyte again. The quadruple-ion battery achieves a high working voltage of 4.5 V vs. $Li^+/AlLi_x$, attributed to the intercalation of dual anions of AlF_4^- and PF_6^-. The rate capability of the quadruple-ion battery at various current densities has also been displayed in Figure 10.21b. The battery exhibits superior rate performance at the charge current density changing from 200 to 1000 mA g^{-1}. Moreover, when the current rate was set back, the discharge capacity can recover to the original value, indicating extraordinarily rate reversibility. However, it should be noted that initial charge–discharge curves displayed the specific capacity declined when the current density increased

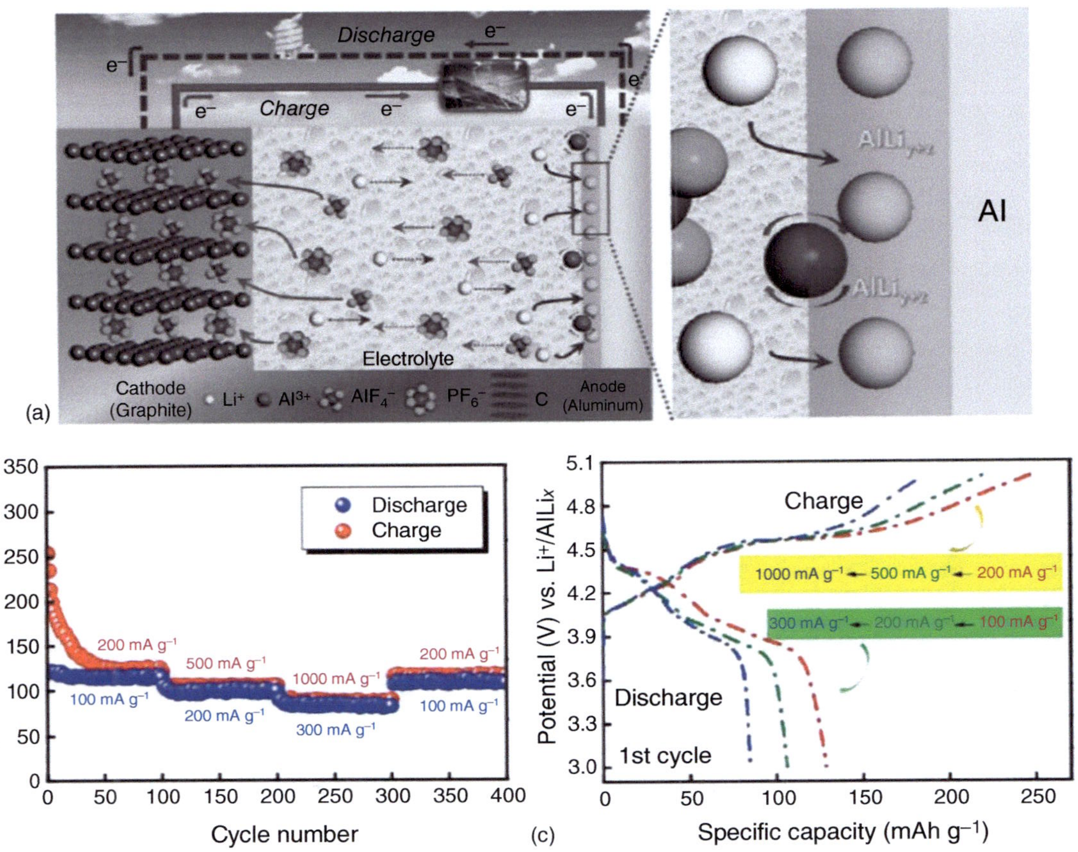

Figure 10.21 (a) Schematic illustration of the quadruple-ion battery during the charging and discharging process. (b) The rate capability at various current densities. (c) The initial charge–discharge curves at current densities varying from low to high. Source: Wang et al. [123]. Reproduced with permission of Wiley-VCH.

owing to the incomplete intercalation of hybrid anions at a higher current density (Figure 10.21c).

Theoretically, other multi-ion configurations based on various cations and anions participating in the electrochemical reactions can also be expected. Multi-ion strategies show significant feasibility to facilitate the development of the emerging battery community and provide a novel pathway to tackle the critical issues involved in the emerging rechargeable batteries.

10.3 Summary and Perspective

DIBs, a novel battery concept just emerged in recent years, have been in the focus of research and considered as a potential alternative to LIBs for energy storage. This chapter comprehensively introduces the DIB technology in terms of cathode materials, anode materials, and electrolytes as well as multi-ion batteries. Compared to the conventional metal-ion batteries based on the "rocking-chair" reaction mechanism, DIBs display a mechanism involving the storage of cations and anions in anode and cathode materials, respectively, resulting in a high operating voltage and energy density. To date, numerous efforts have been undertaken on developing novel electrode materials and electrolytes, and investigating the working mechanisms in DIBs to achieve better electrochemical performance. However, there exist some limitations and challenges which hinder the commercial application of DIBs for energy storage. Some suggestions will be proposed in the following paragraphs, which could be useful for DIBs application in the future.

The first thing is the research and development of electrode materials with high capacity and structural stability. It becomes the short board in DIBs when compared with electrode materials, which were studied deeply for decades in LIB. On the one hand, strategies such as pre-doping with heteroatoms exfoliated with suitable agents to obtain larger interlayer spacing or to functionalize with some active groups can be carried out to improve the performance of commonly used host electrode materials. On the other hand, novel cathode materials/composites benefiting for anion intercalation besides graphite should be explored as candidates for DIBs. Some parameters will be paid attention, such as electronic conductivity for charge transfer, microstructure (surface area and pore size distribution) for anions transportation, and so on.

The second concern is the selection of electrolytes. Especially, the electrolyte serves as active species in DIBs, and therefore more electrolyte and much thicker separator should be needed compared with LIB. Increasing the concentration of active salt in electrolyte would reduce the solvent used, which can improve the energy density of DIBs. Thus, seeking for high-concentration electrolytes with low-cost solvents, salts, and additives should be a good choice. In addition, the anions with smaller size and high oxidative stability can be beneficial to intercalate into cathode materials. Furthermore, electrolytes with multivalent anions would be another solution to enhance the capacity as they can provide more electrons than the monovalent anions.

The next is the suggestion of possible future direction of DIBs. The replacement of scarce Li element in electrode materials and electrolytes with earth-abundant elements (i.e. Na^+, K^+, Mg^{2+}, Ca^{2+}, Al^{3+}) is encouraged. The multivalent cations (Mg^{2+}, Ca^{2+}, Al^{3+}) are more promising as they can supply more electrons to insert into anode materials than monovalent cations (Na^+, K^+). Besides, it should also be noted that the rapidly developed wearable electronics required a flexible high-performance energy storage system to meet the power demands of the continuously emerging new devices. Owing to the high working potential and energy density, flexible DIBs, especially flexible solid-state DIBs, will be one of the promising flexible energy storage candidates for wearable electronics.

We believe that this DIB technology will be an option for sustainable stationary energy storage in the future with unremitting efforts on improvements in energy density, lifetime, and safety.

Acknowledgments

The authors gratefully acknowledge financial supports from the Key-Area Research and Development Program of Guangdong Province (No. 2019B090914003), National Natural Science Foundation of China (Nos. 51972329, 51822210, and 52061160484), Science and Technology Planning Project of Guangdong Province (2019TX05L389), Shenzhen Science and Technology Planning Project (No. JCYJ20200109115624923), and China Postdoctoral Science Foundation (No. 2020M672875).

References

1 Winter, M., Barnett, B., and Xu, K. (2018). Before Li ion batteries. *Chem. Rev.* 118: 11433–11456.
2 Goodenough, J.B. (2013). Evolution of strategies for modern rechargeable batteries. *Acc. Chem. Res.* 46: 1053–1061.
3 Goodenough, J.B. and Kim, Y. (2010). Challenges for rechargeable Li batteries. *Chem. Mater.* 22: 587–603.
4 Ji, B., Zhang, F., Wu, N., and Tang, Y. (2017). A dual-carbon battery based on potassium-ion electrolyte. *Adv. Energy Mater.* 7: 1700920.
5 Zhang, X., Tang, Y., Zhang, F., and Lee, C.-S. (2016). A novel aluminum-graphite dual-ion battery. *Adv. Energy Mater.* 6: 1502588.
6 Wang, M. and Tang, Y. (2018). A review on the features and Progress of dual-ion batteries. *Adv. Energy Mater.* 8: 1703320.
7 Rüdorff, W. and Hofmann, U. (1938). Uber Graphitsalze. *Z. Anorg. Allg. Chem.* 238: 1–50.
8 McCullough, F.P., Levine, C.A., and Snelgrove, R.V. (1989). Secondary battery. US Patent 4,830,938.
9 Carlin, R.T., De Long, H.C., Fuller, J., and Trulove, P.C. (1994). Dual intercalating molten electrolyte batteries. *J. Electrochem. Soc.* 141: L73–L76.

10 Seel, J.A. and Dahn, J.R. (2000). Electrochemical intercalation of PF_6^- into graphite. *J. Electrochem. Soc.* 147: 892–898.

11 Placke, T., Bieker, P., Lux, S.F. et al. (2012). Dual-ion cells based on anion intercalation into graphite from ionic liquid-based electrolytes. *Z. Phys. Chem.* 226: 391–407.

12 Read, J.A., Cresce, A.V., Ervin, M.H., and Xu, K. (2014). Dual-graphite chemistry enabled by a high voltage electrolyte. *Energy Environ. Sci.* 7: 617–620.

13 Rothermel, S., Meister, P., Schmuelling, G. et al. (2014). Dual-graphite cells based on the reversible intercalation of bis(trifluoromethanesulfonyl)imide anions from an ionic liquid electrolyte. *Energy Environ. Sci.* 7: 3412–3423.

14 Wang, M., Jiang, C., Zhang, S. et al. (2018). Reversible calcium alloying enables a practical room-temperature rechargeable calcium-ion battery with a high discharge voltage. *Nat. Chem.* 10: 667–672.

15 Liu, Q., Wang, H., Jiang, C., and Tang, Y. (2019). Multi-ion strategies towards emerging rechargeable batteries with high performance. *Energy Storage Mater.* 23: 566–586.

16 Li, N., Zhang, F., and Tang, Y. (2018). Hierarchical T-Nb_2O_5 nanostructure with hybrid mechanisms of intercalation and pseudocapacitance for potassium storage and high-performance potassium dual-ion batteries. *J. Mater. Chem. A* 6: 17889–17895.

17 Wu, S., Zhang, F., and Tang, Y. (2018). A novel calcium-ion battery based on dual-carbon configuration with high working voltage and Long cycling life. *Adv. Sci. (Weinh)* 5: 1701082.

18 Zhu, H., Zhang, F., Li, J., and Tang, Y. (2018). Penne-like MoS_2/carbon nanocomposite as anode for sodium-ion-based dual-ion battery. *Small* 14: e1703951.

19 Wu, M., Liao, J., Yu, L. et al. (2020). 2020 roadmap on carbon materials for energy storage and conversion. *Chem. Asian J.* 15: 995–1013.

20 Zhou, X., Liu, Q., Jiang, C. et al. (2020). Strategies towards low-cost dual-ion batteries with high performance. *Angew. Chem. Int. Ed. Engl.* 59: 3802–3832.

21 Rüdorff, W. (1940). Kristallstruktur der Säureverbindungen des Graphits. *Z. Phys. Chem.* 45B: 42–68.

22 Daumas, N. and Herold, A. (1969). Relations between phase concept and reaction mechanics in graphite insertion compounds. *C. R. Acad. Sci.* 268: 373.

23 Placke, T., Rothermel, S., Fromm, O. et al. (2013). Influence of graphite characteristics on the electrochemical intercalation of bis(trifluoromethanesulfonyl) imide anions into a graphite-based cathode. *J. Electrochem. Soc.* 160: A1979–A1991.

24 Kravchyk, K.V. and Kovalenko, M.V. (2019). Rechargeable dual-ion batteries with graphite as a cathode: key challenges and opportunities. *Adv. Energy Mater.* 9: 1901749.

25 Li, W.-H., Ning, Q.-L., Xi, X.-T. et al. (2019). Highly improved cycling stability of anion de-/intercalation in the graphite cathode for dual-ion batteries. *Adv. Mater.* 31: 1804766.

26 Deunf, É., Moreau, P., Quarez, É. et al. (2016). Reversible anion intercalation in a layered aromatic amine: a high-voltage host structure for organic batteries. *J. Mater. Chem. A* 4: 6131–6139.

27 Rodríguez-Perez, I.A., Zelang, J.Z., Waldenmaier, P.K. et al. (2016). A hydrocarbon cathode for dual-ion batteries. *ACS Energy Lett.* 1: 719–723.

28 Speer, M.E., Kolek, M., Jassoy, J.J. et al. (2015). Thianthrene-functionalized polynorbornenes as high-voltage materials for organic cathode-based dual-ion batteries. *Chem. Commun.* 51: 15261–15264.

29 Fan, L., Liu, Q., Xu, Z., and Lu, B. (2017). An organic cathode for potassium dual-ion full battery. *ACS Energy Lett.* 2: 1614–1620.

30 Hao, J., Li, X., Song, X., and Guo, Z. (2019). Recent progress and perspectives on dual-ion batteries. *EnergyChem* 1: 100004.

31 Sui, Y., Liu, C., Masse, R.C. et al. (2020). Dual-ion batteries: the emerging alternative rechargeable batteries. *Energy Storage Mater.* 25: 1–32.

32 Rajesh, M., Dolhem, F., Davoisne, C., and Becuwe, M. (2020). Reversible anion insertion in molecular phenothiazine-based redox-active positive material for organic ion batteries. *ChemSusChem* 13: 2364–2370.

33 Lee, M., Hong, J., Lee, B. et al. (2017). Multi-electron redox phenazine for ready-to-charge organic batteries. *Green Chem.* 19: 2980–2985.

34 Li, Q., Qiao, Y., Guo, S. et al. (2018). Both cationic and anionic co-(de)intercalation into a metal-oxide material. *Joule* 2: 1134–1145.

35 Ferey, G., Millange, F., Morcrette, M. et al. (2007). Mixed-valence Li/Fe-based metal-organic frameworks with both reversible redox and sorption properties. *Angew. Chem. Int. Ed.* 46: 3259–3263.

36 Aubrey, M.L. and Long, J.R. (2015). A dual-ion battery cathode via oxidative insertion of anions in a metal-organic framework. *J. Am. Chem. Soc.* 137: 13594–13602.

37 Dühnen, S., Nölle, R., Wrogemann, J. et al. (2019). Reversible anion storage in a metal-organic framework for dual-ion battery systems. *J. Electrochem. Soc.* 166: A5474–A5482.

38 Zhang, Z., Hu, X., Zhou, Y. et al. (2018). Aqueous rechargeable dual-ion battery based on fluoride ion and sodium ion electrochemistry. *J. Mater. Chem. A* 6: 8244–8250.

39 Yao, H.R., You, Y., Yin, Y.X. et al. (2016). Rechargeable dual-metal-ion batteries for advanced energy storage. *Phys. Chem. Chem. Phys.* 18: 9326–9333.

40 Walter, M., Kovalenko, M.V., and Kravchyk, K.V. (2020). Challenges and benefits of post-lithium-ion batteries. *New J. Chem.* 44: 1677–1683.

41 Placke, T., Fromm, O., Lux, S.F. et al. (2012). Reversible intercalation of bis(trifluoromethanesulfonyl)imide anions from an ionic liquid electrolyte into graphite for high performance dual-ion cells. *J. Electrochem. Soc.* 159: A1755–A1765.

42 Zhang, X.Y., Wang, A.X., Liu, X.J., and Luo, J.Y. (2019). Dendrites in lithium metal anodes: suppression, regulation, and elimination. *Acc. Chem. Res.* 52: 3223–3232.

43 Wu, L.N., Peng, J., Sun, Y.K. et al. (2019). High-energy density Li metal dual-ion battery with a lithium nitrate-modified carbonate-based electrolyte. *ACS Appl. Mater. Interfaces* 11: 18504–18510.

44 Xi, X.T., Li, W.H., Hou, B.H. et al. (2019). Dendrite-free lithium anode enables the lithium//graphite dual-ion battery with much improved cyclic stability. *ACS Appl. Energy Mater.* 2: 201–206.

45 Xi, X.T., Feng, X., Nie, X.J. et al. (2019). Dendrite-free deposition on lithium anode toward long-life and high-stable Li//graphite dual-ion battery. *Chem. Commun.* 55: 8406–8409.

46 Xu, X., Lin, K., Zhou, D. et al. (2020). Quasi-solid-state dual-ion sodium metal batteries for low-cost energy storage. *Chem* 6: 902–918.

47 Kraychyk, K.V., Bhauriyal, P., Piveteau, L. et al. (2018). High-energy-density dual-ion battery for stationary storage of electricity using concentrated potassium fluorosulfonylimide. *Nat. Commun.* 9: 4469.

48 Tang, Y., Liu, C.X., Zhu, H.R. et al. (2020). Ion-confinement effect enabled by gel electrolyte for highly reversible dendrite-free zinc metal anode. *Energy Storage Mater.* 27: 109–116.

49 Kravchyk, K.V., Seno, C., and Kovalenko, M.V. (2020). Limitations of chloroaluminate ionic liquid anolytes for aluminum-graphite dual-ion batteries. *ACS Energy Lett.* 5: 545–549.

50 Pan, W., Wang, Y., Zhang, Y. et al. (2019). A low-cost and dendrite-free rechargeable aluminium-ion battery with superior performance. *J. Mater. Chem. A* 7: 17420–17425.

51 Zhao, Y., Adair, K.R., and Sun, X.L. (2018). Recent developments and insights into the understanding of Na metal anodes for Na-metal batteries. *Energy Environ. Sci.* 11: 2673–2695.

52 Li, M., Lu, J., Ji, X.L. et al. (2020). Design strategies for nonaqueous multivalent-ion and monovalent-ion battery anodes. *Nat. Rev. Mater.* 5: 276–294.

53 Aurbach, D., Skaletsky, R., and Gofer, Y. (1991). The electrochemical-behavior of calcium electrodes in a few organic electrolytes. *J. Electrochem. Soc.* 138: 3536–3545.

54 Ponrouch, A., Frontera, C., Barde, F., and Palacin, M.R. (2016). Towards a calcium-based rechargeable battery. *Nat. Mater.* 15: 169–172.

55 Wang, D., Gao, X.W., Chen, Y.H. et al. (2018). Plating and stripping calcium in an organic electrolyte. *Nat. Mater.* 17: 16–20.

56 Rajagopalan, R., Tang, Y.G., Ji, X.B. et al. (2020). Advancements and challenges in potassium ion batteries: a comprehensive review. *Adv. Funct. Mater.* 30: 1909486.

57 Wang, K., Wang, N., Li, X.D. et al. (2019). In-situ preparation of ultrathin graphdiyne layer decorated aluminum foil with improved cycling stability for dual-ion batteries. *Carbon* 142: 401–410.

58 Liu, Y., Hudak, N.S., Huber, D.L. et al. (2011). In situ transmission electron microscopy observation of pulverization of aluminum nanowires and evolution of the thin surface Al_2O_3 layers during lithiation-delithiation cycles. *Nano Lett.* 11: 4188–4194.

59 Hamon, Y., Brousse, T., Jousse, F. et al. (2001). Aluminum negative electrode in lithium ion batteries. *J. Power Sources* 97-8: 185–187.

60 McAlister, A.J. (1982). The Al–Li (aluminum–lithium) system. *Bull. Alloy Phase Diagrams* 3: 177–183.

61 Zhang, W.J. (2011). A review of the electrochemical performance of alloy anodes for lithium-ion batteries. *J. Power Sources* 196: 13–24.

62 Tong, X.F., Zhang, F., Chen, G.H. et al. (2018). Core-shell aluminum@carbon nanospheres for dual-ion batteries with excellent cycling performance under high rates. *Adv. Energy Mater.* 8: 1701967.

63 Tong, X.F., Zhang, F., Ji, B.F. et al. (2016). Carbon-coated porous aluminum foil anode for high-rate, Long-term cycling stability, and high energy density dual-ion batteries. *Adv. Mater.* 28: 9979–9985.

64 Huang, Y.G., Lin, X.L., Pan, Q.C. et al. (2016). Al@C/expanded graphite composite as anode material for lithium ion batteries. *Electrochim. Acta* 193: 253–260.

65 Zhang, G., Ou, X.W., Cui, C.Y. et al. (2019). High-performance cathode based on self-templated 3D porous microcrystalline carbon with improved anion adsorption and intercalation. *Adv. Funct. Mater.* 29: 1806722.

66 Xie, D.H., Zhang, M., Wu, Y. et al. (2020). A flexible dual-ion battery based on sodium-ion quasi-solid-state electrolyte with long cycling life. *Adv. Funct. Mater.* 30: 1906770.

67 Sultana, I., Ramireddy, T., Rahman, M.M. et al. (2016). Tin-based composite anodes for potassium-ion batteries. *Chem. Commun.* 52: 9279–9282.

68 Obrovac, M.N. and Chevrier, V.L. (2014). Alloy negative electrodes for Li-ion batteries. *Chem. Rev.* 114: 11444–11502.

69 Hume-Rothery, W. (1928). The system sodium-tin. *J. Chem. Soc.*: 947–963.

70 Sangster, J. and Bale, C.W. (1998). The K-Sn (potassium-tin) system. *J. Phase Equilib.* 19: 67–69.

71 Ohno, A., Kozlov, A., Arroyave, R. et al. (2006). Thermodynamic modeling of the Ca-Sn system based on finite temperature quantities from first-principles and experiment. *Acta Mater.* 54: 4939–4951.

72 Sheng, M.H., Zhang, F., Ji, B.F. et al. (2017). A novel tin-graphite dual-ion battery based on sodium-ion electrolyte with high energy density. *Adv. Energy Mater.* 7: 1601963.

73 Xu, Y.H., Zhu, Y.J., Liu, Y.H., and Wang, C.S. (2013). Electrochemical performance of porous carbon/tin composite anodes for sodium-ion and lithium-ion batteries. *Adv. Energy Mater.* 3: 128–133.

74 Liu, Y.C., Zhang, N., Jiao, L.F. et al. (2015). Ultrasmall Sn nanoparticles embedded in carbon as high-performance anode for sodium-ion batteries. *Adv. Funct. Mater.* 25: 214–220.

75 Ji, B.F., Zhang, F., Song, X.H., and Tang, Y.B. (2017). A novel potassium-ion-based dual-ion battery. *Adv. Mater.* 29: 1700519.

76 Chan, C.K., Peng, H.L., Liu, G. et al. (2008). High-performance lithium battery anodes using silicon nanowires. *Nat. Nanotechnol.* 3: 31–35.

77 Yuan, S., Zhu, Y.H., Li, W. et al. (2017). Surfactant-free aqueous synthesis of pure single-crystalline SnSe nanosheet clusters as anode for high energy-and power-density sodium-ion batteries. *Adv. Mater.* 29: 1602469.

78 Zhu, Y.H., Yang, X.Y., Liu, T., and Zhang, X.B. (2020). Flexible 1D batteries: recent progress and prospects. *Adv. Mater.* 32: 1901961.

79 He, S.G., Wang, S.F., Chen, H.D. et al. (2020). A new dual-ion hybrid energy storage system with energy density comparable to that of ternary lithium ion batteries. *J. Mater. Chem. A* 8: 2571–2580.

80 Jiang, C.L., Xiang, L., Miao, S.J. et al. (2020). Flexible interface design for stress regulation of a silicon anode toward highly stable dual-ion batteries. *Adv. Mater.* 32: 1908470.

81 Zhou, J., Zhou, Y., Zhang, X. et al. (2020). Germanium-based high-performance dual-ion batteries. *Nanoscale* 12: 79–84.

82 Shao, Y.Y., Gu, M., Li, X.L. et al. (2014). Highly reversible Mg insertion in nanostructured Bi for Mg ion batteries. *Nano Lett.* 14: 255–260.

83 Zhang, W.C., Mao, J.F., Li, S.A. et al. (2017). Phosphorus-based alloy materials for advanced potassium-ion battery anode. *J. Am. Chem. Soc.* 139: 3316–3319.

84 Luo, Y.Q., Tang, Y.J., Zheng, S.S. et al. (2018). Dual anode materials for lithium- and sodium-ion batteries. *J. Mater. Chem. A* 6: 4236–4259.

85 Heidrich, B., Heckmann, A., Beltrop, K. et al. (2019). Unravelling charge/discharge and capacity fading mechanisms in dual-graphite battery cells using an electron inventory model. *Energy Storage Mater.* 21: 414–426.

86 Bommier, C. and Ji, X.L. (2018). Electrolytes, SEI formation, and binders: a review of nonelectrode factors for sodium-ion battery anodes. *Small* 14: 1703576.

87 Hu, X.L., Zhang, W., Liu, X.X. et al. (2015). Nanostructured Mo-based electrode materials for electrochemical energy storage. *Chem. Soc. Rev.* 44: 2376–2404.

88 Liu, Y.J., Hu, X., Zhong, G.B. et al. (2019). Layer-by-layer stacked nanohybrids of N,S-co-doped carbon film modified atomic MoS_2 nanosheets for advanced sodium dual-ion batteries. *J. Mater. Chem. A* 7: 24271–24280.

89 Zhang, X., Zhao, R.F., Wu, Q.H. et al. (2017). Petal-like MoS_2 nanosheets space-confined in hollow mesoporous carbon spheres for enhanced lithium storage performance. *ACS Nano* 11: 8429–8436.

90 Cui, C.Y., Wei, Z.X., Xu, J.T. et al. (2018). Three-dimensional carbon frameworks enabling MoS_2 as anode for dual ion batteries with superior sodium storage properties. *Energy Storage Mater.* 15: 22–30.

91 Park, G., Gunawardhana, N., Lee, C. et al. (2013). Development of a novel and safer energy storage system using a graphite cathode and Nb_2O_5 anode. *J. Power Sources* 236: 145–150.

92 Gunawardhana, N., Park, G.J., Thapa, A.K. et al. (2012). Performance of a graphite (KS-6)/MoO_3 energy storing system. *J. Power Sources* 203: 257–261.

93 Thapa, A.K., Park, G., Nakamura, H. et al. (2010). Novel graphite/TiO_2 electrochemical cells as a safe electric energy storage system. *Electrochim. Acta* 55: 7305–7309.

94 Zhu, J.J., Xu, Y.T., Fu, Y.J. et al. (2020). Hybrid aqueous/nonaqueous water-in-bisalt electrolyte enables safe dual ion batteries. *Small* 16: 1905838.

95 Wrogemann, J.M., Kunne, S., Heckmann, A. et al. (2020). Development of safe and sustainable dual-ion batteries through hybrid aqueous/nonaqueous electrolytes. *Adv. Energy Mater.* 10: 1902709.

96 Yan, T., Ding, R., Ying, D.F. et al. (2019). An intercalation pseudocapacitance-driven perovskite NaNbO3 anode with superior kinetics and stability for advanced lithium-based dual-ion batteries. *J. Mater. Chem. A* 7: 22884–22888.

97 Yan, T., Ding, R., Ying, D.F. et al. (2020). A high tap density perovskite $NaTaO_3$ nanocrystal anode driven by pseudocapacitive conversion/intercalation hybrid mechanisms for advanced Li-ion/dual-ion storage. *J. Mater. Chem. A* 8: 4747–4752.

98 Sui, L., Shi, X., Deng, T. et al. (2019). Integrated Co_3O_4/carbon fiber paper for high-performance anode of dual-ion battery. *J. Energy Chem.* 37: 7–12.

99 Wang, S., Tu, J., Xiao, J. et al. (2019). 3D skeleton nanostructured Ni_3S_2/Ni foam@RGO composite anode for high-performance dual-ion battery. *J. Energy Chem.* 28: 144–150.

100 Zheng, M.B., Tang, H., Li, L.L. et al. (2018). Hierarchically nanostructured transition metal oxides for lithium-ion batteries. *Adv. Sci.* 5: 1077592.

101 Obrezkov, F.A., Shestakov, A.F., Traven, V.F. et al. (2019). An ultrafast charging polyphenylamine-based cathode material for high rate lithium, sodium and potassium batteries. *J. Mater. Chem. A* 7: 11430–11437.

102 Yao, X.H., Ke, Y.J., Ren, W.H. et al. (2019). Defect-rich soft carbon porous nanosheets for fast and high-capacity sodium-ion storage. *Adv. Energy Mater.* 9: 1803260.

103 Xu, K. (2004). Nonaqueous liquid electrolytes for lithium-based rechargeable batteries. *Chem. Rev.* 104: 4303–4417.

104 Lee, Y., Lee, J., Kim, H. et al. (2016). Highly stable linear carbonate-containing electrolytes with fluoroethylene carbonate for high-performance cathodes in sodium-ion batteries. *J. Power Sources* 320: 49–58.

105 Wang, H.G., Wang, H.D., Si, Z.J. et al. (2019). A bipolar and self-polymerized phthalocyanine complex for fast and tunable energy storage in dual-ion batteries. *Angew. Chem. Int. Ed.* 58: 10204–10208.

106 Miyoshi, S., Nagano, H., Fukuda, T. et al. (2016). Dual-carbon battery using high concentration $LiPF_6$ in dimethyl carbonate (DMC) electrolyte. *J. Electrochem. Soc.* 163: A1206–A1213.

107 Heckmann, A., Thienenkamp, J., Beltrop, K. et al. (2018). Towards high-performance dual-graphite batteries using highly concentrated organic electrolytes. *Electrochim. Acta* 260: 514–525.

108 Song, C.H., Li, Y.P., Li, H. et al. (2019). A novel flexible fiber-shaped dual-ion battery with high energy density based on omnidirectional porous Al wire anode. *Nano Energy* 60: 285–293.

109 Shin, K., Zhang, F., Ou, X.W. et al. (2019). Hybridizing anions towards fast diffusion kinetics for tri-ion batteries with significantly improved rate capability and cycling life. *J. Mater. Chem. A* 7: 10930–10935.

110 Fan, H., Gao, J., Qi, L., and Wang, H. (2016). Hexafluorophosphate anion intercalation into graphite electrode from sulfolane/ethylmethyl carbonate solutions. *Electrochim. Acta* 189: 9–15.

111 Holoubek, J., Yin, Y.J., Li, M.Q. et al. (2019). Exploiting mechanistic solvation kinetics for dual-graphite batteries with high power output at extremely low temperature. *Angew. Chem. Int. Ed.* 58: 18892–18897.

112 Koketsu, T., Ma, J.W., Morgan, B.J. et al. (2020). Exploiting cationic vacancies for increased energy densities in dual-ion batteries. *Energy Storage Mater.* 25: 154–163.

113 Fan, J.X., Zhang, Z.X., Liu, Y.H. et al. (2017). An excellent rechargeable PP14TFSI ionic liquid dual-ion battery. *Chem. Commun.* 53: 6891–6894.

114 Placke, T., Heckmann, A., Schmuch, R. et al. (2018). Perspective on performance, cost, and technical challenges for practical dual-ion batteries. *Joule* 2: 2528–2550.

115 Zhang, F., Ji, B., Tong, X. et al. (2016). A dual-ion battery constructed with aluminum foil anode and mesocarbon microbead cathode via an alloying/intercalation process in an ionic liquid electrolyte. *Adv. Mater. Interfaces* 3: 1600605.

116 Zhang, Y.D., An, Y.F., Yin, B. et al. (2019). A novel aqueous ammonium dual-ion battery based on organic polymers. *J. Mater. Chem. A* 7: 11314–11320.

117 Jiang, H., Wei, Z.X., Ma, L. et al. (2019). An aqueous dual-ion battery cathode of Mn_3O_4 via reversible insertion of nitrate. *Angew. Chem. Int. Ed.* 58: 5286–5291.

118 Suo, L.M., Borodin, O., Gao, T. et al. (2015). "Water-in-salt" electrolyte enables high-voltage aqueous lithium-ion chemistries. *Science* 350: 938–943.

119 Wang, F., Borodin, O., Ding, M.S. et al. (2018). Hybrid aqueous/non-aqueous electrolyte for safe and high-energy Li-ion batteries. *Joule* 2: 927–937.

120 Kondo, Y., Miyahara, Y., Fukutsuka, T. et al. (2019). Electrochemical intercalation of bis(fluorosulfonyl)amide anions into graphite from aqueous solutions. *Electrochem. Commun.* 100: 26–29.

121 Jiang, C., Fang, Y., Zhang, W. et al. (2018). A multi-ion strategy towards rechargeable sodium-ion full batteries with high working voltage and rate capability. *Angew. Chem. Int. Ed.* 57: 16370–16374.

122 Lang, J., Jiang, C., Fang, Y. et al. (2019). Room-temperature rechargeable Ca-ion based hybrid batteries with high rate capability and Long-term cycling life. *Adv. Energy Mater.* 9: 1901099.

123 Wang, S., Jiao, S., Tian, D. et al. (2017). A novel ultrafast rechargeable multi-ions battery. *Adv. Mater.* 29: 1606349.

Index

Battery Technologies: Materials and Components, First Edition. Edited by Jianmin Ma.

p

q

r

s